REPRODUCTION IN THE FEMALE MAMMAL

Published Reports of Previous Easter Schools in Agricultural Science

SOIL ZOOLOGY
Edited by D. K. McE. Kevan
(Butterworths: London, 1955)

THE GROWTH OF LEAVES
Edited by F. L. Milthorpe
(Butterworths: London, 1956)

CONTROL OF THE PLANT ENVIRONMENT
Edited by J. P. Hudson
(Butterworths: London, 1957)

NUTRITION OF THE LEGUMES
Edited by E. G. Hallsworth
(Butterworths: London, 1958)

THE MEASUREMENT OF GRASSLAND PRODUCTIVITY
Edited by J. D. Ivins
(Butterworths: London, 1959)

DIGESTIVE PHYSIOLOGY AND NUTRITION OF
THE RUMINANT
Edited by D. Lewis
(Butterworths: London, 1960)

NUTRITION OF PIGS AND POULTRY
Edited by J. T. Morgan and D. Lewis
(Butterworths: London, 1961)

ANTIBIOTICS IN AGRICULTURE
Edited by M. Woodbine
(Butterworths: London, 1962)

THE GROWTH OF THE POTATO
Edited by J. D. Ivins and F. L. Milthorpe
(Butterworths: London, 1963)

EXPERIMENTAL PEDOLOGY
Edited by E. G. Hallsworth and D. V. Crawford
(Butterworths: London, 1964)

THE GROWTH OF CEREALS AND GRASSES
Edited by F. L. Milthorpe and J. D. Ivins
(Butterworths: London, 1965)

REPRODUCTION

IN THE

FEMALE MAMMAL

*Proceedings of the Thirteenth Easter School in
Agricultural Science, University of Nottingham, 1966*

Edited by

G. E. Lamming

*Professor of Agriculture,
University of Nottingham*

and

E. C. Amoroso

*Professor of Physiology,
Royal Veterinary College,
University of London*

Springer Science+Business Media, LLC

Suggested U.D.C. *No.* 591.16

ISBN 978-1-4899-6187-7   ISBN 978-1-4899-6377-2 (eBook)
DOI 10.1007/978-1-4899-6377-2

CONTENTS

CONTENTS

VI. ARTIFICIAL CONTROL OF REPRODUCTION

VII. HORMONAL CONTROL OF UTERINE REACTIONS

VIII. THE SYMPOSIUM IN PERSPECTIVE

PREFACE

EACH generation has supporters of the Malthusian theory, that increases in world population must inevitably lead to an increase in those areas where human populations suffer malnutrition and poverty. Alternatively, some authorities argue that improved techniques for the production, processing and distribution of food, together with effective methods of birth control and education in their use will provide counteracting forces and thus maintain the *status quo*. Whatever argument is supported, a basic knowledge of the physiology of reproduction is essential to achieve control of reproduction in man and his domestic animals.

Many of the past Easter Schools have been concerned with the basic and applied sciences affecting techniques of food production. The Thirteenth Easter School was organized to provide a review of current knowledge on the physiology and biochemistry of reproductive processes in the female and, in particular, to provide a forum where recent information on the control of reproduction in both the human and domesticated species could be reviewed and discussed.

The objectives of the Symposium were to consider the basic processes of reproduction, beginning with the neuro-humoral control of hypophysial function and proceeding via the hypophysis and the gonads and to study those reactions of the uterus which influence the activity of spermatozoa, fertilized and unfertilized ova, the early development of the embryo and the sequence of processes leading to implantation. In view of the wide current interest in the interrelationships between the hypophysis, the corpus luteum and the uterus, a special session was organized to consider these aspects. We felt it was important not to exclude consideration of reproduction in poultry, notwithstanding the Symposium title, for not only are the comparative aspects of reproduction interesting, but also, the numerically small band of enthusiasts who have used poultry as a laboratory species have contributed significantly to our knowledge of reproductive physiology.

To what avail our knowledge, unless it is eventually put to good use? The control of ovulation in the human, whether to limit or enhance fertility, was considered simultaneously with data from experiments designed to synchronize oestrus in animals or to stimulate those with seasonal breeding activity to mate during the normal seasonal anoestrous period.

The Symposium brought together speakers from many countries and from several disciplines, including physiologists and biochemists as well as medical and veterinary personnel involved in clinical practice. The programme was planned to co-ordinate with that of the 14th Easter Symposium on 'The Growth and Development of Mammals'. An arbitrary division was made between the two symposia whereby the early development of the embryo and implantation was considered here, while post-implantation and postnatal development is considered in the following symposium.

Many considered that our original aim of studying a limited field in depth formed a desirably precise objective. In the event, the subject proved more extensive than anticipated, a reflection of the tremendous expansion which has occurred in research into the physiology of reproduction. As the discussions proceeded it became clear that a more complete knowledge of the basic physiology of reproduction is required before man may achieve control over his own rate of procreation or that of his domestic animals.

G. E. L.
E. C. A.

ACKNOWLEDGEMENTS

THE Thirteenth Easter School owed much of its success to those who presented papers and those who participated in the discussions which followed. Indeed, the response to invitations to present papers was sufficiently encouraging that, in order to allow sufficient time for discussion on the last day of the Symposium, Dr. Haynes' paper was withdrawn from the programme, but is published in the Proceedings without discussion. We were delighted that Professor Dainton, the new Vice-Chancellor of the University of Nottingham, was able to welcome delegates and to open the Conference.

Grateful thanks are due to those who acted as Chairmen of the sessions; Professor Folley, Dr. Loraine, Messrs. Peter Williams and Rowson and Professors Melampy, Nalbandov and Amoroso.

The attendance of Dr. Van Rees was arranged under the Foreign University Exchange Scheme of the British Council. Professor Nalbandov was financed by funds supplied by the British Egg Marketing Board, and Professor Hansel and Dr. Chang by funds from the Milk Marketing Board. Professor Robinson was brought from Australia by G. D. Searle and Co. Ltd. and Professor Rothchild from the United States by Syntex Pharmaceuticals Ltd.

The University of Nottingham also expresses gratitude to the following organizations for financial assistance towards expenses of overseas speakers:

> Pig Industry Development Authority
> The British Society of Animal Production
> Imperial Chemical Industries Ltd.
> E. R. Squibb and Sons Ltd.

Finally, we thank the staff of the School for their assistance in organizing the Symposium.

I. HYPOTHALAMIC CONTROL OF REPRODUCTIVE PROCESSES

CONTROL OF FOLLICULAR GROWTH AND OVULATION

B. T. DONOVAN

*Department of Neuroendocrinology, Institute of Psychiatry,
De Crespigny Park, London, S.E.5*

IN THE following account attention will be directed mainly to the part played by the nervous system in the control of follicular development and ovulation. Discussion will be limited to certain aspects of the events occurring during infancy, to the regulation of the onset of the breeding season, and to the triggering of ovulation. Since the neural regulation of gonadotrophin secretion and of the mammalian sexual cycle has been described in detail in a number of recent reviews (Harris, 1955; Greep, 1961; Everett, 1961, 1964; Szentá-gothai, Flerkó, Mess and Halász, 1962; Flerkó, 1963; Schreiber, 1963; Sawyer, 1964; Donovan, 1966; Harris and Campbell, 1966) encyclopaedic coverage will not be attempted.

GONADAL-HYPOPHYSIAL INTERACTION DURING INFANCY

Gonadal activity and pituitary function are in a delicate state of equilibrium during infancy (Greep, 1961; Everett, 1964; Donovan and van der Werff ten Bosch, 1965). Thus, both hypophysectomy and ovariectomy arrest the growth of the genital tract. Gonado-trophic potency of the hypophysis increases following removal of the gonads, as in adults, and this change can be corrected by administra-tion of very small amounts of oestrogen. Again, as in adults, cytological changes appear in the hypophysis of infantile rats after spaying which can be prevented by minute amounts of oestrogen. Further, treatment of intact infantile rats with sex hormone has been shown to inhibit gonadal development or to advance puberty (Ramirez and Sawyer, 1965). The pituitary gland and ovaries therefore interact during infancy in a similar manner to that seen during adult life and, since the sensitivity of the ovaries to gonado-trophin does not appear to increase once an initial phase of re-fractoriness is completed, sexual maturation can only be attributed

3

to a rise in the output of gonadotrophin by the hypophysis. Yet th
primary cause of the increased output remains uncertain.

Despite the apparent inactivity of the gonad-hypophysial axis i
infancy a variety of events can now be outlined which indicate tha
the control of gonadal function before sexual maturation is some
what more complex than had previously been imagined. Thus, i
has become clear that the mode of function of the adult hypothalamo·
hypophysial system is determined in the rat during the first few days
of post-natal life. In the mature female ovulation occurs at regular
intervals, but if the infant is treated with gonadal steroids soon after
birth, then the sexual rhythm is abolished and the rat fails to ovulate
when maturity is reached. From the point of view of the control of
gonadotrophin secretion the female has been masculinized. Initially
this effect was brought about by transplanting testicular tissue into
new-born female rats (Pfeiffer, 1936), but it can be obtained more
easily by the administration of testosterone or other steroids and
much literature has been written on this topic (*see* Harris, 1964;
Donovan and van der Werff ten Bosch, 1965; Jacobsohn, 1965;
Mayer, Thévenot-Duluc and Burin, 1965).

The fundamental mechanism controlling pituitary function is considered to be cyclic in character in both male and female and is
rendered acyclic by the action of androgen during the first few days
after birth. When the supply of androgen is removed, as by
castration of the new-born male, continued cyclic function on the
part of the hypophysis is maintained, as demonstrated by the regular
occurrence of follicular development and ovulation in ovarian tissue
grafted into the anterior chamber of the eye (Pfeiffer, 1936) or elsewhere (Yazaki, 1960; Kawashima, 1960; Harris, 1964; Gorski and
Wagner, 1965). From the present point of view two features of this
process of sexual differentiation merit particular attention. The
first is that the organization of the system controlling gonadotrophin
secretion in the male, or the conversion of that of the female to the
male type by androgen, takes place during a well-defined period of
time. Androgen given to infantile female rats causes constant follicular activity after puberty only when administered during the first
5 or 6 days after birth, although treatment with massive doses of
hormone may extend this term. Castration must be performed
during the first 2 days after delivery, preferably during the first 24 h,
in order to preserve cyclic gonadotrophin secretion (Harris, 1964).
Outside this critical period in the female comparable treatment with
androgen is ineffective and the capacity for cyclic pituitary function
is retained. Large doses of androgen injected into pregnant rats

have failed to modify subsequent ovarian function in the young, even though sufficient hormone reaches them to cause masculinization of the genitalia (Swanson and van der Werff ten Bosch, 1964; Cagnoni, Fantini and Morace, 1964; Kincl and Maqueo, 1965). It may be that small amounts of progesterone can exert a protective or antagonistic action toward androgen for testosterone propionate has failed to modify pituitary function in rats provided also with progesterone (Cagnoni, Fantini, Morace and Ghetti, 1965; Kincl and Maqueo 1965). It is also possible that the hypothalamo-hypophysial system is not ready for masculinization during foetal life in the rat. As yet, there is little information on the existence of this phenomenon in other species, largely because it probably falls within the gestation period. Attempts to reproduce this effect in guinea-pigs by post-natal treatment with testosterone have failed (Barry, Lefranc and Léonardelli, 1964) but pregnant guinea-pigs given testosterone delivered young in which the genitalia were masculinized. These females displayed components of male sexual behaviour on reaching maturity (Phoenix, Goy, Gerall and Young, 1959).

The second feature to be considered concerns the part played by the brain. In his pioneer work on the effects of gonadectomy and gonad transplantation in new-born rats, Pfeiffer (1936) concluded that it was the pituitary gland that became male or female in type. However, the later studies of Harris and Jacobsohn (1952) in the rat, and Martinez and Bittner (1956) in the mouse, revealed that the transplanted hypophysis of the male could function in a cyclic manner when in contact with the hypothalamus of a hypophysectomized female. Further, Segal and Johnson (1959) found that the pituitary glands of females, failing to ovulate after puberty following post-natal treatment with testosterone, secreted the appropriate hormones for the resumption of normal oestrous cycles and pregnancy when transferred to the sella turcica of hypophysectomized normal females. Thus the pattern of hormone secretion is not determined by the pituitary gland but seems to be governed by the brain. Appropriately, the effect of androgens in female rats can be prevented by simultaneous treatment with reserpine. Chlorpromazine and reserpine can block the action of endogenous androgens on the brain of very young males, for ovulation occurred in ovaries later implanted into treated castrated males (Kikuyama, 1961, 1962; Kawashima, 1964). Since the pattern of sexual behaviour is also modified in female rats when the normal rhythm of gonadotrophin release fails to set in, and cannot be restored by appropriate dosage of the spayed animal with sex hormones (Barraclough and Gorski, 1962; Harris and Levine, 1965), there is little doubt that androgens affect the subsequent pattern of gonadotrophin secretion by an action on the brain.

One widely discussed hypothesis concerning the mode of action of androgens (Barraclough and Gorski, 1961; Gorski and Wagner, 1965) takes as its starting point the postulate that there are two functionally distinct regions of the diencephalon concerned with the control of ovulation: a pre-optic (suprachiasmatic) area located anteriorly which is concerned with the cyclic release of luteinizing hormone (LH) and is influenced by exteroceptive factors and gonadal hormones, and a median eminence area which promotes a basal secretion of LH sufficient for the steady secretion of oestrogen. The pre-optic area is considered to operate cyclically in both sexes at birth but is rendered acyclic by the androgen produced in the infantile male testis. Exogenous androgen given to neonatal females acts likewise so that only the basal secretion of LH, maintained by the median eminence centre, continues past puberty. This concept will be recalled at several points later on.

The critical period in the organization of reproductive function correlates very well with a special period in the differentiation of the brain when the neurones of the cerebral cortex approach maturity—as judged by the nuclear volume, presence of Nissl substance and the development of dendritic processes (Himwich, 1962). On the basis of anatomical data and the appearance of the EEG this lies between 4 and 10 days after birth in the rat and around days 44–46 of gestation in the guinea-pig. It is, of course, unlikely that the span of effective action of androgen on pituitary function is related to changes in cerebral cortical development but there is little doubt that rapid organization of lower centres is also in progress. Information concerning these events is fragmentary but, in the hamster for example, the supra-optic nuclei of the hypothalamus do not appear before the third to the fifth day after birth. Differentiation of the pre-optic area is not completed until the tenth day, when an adult configuration of the hypothalamic nuclei is attained (Auer, 1951). Perhaps during these formative days the hypothalamus and associated structures are peculiarly sensitive to steroids. Oestrogen is known to affect the development of sensitivity to electroshock when given during the first week after birth (Heim and Timiras, 1963) and can act like an androgen upon sexual differentiation of the brain (Takewaki, 1962). Significantly, it has been learned recently that oestrogen given from days 6–10 post-natally increased the rate of myelination of the axons of the hypothalamic neurones (Curry and Heim, 1966). It is thus not surprising that attempts to modulate the pituitary function of female rats by treatment with testosterone during foetal life, probably before the anatomical substrate in the brain is laid down, are unavailing.

After the lapse of a week or so from birth it might appear, in the rat, that a balance is struck between the secretion of gonadotrophin and gonadal hormone which is not upset until near puberty. However, stimuli of many kinds can alter the output of gonadotrophin and almost all seem to act through the nervous system. Light provides an obvious example and there is general agreement that exposure to prolonged illumination advances the onset of puberty in the mouse or rat and that the first ovulation can be delayed by keeping the animals in the dark or under short-day conditions (Whitaker, 1936; Luce-Clausen and Brown, 1939; Fiske, 1941; Jöchle, 1956). Other factors are listed and described by Donovan and van der Werff ten Bosch (1965). As previously mentioned, sex hormones given in infancy affect the timing of puberty. Frank, Kingery and Gustavson (1925) first found that a brief period of treatment of infantile rats with oestrogen advanced true puberty and this has been confirmed by Ramirez and Sawyer (1965). In both studies, oestrous cycles followed vaginal opening and continued after withdrawal of the steroid so that it is reasonable to suppose that, in these studies, oestrogen hastened maturation of the system controlling the release of gonadotrophin. However, all the components of the apparatus are ready to function long before puberty. The pituitary gland is sensitive to excitation in infancy, for the release of LH and ovulation can be induced by single injections of oestrogen (Hohlweg, 1934; Everett, 1961) and the controlling influence of the brain is demonstrable. Pregnant mare serum (PMS) can cause ovulation in immature female rats, provided the hypophysis is present, and the induced ovulation can be prevented by barbiturate sedation applied between 2 p.m. and 4 p.m. but not at other times (Strauss and Meyer, 1962; McCormack and Meyer, 1962). Other drugs acting on the nervous system, including serotonin, atropine, SKF-501, and chlorpromazine prevent PMS-induced ovulation in immature rats (O'Steen, 1964; Quinn and Zarrow, 1964; Coppola, Leonardi and Ringler, 1966). In addition, it has been found that progesterone will facilitate ovulation in PMS-primed rats (McCormack and Meyer, 1963, 1964) and that the action of progesterone can be blocked by barbiturates. Interestingly, oestradiol, corticosterone and testosterone failed to facilitate ovulation when given under the

7

same conditions, but other progestational compounds were effective in this regard (McCormack and Meyer, 1965). The hypothalamic areas concerned in the response of the 28-day-old rat to PMS have been investigated by the placement of lesions in various locations 2 days before giving the gonadotrophin (Quinn and Zarrow, 1965). Lesions in the medial pre-optic area, basal-medial pre-optic area, or ventral medial anterior hypothalamic area prevented ovulation but damage to other parts of the hypothalamus was not depressant. Lesions in the anterior hypothalamus also block the ovulation expected after oestrogen administration to immature rats but here a collateral action on the hypophysis is suggested (Döcke and Dörner, 1965). Observations of this kind illustrate very clearly the type of interaction between gonadal hormones and the nervous system that is operative long before puberty and reveal the occurrence of cyclic activity on the part of the infantile hypothalamus.

Abnormalities in the region of the hypothalamus of children have long been associated with the occurrence of sexual precocity and many theories implicating the neural control of gonadotrophin secretion have been advanced to account for the condition (Donovan and van der Werff ten Bosch, 1965). Better control of the extent and location of the damage to the brain can be achieved experimentally and it has been found that lesions placed in the hypothalamus of infantile rats hasten vaginal opening (Donovan and van der Werff ten Bosch, 1956a, 1959a; Bogdanove and Schoen, 1959; Krejci and Critchlow, 1959; Elwers and Critchlow, 1960; Gellert and Ganong, 1960; Corbin and Schottelius, 1960; Schiavi, 1964). Despite the large number of studies undertaken it has not proved possible to pinpoint the hypothalamic structure that must be damaged to advance sexual development and the effect is attributed to interference with the feed-back of gonadal hormones on the brain which favours the secretion of more gonadotrophin by the pituitary gland.

True puberty in the female is marked by the occurrence of ovulation. This means that the secretion of LH has passed beyond threshold levels and, in turn, might indicate that the mechanisms concerned with the timing of puberty centre around the control of this hormone, with the discharge of LH being increased against the background of a constant limited secretion of FSH. Some clinical information is available in support of this view but experimental studies in animals are lacking. Measurement of the urinary secretion of gonadotrophin in one group of children has shown that the elimination of FSH in girls of 10–11, 12–13, and 14–15 years of age falls steadily, while that of LH rises (Brown, 1958). Detailed analyses of the changes in the gonadotrophin content of the hypophyses of children as growth proceeds have not been made although both FSH and LH are present in small amounts. The FSH concentration of the glands of children has been reported to be about one-fortieth of that of an adult (Witschi, 1961) and that of LH in infants to be about one-fifth of that of young women (Ryan, 1962), but little weight can be attached to these ratios. The variations in the gonadotrophin content of the hypophysis of the immature rat have been measured on a number of occasions (Clark, 1935; McQueen-Williams, 1935; Lauson, Golden and Sevringhaus, 1939; Hoogstra and Paesi, 1955; Ramirez and McCann, 1963) and it seems that adult concentrations are reached before puberty although Hoogstra and Paesi (1955) found that little LH was present in young rats and that the concentration of FSH was higher than that of adult females. On the other hand, Ramirez and McCann (1963) reported that LH was present in about the same amount in both infantile and adult females. The plasma content of LH rose markedly after spaying and could be reduced to normal values by supplying oestradiol benzoate daily at a dose related to body weight which was two or three times less than that required by adults.

Although experimental studies employing oestradiol or progesterone, or using the techniques of spaying or ovarian transplantation, substantiate the existence of a gonad pituitary relationship in infancy, the gonadal hormones concerned cannot yet be equated with those of the adult. It remains possible that the spectrum of steroids secreted by the ovary changes during development so that the type of restraint exerted on the hypothalamo-hypophysial system becomes altered in character. Stegmann (1959), Maekawa (1960), Kimura (1960) and Donovan and O'Keeffe (1966) have shown in experiments involving the transfer of ovarian tissue to spleen or to kidney that up to the time of puberty the liver is unable to inactivate gonadal hormones. The hormone produced by the intrasplenic graft passed through the liver to maintain the growth of the uterus. On the other hand, when pure oestradiol was used in place of autografts of ovarian tissue, inactivation of the oestrogen was observed from the earliest age studied, 21 days (Donovan, O'Keeffe and O'Keeffe, 1966), so that the endogenous gonadal hormones appear to differ markedly from oestradiol and a possible change in liver function at the time of puberty may be discounted. However, no direct information based on the isolation of steroids from the ovarian venous blood of immature animals is yet available and the problem awaits a satisfactory answer.

It will be evident that the reasons underlying the increased output of gonadotrophin seen at puberty are not well understood. According to Harris (1954), Donovan and van der Werff ten Bosch (1959a) and Ramirez and McCann (1963) this follows from a reduction in the sensitivity of the hypothalamus to inhibition by gonadal hormone. There is no doubt that the secretion of gonadotrophin is more readily inhibited by oestrogen in immature rats than in adults (*see* Donovan and van der Werff ten Bosch, 1965) and a drive on the part of the brain toward the secretion of gonadotrophin which is opposed by the depressant effect of gonadal hormone is easily envisaged. Nevertheless, it is apparent that more needs to be learned concerning gonadotrophin output during infancy and the kind of hormone secreted by the ovary during this phase of life. If the latter differs significantly from oestradiol, then much of present-day thinking will need revision. Study of the developmental changes taking place in the brain during infancy will also prove rewarding. It will also not have escaped notice that much of the previous discussion has been devoted to work on the rat and generalizations based on findings in this species may prove hazardous.

CONTROL OF THE ONSET
OF THE BREEDING SEASON

It is perhaps artificial to separate discussion of the factors controlling the onset of the breeding season from those governing the timing of puberty, for they are closely similar. However, mention of the breeding season signifies that emphasis will be placed on the environmental regulation of gonadotrophin secretion.

There can be little question that light provides the most important of a number of cues that may be utilized in relating changes in reproductive function to changes in the environment. Numerous examples are given in a recent article by Ortavant, Mauleon and Thibault (1964). In the absence of such cues an inherent rhythm (Bissonnette, 1938; Thomson, 1954) may become manifest but it is extremely difficult to eliminate the possibility that normally unimportant changes in noise, temperature, humidity, odours or dietary constituents have gained unwonted influence. The development of oestrus in sheep is normally favoured or accelerated by exposure to short days, or more accurately, to a ratio of light to darkness in which the dark period is about twice as long as the light. Even under unfavourable conditions, however, the appearance of oestrous cycles cannot be delayed indefinitely (Farner, 1961). It is also wise to recall that in certain strains of sheep the breeding season extends into the spring and summer so that a second crop of lambs can be obtained. Environmental factors would seem to be of less importance in these breeds. The ferret provides an oft-studied example of a species in which the onset of the breeding season is controlled by lengthening days. Numerous experiments involving manipulation of the environmental lighting have been performed (*see* Hammond, 1954; Amoroso and Marshall, 1960; Farner, 1961; Fraps, 1962) and the most effective sequences of light and darkness determined. In this way it has been discovered that 4 h of light daily (when given in the sequence 2 h light, 10 h dark, 2 light, 10 dark) will suffice to induce sexual development in the ferret. Yet surprisingly little is known about the physiological mechanisms involved in the action of light. It is thus not surprising that there has been much argument over the relative importance of light versus darkness in the control of pituitary function, for the processes taking place after stimulation of the retina are not understood.

Under normal circumstances light acts upon gonadotrophin secretion through the eyes and optic nerves, for in the ferret blinding prevents the ovarian response to prolonged illumination (Marshall and Bowden, 1934; Bissonnette, 1936, 1938; Thomson, 1954). It might be expected that from the eyes the neural stimulus traverses the classic visual pathway but this has not been proved in the ferret (Clark, McKeown and Zuckerman, 1939; Jefferson, 1940) or in other species (Critchlow, 1963) and all that is certain is that the stimulus reaches the hypothalamus. This is evident from studies of the electrical responses to photic stimuli, which appear after latent periods that indicate that a multi-neurone pathway is involved (Ingvar and Hunter, 1955; Massopust and Daigle, 1961; Abrahams, Hilton and Malcolm, 1962). The visual responses in the hypothalamus and pre-optic area in cats following photic stimulation of the eyes have been described recently by Feldman (1964). In the anterior hypothalamus and pre-optic area potentials of short (8–10 msec) latency were recorded which were resistant to deep levels of anaesthesia and anoxia, whereas the responses in the posterior hypothalamus were of long latency and abolished by small amounts of barbiturate. This differential response implies that the visual projections to the anterior hypothalamus are more direct, and possibly of greater significance, than those to the posterior region. The short latency responses were always prominent in the region of the supra-optic nuclei and show an intriguing correlation with the changes in neurosecretory activity and cell size taking place in those

11

nuclei in rats exposed to continuous illumination (Fiske and Greep, 1959; Flament-Durand, 1965a). Such changes can be accounted for as indirect consequences of the changes in hormone secretion following upon prolonged illumination, but this possibility seems less likely in view of the electrophysiological studies.

It has often been suggested that direct communications between the optic chiasma and hypothalamus exist. Critchlow (1963), for example, gives an extensive list of papers describing the pathways but since other anatomists, using improved neurohistological methods, have failed to find these fibres (Hess, 1958; Hayhow, 1959; Hayhow, Webb and Jervie, 1960; Szentágothai, Flerkó, Mess and Halász, 1962; Nauta, 1963) there is little agreement on this question. Functional studies are unhappily few but Critchlow and de Groot (1960; *see also* Critchlow, 1963) used prolonged illumination to induce constant oestrus in rats and found that 4 animals with proven bilateral transection of the optic tracts responded to the stimulus. In contrast, small midline lesions in the suprachiasmatic region interfered with the effects of continuous illumination: some rats failed to show prolonged vaginal cornification while others responded only partially or erratically. Although these lesions failed to implicate a specific structure, blockade of the light response was most often associated with destruction of the median peak of fibres located dorsally on the optic chiasma between the suprachiasmatic nuclei.

A comparatively long delay is normally observed between the time that an animal is exposed to long days and the development of oestrus. This is usually about a month in the ferret, yet only a fraction of this interval is needed for the gonadotrophin released to act on the ovaries to cause follicular development and oestrogen secretion. The processes going on within the hypothalamus are seemingly slow, but may indicate that the stimulus of light can act only for a short period of time each day and that the effects are cumulative. With the current interest in circadian rhythms (Whipple, 1964; Wolfson, 1965), it is perhaps worth pointing out that the sensitivity of the mechanism to excitation may not be constant and may fluctuate during the day. Exposure to light for one or more short periods encompassing the time, or times, of greatest sensitivity may bring about the release of gonadotrophin just as rapidly as prolonged illumination, and so account in the ferret for the stimulative action of 4 h light daily. There is evidence (in the ferret) for long term changes in the sensitivity of the hypothalamic mechanism of anoestrous females to stimulation. When exposure to long days was begun at different times during the summer, the latent period before vulval swelling was detected varied, and a striking effect was observed in animals included in the experiment by the end of August. Some females became oestrous unusually rapidly, within 2 weeks, while others in the same group failed to come into heat for several years despite continuous exposure to prolonged illumination (Donovan, unpublished). A period of short-day lighting was required by the latter animals before they would respond to renewed exposure to long days. If, as Bissonnette (1938) and Thomson (1954) suggest, an inherent rhythm in sexual function exists, then this finding indicates that light can inhibit gonadotrophic hormone secretion on occasion and that the degree of excitability of the hypothalamus is of importance. Toward the end of August the mechanism may be triggered easily, or inhibited readily, by prolonged illumination.

In work with anoestrous female ferrets outside the breeding season, Donovan and van der Werff ten Bosch (1956b, 1959b) found that lesions placed in the anterior hypothalamus led to the secretion of gonadotrophin and early onset of oestrus. Here the ovaries were in a 'resting' condition and the lesions exerted their effect against a background of minimal secretion of gonadal hormone. The specificity of this effect has been questioned by Herbert and Zuckerman (1958) in the light of their observations that lesions placed elsewhere in the brain, or blank operations, also hastened the onset of oestrus. As yet it has proved difficult to reconcile these divergent views, particularly as interference with regions of the brain outside the hypothalamus has not, in our hands, promoted the secretion of gonadotrophin in ferrets (Donovan, 1960), and no changes in endocrine status have been detected in patients undergoing thalamic surgery (Odell, van Buren and Hertz, 1962). Van der Werff ten Bosch (1963) suggested that the prior history of exposure to controlled illumination in laboratory ferrets may be of great importance in this regard and that care should be exercised in the selection of animals for studies of this kind. Nevertheless, it remains unquestionable that lesions placed in the hypothalamus of the ferret accelerate sexual development. Donovan and van der Werff ten Bosch (1956b, 1959b) attributed this result to interference with the feed-back action of gonodal hormone on the brain, but in order to substantiate this view it is necessary to establish the existence of pituitary-gonadal interaction during the anoestrous period. Some progress to this end has been made in that the removal of one ovary is followed by compensatory hypertrophy of the other, whether the operation is performed during oestrus or during sexual quiescence, and that overdosage with oestrogen causes ovarian atrophy (Dono-van, 1964).

OVULATION

The neuro-endocrine interactions that underlie much of repro-ductive function are excellently illustrated in the control of ovula-tion, where the role of the nervous system in species such as the rabbit, cat, ferret and mink has been emphasized by the use of the term 'reflex ovulation', for ovulation does not normally occur in the absence of the male. In other species including the rat, mouse, guinea-pig, sheep, pig and cow, so-called 'spontaneous ovulation' occurs at regular intervals, but here too ovulation is under close neural control; the event is far from unprompted (Aron, Asch, Asch, Roos and Luxembourger, 1965). The hypothalamus in both types responds to appropriate excitation by provoking gonadotrophin release but in the cyclic ovulators the sensitivity of the hypothalamus varies according to the stage of the cycle and to the time of day. As might be expected, the activities of oestrogen and progesterone are of prime importance in adjusting the responsiveness of the hypo-thalamus to excitation and many of the effects of these hormones in causing or inhibiting ovulation are explicable on this basis.

Ovulation can be induced by electrical stimulation of similar parts of the brain in both cyclic and non-cyclic species. Most studies have been made in rabbits, where excitation of a broad area of the hypo-thalamus, from the medial pre-optic region to the mamillary bodies, has proved effective. Comparable experiments in the rat have been performed after pharmacological blockade of the expected ovulation, or under conditions (dioestrus, pregnancy, pseudopregnancy or per-sistent oestrus) in which ovulation does not normally occur (Everett, 1961, 1964; Sawyer, 1964). Inflammatory electrolytic lesions in which a deposit of iron provides a focus of irritation are valuable for mapping studies (Everett, Radford and Holsinger, 1964; Everett, 1964) and it has been postulated that the ovulatory mechanism employs a diffuse system of neurones originating throughout the septal complex which converge as they enter the medial pre-optic area and anterior hypothalamus and assume a restricted basal loca-tion on reaching the tuber cinereum. Extra-hypothalamic struc-tures undoubtedly participate in the control of ovulation in that stimulation of the cerebral cortex, temporal lobes, or amygdaloid nuclei, has caused follicular rupture in rats, rabbits and cats. Lesions placed in the midbrain reticular formation have been found to block ovulation but damage to the temporal lobe nuclei is essentially ineffective. It seems likely that structures outside the hypo-thalamus can modulate the output of LH but are not essential in the regulation of pituitary function (Everett, 1964; Sawyer, 1964; Donovan, 1966; Harris and Campbell, 1966).

Little is known of the capacity of the hypothalamo-hypophysial complex to control gonadotrophin secretion when isolated from the rest of the brain. Woods (1962) found that gonadotrophin secretion continued after preparation of an hypothalamic island in cats and rats, although the pattern of release was altered in that discharge of sufficient LH for ovulation seldom occurred. Surgical isolation of the hypothalamus from the rest of the brain in rats was achieved by de Groot (1962, 1965). His animals tended to have normal vaginal cycles and possessed ovaries with corpora lutea that were within the customary range of size. Gonadal weight fell in other rats in which the hypothalamus was deafferented and oestrous cycles did not con-tinue (Halász and Pupp, 1965). In some females the vaginal smears were almost constantly cornified and at autopsy the ovaries contained follicles without corpora lutea, while in others, oestrous vaginal smears were recorded occasionally and persistent corpora lutea were found in the ovaries. There was no apparent difference in the extent of the deafferented region in either group of rats and no explanation for these diverse results is yet available. From the material currently to hand it seems that the hypothalamic island is deprived of the influence of the pre-optic area. This is the region considered to provide a cyclic drive to the release of ovulatory amounts of gonadotrophin and to be reciprocally regulated by the blood level of circulating gonadal steroids (Barraclough and Gorski, 1961; Barraclough, Yrarrazaval and Hatton, 1964; Gorski and Wagner, 1965). On this basis rhythmic function on the part of an hypothalamic island may not be anticipated and for this reason the findings cited above are most intriguing.

Studies of the functional capacity of hypothalamic islands are particularly valuable in the light of the concept that the hypo-thalamus contains a 'hypophysiotrophic area'. As was first shown in female rats by Halász, Pupp and Uhlarik (1962), pituitary tissue introduced into the hypothalamus of rats which were later hypophy-sectomized could support endocrine function. The hypophysio-trophic area extended up to 1 mm on either side of the midline and included the whole of the arcuate nuclei, the ventral part of the anterior paraventricular nuclei and the parvocellular region of the retrochiasmatic area. It corresponds fairly well with the region from which the tubero-hypophysial tract originates. Within this

15

well circumscribed area the pars distalis tissue contained periodic acid-Schiff positive basophils and could support near normal ovarian weight and structure. Later, it was reported that essentially normal oestrous cycles with ovulation reappeared in hypophysectomized rats with pituitary grafts located in the hypophysiotrophic area and that compensatory hypertrophy of the remaining ovary occurred after unilateral ovariectomy (Halász, Pupp, Uhlarik and Tima, 1965). Thus it would seem that the releasing factors necessary for normal gonadotrophin secretion are produced in or about the hypophysiotrophic area so that pituitary grafts located therein can be appropriately activated. This conclusion is not entirely supported by the parallel work of Flament (1964) and Flament-Durand (1965b, 1966). Although the existence of an hypophysiotrophic area in which regranulation of basophil cells occurred was confirmed, and emphasis placed on the greater influence of the arcuate nuclei than other hypothalamic structures, only 8 of 13 animals with pituitary grafts in the hypophysiotrophic area resumed regular cycles 15–65 days after hypophysectomy. Six of these 8 animals mated on several occasions without becoming pregnant or pseudopregnant. At autopsy the ovaries contained follicles but lacked corpora lutea so that it seemed that insufficient LH was being secreted for ovulation. The discordance between morphology and function in the grafted pituitary tissue suggested that the synthesis of gonadotrophic hormone was being supported though release of the hormone was disturbed. Perhaps this is due to the LH produced by the grafts, for in recent months the suggestion that LH may exert a direct feedback action on the hypothalamus has gained considerable support (Szontágh and Uhlarik, 1964; Ramirez and Sawyer, 1965; Chowers and McCann, 1965; Corbin and Cohen, 1966; David, Fraschini and Martini, 1966).

It seems, then, that the hypothalamus can function with some degree of autonomy and a question arises concerning the mode of action of extrinsic factors in controlling follicular development and ovulation. The activities of the limbic system in this regard have been reviewed recently by de Groot (1965), who reaffirms that electrical stimulation of the amygdaloid nuclei, septum complex or pre-optic area can cause ovulation whereas lesions in the amygdaloid area, habenula or other parts of the brain of adult animals have little effect on gonadal activity. In view of the generally positive effect of stimuli, the hypothalamus together with the pituitary gland is compared to a 'homeostat', or a neuronal aggregate capable of steadily controlling the changes in the internal environment created by exteroceptive stimuli, by alterations in blood hormone levels, or by impulses from extrahypothalamic portions of the nervous system. This suggestion is an attractive one and leads to the concept that many circuits may operate to alter the setting of the homeostatic mechanism. To some extent this is a more complex way of saying that the hypothalamus is exerting an integrating action on the variety of stimuli reaching it and not merely acting as a relay station for impulses directed to the pituitary gland.

The hypophysial portal vessels running in the pituitary stalk provide the functional link between the hypothalamus and the hypophysis. Disruption of this connection by pituitary stalk section or by transplantation of the hypophysis away from the sella turcica profoundly depresses the secretion of FSH and, presumably, LH (*see* Donovan, 1966; Harris and Campbell, 1966; Jacobsohn, 1966) for the cells of the pars distalis are deprived of the releasing factors liberated into the primary plexus of the portal vessels by nerve fibres ending in the median eminence. The isolation and activities of the neurohumoral agents affecting gonadotrophin secretion are described elsewhere (McCann, page 55).

Few detailed studies of the feed-back action of endogenous gonadal hormones in species experiencing a long period of oestrus have been made, despite the remarkable combination of levels of gonadotrophin and oestrogen secretion that are considered incompatible in rats and mice. Removal of one ovary causes compensatory hypertrophy of the other (Bond, 1906; Carmichael and Marshall, 1908; Hammond, 1925) so that the products of hypophysis and ovary normally are in balance. It further seems that the ovaries themselves ensure that the gonadotrophin secreted by the hypophysis is follicular in character, for when the blood oestrogen level is reduced in the rabbit by transplantation of the ovary to the mesentery or spleen (Mayer and Soumireu, 1948; Koulischer, 1960) luteal tissue appears in the grafts. It is possible that the ovarian changes reflect an over-secretion of FSH (Alloiteau, 1956) but an entirely satisfactory explanation has yet to come.

There is now little doubt that the feed-back action of gonadal hormones on gonadotrophin secretion is mediated by the brain. The blockade of ovulation by drugs known to act on the nervous system provided early evidence for this conclusion (Sawyer, Markee and Hollinshead, 1947) and further point was given to these results by the finding that the advancement of ovulation by oestrogen or progesterone in rats could be blocked by anticholinergic and anti-adrenergic drugs (Everett and Sawyer, 1949; Sawyer, Everett and

17

Markee, 1949). Many examples based on events in the rat and rabbit are given by Everett (1961, 1964) and Sawyer (1964). Progesterone-induced ovulation in the cow has been blocked by atropine (Hough, Bearden and Hansel, 1955) and normal ovulation in the ewe delayed by chlorpromazine or barbiturate anaesthesia (Robertson and Rakha, 1965; Radford, 1966).

Greater precision on the locus of action of gonadal hormone on the brain can be achieved by the local implantation of small amounts of steroid in selected areas. The rate of release from the depot must be small enough to avoid escape of detectable amounts of hormone into the systemic circulation and to avoid affecting large areas within the brain. At first, fragments of ovarian tissue were implanted into the hypothalamus of rats, with implants of liver tissue being made as controls, and it was found that only grafts of ovarian tissue located in the anterior hypothalamus caused a significant fall in the weight of the uterus (Flerkó and Szentágothai, 1957). This work was followed by experiments in which oestrogen alone was used. Lisk (1960) found that the application of oestrogen to the arcuate-mamillary nuclei area of the hypothalamus caused long periods of dioestrus, loss of ovarian weight and the appearance of the 'wheel' figures in the nuclei of ovarian interstitial cells that are usually linked with a deficiency of LH. Implants made into the suprachiasmatic area exerted little effect. The rise in the plasma concentration of LH expected after ovariectomy in the rat was prevented by the implantation of small amounts of oestradiol into the basal tuberal hypothalamus, but not by the placement of steroid in the suprachiasmatic region or globus pallidus (Ramirez, Abrams and McCann, 1964), while compensatory ovarian hypertrophy was suppressed by implantation of oestradiol into the anterior hypothalamus or mamillary bodies but not in the pre-optic region, lateral hypothalamus or amygdaloid area (Littlejohn and de Groot, 1963). Corresponding information is available for the rabbit, where ovarian atrophy followed the introduction of oestradiol benzoate into the posterior median eminence–basal tuberal region (Davidson and Sawyer, 1961). Again, in treated animals the plasma concentration of LH failed to rise after spaying and the pituitary content of this hormone fell (Kanematsu and Sawyer, 1964). The results in rat and rabbit differ in that oestrogen applied to the mamillary bodies or anterior median eminence in the rabbit failed to alter the hypophysial content of LH.

Since oestradiol applied directly to the anterior pituitary gland was also effective in preventing a rise in the plasma concentration of

LH after ovariectomy (Ramirez, Abrams and McCann, 1964) and is known to modify the cytological changes seen in the hypophysis after spaying (Bogdanove, 1963) it seems clear that oestrogen can act directly on the cells of the pars distalis. Bogdanove (1964) has argued that many of the results of experiments involving local implantation of sex hormone can be explained on this basis—that the hormone diffuses into the hypophysial portal system and is better dispersed through the pituitary gland, where modification in the secretion of gonadotrophin occurs. Nevertheless, the inhibition of gonadotrophin secretion exerted with application of sex hormone outside the median eminence (e.g. to the mamillary bodies) is hardly explicable in this way, particularly as the implants may be very small and unilateral.

Experimentally, suspension of ovulation and a state of continuous vaginal cornification can be induced in the rat in a variety of ways (Takewaki, 1962). The term 'persistent oestrus', which is often applied to this condition, is a misnomer, for such rats are not in a psychic state of heat and seldom mate. A period of treatment with progesterone is normally required before mating will occur (Everett, 1940, 1943; Kempf, 1950; Greer, 1953), but the term is firmly established and cannot easily be discarded. The ovaries of such rats are follicular in character, lack corpora lutea, and are comparable with those of oestrous rabbits, ferrets and cats. Further, ovulation can be induced by means that are effective in causing follicular rupture in the intact animal (progesterone, or electrical stimulation of the median eminence). The action of many procedures causing persistent oestrus is explicable to a large extent on the basis of interference with the feed-back action of gonadal steroids, or, in the words of Everett (1964): 'the absence of corpora lutea . . . results from failure of the neural mechanism which under normal circumstances spontaneously activates the ovulation inducing quota of LH, a mechanism whose operation depends in part on the feed-back action of gonadal steroids.' The demonstration that oestrogen exerts an inhibitory effect on gonadotrophin secretion when implanted into the basal tuberal part of the hypothalamus need not imply that actions exerted elsewhere (perhaps in the pre-optic area) are unimportant.

Despite the similarity in ovarian function of rats brought into persistent oestrus by androgen treatment in infancy ('androgen blocked'), by constant illumination or by hypothalamic lesions, the mechanisms involved are not necessarily identical. Caution is indicated by the difficulties experienced in inducing corpus luteum formation in androgen-blocked rats by progesterone (Barraclough and Gorski, 1961). Ovulation has followed electrical stimulation of the hypothalamus in the neighbourhood of the arcuate or ventromedian nuclei in animals in which the storage of gonadotrophin by the hypophysis was promoted by prior treatment with progesterone (Barraclough and Gorski, 1961). A further complication is introduced by the finding that the ovaries of androgen-treated rats secrete, not only oestrogen, but androgen (Kawashima, 1960; Goldzieher and Axelrod, 1963; Weisz, Matsuyama, Self and Lloyd, 1964). Androgen is known to abolish cyclic processes in the hypothalamus, as is seen from the treatment of infantile rats (page 4) and from the fact that ovaries transferred to castrated males show follicular development only and seldom ovulate. Nevertheless, under special circumstances stimulation of the pre-optic area or median eminence of the male rat has been shown to cause an ovulatory discharge of gonadotrophin (Moll and Zeilmaker, 1966; Quinn, 1966). As in the case of reflexly ovulating constantly oestrous species (page 19) the release of LH as shown by luteinization of the ovaries, can be elicited in persistent oestrous rats by disturbance of the pattern of gonadal hormone feeding back at the hypothalamic level. Thus transplantation of the ovaries to the spleen is followed by intense luteinization (Desclin, 1954; Noumura, 1958). This effect may be attributed to a complete loss of the feed-back action of oestrogen, with consequent oversecretion of gonadotrophin.

In contrast to the wealth of knowledge concerning the interplay between the hypophysis and the ovaries, in which the hypothalamus exercises such an important role, little is known about the changes in hypothalamic function brought about by steroid hormones. The size of the nuclei (Szentágothai *et al.*, 1962) or nucleoli (Lisk and Newlon, 1963; Ifft, 1964) of cell groups of the hypothalamus varies in response to gonadectomy or gonadal hormone administration but the changes have proved difficult to interpret. Thus both spaying and oestrogen administration reduced the size of the nucleoli of the arcuate cell group although the volume of the nuclei of these cells was increased. The hypothalamus of the cyclic rat is not constantly responsive to afferent stimuli for in the Duke University strain of animals the stimulus for ovulation operates only between 2 p.m. and 4 p.m. Although the neurogenic stimulus is transmitted to the hypophysis once every 4 or 5 days, there is, nevertheless, a daily rhythm in the sensitivity of the hypothalamus to excitation. This is evident from experiments involving the advancement of ovulation by progesterone and from the delay in ovulation caused by repetitive daily blockade by barbiturate (Everett, 1961, 1964; Sawyer, 1964). The diurnal rhythm is manifest long before sexual maturation, for Strauss and Meyer (1962) found that barbital sodium could block PMS-induced ovulation when given at 2 p.m., but not when supplied at 4 p.m. Confirmatory findings have been reported in 24-day-old rats by McCormack and Meyer (1962) with barbiturate treatment and in 18-day-old rats with progesterone administration (Mc-Cormack and Meyer, 1964). In adult rats the duration of the stimulus required for ovulation has been measured fairly accurately and lasts about half an hour, with the release of LH proceeding simultaneously with excitation of the hypothalamus (Everett, 1961, 1964). For rabbits it has been shown that an increased output of progesterone in the ovarian venous blood can be traced as early as 10 min after mating (Hilliard, Endröczi and Sawyer, 1961; Hilliard, Archibald and Sawyer, 1963) but diurnal variations in the sensitivity of the mechanism have not been reported. The release of gonadotrophin after coitus lasts for about 2 h (Hilliard, Hayward and Sawyer, 1964). Both oestrogen and progesterone lower the threshold to EEG arousal and the EEG after-reaction to coitus (Kawakami and Sawyer, 1959; Sawyer, 1963) and electrical changes in the lateral hypothalamus induced by mechanical stimulation of the cervix of the cat have been traced by Porter, Cavanaugh, Critchlow and Sawyer (1957). A sharply differentiated EEG response to stimulation of the cervix uteri has been recorded in the lateral pre-optic and lateral hypothalamic areas of rats (Barraclough, 1959, 1960) with less well defined changes occurring after oestrogen and stimulation of the vagina (Law and Sackett, 1965). Changes in the threshold to electrically-induced seizures upon treatment with gonadal steroids are also known to occur (Woolley and Timiras, 1962a, b; Heim and Timiras, 1963).

Studies of the electrical activity of single cells in the hypothalamus of the rat and their changing response to alterations in the hormonal background and to stimulation of the genitalia have been made by Barraclough and Cross (1963) and Cross and Silver (1965). Little specificity to genital stimuli (probing of the cervix) was observed in that the majority of neurones reacting were also sensitive to more than one other stimulus modality (pain, cold, smell). The responsiveness of the neurones changed during the oestrous cycle and there was a trend toward a reduction in excitability during oestrus. Progesterone given intravenously depressed the responsiveness to genital stimuli but this effect was short-lived and had disappeared an hour after injection. In pseudopregnant rats, too, the number of hypo-thalamic neurones excited by cervical probing was reduced; removal of the ovaries immediately before recording was begun led to a marked increase in the percentage of neurones excited by vaginal probing. Fewer experiments involving the administration of oestro-gen have been performed, but these indicate that this type of hor-mone exercises a predominantly inhibitory influence on the response of hypothalamic neurones to manipulation of the cervix (Cross, 1964).

REFERENCES

Abrahams, V. C., Hilton, S. M. and Malcolm, J. L. (1962). *J. Physiol., Lond.* **164,** 1–16

Alloiteau, J. J. (1956). *C.r. Séanc. Soc. Biol.* **150,** 250–251

Amoroso, E. C. and Marshall, F. H. A. (1960). In: *Marshall's Physiology of Reproduction* Vol. 1, Part 2, 707–831. Ed. by A. S. Parkes. London; Longmans Green

Aron, C., Asch, G., Asch, L., Roos, J. and Luxembourger, M. M. (1965). *Path. Biol., Paris* **13,** 603–614

Auer, J. (1951). *J. comp. Neurol.* **95,** 17–41

Barraclough, C. A. (1959). XXI *Int. physiol. Congr. Buenos Aires.* Abst. Commun. p. 28

— (1960). *Anat. Rec.* **136,** 159

— and Cross, B. A. (1963). *J. Endocr.* **26,** 339–359

— and Gorski, R. A. (1961). *Endocrinology* **68,** 68–79

— — (1962). *J. Endocr.* **25,** 175–182

— Yrarrazaval, S. and Hatton, R. (1964). *Endocrinology* **75,** 838–845

Barry, J., Lefranc, G. and Léonardelli, J. (1964). *C.r. Séanc. Soc. Biol.* **158,** 2082–2084

Bissonnette, T. H. (1936). *Anat. Rec.* **64,** Suppl. 3, 89–90

— (1938). *Res. Publs Ass. Res. nerv. ment. Dis.* **17,** 361–376

Bogdanove, E. M. (1963). *Endocrinology* **73,** 696–712

— (1964). *Vitam. Horm. Lpz.* **22,** 205–260

— and Schoen, H. C. (1959). *Proc. Soc. exp. Biol. Med.* **100,** 664–669

Bond, C. J. (1906). *Br. med. J.* ii, 121–127

Brown, P. S. (1958). *J. Endocr.* **17,** 329–336

Cagnoni, M., Fantini, F. and Morace, G. (1964). *Rass. Neurol. veg.* **18,** 275–284

— — — and Ghetti, A. (1965). *J. Endocr.* **33,** 527–528

Carmichael, E. S. and Marshall, F. H. A. (1908). *J. Physiol., Lond.* **36,** 431–434

Chowers, I. and McCann, S. M. (1965). *Endocrinology* **76,** 700–708

Clark, H. M. (1935). *Anat. Rec.* **61,** 175–192

Clark, W. E. le Gros, McKeown, T. and Zuckerman, S. (1939). *Proc. R. Soc. B.* **126,** 449–468

CONTROL OF FOLLICULAR GROWTH AND OVULATION

Coppola, J. A., Leonardi, R. G. and Ringler, I. (1966). *J. Reprod. Fert.* **11,** 65–71

Corbin, A. and Cohen, A. I. (1966). *Endocrinology* **78,** 41–46

— and Schottelius, B. A. (1960). *Proc. Soc. exp. Biol. Med.* **103,** 208–210

Critchlow, V. (1963). In: *Advances in Neuroendocrinology* 377–402. Ed. by A. V. Nalbandov. Urbana; University of Illinois Press

— and de Groot, J. (1960). *Anat. Rec.* **136,** 179

Cross, B. A. (1964). *Symp. Soc. exp. Biol.* **18,** 157–193

— and Silver, I. A. (1965). *J. Endocr.* **31,** 251–263

Curry, J. J. and Heim, L. M. (1966). *Nature, Lond.* **209,** 915–916

David, M. A., Fraschini, F. and Martini, L. (1966). *Endocrinology* **78,** 55–60

Davidson, J. M. and Sawyer, C. H. (1961). *Acta endoc., Copenh.* **37,** 385–393

Desclin, L. (1954). *C.r. Séanc. Soc. Biol.* **148,** 187–189

Döcke, F. and Dörner, G. (1965). *J. Endocr.* **33,** 491–499

Donovan, B. T. (1960). *Mem. Soc. Endocr.* **9,** 1–15

— (1964). *Anat. Rec.* **148,** 277

— (1966). In: *The Pituitary Gland* Vol. 2, 49–98. Ed. by G. W. Harris and B. T. Donovan. London; Butterworths

— and O'Keeffe, M. C. (1966). *J. Endocr.* **34,** 469–478

— — O'Keeffe, H. T. (1966). *Proc. 2nd Int. Congr. Hormonal Steroids,* 365–366

— and van der Werff ten Bosch, J. J. (1956a). *Nature, Lond.* **178,** 745

— — (1956b). *J. Physiol., Lond.* **132,** 57P–58P

— — (1959a). *J. Physiol., Lond.* **147,** 78–92

— — (1959b). *J. Physiol., Lond.* **147,** 93–108

— — (1965). *Physiology of Puberty.* Monographs of the Physiological Society, No. 15. London; Edward Arnold

Elwers, M. and Critchlow, V. (1960). *Am. J. Physiol.* **198,** 381–385

Everett, J. W. (1940). *Endocrinology* **27,** 681–686

— (1943). *Endocrinology* **32,** 285–292

— (1961). In: *Sex and Internal Secretions* 3rd edn. 1, 497–555. Ed. by W. C. Young. Baltimore; Williams & Wilkins

— (1964). *Physiol. Rev.* **44,** 373–431

— Radford, H. M. and Holsinger, J. (1964). *Proc. 1st Int. Congr. Hormonal Steroids,* 1, 235–246

— and Sawyer, C. H. (1949). *Endocrinology* **45,** 581–595

Farner, D. S. (1961). *A. Rev. Physiol.* 23, 71–96

Feldman, S. (1964). *Ann. N.Y. Acad. Sci.* **117,** 53–66

Fiske, V. M. (1941). *Endocrinology* **29,** 187–196

— and Greep, R. O. (1959). *Endocrinology* **64,** 175–185

Flament, J. (1964). *C.r. hebd. Séanc. Acad. Sci., Paris* **259,** 4376–4378

Flament-Durand, J. (1965a). *Annls Endocr.* **26,** 609–613

— (1965b). *Endocrinology* **77,** 446–454

— (1966). *Annls Soc. r. Sci. méd. nat. Brux.* **19,** 1–120

Flerkó, B. (1963). In: *Advances in Neuroendocrinology* 211–224. Ed. by A. V. Nalbandov. Urbana; University of Illinois Press

Flerkó, B. and Szentágothai, J. (1957). *Acta endocr., Copenh.* **26,** 121–127

Frank, R. T., Kingery, H. M. and Gustavson, R. (1925). *J. Am. med. Ass.* **85,** 1558–1559

Fraps, R. M. (1962). In: *The Ovary* 2, 317–379. Ed. by S. Zuckerman. London; Academic Press

Gellert, R. J. and Ganong, W. F. (1960). *Acta Endocr., Copenh.* **33,** 569–576

Goldzieher, J. W., Axelrod, L. R. (1963). *Fert. Steril.* **14,** 631–653

Gorski, R. A. and Wagner, J. W. (1965). *Endocrinology* **76,** 226–239

Greep, R. O. (1961). In: *Sex and Internal Secretions* 1, 240–301. Ed. by W. C. Young. London; Baillière, Tindall and Cox

Greer, M. A. (1953). *Endocrinology* **53,** 380–390

de Groot, J. (1962). In: *Proc. XXII Int. Congr. Physiological Sciences* 1, 623–624

— (1965). In: *Sex and Behavior* 496–505. Ed. by F. A. Beach. New York; John Wiley

Halász, B. and Pupp, L. (1965). *Endocrinology* **77,** 553–562

— — and Uhlarik, S. (1962). *J. Endocr.* **25,** 147–154

— — — and Tima, L. (1965). *Endocrinology* **77,** 343–355

Hammond, J. (1925). *Reproduction in the Rabbit.* Edinburgh; Oliver and Boyd

Hammond, J. Jnr. (1954). *Vitam. Horm., Lpz.* **12,** 157–206

Harris, G. W. (1954). In: *Biochemistry of the Developing Nervous System* 431–442. New York; Academic Press

— (1955). *Neural Control of the Pituitary Gland.* London; Edward Arnold

— (1964). *Endocrinology* **75,** 627–648

— and Campbell, H. J. (1966). In: *The Pituitary Gland* 2, 99–165. Ed. by G. W. Harris and B. T. Donovan. London; Butterworths

— and Jacobsohn, D. (1952). *Proc. R. Soc. B.* **139,** 263–276

— and Levine, S. (1965). *J. Physiol., Lond.* **181,** 379–400

Hayhow, W. R. (1959). *J. comp. Neurol.* **113,** 281–314

— Webb, C. and Jervie, A. (1960). *J. comp. Neurol.* **115,** 187–215

Heim, L. M. and Timiras, P. S. (1963). *Endocrinology* **72,** 598–606

Herbert, J. and Zuckerman, S. (1958). *J. Endocr.* **17,** 433–443

Hess, A. (1958). *J. comp. Neurol.* **109,** 91–115

Hilliard, J., Archibald, D. and Sawyer, C. H. (1963). *Endocrinology* **72,** 59–66

— Endröczi, E. and Sawyer, C. H. (1961). *Proc. Soc. exp. Biol. Med.* **108,** 154–156

— Hayward, J. N. and Sawyer, C. H. (1964). *Endocrinology* **75,** 957–963

Himwich, W. A. (1962). *Int. Rev. Neurobiol.* **4,** 117–158

Hohlweg, W. (1934). *Klin. Wschr.* **13,** 92–95

Hoogstra, M. J. and Paesi, F. J. A. (1955). *Acta physiol. pharmac. Néerl.* **4,** 395–406

Hough, W. H., Bearden, H. J. and Hansel, W. (1955). *J. Anim. Sci.* **14,** 739–745

Ifft, J. D. (1964). *Anat. Rec.* **148,** 599–603

CONTROL OF FOLLICULAR GROWTH AND OVULATION

Ingvar, D. H. and Hunter, J. (1955). *Acta physiol. scand.* **33,** 194–218
Jacobsohn, D. (1965). *Acta Univ. lund.* Section II, No. 17
— (1966). In: *The Pituitary Gland* 2, 1–21. Ed. by G. W. Harris and
    B. T. Donovan. London; Butterworths
Jefferson, J. M. (1940). *J. Anat.* **75,** 106–134
Jöchle, W. (1956). *Endokrinologie* **33,** 129–138
Kanematsu, S. and Sawyer, C. H. (1964). *Endocrinology* **75,** 579–585
Kawakami, M. and Sawyer, C. H. (1959). *Endocrinology* **65,** 652–668
Kawashima, S. (1960). *J. Fac. Sci. Tokyo Univ.* Sect. IV **9,** 117–125
— (1964). *Annotnes zool. jap.* **37,** 79–85
Kempf, R. (1950). *Archs Biol. Paris* **61,** 501–594
Kikuyama, S. (1961). *Annotnes zool. jap.* **34,** 111–116
— (1962). *Annotnes zool. jap.* **35,** 6–11
Kimura, T. (1960). *J. Fac. Sci. Tokyo Univ.* Sect. IV **9,** 127–135
Kincl, F. A. and Maqueo, M. (1965). *Endocrinology* **77,** 859—862
Koulischer, L. (1960). *Annls. Endocr.* **21,** 314–320
Krejci, M. E. and Critchlow, V. (1959). *Anat. Rec.* **133,** 300
Lauson, H. D., Golden, J. B. and Sevringhaus, E. L. (1939). *Am. J. Physiol*
    125, 396–404
Law, O. T. and Sackett, G. P. (1965). *Neuroendocrinology* **1,** 31–44
Lisk, R. D. (1960). *J. exp. Zool.* **145,** 197–207
— and Newlon, M. (1963). *Science, N.Y.* **139,** 223–224
Littlejohn, B. M. and de Groot, J. (1963). *Fedn Proc. Fedn Am. Socs exp. Biol.*
    22, 571
Luce-Clausen, E. M. and Brown, E. F. (1939). *J. Nutr.* **18,** 551–562
McCormack, C. E. and Meyer, R. K. (1962). *Proc. Soc. exp. Biol. Med.* **110,**
    343–346
— — (1963). *Gen. comp. Endocr.* **3,** 300–307
— — (1964). *Endocrinology* **74,** 793–799
— — (1965). *Fert. Steril.* **16,** 384–392
McQueen-Williams, M. (1935). *Proc. Soc. exp. Biol. Med.* **32,** 1051–1052
Maekawa, K. (1960). *Endocr. jap.* **7,** 53–56
Marshall, F. H. A. and Bowden, F. P. (1934). *J. exp. Biol.* **11,** 409–422
Martinez, C. and Bittner, J. J. (1956). *Proc. Soc. exp. Biol. Med.* **91,** 506–509
Massopust, L. C. and Daigle, H. J. (1961). *Expl Neurol.* **3,** 476–486
Mayer, G. and Soumireu, J. (1948). *C.r. Séanc. Soc. Biol.* **142,** 964–967
— Thévenot-Duluc, A. J. and Burin, P. (1965). *Path.-Biol., Paris* **13,** 989–
    1002
Moll, J. and Zeilmaker, G. H. (1966). *Acta endocr. Copenh.* **51,** 281–289
Nauta, W. J. H. (1963). In: *Advances in Neuroendocrinology* 5–21. Ed. by
    A. V. Nalbandov. Urbana; University of Illinois Press
Noumura, T. (1958). *J. Fac. Sci. Tokyo Univ.* Sect. IV **8,** 317–335
Odell, W. D., van Buren, J. M. and Hertz, R. (1962). *J. clin. Endocr.*
    Metab. **22,** 1262–1265
Ortavant, R., Mauleon, P. and Thibault, C. (1964). *Ann. N.Y. Acad. Sci.*
    117, 157–192

REPRODUCTION IN THE FEMALE MAMMAL

O'Steen, W. K. (1964). *Endocrinology* **74**, 885–888
Pfeiffer, C. A. (1936). *Am. J. Anat.* **58**, 195–225
Phoenix, C. H., Goy, R. W., Gerall, A. A. and Young, W. C. (1959). *Endocrinology* **65**, 369–382
Porter, R. W., Cavanaugh, E. B., Critchlow, B. V. and Sawyer, C. H. (1957). *Am. J. Physiol.* **189**, 145–151
Quinn, D. L. (1966). *Nature, Lond.* **209**, 891–892
—— and Zarrow, M. X. (1964). *Endocrinology* **74**, 309–313
—— —— (1965). *Endocrinology* **77**, 255–263
Radford, H. M. (1966). *J. Endocr.* **34**, 135–136
Ramirez, V. D., Abrams, R. M. and McCann, S. M. (1964). *Endocrinology* **75**, 243–248
—— and McCann, S. M. (1963). *Endocrinology* **72**, 452–464
—— and Sawyer, C. H. (1965). *Endocrinology* **76**, 1158–1168
Robertson, H. A. and Rakha, A. M. (1965). *J. Endocr.* **32**, 383–386
Ryan, R. J. (1962). *J. clin. Endocr. Metab.* **22**, 300–303
Sawyer, C. H. (1963). In: *Advances in Neuroendocrinology* 444–457. Ed. by A. V. Nalbandov. Urbana; University of Illinois Press
—— (1964). In: *Gonadotropins* 113–159. Ed. by H. H. Cole. San Francisco; W. H. Freeman and Company
—— Everett, J. W. and Markee, J. E. (1949). *Endocrinology* **44**, 218–233
—— Markee, J. E. and Hollinshead, W. H. (1947). *Endocrinology* **41**, 395–402
Schiavi, R. C. (1964). *Am. J. Physiol.* **206**, 805–810
Schreiber, V. (1963). *Hypothalamo-Hypophysial System;* Prague Publishing House of the Czechoslovak Academy of Sciences
Segal, S. J. and Johnson, D. C. (1959). *Archs Anat. Microsc.* **48**, 261–273
Stegmann, H. (1959). *Gynaecologia* **148**, 262–268
Strauss, W. F. and Meyer, R. K. (1962). *Science, N.Y.* **137**, 860–861
Swanson, H. E. and van der Werff ten Bosch, J. J. (1964). *Acta endocr., Copenh.* **47**, 37–50
Szentágothai, J., Flerkó, B., Mess, B. and Halász, B. (1962). *Hypothalamic Control of the Anterior Pituitary.* Budapest; Akademiai Kiado
Szontágh, F. E. and Uhlarik, S. (1964). *J. Endocr.* **29**, 203–204
Takewaki, K. (1962). *Experientia* **18**, 1–6
Thomson, A. P. D. (1954). *Proc. R. Soc. B.* **142**, 126–135
Weisz, J., Matsuyama, E., Self, L. W. and Lloyd, C. W. (1964). *Progm. Forty-Sixth Meeting Endocrine Soc.* 25
van der Werff ten Bosch, J. J. (1963). *J. Endocr.* **26**, 113–123
Whipple, H. E. (1964). *Am. N.Y. Acad. Sci.* **117**, 1–645
Whitaker, W. L. (1936). *Proc. Soc. exp. Biol. Med.* **34**, 329–339
Witschi, E. (1961). In: *Human Pituitary Gonadotropins* 349–351. Ed. by A. Albert. Springfield; C. C. Thomas
Wolfson, A. (1965). *Archs. Anat. microsc.* **54**, 579–598
Woods, J. W. (1962). In: *Proc. XXII Int. Congr. Physiological Sciences* **1**, 611–612
Woolley, D. E. and Timiras, P. S. (1962a). *Endocrinology* **70**, 196–209

CONTROL OF FOLLICULAR GROWTH AND OVULATION

Woolley, D. E. and Timiras, P. S. (1962b). *Endocrinology* **71**, 609–617

Yazaki, I. (1960). *Annotnes zool. jap.* **33**, 217–225

DISCUSSION

S. J. FOLLEY (*Reading*)

In opening the discussion nearly all the work quoted has been carried out on the rat. Of importance in this connection is that prolactin secretion is inhibited by a hypothalamic factor in this and other mammals, but in birds the latest evidence is that the hypothalamus produces a factor which promotes prolactin secretion. This points to the necessity of widening studies to encompass as many species as possible and I would like to ask Dr. Donovan if there are any studies on the masculinization of the hypothalamus in any other species; primates for instance?

DONOVAN

No successful work on primates has been published as far as I am aware. Work is proceeding in the Oxford Department of Anatomy where attempts are being made to breed monkeys for studies with animals of known ages. Of course, the difficulty lies in overcoming the placental barrier or, if you put the hormone into the foetus, of ensuring purely hypothalamic masculinization and avoiding masculinization of the genitalia.

W. M. HANSEL (*Ithaca, U.S.A.*)

I would like to ask a question about the androgen-sterilized rats. You have quite logically ascribed the phenomenon to changes which occur during the critical period of organization of certain nuclei in the hypothalamus. Many workers, however, have produced semi-sterile rats, for example, rats which undergo half a dozen oestrous cycles after they mature and then show constant vaginal cornification. This does not seem to fit your explanation.

DONOVAN

I think this can be reconciled with my view provided you accept that the androgen is exerting a partial or incomplete effect on, let us say, the hypothalamus, although we do not really know where the androgen acts. It is certainly true that rats will reach puberty at approximately the normal time, show a few cycles and then enter persistent oestrus. It has been suggested, for example, that this might be an accelerated ageing of the hypothalamus. We can say that, as the hypothalamus is concerned in rhythmic fluctuations in gonadal hormonal levels, it gradually becomes refractory to subsequent stimulation, but this is pure guesswork.

I. ROTHCHILD (*Cleveland, U.S.A.*)

I think that the explanation in relation to ageing is probably the best one, since it has been demonstrated that the same incidence of anovulation occurs with small doses of hormone as with high doses, but at a much later age.

DONOVAN

One of the reasons for discussing the excretion of FSH and LH was that I hoped someone would come forward with more up-to-date information. It seems very surprising that while some work was done on gonadotrophin excretion in children in the 1930s and early 1940s little has been done since then. Is this due to a lack of interest or lack of technique? Brown published some work in 1958 but the difficulty in all the early work is that children of known ages were used. While the obvious course may be to compare children of 10, 11 and 12 years of age physiologically this is wrong. What one should do is to grade the children according to developmental age, and then see at which stage of puberty they are at.

J. R. CLARK (*Oxford*)

I would like to ask what is the nature of the evidence concerning the time taken for a light impulse to pass from the retina to the hypothalamus?

DONOVAN

This is evidence from electrophysiological studies where, for example, micro-electrodes are inserted into the hypothalamus and measurements made of the time taken for a flash of light to cause evoked potentials within the hypothalamus. This can be done in several ways. In one recent study, electrodes were placed in the anterior and posterior hypothalamus of cats and the evoked potential could be recorded in the anterior before the posterior hypothalamus. This would imply that the pathway to the anterior hypothalamus is shorter or has fewer synapses than the pathway to the posterior hypothalamus.

CLARK

So a single flash of light produced a change that could be recorded?

DONOVAN

Yes, that is so.

R. A. WELSH (*Cambridge*)

Could you give us some of the evidence that pituitary gonadotrophins feed-back directly on the hypothalamus?

DONOVAN

Yes. Most recently it has been found that local implantation of LH, like the local implantation of oestradiol, will depress the pituitary content of LH, and will also I believe depress LH output in spayed animals. This work is based on the local implantation of micro amounts of LH within the hypothalamus itself. Dr. McCann may be able to elaborate on this.

S. M. McCANN (*Texas, U.S.A.*)

This is correct. It is also claimed that it is effective in spayed animals. I might add that we have had a great deal of trouble in reproducing this effect. In nine experiments we have only two positive results and we do not really know the reason for this discrepancy. I have an open mind on this matter at the moment.

P. J. HEALD (*London*)

Could Dr. Donovan elaborate on the concept of how much oestrogen and progesterone might be getting into the pre-optic areas in the infant and adult rat? This should be measurable not by implantation but by measurement of uptake by radio-labelled hormones. Has this been done?

28

Donovan

No. A problem here has been the development of a blood-brain barrier which is not as fully developed in infantile animals as it is in adults. This is one factor which might be concerned in the critical periods outlined. During this time it may be a lot easier for gonadal hormones to reach the hypothalamus. Certainly they reach the hypothalamus of the infantile animal but they may reach a greater concentration more readily because of the lack of development of the blood-brain barrier. I should add, however, that there are some who say that such a blood-brain barrier does not exist.

THE NEUROLOGIC BASIS FOR THE ANOVULATION OF THE LUTEAL PHASE, LACTATION AND PREGNANCY

IRVING ROTHCHILD

Professor of Reproductive Biology, Department of Obstetrics and Gynecology, Western Reserve University School of Medicine, Cleveland, Ohio, U.S.A.

INTRODUCTION

OVULATION in the mammal implies not just the rupture of the follicle and release of the ovum, but the entire process of growth and maturation of the Graafian follicle up to and including its rupture. In this sense, ovulation can be regarded as the expression of the most basic relationship which the ovary bears to its total environment. We can try to understand this relationship by studying the process of spontaneous ovulation and conditions which prevent it. Studies of anovulatory conditions are especially valuable and complementary to a direct study of ovulation.

The primary object of this paper is to discuss the causes of anovulation during the luteal phase, lactation and pregnancy, with emphasis on the former and very little concerning the latter. My research interests have been concerned with the first two, the causes of which are comparable, while the anovulation of pregnancy seems to be a special case and not directly comparable with the other two.

We already know that the anovulation of the luteal phase is caused by progesterone, that the anovulation of lactation is caused by suckling, and that the anovulation of pregnancy (where the cause is other than that of the luteal phase itself) is caused by oestrogens and progestagens. We should like to know in addition: (*a*) the specific neural mechanisms by which progesterone prevents ovulation; (*b*) whether these are similar to or different from suckling; and (*c*) the neural mechanism by which continued secretion of large amounts of oestrogens and progestogens prevents ovulation.

THE PROCESS OF OVULATION

Ovulation occurs as the last of a series of orderly changes in organization of the Graafian follicle. These changes are dependent on a progressive increase in the secretion of the folliculotrophic* hormones of the pituitary, which is influenced to an important degree by the ovarian steroids. Although the phase of growth, development and hormone secretion by the Graafian follicle preceding these changes may take place in the presence of an unchanging rate of secretion of FSH and LH, the events in the follicle which lead inevitably to ovulation occur only following an exponential increase in secretion rate of the folliculotrophins, the most important of which is LH. What we commonly call the ovulatory discharge of LH is probably only the terminal portion of the curve expressing the changing rate of its pre-ovulatory secretion.

I believe the increase in rate of secretion of LH is the result of a sequential diminution of neural inhibition over a primary, and perhaps intrinsically active neural stimulation to the pituitary for LH secretion. The neural inhibition may act more as a brake limiting the amount of stimulation allowed to reach the pituitary, than by depressing the degree of stimulation itself. The strength of this inhibition is lowered (or is diverted away from the primary neural stimulus to the pituitary) by a combination of factors, among which the most important are a rising rate of oestrogen secretion, a relatively terminal secretion of progesterone, and some event or events connected with the animal's external environment (*Figure 1*; Rothchild, 1965). This means that the inhibitory influence is not that which is exerted on the secretion of the folliculotrophins by the long-term action of oestrogens. The reason for postulating its existence is that the rate of secretion of the folliculotrophins during the major part of the follicular phase of the cycle is much less than the maximal possible. Since the folliculotrophin secretion rate does rise to an almost maximal one during the pre-ovulatory stage of the cycle, and since this stage is also characterized by a rising rate of oestrogen secretion, some inhibitory factor other than oestrogen must maintain folliculotrophin secretion at less than the maximum rate during all but the late stages of follicle growth.

Anovulation can be due to factors which reduce stimulation of the ovary, or to factors which make the ovary unable to respond to stimulation. To examine the anovulation of the luteal phase, pregnancy and lactation we must deal, essentially, with the first category of factors. Because of the dependency of the pituitary on the central nervous system (CNS), this concept of ovulation allows us to say that anovulation due to changes in pituitary activity may be

---

* The term 'folliculotrophin' is used in a generic sense to describe either FSH or LH, or combinations of these two hormones; for justification *see* Rothchild (1960).

due to either of two general mechanisms, that which acts *primarily* to inhibit maximal-rate secretion of LH, and that which acts *primarily* to reduce the state of activity in the stimulatory component of the system controlling folliculotrophin secretion. The first mechanism would leave the stimulatory element of the regulatory system essentially intact and potentially able to function at maximal

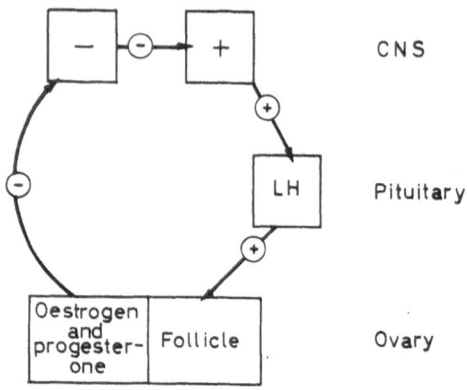

Figure 1. The basic mechanism of ovulation

Symbols: CNS—central nervous system, in particular the hypothalamus; *LH*—the secretion of LH by the anterior pituitary; +stimulation; −inhibition. The diagram attempts to show the most essential elements of the process through which maximal rate LH secretion is effected, i.e., by a depression of neural inhibitory influence—through the effects of oestrogens and progesterone—on a neural stimulatory influence which causes the pituitary to secrete LH at a maximal rate (*see* text for further details).

capacity, while the second would severely reduce its potential capacity to function. Not included in these considerations is the possibility that a reduction in folliculotrophin secretion may be due to causes which primarily affect the pituitary, rather than the neural systems which regulate its activity. I have omitted this because the evidence is almost overwhelmingly in favour of neural factors as primary causes of the anovulation of the luteal phase and lactation; what I have to say about pregnancy will not be affected by this omission.

THE ANOVULATION OF THE LUTEAL PHASE

Progesterone apparently inhibits ovulation by preventing the terminal maximal-rate secretion of LH, but it does not inhibit a basal rate of folliculotrophin secretion, and may actually increase the potential capacity of the pituitary for maximal-rate secretion. The evidence, in brief, for this viewpoint (Rothchild, 1965) is as follows. In the presence of a continual secretion or administration of progesterone, follicle growth to but not including pre-ovulation is not prevented. Oestrogens are secreted at an appreciable, though less than maximal, rate and follicle sensitivity to folliculotrophins is not significantly impaired. Folliculotrophic potency of the pituitary increases, and in women a basal rate of excretion of folliculotrophins is maintained. Ovulation can be induced in some mammals by the administration of oestrogens.

Since progesterone probably inhibits ovulation by acting on the neural influences which regulate folliculotrophin secretion, we would say that progesterone probably activates a neural inhibitory influence whose major site of action is on that neural process which promotes the exponentially rising rate of LH secretion. We know little about the specific nature of this effect, or the areas of the CNS which are involved in it, apart from the fact that a threshold-raising effect may be one aspect (Kawakami and Sawyer, 1959) and that a location in the hypothalamus may be another (Everett, 1964). We do know that the luteal phase itself, or the anovulation caused by chronic treatment with progesterone, has certain characteristics, and is accompanied by certain interesting and somewhat specific changes in physiological activity. Examination of these might clarify the specific nature of the ovulation-inhibiting effect. The problems can be stated as follows:

(a) Is the inhibitory influence exerted by progesterone the same as the one which normally controls maximum folliculotrophin secretion? (b) What is the connection between the way progesterone inhibits ovulation and the process which increases the folliculotrophic potency of the pituitary? (c) Is the inhibition over ovulation an ephemeral phenomenon, like many of the other effects of progesterone on neural mechanisms, and if so, what connection does this have with ovulation-inhibition? (d) How is ovulation-inhibition connected with the facilitation of ovulation during the period following withdrawal of progesterone? (e) What is the connection between the fact that progesterone inhibits ovulation and the facts that progesterone stimulates appetite, depresses motor activity and sexual receptivity, elevates body temperature, and maintains the secretion of prolactin? Each of these is considered in turn.

(a) *The identity of the neural inhibition of LH release caused by progesterone*

A basic premise is that the neural inhibition which acts normally to maintain a less than maximal-rate secretion of LH, and the neural inhibition associated with certain types of anovulatory disturbances in rats (Rothchild, 1965) are essentially the same, though different

2+   33

in degree. If the occurrence of ovulation means that the degree of normal inhibition over folliculotrophin secretion can be lowered in response to certain stimuli, then there must be certain limits within which the strength of this inhibitory influence may vary. This implies that certain stimuli or conditions can increase the strength of this inhibition above the normal range. When we call one range of degrees of inhibitory activity 'normal', we mean that, in the interval between ovulations, these degrees of inhibition can maintain a less than maximal-rate secretion of folliculotrophins, and that these degrees of inhibition can be lowered sufficiently for ovulation to occur, by factors which are *also* normal. Furthermore, we assume that what causes such 'normal' inhibition, or reduces its strength, is a complex of factors, arising from the animal's external as well as internal environment which impinges in various ways upon the CNS. (For example, tonic activity in the cerebral cortex could be part of this inhibitory influence, since the induction of cortical spreading depression in rats leads to a large-scale release of LH (Taleisnik and Caligaris, 1962; Taleisnik, Caligaris and De Olmos, 1962).) The inhibition which is responsible for several varieties of anovulation of the 'constant oestrus' type (Rothchild, 1965, page 259) may arise, therefore, from what are essentially the same neural complexes and activities which are responsible for the normal inhibition, but the inhibition is stronger either because the activating environmental factors have also increased in strength, or because other factors have raised the sensitivity of the neural inhibitory influence to what would otherwise be a normal environment.

I do not think that progesterone acts through this 'normal' neural inhibition to prevent ovulation, but rather that it activates an inhibitory influence which may be of more limited or special character than the normal one. Some of the reasons for this impression follow.

Progesterone, together with oestrogen, seems to *diminish* the normal inhibition over maximal-rate LH release, and thus effects ovulation. I am well aware that the chronic effect of progesterone is the exact opposite of its initial effect on several physiologic processes (Rothchild, 1965), but it makes better physiologic sense to suppose that the reason for the change from an ovulation-inducing effect to an ovulation-inhibiting one amounts to the difference between reducing one source and increasing *another* source (and kind) of neural inhibitory activity, than that the difference is between one kind of effect (inhibition) and another (stimulation) on the *same* kind of neural mechanism.

Another reason is that when the 'normal' inhibition becomes high enough to cause anovulation, it is not diminished by oestrogen treatment and may even be increased by it; that is to say, oestrogen treatment does not induce ovulation in the 'constant oestrus' type of anovulatory conditions. Oestrogen treatment during progesterone-induced anovulation does lead to ovulation (Everett, 1945; Kidder, Casida and Grummer, 1955; Klein, 1947) and in fact, in the type of 'constant oestrus' anovulation which occurs in the DA strain of rats, the *administration* of progesterone permits oestrogens to induce ovulation (Everett, 1950). Furthermore, progesterone counteracts, to varying degrees, the conditions of the 'constant oestrus' anovulatory state itself. I have discussed this phenomenon in detail elsewhere (Rothchild, 1965, pages 253–255; 259–261) but a few examples might be appropriate here. Small daily doses of progesterone delay the onset of anovulation in young DA rats, and spontaneous ovulatory cycles commonly follow induced ovulation and pseudopregnancy in rats of the DA strain or in rats made anovulatory by anterior hypothalamic lesions. Even the induction of the anovulatory state in rats by treatment with androgen in early life can be prevented by treatment with progesterone (Cagnoni, Fantini, Morace and Ghetti, 1965; Kincl and Maqueo, 1965). If progesterone inhibited ovulation by acting on the same neural inhibitory activity that was responsible for the abnormal anovulatory state, one would not expect these findings.

A third reason is that the typical accompaniments of progesterone-induced anovulation—i.e., depressed motor activity and sexual receptivity, and increased appetite, body temperature, and prolactin secretion—are not the usual accompaniments of the abnormal anovulatory states. For example, rats made anovulatory by treatment with androgen in early life show almost as much running activity as do normal ones (Kennedy, 1964b). Only in the anovulation of old age in rats are some of these characteristics present; this similarity does not necessarily weaken my point and may even strengthen it.

(*b*) *The connection between ovulation inhibition and the increase in folliculo-trophic potency of the pituitary*

The luteal phase increase in folliculotrophic potency of the pituitary (Rothchild, 1965, pages 231–236) may not be due directly to progesterone, but by delaying ovulation (i.e., by preventing maximal-rate release of the folliculotrophins) progesterone may permit the process which is directly responsible for the increase to continue (Rothchild, 1965, page 242). However, since the problem is by no means solved we have been carrying out some rather simple experiments. One involves comparing the increase of folliculotrophins with that which follows castration; another deals with the effect of oestrogen administration. So far, we have followed only the changes in LH potency of the pituitary and have not yet investigated how these changes are related to either release or production rate, so their interpretation must obviously be made with reservations.

LH was measured by the ovarian ascorbic acid depletion (OAAD) test, using a 4 h interval between injection of test material and autopsy. Each ovary was assayed separately for its ascorbic acid content. A 4-point assay design was used. The two LH standard doses (NIH-LH-S1) were 0·4 μg and 1·6 μg/rat and the pituitaries were assayed as a dry powder suspended in saline, at a dose of 100 μg or 400 μg/rat. Five rats were used for each dose of standard or unknown and all injections were made intravenously between 08.00 and 10.00 h. Potency was calculated according to the statistical procedure described by Bliss (1956) and expressed as microgramme equivalents of LH standard per milligramme of dry pituitary. The results, which are preliminary only, are shown in *Figure 2*.

The rate at which LH accumulates in the pituitary during the first 8 days of pseudopregnancy is the same at which it increases during an equivalent period of time following castration on day 1 of dioestrus (*Figure 2, a*). If the increase in both conditions expresses the same aspect of the secretion process as a whole (and it may not do so), it would be certain that the luteal phase increase need not be due to progesterone, since it occurs even in its absence. The increase also does not seem to be due to a relative lack of oestrogen since oestrogens are certainly secreted in pseudopregnant rats, and the administration of 5·0 μg of oestradiol daily—which markedly depresses pituitary LH potency in intact rats after 28 days of treatment (Rothchild and Schwartz, 1965)—had no observable effect on LH potency in either the castrated or the pseudopregnant rats during the first 6–8 days after ovulation (*Figure 2b, c* and *d*).

What these results may mean is that the factor which stimulates the pituitary to synthesize LH (and probably both folliculotrophins) can act apparently quite independently of oestrogen levels for an appreciable period of time. It also does not seem to be directly dependent on progesterone, and the continued rise of pituitary folliculotrophic potency during the luteal phase, therefore, may really be an expression of only the delay of ovulation.

(c) *The ephemerality of the progesterone-induced block of ovulation*

The effects of progesterone on several phenomena that are either

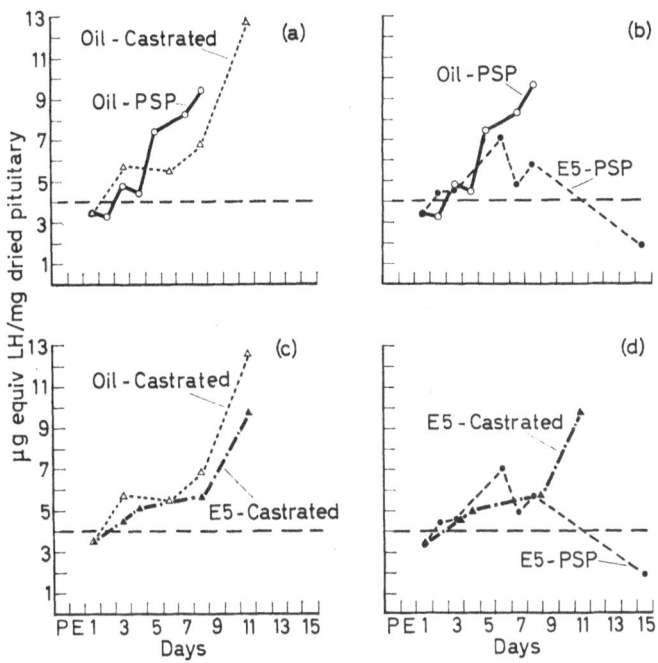

Figure 2. Changes in pituitary LH potency during pseudopregnancy, or after castration or oestrogen treatment

Symbols: P—day of pro-oestrus; E—day of oestrus; *abscissae*—days after day of oestrus (i.e., day 1 is first day of dioestrus). In all cases, treatment with oil or oestradiol was begun on day 1; castration was also done on this day. *Ordinates*: pituitary LH potency expressed as μg equivalent of standard LH/mg of dried pituitary. The broken horizontal line in each figure represents the potency of the pituitary of the cyclic rat at pro-oestrus (the highest point in the cycle); the lowest point in the cycle (not shown) occurs on the day of oestrus, and is almost always about 50 per cent of the pro-oestrus value (Schwartz and Bartosik, 1962). *PSP*—pseudopregnant rats. *E5*—rats treated with 5·0 μg of oestradiol daily. (a). Oil-treated castrated rats' pituitaries compared with those of oil-treated pseudo-pregnant rats; (b). Oil-treated pseudopregnant rats' pituitaries compared with those of oestrogen-treated pseudopregnant rats; (c). Oil-treated castrated rats' pituitaries compared with those of oestrogen-treated castrated rats; (d). Oestrogen-treated castrated rats' pituitaries compared with those of oestrogen-treated pseudopregnant rats (for further details *see* text).

clearly or very probably CNS-dependent have a strong tendency to gradually diminish and eventually to disappear (Rothchild, 1965, pages 243–249). This is true of the anaesthetic effect, the temperature elevating effect, the depression of sexual receptivity (and its correlate of raising the EEG arousal threshold to reticular formation stimulation (Kawakami and Sawyer, 1959)), and the increase in the threshold for electric shock-induced convulsions. Progesterone's stimulating effect on prolactin secretion (Rothchild and Schwartz, 1965) and the ovulation-inhibiting effect may also be ephemeral. Wolfe (1946) found that continual treatment of immature rats with progesterone prevented follicular luteinization for 25 days but not for periods longer than 90 days. Kanematsu and Sawyer (1965) showed that implantation of a norethisterone pellet in the hypo-thalamus prevented coitus-induced ovulation in rabbits for up to 4 weeks but not for 8 weeks. In my own laboratory we found that adult pseudopregnant rats injected daily with either 2 or 5 mg of progesterone apparently ovulated between the third and fourth weeks of treatment (Rothchild and Schwartz, 1965). Similar find-ings have been made in adult castrated rats bearing autotransplanted ovarian tissue beneath their kidney capsules (Kaufman and Roth-child, 1966; and unpublished findings).

The fact that ovulation-inhibition by progesterone is temporary is in contrast, again, to the kind of inhibition responsible for the 'con-stant oestrus' types of anovulation. In the latter, ovulation virtually never occurs spontaneously. It is interesting that the anovulation of lactation also tends to be ephemeral, and in certain respects, though by no means all, it resembles the anovulation caused by progesterone. This suggests that the type of neural process inhi-bited or depressed by progesterone, and perhaps by suckling, is one which has a relatively high potentiality for being activated by other factors. For example, it may be one which is kept in an active state by stimuli coming to it from many other parts of the CNS, so that these tend, eventually, to override the effect of progesterone.

(d) *The connection between the progesterone-induced inhibition of ovulation and the facilitation of ovulation following progesterone withdrawal*

The facilitatory effect on ovulation of discontinuation of pro-gesterone secretion or treatment bears a strong resemblance to the way progesterone affects sexual receptivity in sheep and maternal nest-building in rabbits (Rothchild, 1965, pages 245–248). Pro-gesterone prevents the expression of these behaviours but the full pattern is crucially dependent on the animal having been under its influence beforehand. Although ovulation is not so crucially dependent on a period of pre-treatment with progesterone this is probably a minor difference. All three phenomena are inhibited during periods of continual progesterone secretion or treatment.

They occur at a fairly uniform interval following withdrawal of progesterone, and unfold as a whole in a much more regular and complete way than would be the case in the absence of previous progesterone.

The combination of inhibition of a behaviour pattern *during*, with a stereotyped expression of the pattern *after* a period of progesterone secretion, suggests a process in which all the participating elements are brought to a state of maximum readiness for interaction, but in which a crucial set of elements is not engaged with the others, so that the complete process cannot occur. What strikes me as the factor common to these two behaviour patterns and the process of ovulation is a positive feed-back system of regulation; each process builds up to a climax and then is resolved into another kind of activity. If we could imagine progesterone acting to hold back one essential connecting element in a circular system of mutual stimulations, it would be easy to understand why the process goes so well once the block is removed.

One aspect of the post-progesterone facilitation of ovulation suggests that prolactin secretion may be involved in it. When pseudopregnancy is induced in certain types of anovulatory rats there is a tendency, before the anovulatory state returns, for several spontaneous ovulations to follow the pseudopregnancy. This may occur because the neural inhibition over prolactin secretion (Rothchild, 1965, pages 289–290) is brought to a lower level of activity by progesterone. In general, in anovulation induced by constant light and in that induced by treatment with androgens in early life, there seems to be a direct relationship between the degree of neural inhibition over maximal-rate LH release and that over prolactin secretion. For example, ovulation is readily induced in rats showing a light-dependent anovulation but is induced with difficulty in those showing the anovulation caused by treatment with androgen or oestrogen in early life (Rothchild, 1965, pages 252–255, 257–258). Similarly, pseudopregnancy frequently follows mating-induced ovulation in the former types, while mating only rarely induces either ovulation or pseudopregnancy in the latter (Segal and Johnson, 1959; Gorski, 1963; Swanson and van der Werff ten Bosch, 1964). When pseudopregnancy does occur, there is an immediate return to the anovulatory state (Aschheim, 1965). In the castrated male rat bearing an ovarian autotransplant there is also evidence that neither prolactin secretion nor maximal-rate LH release is readily induced (Zeilmaker, 1963). It might be that the facilitation of prolactin secretion continues into the post-pseudopregnancy period long

39

enough to allow a small amount of progesterone to be secreted by the corpora lutea of the cycle, and thus permits ovulations to continue, at least until the neural inhibition over prolactin secretion is fully restored (Everett, 1944). In the *normal* mammal, one might assume that the degree of neural inhibition over prolactin secretion is never as great as it is in the anovulatory state, and that the amount of prolactin secreted under the influence of progesterone plays an important but undefined role in facilitating ovulation.

(e) The connection between ovulation-inhibition by progesterone and the effects of progesterone on appetite, motor activity, sexual receptivity, body temperature, and prolactin secretion

The inhibition of ovulation by progesterone, which we have defined as an inhibition of maximal-rate LH release, is frequently accompanied by some or all of the following phenomena: an increase in appetite and related processes (Brobeck, Wheatland and Strominger, 1947), an increase in body temperature (Brobeck *et al.*, 1947), indirect evidence for an increase in prolactin secretion (Rothchild, 1965, pages 291–295), a depression of motor activity (Brobeck *et al.*, 1947; Richter, 1956), and a depression or absence of sexual receptivity (Dempsey, Hertz and Young, 1936; Dziuk, 1960; Goy, Phoenix and Young, 1964). During the oestrogen-dominated phase of the cycle and particularly from pre-ovulation up to and including ovulation, the activity of these modalities is opposite. (Table 1.)

Table 1

A comparison between pro-oestrous–oestrous phase of the cycle and the luteal phase with respect to the state of activity of several behavioural or other physiological processes

	Pro-oestrus-Oestrus (*oestrogen-dominated*)	*Luteal phase* (*progesterone-dominated*)
Motor activity (running, playing, etc.)	High	Low
Sexual activity (receptivity, coitus, etc.)	High	Low
Appetite (appetite, feeding, food-seeking, etc.)	Low	High
Body temperature	Low	High
LH secretion rate	Rising to maximum	Held at basal rate
Prolactin secretion rate	Low	High

Since each process is controlled to an important extent by the hypothalamus, the *pattern* of change in the level of their activities in relation to the ovulation cycle could be an important clue, not only to the neural mechanism which generally controls any one or all of them, but to the specific neural mechanism through which progesterone inhibits ovulation.   The hypothalamic controls over appetite (Mayer, 1957; Anand, 1961; Morgane, 1961a, b; Miller, 1965); body temperature (von Euler, 1961), and sexual receptivity (Goy and Phoenix, 1963), apparently include a primary stimulatory component which is under inhibitory control by another hypothalamic area.   For motor activity (Kennedy, 1964a, b) and for LH release (Everett, 1964) there is evidence for stimulatory control, and although there is probably also some inhibitory control over these processes, the evidence so far is still indirect.   For prolactin secretion there is good evidence that the primary hypothalamic control is inhibitory (Rothchild, 1965, pages 289–290).

A change in the level of activity of any modality can therefore be defined in terms of the necessary change which may occur in the control system (e.g., increase in appetite results from depression of the activity of the inhibitory component in appetite control). Similarly, for progesterone to affect each modality as it does, the ultimate change in any control system can be defined (e.g., for progesterone to depress motor activity the stimulatory component of motor activity control must be depressed).   This way of defining the problem of how ovulation-inhibition may be related to these other effects is shown in Table 2, and we can now proceed to the question of how the overall pattern might be the result of effects on the control systems.   There are two general possibilities. The first is that progesterone might affect each control system directly; the second is that the control systems are connected together in series and that by acting on only one of them directly, progesterone might also affect the others by the influence the affected control system exerts on them.   Let us examine the first possibility.

To induce the effects described, progesterone would have to act either as an inhibitor or as an excitant.   If it acted directly on the control systems of each process, it would have to affect either the inhibitory or the stimulatory element of each set of controls.   It probably does not act as an excitant, for to do so it could influence only the stimulatory component over appetite in the far-lateral area of the hypothalamus (Morgane, 1961a, b), and this area seems to be intrinsically active, its primary system of control being through modulation of the inhibitory activity arising in the ventromedial

Table 2

An attempt to describe the nature of the hypothalamic control over certain behavioural or other physiologic processes which are affected by a continuous secretion (or administration) of progesterone

Physiologic activity or process	Nature of the ultimate or primary hypothalamic control	Nature of the modifying or secondary hypothalamic control	Change in activity of the process under the influence of progesterone	Probable change in the control system required to produce the change in activity
Motor activity	stimulatory	inhibitory	decreased	inhibition of stimulation
Sexual receptivity	stimulatory	inhibitory	decreased	inhibition of stimulation
Appetite	stimulatory	inhibitory	increased	inhibition of inhibition
Body temperature (heat-losing processes) *	stimulatory*	inhibitory	decreased*	inhibition of stimulation*
Maximal-rate LH secretion	stimulatory	inhibitory	decreased	inhibition of stimulation
Prolactin secretion	inhibitory	(both inhibitory and stimulatory?)	increased	inhibition of inhibition

* Progesterone causes a rise in body temperature, which is assumed to result from *decreased* activity of the hypothalamic influences promoting heat-loss.

nucleus (Mayer, 1957; Morgane, 1961b). As an excitant, progesterone would have to affect the inhibitory components of the controls over motor activity, heat-losing processes, sexual receptivity, and maximal-rate LH release. This would imply, as far as the latter is concerned, that progesterone would increase the 'normal' inhibition, a conclusion I have tended to avoid (*see* above). Furthermore, in order to stimulate prolactin secretion, progesterone acting as an excitant would have to work through another inhibitory element, rather than the one directly controlling the pituitary.

If progesterone acted as an inhibitor, it would have to affect the inhibitory elements of the controls over appetite and prolactin secretion, and the stimulatory elements of the controls over motor activity, sexual receptivity, heat-losing processes, and maximum-rate LH release. There are many indications that progesterone acts on the CNS as an inhibitor (Rothchild, 1965, pages 243–249) and there is no reason to think that its action cannot be directed to the inhibitory component of one system and to the stimulatory component of another.

These explanations may be correct, but in my opinion they do not answer the puzzle of why the inhibitory effects of progesterone on ovulation are so frequently associated with the particular combination of changes I have already described in appetite, motor and sexual activity, body temperature and prolactin secretion.

One clue to the puzzle, as I see it, is that progesterone increases appetite, and another is that it stimulates prolactin secretion. The neural control system for appetite consists of at least the ventromedial nucleus and the far-lateral area, the latter functioning as the centre for stimulation of the appetite drive itself, as well as for its associated processes (such as food seeking, ingestion, etc.), while the ventromedial nucleus acts on the far-lateral area as an inhibitory influence (*see* references above). Somewhat similarly, the prolactin-secretion control system consists of an intrinsic capacity within the pituitary to secrete prolactin, and an inhibitory influence within an unknown area of the hypothalamus. It is remarkable that factors which increase appetite and its associated processes, as well as certain ones which stimulate prolactin secretion, tend to be accompanied by some or all of the changes in the other processes that also occur under the influence of progesterone. Feeding is closely associated in rats with a marked decrease in motor activity (Hollifield and Parson, 1957; Stevenson and Rixon, 1957; Kennedy and Mitra, 1963a) and an increase in body temperature (Brobeck, 1948; Stevenson and Rixon, 1957). The hyperphagia induced in rats by lesions in the ventromedial nucleus (Brobeck, 1946) is accompanied by a decreased motor activity and absence of sexual receptivity (Kennedy and Mitra, 1963a). Lactation itself is associated with increased appetite, decreased motor activity and decreased sexual receptivity (Kennedy, 1953; Richter, 1956), as well as possibly, increased body temperature. The increased appetite shown by rats in old age (Kennedy, 1953) is frequently accompanied by anovulation and signs of prolactin secretion (Aschheim, 1961, 1962; Aschheim and Pasteels, 1963). Lesions, presumably in the ventromedial nucleus, prevent ovulation in rabbits in response to injection of copper salts (Kurachi and Suchowsky, 1958), and in rats, frequently, though not always, lead to anovulation and decreased sexual receptivity (Greer, 1953; Barnet and Mayer, 1954; Van Dyke, Simpson, Lepkovsky, Konoff and Brobeck, 1957) as well as hyperphagia (Kennedy and

Mitra, 1963a). Treatment of mice with gold thioglucose—which damages the ventromedial nucleus (Mayer and Marshall, 1956; Heatherington and Ranson, 1942; Wagner and de Groot, 1963)—causes an increase in appetite, anovulation of the 'constant oestrus' type and signs of prolactin secretion (Liebelt, Sekiba and Taylor, 1961; Browning and Kwan, 1964; McBurney, Leibelt and Perry, 1965).

The concomitance of inhibition of ovulation with these effects, and especially with increase of appetite, may be due therefore to the possibility that progesterone acts primarily on the control system over appetite, specifically, by depressing the activity of at least part of the ventromedial nucleus. The consequent release of the far-lateral area from inhibition would allow it to stimulate appetite, but through its connections with the control systems of the other modalities, on which it would act as an *inhibitory* influence on the primary control elements, its other effects would be manifested. Since the primary controls are stimulatory for motor activity, sexual receptivity, heat-losing processes and maximal-rate LH release, and inhibitory for prolactin secretion, the effects would be a decrease in motor and sexual activity, increase in body temperature, prevention of maximal-rate LH release, and increase in prolactin secretion (*Figure 3*). An alternative possibility is that far-lateral area activity as a whole may be inhibitory. It may therefore affect appetite as it does, not by stimulating the neural mechanisms that directly control food seeking, etc. but by depressing an unknown inhibitory activity over such processes. The results of Knigge and Hays (1963), cited below, should remind us that such multiple levels of inhibition are possible.

Some findings do not seem to agree with this hypothesis. A lesion of the ventromedial nucleus in rats depressed motor and sexual activity and increased food intake but it did not usually prevent ovulation (Kennedy and Mitra, 1963a). In guinea-pigs, destruction of the lateral hypothalamic area resulted in anovulation (Leonardelli and Barry, 1964). Norethisterone, implanted in the posterior median eminence area, in rabbits, prevented ovulation in response to coitus (Kanematsu and Sawyer, 1965). These results, however, do not necessarily invalidate the explanation.

Kennedy and Mitra's (1963a) results may mean that a lesion in the ventromedial nucleus and the effects on the ventromedial nucleus of progesterone are not necessarily the same, since the effects of a lesion may be much less specific than those of a chemical (Miller, 1965). Their results may also mean that the postulated inhibitory effect of far-lateral area activity on the control system for maximal-rate LH release is not direct but may include synapses with certain neurons in the ventromedial nucleus. Alternatively, the ventromedial nucleus area may include the required fibres of passage.

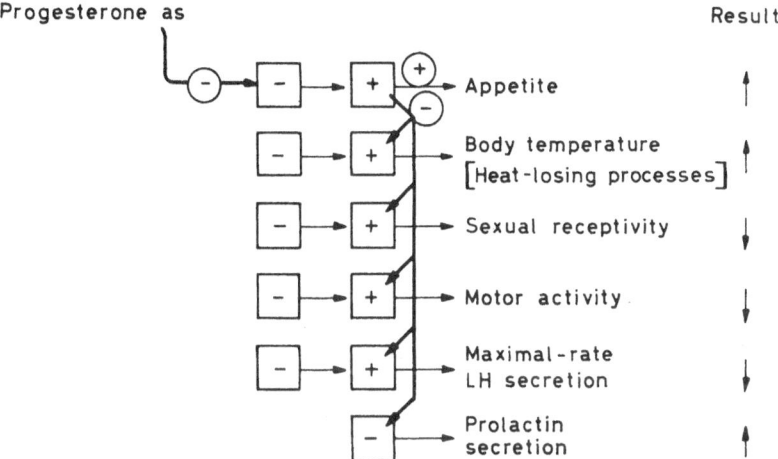

Figure 3. A symbolic representation of how progesterone may inhibit ovulation

Symbols: The hypothalamic neural controls for each modality (appetite, heat-losing processes, sexual receptivity, etc.) are assumed to consist at least of a primary stimulating influence (box with + sign), which is kept under tonic inhibition (box with − sign); the primary controlling influence for prolactin secretion, however, is an inhibitory one. Progesterone is assumed to depress the inhibitory element of the appetite-controlling system, and thus to free the stimulatory element of this system from restraint. The stimulatory element over appetite is assumed to feed inhibitory impulses to the primary controlling element of the other modalities, and thus to bring about the changes in each modality represented under 'Result', where an arrow pointing downward represents a decrease, and an arrow pointing upward an increase, in the level of activity of the modality. The concomitance of ovulation-inhibition with increase in appetite, increase in body temperature, decrease in sexual receptivity, decrease in body motor activity, and stimulation of prolactin secretion, may therefore be explained by this hypothesis (*see* text for further details).

Such a possibility should not be discounted, since other evidence for multiple levels of inhibitory control over pituitary activity can be found. For example, the increased secretion of ACTH in response to ether or haemorrhage can be blocked by lesions in the midbrain reticular formation or the amygdala, but if an additional lesion is placed in the hippocampus, the block is removed (Knigge and Hays, 1963). Such results indicate that the hippocampus can prevent ACTH secretion, but can be inhibited in turn by either the midbrain reticular formation or the amygdala, or both.

Furthermore, the lack of absolute correlation between the evidence for unrestrained far-lateral area activity (e.g. increased appetite) and anovulation must be compared with the findings cited above that lesions in the area of the ventromedial nucleus have frequently been associated with anovulation, and with the statement by Kennedy and Mitra (1963a) in their discussion, that underfeeding of their lesioned rats, as well as underfeeding of intact rats, leads to an increased incidence of constant vaginal cornification, i.e., anovulation. This is an extremely interesting observation, because it suggests the possibility, at least, that the far-lateral area can be excited by withdrawal of food, even after the normal restraint exerted on it by the ventromedial nucleus has been removed and, of course, this observation agrees with the hypothesis that far-lateral area activity can inhibit maximal-rate LH release. It is also very interesting that Morgane (1961b) in discussing the effects on appetite of lesions in the median forebrain bundle remarks that just the passage of the electrode (without current) through the cerebral cortex led to aphagia and adipsia lasting for about 24–36 h. This, of course, could indicate that the ventromedial nucleus was activated by the injury to the cortex, and the far-lateral area depressed as a result. Since a cortical spreading depression can be induced by similar handling of the cortex, and is associated with excellent evidence for maximal-rate LH release (Taleisnik and Caligaris, 1962), the two sets of findings complement each other and agree with the hypothesis.

The results of Leonardelli and Barry (1964) are not conclusive, since their lesions may not have included the area involved in appetite control, and the animals were not followed long enough to be sure that the anovulation was a permanent effect. Kanematsu and Sawyer's (1965) findings indicate only that the median eminence area can also be affected directly by a progestogen, but it is also possible that norethisterone does not behave in all respects as does progesterone. It is interesting that (according to their own unpublished findings) progesterone implanted in the same area did not inhibit ovulation.

The pattern of change in these processes which we see during the pro-oestrus-oestrus phase of the cycle could also agree with this hypothesis if we assume that, in the absence of a continual secretion of progesterone, oestrogen *stimulates* ventromedial nucleus activity. The inhibitory influence of the ventromedial nucleus over appetite, through its effects on the far-lateral area, could be associated with the other changes, i.e., increased motor and sexual activity (a beautiful example is the fact that seals do not feed during the entire rutting season), decreased body temperature, and an increasing rate of LH secretion. These influences could occur either because the ventromedial nucleus itself activates the stimulatory component of the control systems for these processes, or because the control systems are removed from the inhibitory influence of far-lateral area activity. It is interesting that oestrogen treatment, which stimulates running as well as sexual activity in intact rats, does not do so after destructive lesions in the ventromedial nucleus (Kennedy, 1964a).

The hypothesis may fit with the potentiation of folliculotrophin secretion during, and with the facilitation of ovulation following periods of continuous progesterone secretion. If the inhibitory influence of the far-lateral area is directed only to the maximal-rate release process (for LH), it would not interfere with the stimulus to folliculotrophin synthesis, while progesterone would still be permitted to act on the process that prepares the pituitary for maximal-rate LH release. Work in my laboratory has shown that in *prepubertal* rats, progesterone treatment does not increase body weight as it does in adults, nor does it cause a facilitation of ovulation. The lack of effect on body weight should not be surprising, since the ventromedial nucleus is apparently inactive in young growing rats (Kennedy, 1957). The lack of facilitation of ovulation agrees with the same explanation, since if this effect depended on the activity of the far-lateral area holding back all but the terminal stage of LH secretion, such a process would already be operating at a maximal rate in the prepubertal rat, because of the absence at this age of inhibitory activity in the ventromedial nucleus (Kennedy, 1957). Ovulation would be facilitated in such rats by the process which stimulates folliculotrophin secretion, and as I have already indicated (*Figure 2*), progesterone does not seem to affect this process.

It may also be worth pointing out at this point, that the prepubertal rat does not ovulate and its pituitary, like that of the pseudo-pregnant rat, has a high folliculotrophic potency; its motor activity is very low compared with the adult and is not increased by oestrogens (Kennedy and Mitra, 1963b), and its appetite and food consumption are maximal and cannot be increased by lesions of the ventromedial nucleus (Kennedy, 1957). At puberty, however, it ovulates, develops a restraint over appetite (Kennedy, 1957), and increases its running activity in response to oestrogen. I am aware of no information on body temperature changes or on prolactin secretion before and during puberty, but the changes in the modalities we have discussed certainly agree with a process in which activity in the ventromedial nucleus increases and activity in the far-lateral area decreases in association with the transition to sexual maturity, i.e., with a decrease in inhibition over folliculotrophin secretion.

THE ANOVULATION OF LACTATION

I have postulated (Rothchild, 1960) that suckling prevents ovulation by depressing the primary neural stimulus to folliculotrophin secretion. This mechanism resembles that by which it stimulates prolactin secretion, by inhibiting the primary neural inhibition over the secretion of prolactin. A part of this theory was that the threshold of response of the folliculotrophin regulating system to suckling must be higher than that of the prolactin-controlling system because the minimal amount of suckling stimulus which will suppress folliculotrophin secretion is higher than the minimal amount which will stimulate prolactin secretion. Also, ovulation cycles eventually reappear with prolonged suckling, although milk secretion is maintained (Bruce, 1958).

The major difference between the effect of suckling and the effect of progesterone in the rat seems to be the way ovulation is inhibited. Suckling depresses the synthesis of the folliculotrophins, while progesterone primarily affects the release process. In mice and hamsters there is some evidence that FSH and LH secretion rates are affected differently by suckling (Greenwald, 1962, 1965) and in women some findings suggest a direct effect—possibly of prolactin—on the ovary (Keetel and Bradbury, 1961). The difference between the effect of progesterone and that of suckling on folliculotrophin secretion, however, tends to diminish with time; that is, the folliculotrophin secretion control system eventually escapes from total suppression, and a phase of anovulation similar to that induced by progesterone apparently precedes resumption of cyclic ovulations. The difference between the effects of progesterone and of suckling on prolactin secretion, on the other hand, may increase with time, since suckling seems to stimulate prolactin secretion as long as it is applied, while progesterone may be able to do so for only a limited time (Rothchild and Schwartz, 1965).

In other respects, however, the effects of suckling and of progesterone resemble each other as well as a syndrome which appears in old female rats. Lactation is associated with an increase of appetite and a decrease in motor activity equivalent to that seen following lesions in the ventromedial nucleus (Kennedy, 1953). The old-age syndrome in rats includes an increase in appetite, a spontaneous secretion of prolactin, and a tendency to anovulation of the constant oestrus type which may eventually progress to ovarian atrophy (*see* references above). The old-age syndrome may be due to a change in the neural controls over appetite, and over prolactin and folliculotrophin secretion, similar to that induced by suckling. Since suckling has the eventual effect of depressing the neural inhibition over prolactin secretion, the neural stimulation over folliculotrophin secretion, and the neural inhibition over appetite, the cause of the old-age syndrome may be the spontaneous development of an inhibitory influence which acts very much like the one that is activated by suckling. If the ventromedial nucleus and far-lateral area act as I have suggested, it is only logical to suppose that the essential cause of the old-age syndrome is a diminution with age of the inhibitory influence of the ventromedial nucleus, and the consequent unrestrained activity of the far-lateral area. It is also logical to suggest therefore (because of the resemblance between the old-age syndrome and the effects of suckling) that the mechanism through which suckling affects ovulation, prolactin secretion, and appetite, is through a depression of activity of the ventromedial nucleus (*Figure 4*;

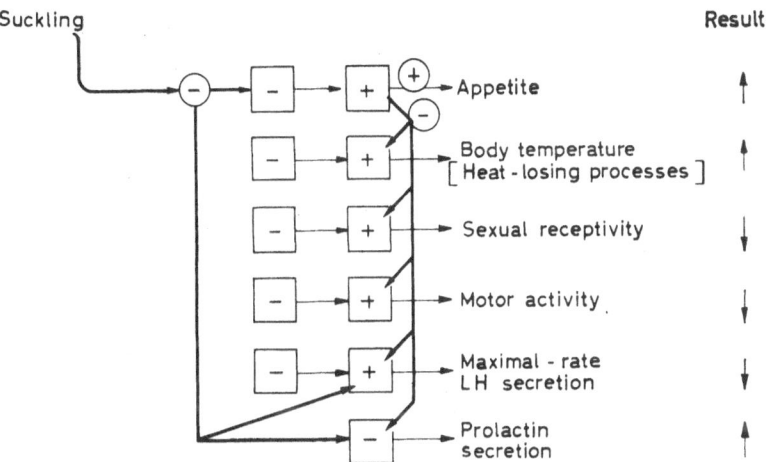

Figure 4. A symbolic representation of how suckling may inhibit ovulation

Symbols and format the same as in *Figure 3*. The major difference between progesterone and suckling in their effects on ovulation and the other modalities may be in the possibility that suckling also directly affects the primary controls over maximal rate LH secretion and prolactin secretion (*see* text for further details).

see also the section entitled 'The Suckling Stimulus' in the paper by Tindal presented at this symposium).

The fact that the pituitary can secrete prolactin and folliculotrophins simultaneously, or that prolactin secretion is not invariably accompanied by an increase of appetite, does not necessarily invalidate this suggestion. Regular ovulation cycles during lactation can reflect, for example, selective breeding for low sensitivity of the system controlling folliculotrophin secretion to the suppressing effect of suckling (e.g. the dairy cow). Galactorrhoea in women, however, can occur quite independently of any increase in appetite or depression of ability to ovulate, but this may only mean that the neural inhibition over prolactin secretion may be lowered by other mechanisms, as well as by the activity of the far-lateral area. The activity of the neural inhibitor over prolactin secretion may be dependent on inputs from various parts of the CNS and peripheral nervous system. For example, removal of the cerebral cortex in rabbits (Beyer and Meno, 1965), or lesions in the midbrain reticular formation (Benedetti, Appeltauer, Reissenweber, Dominguez, Grino and Sas, 1965), as well as a host of other apparently unrelated stimuli, are followed by signs of prolactin secretion in several mammals. The secretion of prolactin, therefore, could follow from a downward change in the amount of neural input supporting the activity of the primary neural inhibitory influence over prolactin secretion. Under conditions in which such inputs themselves are changed, there need be no association between the secretion of prolactin and a change in appetite, ovulation, or any of the other processes that are so characteristically associated with the suckling stimulus.

The efficacy with which suckling maintains prolactin secretion (in comparison at least with the way progesterone does) might mean that even if suckling acted through the ventromedial nucleus and far-lateral area to cause the increase in appetite, decrease in motor activity, anovulation, etc., it might *also* act directly on the primary neural inhibitory influence over prolactin secretion to depress its activity. Such a possibility deserves consideration. The depression of folliculotrophin synthesis, even though it is a relatively short-term effect, also suggests that the suckling stimulus may be directed to more than just the ventromedial nucleus and far-lateral area.

The similarity between the *long-term* effects of progesterone and suckling on folliculotrophin secretion (i.e., the fact that the control system eventually escapes from inhibition), leads to the question of how the ventromedial nucleus—far-lateral area system could be involved in this effect. Is it because the far-lateral area, even in the way it controls appetite, does not remain in a state of high or maximal activity permanently? For example, hyperphagia appears after ventromedial nucleus lesions but eventually the obese rats take in a daily amount of food which maintains a stable body weight, and it is only after periods of partial or complete starvation that they again show hyperphagia (Kennedy, 1951). If all the factors which control far-lateral area activity are such, that even when the area is removed from the restraint exerted on it by the ventromedial nucleus, its level of activity would tend to return gradually to a sub-maximal or a basal level, then the fact that the inhibiting effects of progesterone or of suckling on ovulation are ephemeral is further confirmation of the hypothesis that progesterone and suckling affect ovulation via the far-lateral area.

THE ANOVULATION OF PREGNANCY

The problem may be condensed to one interesting question for which I have no answer. In some animals (e.g. rats, rabbits, mice and hamsters) the anovulation of pregnancy is due essentially to the same factors responsible for anovulation of the luteal phase, the only difference being in the process which prolongs the secretion of progesterone. In others (e.g. dogs and other seasonal breeders of the same type), either progesterone, or a seasonally-related depression of folliculotrophin synthesis, or both, may be responsible. In still others (e.g. women), a very high and prolonged rate of secretion of oestrogens and progesterone evidently suppresses the synthesis as well as release of folliculotrophins. The problems of how progesterone on the one hand, and seasonal changes on the other, affect ovulation have been and are being actively attacked, but the problem which has not yet really been directly investigated is the cause of anovulation in human pregnancy.

There is no evidence that oestrogens exert a negative feed-back effect on folliculotrophin secretion *during the ovulation cycle*, but there is evidence that they exert a positive feed-back effect, during a part of the follicular phase, and *at least*, on the secretion of LH. There is, however, ample evidence that oestrogens can exert a negative feed-back effect on folliculotrophin secretion under conditions which are quite different from those of the ovulation cycle. The real problem then is, if there is a neural mechanism through which the oestrogens inhibit folliculotrophin secretion, or even one in which the oestrogens act directly on the pituitary to accomplish the same effect, in what specific way is this mechanism kept from operating during the conditions of the ovulation cycle, and in what way is it related to the one through which the oestrogens increase the rate of secretion of the folliculotrophins?

SUMMARY

The neural mechanisms through which ovulation is prevented

51

during the luteal phase by progesterone, and during lactation by suckling, have been examined from a comparative point of view. An explanation of the way progesterone inhibits ovulation has been attempted by examining certain phenomena associated with, or characteristic of, progesterone-induced anovulation. These include the increase in pituitary folliculotrophin potency; the ephemerality of the anovulation and of other effects of progesterone (such as anaesthesia, depression of sexual receptivity, elevation of body temperature and depression of sensitivity to electric shock-induced convulsions); the facilitation of ovulation, of sexual receptivity in sheep, and of maternal nest-building in rabbits following progesterone treatment or secretion. We have considered also the association between the effects of progesterone on ovulation, appetite, motor activity, sexual receptivity, body temperature, and prolactin secretion.

The hypothesis was offered that progesterone induces a neural inhibition primarily directed to the system controlling maximal-rate LH release; this inhibitory influence is not the same as that which normally maintains a less than maximal rate of secretion of folliculotrophins during the follicular phase of the ovulation cycle. The source of this progesterone-induced inhibition is the activity of the far-lateral area of the hypothalamus, induced by depressing the inhibitory influence of the ventromedial nucleus over the far-lateral area. In addition to feeding inhibitory impulses into the system controlling folliculotrophin secretion, the far-lateral area is assumed to feed inhibitory impulses to the neural inhibitory control over prolactin secretion and to the stimulatory components of the controls over motor and sexual activity, and heat-losing processes. At the same time it acts as a stimulus to appetite and related processes. The suggestion has also been made, that, with minor differences, the suckling stimulus may act in the same way to prevent ovulation, although its stimulation of prolactin secretion may involve an additional pathway to the primary neural inhibition over prolactin secretion.

ACKNOWLEDGEMENT

Supported in part by U.S.P.H.S. Grant No. HD-00028 from the National Institute of Child Health and Human Development.

REFERENCES

Anand, B. K. (1961). *Physiol. Rev.* **41**, 677–708
Aschheim, P. (1961). *C.r. hebd. Séanc. Acad. Sci., Paris* **253**, 1988–1990
— (1962). *C.r. hebd. Séanc. Acad. Sci., Paris* **255**, 3053–3055

Aschheim, P. (1965). *C.r. hebd. Séanc. Acad. Sci., Paris* **260,** 5627–5630

— and Pasteels, J. J. (1963). *C.r. hebd. Séanc. Acad. Sci., Paris* **257,** 1373–1375

Barnett, R. J. and Mayer, J. (1954). *Anat. Rec.* **118,** 374–375

Benedetti, W. L., Appeltauer, L. C., Reissenweber, N. J., Dominquez, R., Grino, E. and Sas, S. (1965). *Acta Physiol. latinoam.* **15,** 218–220

Beyer, C. and Meno, F. (1965). *Am. J. Physiol.* **208,** 289–291

Bliss, C. I. (1956). *Drug Stand.* **24,** 33–68

Brobeck, J. R. (1946). *Physiol. Rev.* **26,** 541–558

— (1948). *Yale J. Biol. Med.* **20,** 545–552

— Wheatland, M. and Strominger, J. L. (1947). *Endocrinology* **40,** 65–72

Browning, H. C. and Kwan, L. P. (1964). *Tex. Rep. Biol. Med.* **22,** 579–691

Bruce, H. M. (1958). *Proc. R. Soc. (Biol)* **149,** 421–423

Cagnoni, M., Fantini, F., Morace, G. and Ghetti, A. (1965). *J. Endocr.* **33,** 527–528

Dempsey, E. W., Hertz, R. and Young, W. C. (1936). *Am. J. Physiol.* **116,** 201–209

Dziuk, P. J. (1960). *Endocrinology* **66,** 898–900

Everett, J. W. (1944). *Endocrinology* 35, 507–520

— (1945). *Anat. Rec.* **91,** 272–273

— (1950). In: *Progress in Clinical Endocrinology* (S. Soskin, ed.), p. 319. New York; Grune and Stratton

— (1964). *Physiol. Rev.* **44,** 373–431

Gorski, R. A. (1963). *Am. J. Physiol.* **205,** 842–844

Goy, R. W. and Phoenix, C. H. (1963). *J. Reprod. Fert.* **5,** 23–40

— — and Young, W. C. (1964). *Am. Zool.* **4,** 301

Greenwald, G. S. (1962). *Gen. Comp. Endocr.* **2,** 453–457

— (1965). *Endocrinology* **77,** 641–650

Greer, M. (1953). *Endocrinology* **53,** 380–390

Heatherington, A. W. and Ranson, S. W. (1942). *J. Comp. Neurol.* **76,** 475–499

Hollifield, G. and Parson, W. (1957). *J. clin. Invest.* **36,** 1638–1641

Kanematsu, S. and Sawyer, C. H. (1965). *Endocrinology* **76,** 691–699

Kaufman, A. B. and Rothchild, I. (1966). *Acta Endocr. (Copenh.)* **51,** 231–244

Kawakami, M. and Sawyer, C. H. (1959). *Endocrinology* **65,** 652–658

Keetel, W. C. and Bradbury, J. T. (1961). *Am. J. Obstet. Gynec.* **82,** 995–1002

Kennedy, G. C. (1951). *Proc. R. Soc. (Biol)* **44,** 899–904

— (1953). *Proc. R. Soc. (Biol)* **140,** 578–592

— (1957). *J. Endocr.* **16,** 9–17

— (1964a). *J. Physiol.* **172,** 383–392

— (1964b). *J. Physiol.* **172,** 393–399

— and Mitra, J. (1963a). *J. Physiol.* **166,** 395–407

— — (1963b). *J. Physiol.* **166,** 408–418

REPRODUCTION IN THE FEMALE MAMMAL

Kidder, H. E., Casida, L. E. and Grummer, R. H. (1955). *J. Anim. Sci.* **14**, 470–474

Kincl, F. and Maqueo, M. (1965). *Endocrinology* **77**, 859–862

Klein, M. (1947). *J. Endocr.* **5**, xxv–xxvi

Knigge, K. M. and Hays, M. (1963). *Proc. Soc. exp. Biol. Med.* **114**, 67–69

Kurachi, K. and Suchowsky, G. (1958). *Acta Endocr. (Copenh.)* **29**, 27–32

Leonardelli, S. and Barry, J. (1964). *C.r. Séanc. Soc. Biol.* **158**, 1630–1632

Liebelt, R. A., Sekiba, K. and Taylor, H. G. (1961). *Anat. Rec.* **139**, 249

Mayer, J. (1957). *Bull. N.Y. Acad. Med.* **33**, 744–761

— and Marshall, N. B. (1956). *Nature, Lond.* **178**, 1399–1400

McBurney, P. L., Liebelt, R. A. and Perry, J. H. (1965). *Tex. Rep. Biol. Med.* **23**, 737–752

Miller, N. E. (1965). *Science, N.Y.* **148**, 328–338

Morgane, P. J. (1961a). *Science, N.Y.* **133**, 887–888

— (1961b). *J. Comp. Neurol.* **117**, 1–25

Richter, C. P. (1956). *Gestation Trans. Confs. Josiah Macy jr. Fdn* 1955, p. 11

Rothchild, I. (1960). *Endocrinology* **67**, 9–41

— (1965). *Vitams Horm.* **23**, 209–327

— and Schwartz, N. B. (1965). *Acta Endocr. (Copenh.)* **49**, 120–137

Schwartz, N. B. and Bartosik, D. (1962). *Endocrinology* **71**, 756–762

Segal, S. J. and Johnson, D. C. (1959). *Archs Anat. microsc. Morph. exp.* **48** (suppl.), 261–273

Stevenson, J. A. F. and Rixon, R. H. (1957). *Yale J. Biol. Med.* **29**, 575–584

Swanson, H. E. and van der Werff ten Bosch, J. J. (1964). *Acta Endocr. (Copenh.)* **45**, 1–12

Taleisnik, S. and Caligaris, L. (1962). *Experientia* **18**, 578–579

— — and DeOlmos, J. (1962). *Am. J. Physiol.* **203**, 1109–1112

van Dyke, D. C., Simpson, M. E., Lepkovsky, S., Koneff, A. A. and Brobeck, J. R. (1957). *Proc. Soc. exp. Biol. Med.* **95**, 1–5

von Euler, C. (1961). *Pharmac. Rev.* **13**, 361–398

Wagner, J. W. and de Groot, J. (1963). *Proc. Soc. exp. Biol. Med.* **112**, 33–37

Wolfe, J. M. (1946). *Am. J. Anat.* **79**, 199–239

Zeilmaker, G. H. (1963). *Acta Endocr. (Copenh.)* **43**, 246–254

REGULATION OF GONADOTROPHIN AND PROLACTIN SECRETION BY HYPOTHALAMIC NEUROHUMORAL FACTORS

S. M. McCANN, J. ANTUNES-RODRIGUES,
S. WATANABE, A. RATNER and A. P. S. DHARIWAL

*Department of Physiology, University of Texas Southwestern Medical School,
Dallas, Texas, U.S.A.*

INTRODUCTION

THE SECRETION of gonadotrophins and prolactin by the adenohypophysis is influenced by a variety of environmental and hormonal factors. There are many examples of neurally mediated environmental influences. Exposure of rats to constant lighting results in the development of constant vaginal oestrus associated with a blockade of cyclic ovulation. Lactation, presumably operating via the stimulus of nursing, augments prolactin release and inhibits secretion of follicle stimulating hormone (FSH) and luteinizing hormone (LH) (McCann and Ramirez, 1964; Everett, 1964).

The gonadal steroids exert complex feed-back effects on the secretion of these three hormones. Oestrogen is a potent inhibitor of FSH and LH secretion in ovariectomized animals, but the rising levels of oestrogens from the pre-ovulatory follicle appear to be involved in stimulating the release of LH which provokes ovulation. Similarly, progesterone can also inhibit LH secretion in gonadectomized rats, particularly if given after oestrogen priming, and yet, it can be shown to stimulate LH secretion when given during the pre-ovulatory phase of the oestrous cycle. Oestrogen in sufficient dosage stimulates prolactin secretion as evidenced by the development of a pseudopregnancy-like syndrome (Ramirez and McCann, 1964).

Current evidence overwhelmingly supports the concept that the hypothalamus plays a key role in mediating both environmental and hormonal influences on the secretion of gonadotrophins and prolactin. The most direct evidence has been provided by classical approaches such as electrical stimulation or ablation of various hypothalamic areas. Hypothalamic stimulation can evoke ovulation in both rat (Everett, 1964) and rabbit (Harris, 1960), which is accompanied by elevated release of LH (McCann, Ramirez and

55

Abrams, 1964). Conversely, hypothalamic lesions in the median eminence of the tuber cinereum reduce secretion of both FSH (Igarashi and McCann, 1964a) and LH (Taleisnik and McCann, 1961). By contrast, hypothalamic lesions or grafting the pituitary to a site at a distance from the hypothalamus have the converse effect on prolactin secretion and cause it to increase (Everett, 1954; Nikitovitch-Winer and Everett, 1958; McCann and Friedman, 1960; Haun and Sawyer, 1960). Even when the pituitary is incubated *in vitro*, prolactin is secreted at a high rate (Talwalker, Ratner and Meites, 1963). Thus, the hypothalamus has a net stimulatory effect on secretion of FSH and LH on the one hand, and a net inhibitory effect on the secretion of prolactin on the other.

Most workers agree that the anterior lobe of the pituitary is devoid of a secretory nerve supply. An alternative pathway by which neural control over this portion of the gland could be mediated was provided by the discovery of the hypophyseal portal system of veins. They take their origin in the median eminence and pituitary stalk and carry capillary blood from these two regions down the pituitary stalk to the sinusoids of the anterior lobe. These portal vessels provide a route by which specific chemotransmitter agents secreted in the median eminence and pituitary stalk might trigger release of particular hormones from the anterior lobe (Harris, 1960). This leads to the concept of the neurohumoral control of adenohypophysis.

A variety of experiments have been designed to test the hypothesis of neurohumoral control of the pituitary by means of the portal vessels. The pituitary stalk section, which should interrupt the portal circulation and eliminate any direct secretomotor nerves to the anterior lobe, has resulted in variable effects on anterior pituitary function unless regeneration of the portal vessels has been prevented by means of a mechanical barrier such as a piece of wax paper inserted between the cut ends of the stalk (Harris, 1960). In this situation a permanent interference with pituitary function results. A similar interference with function follows removal of the pituitary from its normal site and grafting it to another location. Grafts beneath the kidney capsule have been most thoroughly studied (Nikitovitch-Winer and Everett, 1958). As in the case of lesions in the median eminence, the secretion rate of FSH and LH appears to be markedly diminished in this situation, whereas that of prolactin proceeds apace. If the pituitary is removed from its normal location and then replaced under the median eminence, so that revascularization is achieved by the portal vessels, then there is no impairment in function. The most dramatic experiment of this sort was performed by Nikitovitch-Winer and Everett (1958a), who first grafted the gland under the kidney capsule and then regrafted it under the median eminence. The first operation resulted in the typical abnormalities of anterior lobe function which were remedied by the second operation. Vascularization of the gland appears to be similar in either locus, so it is hard to escape the conclusion that the return of function after placement under the median eminence is caused by the delivery of specific neurohumoral agents to the gland via the portal vessels. In the succeeding discussion, we shall summarize the evidence for the existence of specific polypeptides which appear to mediate the hypothalamic regulatory influence over gonadotrophin secretion.

LH-RELEASING FACTOR (LH–RF)

Corticotrophin-releasing factor (CRF) was the first specific factor from the hypothalamus shown to alter anterior pituitary function (Saffran and Schally, 1955; McCann and Haberland, 1959; Royce and Sayers, 1960; Guillemin, 1964). It appeared reasonable to postulate other factors to stimulate the release of gonadotrophins and to inhibit the secretion of prolactin. In 1959 we began an intensive search for a possible LH-releasing action of hypothalamic extracts and employed a new sensitive and specific assay for LH, the ovarian ascorbic acid depletion test of Parlow (1961). In immature female rats, which have been pre-treated with gonadotrophins, minute doses of LH evoke a decline in ovarian ascorbic acid concentration. Other anterior lobe hormones are without effect.

Crude acidic extracts of rat stalk-median eminence (SME) tissue were capable of depleting ovarian ascorbic acid and a dose-response relationship was obtained (*Figure 1*) (McCann, Taleisnik and Friedman, 1960; McCann, 1962). Similar activity was found in ovine or bovine hypothalamic extracts (Ramirez and McCann, 1964; Ramirez, Abrams and McCann, 1964; Dhariwal, Antunes-Rodrigues and McCann, 1965). Since the extract exhibited little or no activity upon injection into hypophysectomized rats, it appeared that contamination with LH could not explain these results and that a release of LH from the pituitaries of the test rats had occurred. Furthermore, heating for 10 min in a boiling water bath inactivated rat pituitary LH while leaving unchanged the activity of the hypothalamic extract (*Figure 2*). The results also could not be explained by the content of known physiologically active substances which were present in the extract, such as vasopressin, oxytocin, substance

P, histamine, serotonin, or epinephrine. It should be pointed out, however, that sufficient doses of vasopressin can deplete ovarian ascorbic acid, apparently by a direct action on the ovary (McCann and Taleisnik, 1960) and can even release LH in one sensitive test system (Ramirez and McCann, 1963).

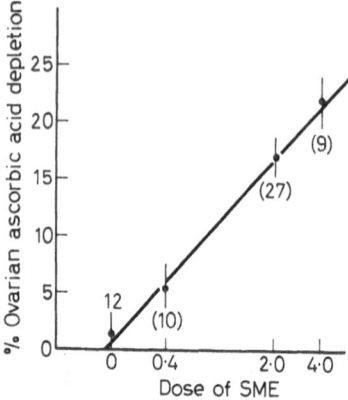

Figure 1. Effect of varying doses of stalk-median eminence extract (SME) on ovarian ascorbic acid depletion. Numbers on the abscissa *refer to the number of hypothalami from which the extract was derived. Except for the control (0), the doses are on a log scale. Numbers in parentheses refer to the number of test animals at a given dose of extract. Points and vertical lines represent the mean and standard error (SEM), respectively* (From S. M. McCann, 1962, by courtesy of the *American Journal of Physiology*)

Similarly prepared extracts of cerebral cortex were ineffective which indicated that the action of hypothalamic extracts was a specific one. Extraction of various hypothalamic areas revealed that the major LH-releasing activity was concentrated in the SME region. There was minimal activity in the overlying ventral hypothalamus (*Figure 3*) (McCann, 1962). At this point the unknown substance responsible for the LH-releasing effect was designated the LH–RF.

Since the immature test rats used in evaluating LH-releasing activity were treated with large doses of gonadotrophins and had relatively low stores of hypophyseal LH (McCann *et al.*, 1960;

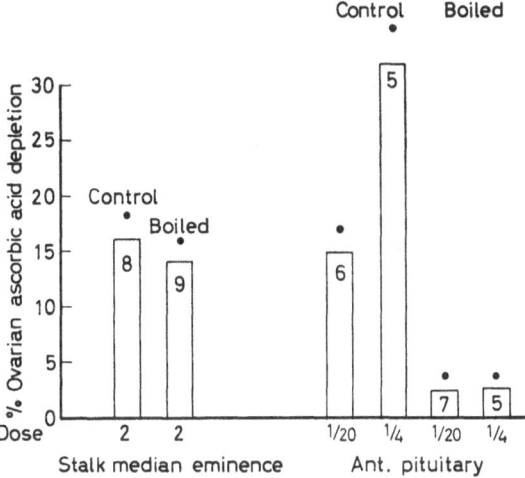

Figure 2. Effect of boiling on the activity of extracts. Numbers on the abscissa refer to the number of stalk-median eminences or anterior pituitaries used. Numbers at the top of each column indicate the number of animals injected; dots indicate the SEM (From S. M. McCann, 1962, by courtesy of the *American Journal of Physiology*)

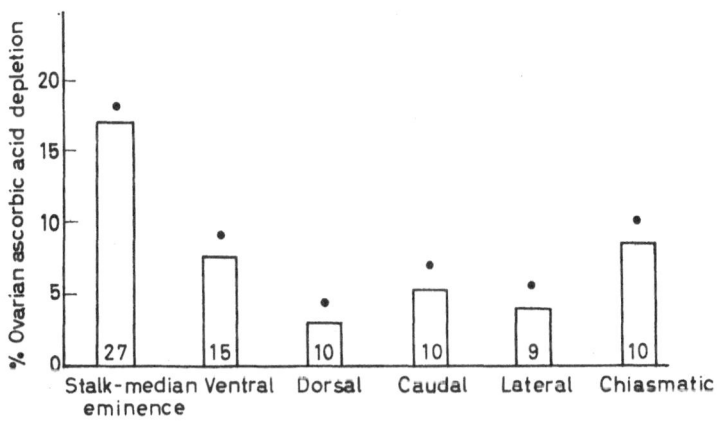

Figure 3. Effect of extracts prepared from various portions of the hypothalamus on ovarian ascorbic acid depletion. In each case the test rat received extract derived from two hypothalami. Numbers refer to the number of test animals used. Dots above bars indicate the SEM (From S. M. McCann, 1962, by courtesy of the *American Journal of Physiology*)

Novella, Alloiteau and Ascheim, 1964), it was advisable to determine if LH-releasing activity could be demonstrated in more physiological circumstances. The LH-releasing action of hypothalamic extracts has now been evaluated in a variety of test animals. In these experiments, the extract was injected intravenously into the test rats which were bled 10 min later from the jugular vein. The effect of the extract on plasma LH activity of the test animals was estimated by the ovarian ascorbic acid depletion test. The factor has been found to be effective in a variety of situations. It is active in normal female rats and in ovariectomized rats in which the release of LH has been inhibited either by administration of gonadal steroids or by lesions in the median eminence (McCann, 1962). This latter observation is important because it indicates that the LH–RF acts directly on the anterior pituitary to release LH. An indirect action via the nervous system would have been blocked by these lesions which interrupted neural control over LH secretion. Further evidence that the LH–RF acts directly on the gland to release LH has been provided by the experiments of Campbell, Feuer and Harris (1964) in rabbits and Nikitovitch-Winer (1962) in rats. They showed that infusion of hypothalamic extract directly into the anterior lobe of the pituitary could evoke ovulation. Systemic administration of the same dose of extract was without effect. Schally and Bowers (1964) have demonstrated an LH-releasing action of hypothalamic extracts on pituitaries incubated *in vitro*, and we have recently confirmed this observation (Watanabe and Ratner, unpublished, 1965). This provides further evidence that the LH–RF acts directly on the hypophysis.

Rather surprisingly, LH–RF failed to elevate plasma LH in untreated ovariectomized rats. Plasma LH is already elevated in the ovariectomized rat because of the elimination of ovarian steroid negative feed-back, so we have postulated that this animal is already responding maximally to release of endogenous LH–RF and is incapable of responding further to exogenous releasing factor (McCann, 1962).

Physiological significance of the LH–RF

It is important to determine if changes in the rate of secretion of the LH–RF are responsible for bringing about the alterations in secretion of LH which occur during the oestrous cycle and after administration of gonadal steroids.

LH secretion fluctuates during the oestrous cycle and reaches a peak at pro-oestrus just prior to ovulation (*Figure 4*) (Ramirez and McCann, 1964b). Pituitary LH begins to decline at this time and

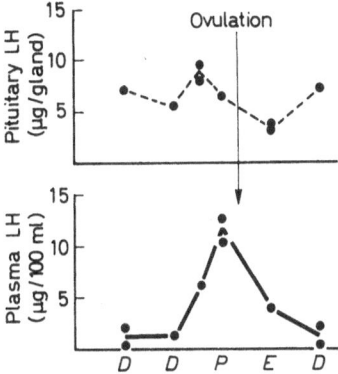

Figure 4. Fluctuations in hypophyseal and plasma LH *during the oestrous cycle of the rat. Data for pituitary* LH *changes are from studies by Dr. Neena B. Schwartz at the University of Illinois. Data for plasma are those of Doctors V. D. Ramirez and S. M. McCann at the University of Pennsylvania School of Medicine. Letters on the* abscissa *refer to the days of the oestrous cycle in the rat: D = dioestrus; P = pro-oestrus; E = oestrus* (From S. M. McCann, 1963, by courtesy of the American Physiological Society)

reaches a minimum after ovulation (Schwartz and Mills, 1961). Accompanying the rise in LH release at pro-oestrus is a fall in content of hypothalamic LH–RF (Table 1) (Chowers and McCann, 1965; Ramirez and Sawyer, 1965). This observation is consistent with the view that LH–RF triggers the pre-ovulatory discharge of LH.

One could also explain the ovulatory discharge of LH by postulating increased responsiveness of the pituitary at pro-oestrus to a constant release of LH–RF. Consequently, we have evaluated the responsiveness of the pituitary to purified LH–RF at various stages of the oestrous cycle. Doses of 0·3 or 0·1 ml of LH–RF were clearly effective in elevating plasma LH within 10 min of their intravenous injection at all stages of the cycle (*Figure 5*) (Antunes-Rodrigues, Dhariwal and McCann, 1966). LH was estimated by the ovarian ascorbic acid depletion produced in recipient test rats. A dose of 0·02 ml was ineffective at all stages of the cycle although there was a suggestion of a positive response during pro-oestrus.

The data to date are consistent with the thesis that an increased release of LH–RF mediates the discharge of LH that triggers ovulation. Conclusive proof will require the demonstration of increased levels of LH–RF in hypophyseal portal blood at this stage of the cycle.

Table 1

Influence of various conditions known to alter LH secretion or the content of stored LH–RF in the hypothalamus

Experimental condition	Plasma LH	Effect on	
		Pituitary LH	Hypothalamic LH–RF
1. Pro-oestrus	increased	decreasing	decreased
2. Testosterone injections*	decreased †	decreased	decreased
3. Oestrogen	decreased †	decreased	no significant change
4. Castration	increased	increased	no significant change
5. Hypothalamic implant of oestrogen or testosterone	decreased †	decreased	decreased
6. Pituitary implant of testosterone	?	no change	no change
7. Pituitary implant of oestrogen	decreased †	no change or decreased	elevated

* 1·0 mg of testosterone propionate s.c./day for 2 weeks.
† Effect on plasma LH was determined in castrated animals.

If LH–RF mediates the ovulatory discharge of LH, what causes the enhanced release of LH–RF? This is presumably brought about by increased levels of circulating oestrogen and progesterone from the maturing ovarian follicles. Everett and others (*see* review by Everett, 1964) have shown that oestrogen or progesterone will advance ovulation when administered at the appropriate stage of the oestrous cycle. We have recently been able to elevate plasma LH by administration of a single subcutaneous injection of progesterone in oil (1·5 mg) during the pre-ovulatory phase of the cycle (Nallar, Antunes-Rodrigues and McCann, 1966) (*Figure 6*). Plasma LH was measured 6 h after injection. Presumably the progesterone acts on the cyclic timing mechanism for gonadotrophin release which is thought to lie in the rostral hypothalamus (Barraclough and Gorski, 1963), and this evokes secretion of LH–RF which in turn stimulates LH secretion.

A hypothalamic site for the inhibitory action of gonadal steroids on gonadotrophin secretion has been suggested from the results of implanting minute amounts of oestrogen or testosterone into the hypothalamus (Lisk, 1960; Davidson and Sawyer, 1961). Oestrogen implants can inhibit LH secretion in gonadectomized rats and can augment prolactin release in intact females (Ramirez *et al.*, 1964;

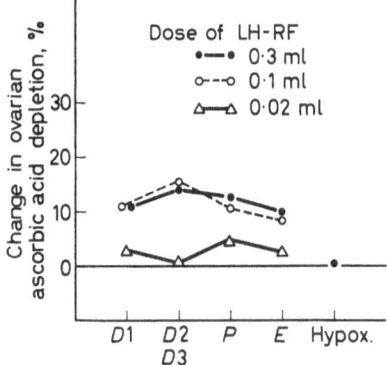

Figure 5. Sensitivity to LH-RF *at various stages of the oestrous cycle of the rat. The increase in plasma LH* 10 min *after iv injection of* LH-RF *is indicated by the percentage change in ovarian ascorbic acid depletion. A significant increase in plasma LH was observed at each stage of the cycle with the* 0·3 *and* 0·1 ml *doses of* LH-RF, *except in the case of hypophysectomized (hypox.) rats which showed no response to the* 0·3 ml *dose. Responses to the* 0·02 ml *dose were not statistically significant* (Antunes-Rodrigues *et al.*, unpublished data)

Ramirez and McCann, 1964a); however, implants of oestrogen in the anterior pituitary are also effective (Bogdanove, 1963; Ramirez *et al.*, 1964). This latter observation clearly establishes that these steroids can alter gonadotrophin and prolactin secretion by an action directly on the pituitary gland. Bogdanove (1963) has stressed the possibility that the results obtained with hypothalamic implants may be due to uptake of steroid in portal vessels with distribution to the pituitary where their effect is mediated. Although this possibility seemed unlikely, it became necessary to employ other approaches to establish a hypothalamic site for the inhibitory effects of gonadal steroids.

If the inhibitory effect of gonadal steroids were exerted primarily on the pituitary, one would predict that large doses of these steroids would block the action of administered LH–RF.   Such is not the case.   Even after pre-treatment of ovariectomized rats for 3 days with massive doses of oestrogen and progesterone, LH–RF was effective in elevating plasma LH.   If anything, it appears that these steroid-blocked rats are supersensitive to the transmitter (Ramirez

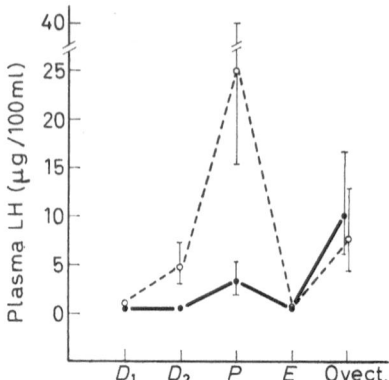

Figure 6.   Plasma LH in oil-injected controls ●——● *and in rats injected with progesterone in oil* ○– – –○.   *Values are those for ovariectomized rats (ovect.) and those exhibiting 4 day oestrous cycles.   The dashed lines connect values for progesterone-treated rats, the solid lines represent the controls.   Vertical lines indicate the 95 per cent confidence limits.   D1 = dioestrus, day 1;   D2 = dioestrus, day 2;   P = pro-oestrus;   E = oestrus*

and McCann, 1963).   These results suggest an inhibitory effect of the steroids on the hypothalamus rather than the pituitary gland itself.

If changes in gonadal steroid titre were accompanied by alterations in the level of LH–RF stored in the hypothalamus, evidence for an effect of these steroids on secretion of LH–RF would be provided. We have been unable to detect an effect of castration on the hypothalamic content of LH–RF (Table 1) (Chowers and McCann, 1965).   Administration of oestrogen also failed to alter the stored LH–RF, but a large dose of testosterone propionate produced a decline in the level of stored factor.   Implants of either oestradiol or testosterone in the median eminence were capable of lowering the content of LH–RF.   Implants of testosterone in the pituitary produced no alteration in the level but oestradiol implants in the pituitary rather surprisingly produced a significant increase in stored LH–RF. A speculative explanation for this latter observation, involving a possible negative feed-back of LH itself on the hypothalamus, has been advanced (Chowers and McCann, 1965). Taken altogether, the observations to date are consistent with an effect of altered levels of gonadal steroids on the content of stored LH–RF.

Direct evidence for an effect of these steroids on LH–RF secretion would be provided if alterations in the level of circulating LH–RF could be produced by administration of gonadal steroids. No detectable LH–RF can be found in plasma of normal rats; however, several months after hypophysectomy the plasma contains a factor which will deplete ovarian ascorbic acid (Nallar and McCann, 1965). The factor vanishes within 24 h of coagulation of the median eminence. Since this procedure would eliminate the site of storage and release of LH–RF, we concluded that the factor circulating in blood of rats several months after hypophysectomy was the LH–RF. Unfortunately, the circulating LH–RF levels are rather low, so that they are not always detectable with present methods. This has hampered our investigation of the factors which alter the level of the circulating LH–RF. Preliminary experiments suggest, however, that the level of circulating LH–RF may be lowered by pre-treatment of the hypophysectomized rats with either LH itself (Nallar, unpublished 1964) or a combination of oestrogen and progesterone (Antunes-Rodrigues, unpublished, 1965). Recently, Frankel, Gibson, Graber, Nelson, Reichert and Nalbandov (1965) have reported an ovarian ascorbic acid depleting activity in hypophysectomized chicken plasma which may also represent circulating LH–RF.

Although more experimentation is needed, it appears highly likely that most changes in LH secretion are mediated by variations in the release of the LH–RF. Some changes may be exerted via direct inhibitory effects of the gonadal steroid on the pituitary gland.

FOLLICLE STIMULATING HORMONE-RELEASING FACTOR (FSH–RF)

It has now been possible to demonstrate a FSH-releasing action of hypothalamic extracts (Igarashi and McCann, 1964a). For most of these experiments a new, sensitive method for assay of FSH was employed, the mouse uterine weight augmentation assay (Igarashi and McCann, 1964), but we have also obtained similar results with the rat ovarian weight augmentation method of Steelman and

Pohley (1953) (Igarashi, Nallar and McCann, 1964). Hypothalamic extracts elevated plasma FSH when injected intravenously into ovariectomized rats in which the release of FSH had been inhibited, either by hypothalamic lesions, or by treatment with both oestrogen and progesterone. Extracts from cerebral cortex were inactive and the results were not caused by contamination of the extract with FSH, itself, or with vasopressin or oxytocin.

Kuroshima, Ishida, Bowers and Schally (1965) have confirmed these results and have also shown, as have Mittler and Meites (1964), that the FSH–RF will increase the release of FSH into the media of pituitaries incubated *in vitro*. Watanabe, Ratner, Dhariwal and McCann (unpublished, 1965) have been able to confirm the *in vitro* FSH–RF activity of rat and sheep hypothalamic extracts (*Figure 7*).

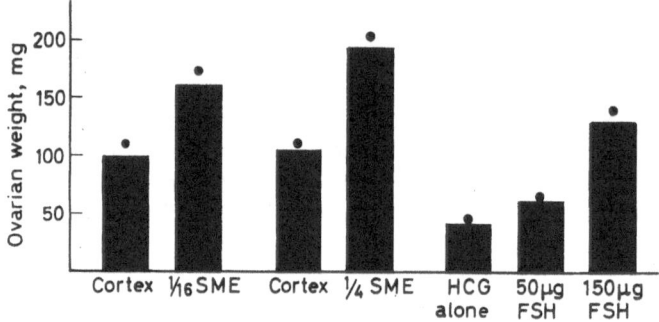

Figure 7. Effect of rat SME extract on the release of FSH *from pituitaries incubated* in vitro. *FSH was assayed by the Steelman-Pohley assay. The three columns at the right of the figure illustrate the response to human chlorionic gonadotrophin (HCG) alone and to 2 doses of ovine FSH standard administered together with* HCG. *The four columns to the left of the figure indicate the amount of* FSH *in the medium of pituitaries incubated* in vitro *for 6 h in the presence of either cerebral cortical extract (cortex) or SME extract. A significant increase in FSH in the medium was observed with either* $\frac{1}{16}$ *or* $\frac{1}{4}$ *of an* SME (Watanabe et al., unpublished data)*

In these latter studies the method of Steelman and Pohley (1953) was used for assay of FSH.

David, Fraschini and Martini (1965) and Ishida, Kuroshima, Bowers and Schally (1965) have reported that hypothalamic extracts can deplete pituitary FSH in normal males and ovariectomized steroid-blocked females, respectively. We have been unable to confirm these observations despite repeated attempts (Antunes and Watanabe, unpublished data, 1964–66); however, it is possible that strain differences or other unknown factors may account for the discrepancy between their results and our own.

PROLACTIN-INHIBITING FACTOR (PIF)

Pituitaries in tissue culture liberate large amounts of prolactin into the culture medium.   They thus behave in a similar manner to the pituitary isolated from hypothalamic influences *in vivo*.   This enhanced release of prolactin can be inhibited by the addition of crude acidic extracts of rat SME to the culture (Talwalker *et al.*, 1963; Pasteels, 1962).   These findings obtained *in vitro* suggest that

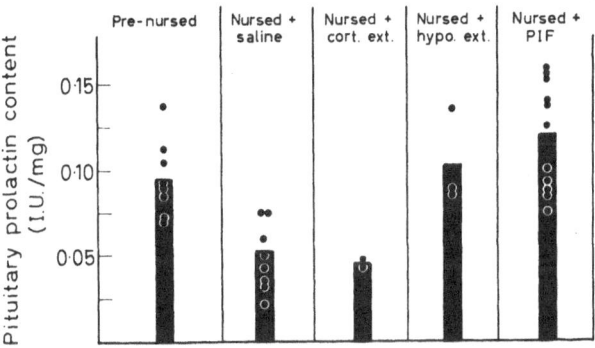

Figure 8.   Ability of hypothalamic extracts to inhibit nursing-induced decline in pituitary prolactin.   Pre-nursed refers to females isolated from their litters for 10 h.   All other groups of females were isolated from their litters for 10 h, and, 1–2 min following i.p. injection of an extract, they were allowed to nurse their litters for 30 min.   Cort. ext. = bovine cerebral cortical extract; Hypo. ext. = bovine SME extract; PIF = purified fractions of ovine SME with PIF activity.   Each point refers to an assay of 4–6 pituitaries by the local crop sac method (Unpublished data of Grosvenor, Dhariwal, Antunes-Rodrigues and McCann)

a PIF may be found in the hypothalamic extracts.   In collaboration with Grosvenor of the University of Tennessee, we have obtained evidence for the inhibitory effect of hypothalamic extracts on pro-lactin release *in vivo* (Grosvenor, McCann and Nallar, 1965). Grosvenor and Turner (1957) had already shown that suckling causes an acute depletion of prolactin from the hypophysis of the lactating rat.   If crude SME extracts of rat or beef origin were injected just prior to suckling, the suckling-induced depletion of prolactin no longer occurred (*Figure 8*). Cerebral cortical extracts were without effect. Thus, both *in vivo* and *in vitro* results point to the existence of a PIF.

STUDIES ON THE CHEMISTRY OF THE FACTORS WHICH ALTER GONADOTROPHIN SECRETION

Considerable information is now available on the chemical nature of the various hypothalamic chemotransmitters (agents) which affect anterior pituitary secretion. FSH–RF, LH–RF and PIF all retain their activity after heating for 10 min in a boiling water bath. By contrast, rat LH is completely inactivated, and bovine and ovine LH are partially inactivated after similar treatment (McCann, 1962; Ramirez and McCann, 1964a; Guillemin, 1964; Igarashi and McCann, 1964a; Grosvenor *et al.*, 1965). The LH–RF is partially or completely inactivated by the proteolytic enzymes, pepsin or trypsin (McCann, 1962; Guillemin, 1964). This observation indicates that intact peptide bonds are required for its activity. Thioglycollate splits the disulphide bridge in oxytocin and vasopressin, thus inactivating the molecules, but this treatment is without influence on LH–RF or hypothalamic CRF (Ramirez and McCann, 1964a). Consequently, it appears likely that the chemical structures of LH–RF and CRF, at least with respect to the disulphide bridge, are dissimilar from that of the known neurohypophyseal polypeptides.

Recently, several attempts have been made to purify the LH–RF, using ovine or bovine extracts as the starting material. Gel filtration on Sephadex G-25 was first used by Porath and Schally (1962) for separation of posterior lobe hormones and has proved to be a good method for purification of the various hypothalamic releasing factors. In our hands, using ovine SME extract and ammonium acetate as the eluting buffer, LH–RF has been eluted from the Sephadex column just prior to the emergence of vasopressin (Dhariwal *et al.*, 1965) (*Figure 9*). Similar results were obtained with bovine or ovine extracts and acetic acid as the eluent (Ramirez, Nallar and McCann, 1964a; Schally and Bowers, 1964; Guillemin, Justisz and Sakiz, 1963). Further purification of LH–RF has been achieved by chromatography on carboxymethylcellulose (CMC) (Dhariwal *et al.*, 1965). After careful desalting of the extract with glacial acetic acid and by several passages through the CMC column, LH-releasing activity was retained on the ion exchanger and was eluted by application of ammonium acetate buffers of increasing pH and ionic strength (*Figure 10*). The LH–RF was eluted from the column just prior to elution of residual vasopressin which still contaminated the fractions obtained from the Sephadex column. The retention of the LH–RF on the CMC column indicates that it is a basic polypeptide. By this

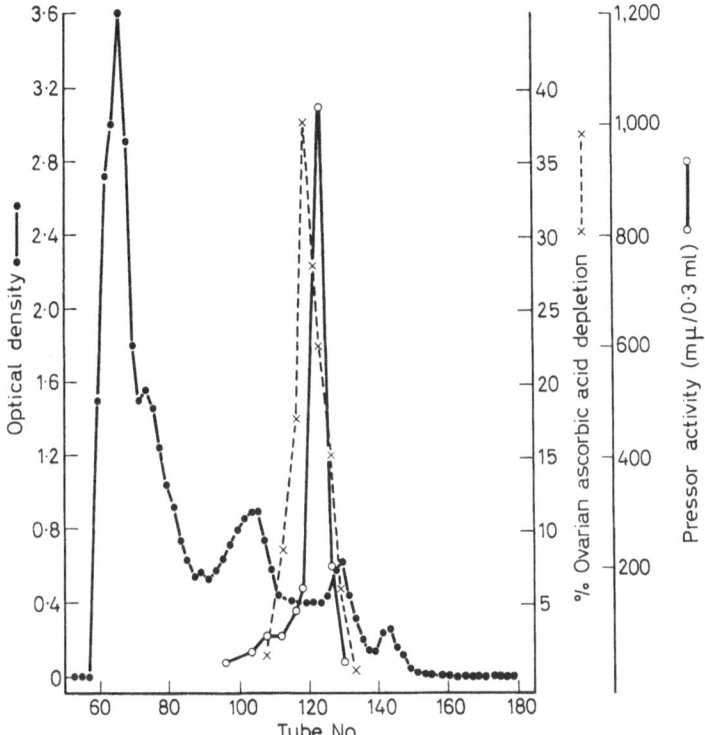

Figure 9. LH-releasing and pressor activity of ovine hypothalamic extracts obtained by gel filtration on Sephadex G-25. Optical density (at 750 mμ) represents peptide concentration as determined by the Folin-Lowry reaction. LH-releasing activity is estimated by ovarian ascorbic acid depletion (From A. P. S. Dhariwal, J. Antunes-Rodrigues and S. M. McCann, 1965, by courtesy of the Editor of *The Soc. exp. Biol. Med.*)

means LH–RF, essentially free of vasopressin (< 5 mU/dose) was obtained, which was active at a dose of < 3 μg of peptide. The highly purified material migrated on paper electrophoresis as a major component with a trailing minor component. It contained the following amino acids: aspartic acid, glutamic acid, threonine, serine, glycine, alanine, phenylalanine, lysine, arginine, traces of proline and histidine. A similar purification of bovine LH–RF has recently been announced by Schally and Bowers (1964a).

Gel filtration on Sephadex is sufficient to separate the FSH–RF from the LH–RF (*Figure 11*) (Dhariwal, Nallar, Batt and McCann, 1965c). In these experiments we employed a tall column of Sephadex G-25 and used ammonium acetate as the eluting buffer.

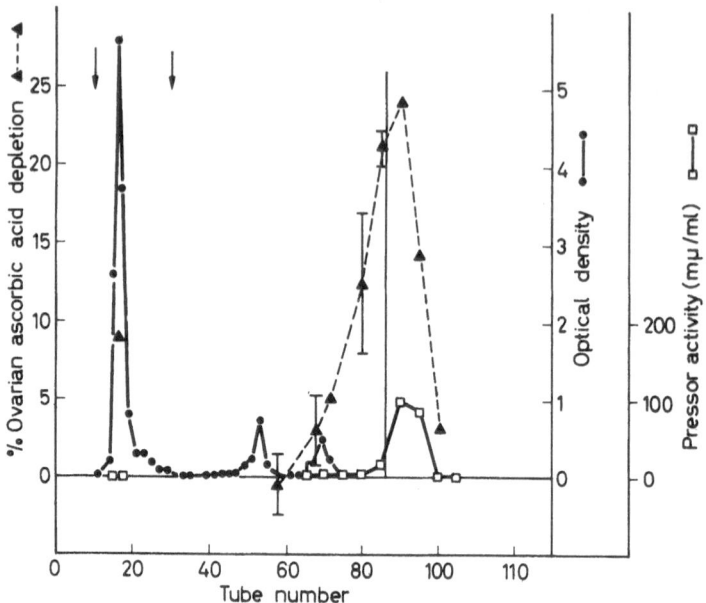

Figure 10. LH-releasing (*percentage ovarian ascorbic acid depletion*) and pressor activity of fractions obtained from the CMC column. *Arrows represent application of gradient to 0·04 M (pH 6·5) ammonium acetate buffer at tube 10 and gradient to 2·0 M (pH 6·5) buffer at tube 30. All tubes to the left of vertical line at tube 87 contain little or no vasopressin. Vertical bars gave standard error where five or more test rats were used* (From A. P. S. Dhariwal, J. Antunes-Rodrigues and S. M. McCann, 1965, by courtesy of the Editor, *The Soc. exp. Biol. Med.*)

Earlier experiments with a shorter column and employing acetic acid as the eluent had failed to separate the two activities (Igarashi *et al.*, 1964). Further chromatography of the active fractions on CMC has resulted in the preparation of highly purified FSH–RF (Dhariwal, Watanabe, Ratner and McCann, 1966a), which was active in the *in vitro* assay system at a dose of < 1 μg of peptide.

Prolactin-inhibiting activity has also been localized in the effluent from the Sephadex column (Dhariwal, Krulich, Antunes-Rodrigues, Chowers, Reeser, Grosvenor and McCann, 1965a, b) (*Figure 8*);

however, to date it has not been possible to separate consistently this activity from the LH–RF (*Figure 12*). The two activities have

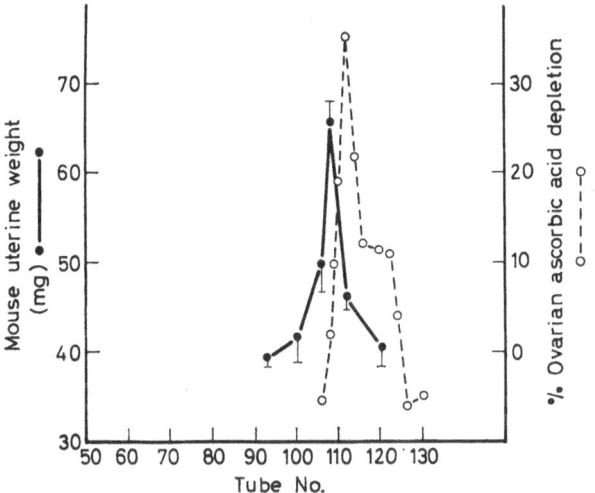

Figure 11. FSH-releasing activity (mouse uterine weight, ●———●) and LH-releasing activity (percentage ovarian ascorbic acid depletion, ○– – –○) of fractions of ovine hypothalamic extract from a column of Sephadex G-25. Mouse uterine weights greater than 45 mg are indicative of significant FSH release. Vertical bars indicate standard error of mean. Ovarian ascorbic acid depletions of greater than 10 per cent indicate LH release (From S. M. McCann and A. P. S. Dhariwal, by courtesy of the Editor, *Trans. New York Academy of Sciences*)

resided in the same tubes although there has been a tendency in two of four runs for the PIF to begin emerging from the column just prior to the LH–RF. Even CMC-purified LH–RF, active at a dose of only a few microgrammes of peptide, possessed prolactin-inhibiting activity as indicated by blockade of nursing-induced depletion of pituitary prolactin. Clearly, a manifold purification of PIF was obtained without achieving a separation from LH–RF. Since there is often a reciprocal relationship between the secretion of LH and prolactin, i.e., after median eminence lesions, in lactation or in pseudopregnancy, it is possible that LH–RF and PIF represent two activities of a single molecule. It will not be possible to resolve this question with certainty until a homogeneous preparation of LH–RF is tested for prolactin-inhibiting activity.

71

Little difficulty has been encountered in separating the gonado-trophin-releasing factors and PIF from the other hypothalamic pituitary stimulating hormones. Gel filtration through Sephadex is sufficient to achieve the separation (*Figure 12*). The order of elution

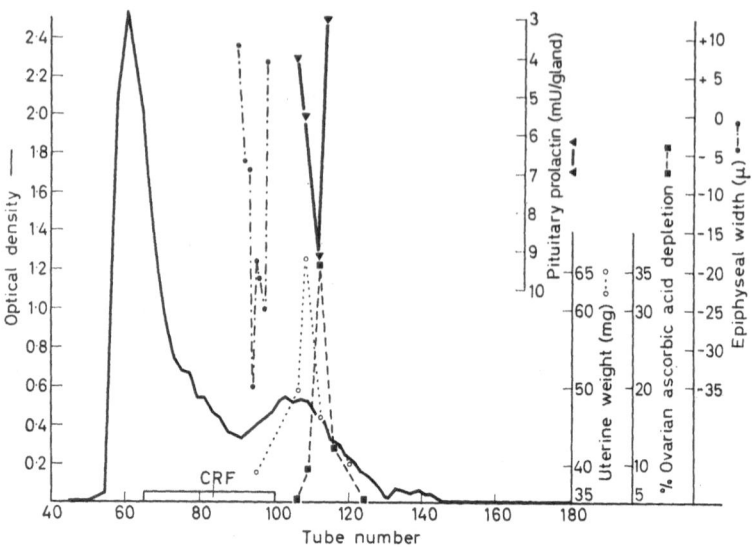

Figure 12. Distribution of releasing factors following gel filtration through Sephadex G-25. Optical density refers to the Folin-Lowry colour and is an indication of peptide content. The bar labelled CRF refers to CRF as indicated by adrenal ascorbic acid depletion in rats with hypothalamic lesions. Epiphyseal cartilage width in microns is an index of growth hormone (GH) content of pituitaries from animals injected with hypothalamic extract. Epiphyseal cartilage width was significantly narrowed with tubes 94–97 which indicates the presence of GH-RF in these tubes. The FSH-releasing zone is indicated by the curve for uterine weight and was maximal at tube 108. LH-releasing activity as indicated by ovarian ascorbic acid depleting activity was maximal at tube 112. Pituitary prolactin content of the pituitaries of recipient animals, an index of prolactin-inhibiting activity, was also maximal at tube 112. These results were obtained in fractionation APSD 1-6

of the various factors has been as follows: CRF (contaminated with ACTH-like activity), growth hormone-releasing factor (GH–RF), melanocyte stimulating-releasing factor (not shown), FSH–RF, PIF and LH–RF (Dhariwal *et al.*, 1965, 1966). LH–RF is followed by vasopressin. Since molecular size is the principal factor affecting mobility on Sephadex, the fact that LH–RF is eluted from these columns just prior to vasopressin suggests that this factor is a small polypeptide slightly larger than vasopressin and oxytocin. The other releasing factors would by this reasoning appear to be somewhat larger molecules than LH–RF. Highly purified CRF and GH–RF also differ from the LH–RF in amino acid composition (Dhariwal et al., unpublished).

By contrast with vasopressin and oxytocin, these highly purified factors (CRF, GH–RF, and LH–RF) had an absence of sulphur-containing amino acids in their molecules. We have already alluded to the resistance of CRF and LH–RF to reduction by thioglycollate which splits the disulphide bridge in the molecules of vasopressin and oxytocin. Consequently, it would appear that the hypothalamic factors which affect pituitary secretion constitute a new family of relatively small, basic polypeptides.

In conclusion, the hypothalamic stimulatory influence on the secretion of FSH and LH and its inhibitory influence on prolactin secretion is mediated by specific polypeptides secreted into the hypophyseal portal vessels. Evidence is mounting to support the view that the variations in output of these three pituitary hormones are brought about mainly through variations in the rate of release of the FSH–RF, LH–RF and PIF. FSH–RF and LH–RF are small polypeptides dissimilar from vasopressin and oxytocin. They can be separated from the other hypothalamic polypeptide hormones involved in the regulation of other adenohypophyseal hormones. The FSH–RF and LH–RF have been separated chemically, but so far PIF and LH–RF activity have been closely associated during purification by gel filtration followed by chromatography on CMC.

ACKNOWLEDGEMENT

Research in the author's laboratory was supported by the following grants: Grant AM-10073 from the National Institutes of Health, Population Council Grant M-6461, and a grant from the Ford Foundation.

REFERENCES

Antunes-Rodrigues, J., Dhariwal, A. P. S. and McCann, S. M. (1966). Proc. Soc. exp. Biol. Med. **122,** 1001–1004

Barraclough, C. A. and Gorski, R. A. (1963). J. Endocr. **25,** 175–182

Bogdanove, E. M. (1963). Endocrinology **72,** 638–642

Campbell, H. J., Feuer, G. and Harris, G. W. (1964). J. Physiol., Lond. **170,** 474–486

Chowers, I. and McCann, S. M. (1965). Endocrinology **76,** 700–708

David, M. A., Fraschini, F. and Martini, L. (1965). C. r. hebd. Séanc. Acad. Sci. Paris **261,** 2249–2251

Davidson, J. and Sawyer, C. H. (1961). *Proc. Soc. exp. Biol. Med.* **107,** 4–7

Dhariwal, A. P. S., Antunes-Rodrigues, J. and McCann, S. M. (1965). *Proc. Soc. exp. Biol. Med.* **118,** 999–1003

— — Krulich, L. and McCann, S. M. (1966). *Proc. Soc. exp. Biol. Med.* **121,** 8–12

— Grosvenor, C., Antunes-Rodrigues, J. and McCann, S. M. (1965a). *Programme of the 47th meeting of the Endocrine Society* p. 84

— Krulich, L., Antunes-Rodrigues, J., Chowers, I., Reeser, F., Grosvenor, C. and McCann, S. M. (1965b). *Proc. Int. Congr. Physiol. Sci.* (Tokyo), p. 252

— Nallar, R., Batt, M. and McCann, S. M. (1965c). *Endocrinology* **76,** 290–294

— Watanabe, S., Ratner, A. and McCann, S. M. (1966a). *Progm. 48th Meeting Endocr. Soc.* (in press)

Everett, J. W. (1954). *Endocrinology* **54,** 685–690

— (1964). *Physiol. Rev.* **44,** 373–431

Frankel, A. I., Gibson, W. R., Graber, J. W., Nelson, D. M., Reichert, L. E. Jr. and Nalbandov, A. V. (1965). *Endocrinology* **77,** 651–657

Grosvenor, C. E. and Turner, C. W. (1957). *Proc. Soc. exp. Biol. Med.* **96,** 723–725

— McCann, S. M. and Nallar, R. (1965). *Endocrinology* **76,** 883–889

Guillemin, R. (1964). *Recent Prog. Horm. Res.* **20,** 89–122

— Justisz, M. and Sakiz, E. (1963). *C.r. hebd. Séanc. Acad. Sci. Paris* **256,** 504–507

Harris, G. W. (1960). In: *Handbook of Physiology* II, 1007–1038. Ed. by J. Field. Baltimore: Williams and Wilkins

Haun, C. K. and Sawyer, C. H. (1960). *Endocrinology* **67,** 270–272

Igarashi, M. and McCann, S. M. (1964). *Endocrinology* **74,** 440–445

— — (1964a). *Endocrinology* **74,** 446–452

— Nallar, R. and McCann, S. M. (1964). *Endocrinology* **75,** 901–907

Ishida, Y., Kuroshima, A., Bowers, C. Y. and Schally, A. V. (1965). *Proc. Int. Congr. Physiol. Sci.* (Tokyo), p. 276

Kuroshima, A., Ishida, Y., Bowers, C. Y. and Schally, A. V. (1965). *Endocrinology* **76,** 614–619

Lisk, R. D. (1960). *Can. J. Biochem. Physiol.* **38,** 1381–1383

McCann, S. M. (1962). *Am. J. Physiol.* **202,** 395–400

— (1963). *Physiol. Physicians* **1,** 12

— and Dhariwal, A. P. S. (1964). *Trans. N.Y. Acad. Sci.* **27,** 39

— and Friedman, H. M. (1960). *Endocrinology* **67,** 597–608

— and Haberland, P. (1959). *Proc. Soc. exp. Biol. Med.* **102,** 319–325

— and Ramirez, V. D. (1964). *Recent Prog. Horm. Res.* **20,** 131–181

— — and Abrams, R. (1964). In: *Hormonal Steroids, Biochemistry and Therapeutics* 1, 251–258. Ed. by L. Martini. New York; Academic Press

— and Taleisnik, S. (1960). *Am. J. Physiol.* **199,** 847–850

— — and Friedman, H. M. (1960). *Proc. Soc. exp. Biol. Med.* **104,** 432–434

Mittler, J. C. and Meites, J. (1964). *Proc. Soc. exp. Biol. Med.* **117,** 309–313
Nallar, R., Antunes-Rodrigues, J. and McCann, S. M. (1966). *Endocrinology* **79,** 907–911
— and McCann, S. M. (1965). *Endocrinology* **76,** 272–275
Nikitovitch-Winer, M. B. (1962). *Endocrinology* **70,** 350–358
— and Everett, J. W. (1958). *Endocrinology* **62,** 522–532
—— (1958a). *Endocrinology* **63,** 916–930
Novella, M., Alloiteau, J. J. and Ascheim, P. (1964). *C. r. hebd. Séanc. Acad. Sci., Paris* **259,** 1953–1956
Parlow, A. F. (1961). In: *Human Pituitary Gonadotrophins* 300–310. Ed. by A. Albert. Springfield, Illinois; C. C. Thomas
Pasteels, J. (1962). *C.r. hebd. Séanc. Acad. Sci., Paris* **254,** 2664–2666
Porath, J. and Schally, A. V. (1962). *Endocrinology* **70,** 738–742
Ramirez, V. D., Abrams, R. and McCann, S. M. (1964). *Endocrinology* **75,** 243–248
— and McCann, S. M. (1963). *Endocrinology* **73,** 193–198
—— (1964). *Endocrinology* **74,** 814–816
—— (1964a). *Am. J. Physiol.* **207,** 441–445
—— (1964b). *Endocrinology* **75,** 206–214
— Nallar, R. and McCann, S. M. (1964a). *Proc. Soc. exp. Biol. Med.* **115,** 1072–1076
— and Sawyer, C. H. (1965). *Endocrinology* **76,** 1158–1168
Royce, P. C. and Sayers, G. (1960). *Proc. Soc. exp. Biol. Med.* **103,** 447–450
Saffran, M. and Schally, A. V. (1955). *Can. J. Biochem. Physiol.* **33,** 408–415
Schally, A. V. and Bowers, C. Y. (1964). *Endocrinology* **75,** 312–320
— and Bowers, C. Y. (1964a). *Endocrinology* **75,** 608–614
Schwartz, N. B. and Mills, J. M. (1961). *Endocrinology* **69,** 844–850
Steelman, S. L. and Pohley, F. M. (1953). *Endocrinology* **53,** 604–616
Taleisnik, S. and McCann, S. M. (1961). *Endocrinology* **68,** 263–272
Talwalker, P. K., Ratner, A. and Meites, J. (1963). *Am. J. Physiol.* **205,** 213–218

DISCUSSION

J. A. Loraine (*Edinburgh*)

I notice in relation to the ascorbic acid depletion data that your results were expressed as percentage depletion of ascorbic acid. Would it have been possible to express these results somewhat differently, for instance in terms of a standard material such as NIH-LH. Would it have been possible to conduct four point assays so that the reliability criteria such as parallelism could have been examined in more detail. The reason I say this, is that, in our experience at least as far as assays of body fluids are concerned, a proportion of such assays using this method when assayed in terms of standard materials are invalid due to the lack of parallelism. Presumably such results would have little meaning from the quantitative point of view.

McCann

For the first results given, where we were measuring pituitary and plasma LH,

75

we did have standards. We have been criticized for not using standards but we have probably recorded more results in terms of LH standard than anyone else in the literature. For LH-RF the proper standard is not LH but purified LH-RF and when we did the early work a standard was not available. It is still not available although we could take some of our own purified fractions and call them standards. In most of the work on LH-RF I was not trying to draw too many conclusions on the quantitative aspect, but was merely demonstrating whether or not it was present. I quite agree that we should use standards but the problem is that with LH-RF the precision of the assay is less than with LH. This is probably because of the introduction of another variable, i.e. the variability of release of LH in the pituitary of test rats. We have tried this work using a four point assay but the precision is not very good; that is why we have not done this routinely. Particularly in the purification work it would have been advantageous to indicate how much purification had been achieved. The difficulty is that nearly all the purified material is used up merely doing quantitative assays, as well as using up time. When we have more precise methods where we can obtain reasonable values without too much effort it certainly should be done.

I. ROTHCHILD (*Cleveland, U.S.A.*)

I would like to make one point about the identity of the prolactin inhibiting factor (PIF) and LH-RF. There are many findings in the literature indicating that they are not the same. This depends on the fact that, in the rat, human-being and cow, ovulation can occur during lactation. In fact, in the rat, appreciable degrees of lactation and therefore probably prolactin secretion can occur, yet the animal can still show a marked increase of luteotrophin in the pituitary after castration. The animal can also ovulate and this must mean that the LH-RF must be secreted and liberated to the pituitary. One would expect the animal to stop lactating if it is inhibiting prolactin secretion but it does not.

McCANN

I agree. I think the difficulty is that we haven't quantitative measurements of the amount of plasma prolactin and LH. Furthermore, it is possible that some interactions may take place at the pituitary level. I still think that it is conceivable that they could be the same in spite of the exceptions you have mentioned. We have been trying to separate them but have not yet been very successful. A single series of experiments has in fact been reported in which purified LH-RF failed to inhibit prolactin *in vitro*. The disadvantage was that the workers did not find the active fraction and it seems to me that if you cannot find this fraction you cannot be sure.

L. E. REICHERT (*Atlanta, Georgia, U.S.A.*)

Have you detected LH-RF by any other assay than the OAAD method? For example, the hyperaemic assay is approximately as sensitive as the OAAD assay and the cholesterol depletion assay is supposedly much more sensitive.

McCANN

We have not done this but have left it to others. Fortunately, they have been able to demonstrate the activity of this material by a variety of methods including ovulation; although no one has done the hyperaemia assay to my knowledge. The reason why we have refrained from ovulation production is that it is not as sensitive as OAAD assay and because others have shown it so well. It is clear that the material produces ovulation but this does not definitely indicate that we are dealing with LH. In hypophysectomized animals one gets better ovulation by mixtures of FSH and LH than with LH alone, but I think this answers the criticism about just relying on OAAD. As far as the cholesterol assay is concerned I understand there are difficulties in interpreting the results.

P. J. HEALD (*London*)

Professor McCann has covered a lot of ground extremely quickly. In one of his early statements he showed that there was a release of LH during the pro-oestrous phase of the cycling rat and that this could be enhanced by progesterone treatment at that particular phase. He then said that it was clear that the endogenous progesterone release stimulated the LH release in the normal cycling rat. This is rather a large jump to make from the experimental evidence he presented. I am not aware of any measurements that have been made of levels of progesterone during the oestrous cycle of the rat and I wonder if Professor McCann would like to elaborate.

McCANN

I showed that there was a release of LH because it was one of the few new things I had to report. I did not mean to imply that the ovulatory discharge of LH was brought about exclusively by progesterone, probably oestrogen is involved also. We have tried to get this effect with oestrogen but haven't succeeded yet but we may not of course be using the right conditions. As to whether or not there is any progesterone or progestational agent in plasma of rats at this stage of the cycle, Everett has been postulating its presence for some time.

ROTHCHILD

I have just finished a review on this subject and there is an enormous amount of indirect evidence to indicate secretion of progesterone in the pre-ovulatory stage, e.g. the condition of the vagina, the uterus, the behavioural aspect of the animal as well as the fact that progesterone will induce ovulation in the rat and others. There are two papers in the literature which, although not entirely in agreement with one another, do indicate that there is an increased secretion of progesterone sometime before ovulation occurs. These are a paper by Talenti, and one by the Japanese group, Eto and co-workers. The latter is a beautifully detailed piece of work on the amount of progesterone in the blood and corpora lutea throughout the cycle, pseudopregnancy, lactation, and in the presence of a transplanted pituitary. This work does show that there is an increase in progesterone before ovulation.

H. KARG (*Munich, Germany*)

I have a question with regard to LH-RF and LH in the ovarian ascorbic acid depletion assay. It is known that the time response to LH reaches a maximum after 4 h with a further change after 8 h. Is there any evidence to indicate how quickly RF acts?

McCANN

We have tried to answer that by directly injecting material in a test rat. If you inject extracts intravenously, an increase in plasma LH within 5 min after injection can be measured. We have not yet tried to get any closer than that, although it is possible from what we know of other releasing factors we might get an increase in a shorter time. Recently in our laboratory Dr. Porter has infused the CRF preparation and obtained the effect within a matter of minutes so these responses appear to be rapid. As to using the 4 h rat test, we have found a lot of difficulty in using the 4 h test in measuring LH-RF mainly because the 4 h test is less sensitive than the 1 h test. That is why we have used the 1 h test in most of the work on the LH-RF. We have used the 4 h test when measuring the pituitary LH because this gives a better precision, although it is less sensitive.

H. A. ROBERTSON (*Aberdeen*)

Can you give any information as to whether LH-RF has any effect on the synthesis of LH as well as on release?

McCANN

It would appear from *in vitro* studies carried out in Meites' Laboratory that the hypothalamic factors effect synthesis as well as release. This does seem to be the case with PIF and GH-RF. We have not done any work on this. As far as LH-RF is concerned I think it also has some effect on synthesis of LH. A very interesting experiment has been reported by Nikitovich-Winer and co-workers who placed pituitary grafts in the kidney capsule and these underwent characteristic cytological changes. The authors then infused median eminence extracts via the renal artery and were able to reconstitute the cytology of the pituitary and also to produce effects on target organs. It can be assumed that there was initially very little LH in the pituitary although this was not checked by assay. There were signs of some storage and stimulation of target organs which would indicate synthesis and release. There is a lot of work to be done here; these are probably not just releasing factors, but are hypophyseal stimulating and inhibitory hormones which affect both synthesis and release. The reason we concentrate on release is because it is easy to measure.

STUDIES ON THE NEUROENDOCRINE CONTROL OF LACTATION

J. S. TINDAL

Physiology Department, National Institute for Research in Dairying, Shinfield, Reading

CONTROL OF THE SECRETION OF ANTERIOR PITUITARY HORMONES

Role of anterior pituitary hormones in lactation

IT HAS been established in those species so far investigated that a functional anterior pituitary is essential for the processes of milk secretion (*see* reviews by Cowie, 1961; Cowie and Folley, 1961; Folley, 1961). Considerable restoration of milk secretion has been achieved in the hypophysectomized or anterior-lobectomized lactating rat with prolactin plus adrenocorticotrophin (ACTH) or adrenal corticosteroids (Cowie, 1957; Bintarningsih, Lyons, Johnson and Li, 1958; Cowie and Tindal, 1961a), while in the hypophysectomized lactating goat partial to complete restoration of lactation was achieved with a combination of prolactin, ox somatotrophin (STH), insulin, tri-iodo-L-thyronine (LT$_3$) and adrenal corticosteroid (Cowie and Tindal, 1960, 1961b; Cowie, Knaggs and Tindal, 1964). Confirmation of our results has been obtained by Gale and Larsson (1963) in goats 'hypophysectomized' by a collimated proton beam. In subsequent studies (Cowie and Tindal, 1965a), we have obtained evidence that the insulin may not be necessary for complete restoration of lactation in the hypophysectomized goat.

Effects of pituitary stalk section and of brain lesions

Severence of the pituitary from its normal connection with the hypothalamus via the hypophysial portal system, by transplantation elsewhere in the body or by pituitary stalk section, results in the virtual cessation of secretion of trophic hormones, except for that of prolactin. For a detailed account of these studies, the recent review by Everett (1964) should be consulted, since only those aspects which have a bearing on lactation will be considered here.

In the rat, when the pituitary is grafted under the renal capsule

79

where it can become well vascularized, it has been shown to secrete sufficient prolactin to maintain milk secretion after hypophysectomy if ACTH is administered (Cowie, Tindal and Benson, 1960), and to initiate milk secretion in the oestrogen-primed rat (Meites and Hopkins, 1960). In the lactating goat after surgical stalk section and the placement of an impervious plate to prevent regeneration of the pituitary portal vessels, milk yield fell to approximately 30 per cent of the pre-operative level. Considerable restoration of milk yield was achieved by administration of STH, LT_3, insulin and adrenal corticosteroid (Cowie, Daniel, Knaggs, Prichard and Tindal, 1964). In this study, the composition of the milk was found to be normal, whether exogenous hormones were being administered or not. Our results did not therefore support the suggestion of Donovan and Van der Werff ten Bosch (1957) that stalk section in the lactating rabbit results in the secretion of milk of poor nutritive value, although a species difference in the response to the operation cannot be entirely ruled out.

Although stalk section together with the insertion of an impervious plate virtually eliminates the possibility of portal vessel regeneration, the technique suffers from the disadvantage of causing severe necrosis of the anterior lobe (Daniel and Prichard, 1958). Another approach has been the complete or partial destruction by various techniques of the median eminence and tuberal region above the pituitary stalk. Surgical damage to this region in the goat (Tverskoy, 1960) and by electrolytic lesions in the rat (McCann, Mack and Gale, 1959) blocked lactation, and in the rat the lactational deficiency was restored to 70–80 per cent by replacement with cortisol (plus oxytocin to achieve milk ejection) and exogenous prolactin was not required (Gale, Taleisnik, Friedman and McCann, 1961). Discrete lesions of the post-tuberal region in the oestrogen-primed ovariectomized rabbit caused activation of the mammary glands, and, in some cases, the onset of copious milk secretion, indicating the secretion or release of considerable quantities of prolactin (Haun and Sawyer, 1960, 1961; Sawyer, Haun, Hilliard, Radford and Kanematsu, 1963; Kanematsu, Hilliard and Sawyer, 1963a). The common site of the effective lesions in these studies was the arcuate and the basal portion of the ventromedial nuclei. Further evidence for the participation of the median eminence in mechanisms controlling prolactin secretion was obtained by lesions in this region in the rat, placed during the oestrous cycle (McCann and Friedman, 1960) or during early gestation (Gale and McCann, 1961) since they resulted in the secretion of sufficient prolactin to support uterine deciduomata and pregnancy, respectively. Recently, the same group have also achieved mammary activation in male rats bearing tuberal lesions (De Voe, Ramirez and McCann, 1966). In comparatively low-yielding lactating goats, median eminence lesions produced either by a proton beam or by radio-frequency energy caused a drop in milk secretion which could be restored by hormonal combinations which omitted prolactin (Gale and Larsson, 1963; Gale, 1963; Gale, 1964).

From the clinical standpoint, these studies present a basis for understanding the occurrence of milk secretion in non-parturient women after section of the pituitary stalk for the alleviation of breast cancer (Ehni and Eckles, 1959), and also certain clinical syndromes, in particular the Chiari–Frommel syndrome, where persistent milk secretion and ovarian atrophy occur in the post-parturient woman. Such a condition is sometimes associated with a tumour in the pituitary stalk or basal hypothalamus, but more frequently with a tumour in the anterior pituitary itself (*see* Folley, 1960; Pasteels, 1963; Jaszmann, 1963; Bercovici and Ehrenfeld, 1963).

In addition to the median eminence, other regions of the hypothalamus may also be concerned with the control of prolactin secretion, since, in the cat with partially regressed mammary glands, surgical damage to the floor of the brain immediately caudal to the optic chiasma resulted in copious milk secretion (Grosz and Rothballer, 1961), while in the rat, lesions in the anterior hypothalamus dorsolateral to the paraventricular nuclei have been reported to produce prolonged dioestrus and to favour the formation of deciduomata after uterine trauma (Flament-Durand and Desclin, 1964). Lesions in this area were shown to prevent lactogenesis in the rat (Averill and Purves, 1963), yet had no effect on established lactation (McCann *et al.*, 1959). In regions more distant from the hypothalamus, medial septal lesions in the rabbit caused a reduction in milk secretion, although the effect may have been secondary to loss of appetite (Cross, 1961b). Destruction of a large region at the mesodiencephalic boundary in the cat was reported to block lactation (Beyer, Mena, Pacheco and Alcarez, 1962b), although in later studies (Beyer, personal communication) it seemed likely that the block was secondary to loss of the milk-ejection reflex. Finally, Beyer and Mena (1965) showed that removal of the telencephalon in oestrogen-primed rabbits caused the onset of milk secretion.

Effects of drugs and steroids

The differentiation and fragmentation of pituitary function has been investigated not only by surgical and destructive means, but also by the systemic administration, or local implantation in the brain, of drugs and steroids. Notable amongst these have been the effects of certain tranquillizing drugs, reserpine and chlorpromazine, which have been shown to cause the release of prolactin and activation of the mammary glands (Kehl, Audibert, Gage and Amarger, 1956; Sulman and Winnik, 1956; Meites, 1957; Sawyer, 1957; Tindal, 1960).

It has been known for a long time that exogenous oestrogen can affect pituitary function, and the first demonstration of the artificial induction of mammary growth and lactation in the ruminant by synthetic oestrogen took place in this laboratory (Folley, Scott Watson and Bottomley, 1940). In recent years, there has been a wealth of reports on the effects of local implants of oestrogen in the brain on trophic hormone secretion by the pituitary. Following their studies on post-tuberal lesions in the rabbit, the Los Angeles group reported that local oestrogen implants in the post-tuberal region caused increased secretion and storage of prolactin, but no release of the hormone. Conversely, when the oestrogen was placed directly in the anterior lobe, prolactin content of the gland fell, and this was followed by activation of the mammary glands in suitably prepared animals (Kanematsu and Sawyer, 1963a, 1963b). In the rat, local oestrogen implants in either the median eminence or the anterior lobe of the pituitary were reported to cause prolactin release, and, on histological assessment, activation of mammary glands (Ramirez and McCann, 1964). There has been some controversy as to whether local oestrogen acts at the level of the median eminence or directly on the pituitary itself. It arose during studies on the control of gonadotrophin secretion, and a full discussion of the arguments would be outside the scope of this paper. Very briefly, it has been proposed that when oestrogen is implanted in the median eminence, it is dissolved and carried by the portal system to the anterior pituitary, and that the most efficient means of introducing oestrogen to all parts of the anterior lobe, therefore, is to put it in the median eminence (*see* Bogdanove, 1963). However, the work of C. H. Sawyer's group (referred to above) has shown a clear differentiation in the type of effect that oestrogen can exert in these two sites. In addition, they demonstrated, on the basis of local implants, that the lactogenic action of reserpine in the oestrogen-primed ovariectomized rabbit was exerted on the basal tuberal region, that is, the same site whose destruction leads to lactogenesis in the suitably prepared rabbit. Reserpine implants placed directly in the anterior pituitary were ineffective, and when reserpine was administered intravenously the integrity of the post-tuberal hypothalamic region was essential for the drug to manifest its lactogenic action (Kanematsu and Sawyer, 1963c; Kanematsu, Hilliard and Sawyer, 1963b).

Studies on the anterior pituitary cultured in vitro *and hypothalamic neurohumours*

The inescapable conclusion from the above studies was that whereas the hypothalamus facilitates the production of most of the trophic hormones, it actively inhibits the secretion and/or release of prolactin. The step forward from a hypothesis to a theory was achieved by the culture of the anterior pituitary gland *in vitro* in an artificial medium. Under these conditions, the anterior pituitary of the rat secreted considerably more prolactin over a period of days than it appeared to contain at the time of removal from the body (Pasteels, 1961a; Meites, Kahn and Nicoll, 1961). Also, the secretion of prolactin was reported to be enhanced by the addition of oestrogen directly to the culture medium (Nicoll and Meites, 1962) or by culturing pituitaries from oestrogen-primed rats (Ratner, Talwalker and Meites, 1963). The direct effect of oestrogen on the cultured pituitary was confirmed by Ben-David, Dikstein and Sulman (1964) but denied by Gala and Reece (1964).

The discovery that the addition of hypothalamic tissue or extracts to the incubation medium could inhibit the secretion of prolactin by the pituitary gland *in vitro* (Pasteels, 1961b, 1963; Talwalker, Ratner and Meites, 1963; Danon, Dikstein and Sulman, 1963) indicated that a hypothalamic neurohumour (designated 'prolactin-inhibiting factor' abbreviation: 'PIF') was responsible for such inhibition and was the first evidence of the nature of the central nervous inhibition of prolactin secretion. The inhibitory action of hypothalamic fragments on the secretion of prolactin *in vitro* was itself inhibited by perphenazine (a phenothiazine derivative related to the tranquilizing drug chlorpromazine) either by adding the drug directly to the incubation medium or by using hypothalami from perphenazine-primed rats (Danon *et al.*, 1963), and reserpine was also found to act in a similar manner (Ratner, Talwalker and Meites, 1965).

No definite evidence is available as to the exact site or sites in the hypothalamus where PIF is secreted or stored. However, the fact that the tuberal lesions causing lactogenesis (reviewed above) and the 'hypophysiotrophic' area of the rat in which pituitary grafts maintain a normal histological appearance (Halász, Pupp and Uhlarik, 1962) share, to a large extent, a common region in the arcuate and basal part of the ventromedial nuclei, suggests that it is in this area that the neurohumours, including PIF, might be found.

As regards the hypothalamic control of the other hormones known to be concerned in the maintenance of lactation, evidence has now been presented for a growth hormone-releasing factor (Deuben and Meites, 1964, 1965; Schally, Steelman and Bowers, 1965; Müller and Pecile, 1965a, 1966; Pecile, Müller, Falconi and Martini, 1965; Krulich, Dhariwal and McCann, 1965) in the basal hypothalamus (Halász, 1965), while the presence of a corticotrophin-releasing factor(s) has been recognized for some time (see Guillemin, 1963; Vernikos-Danellis, 1965).

The pituitary hormones concerned with the maintenance of lactation, therefore, appear to be under the control of neurohumours secreted by the hypothalamus and carried to the pituitary by the hypophysial portal system. However, the mechanisms controlling the pituitary during lactation are presumably more sophisticated than a mere isolation of the anterior pituitary from the central nervous system, and the question still remains as to what central nervous structures and mechanisms are responsible for the fine control of the neurohumours and hence the modulation of pituitary function. The phylogenetically ancient structures known as the limbic system have already been implicated in the feed-back control of gonadotrophin secretion (Kawakami and Sawyer, 1959), and the degeneration studies of Nauta (see Nauta, 1960, 1963) have highlighted the reciprocal interconnections in this system and the probability that it is indeed concerned with pituitary control.

Role of the amygdaloid complex in the control of prolactin secretion

A major structure in the limbic system is the paired amygdaloid complex. The amygdala, like the brainstem reticular formation, receives inputs from all sensory modalities (see Goddard, 1964 for review) and several studies have linked it with pituitary function. Bilateral destruction of the amygdala in the male rat resulted in testicular atrophy (Yamada and Greer, 1960), although the effect may have been secondary to anorexia. Destruction of the basomedial region in young female rats led to precocious ovarian development, suggesting that the structure may be a site for a negative feed-back effect of oestrogen (Elwers and Critchlow, 1960). In the adult female rat, lesions restricted to the central nucleus, which contains many fibres of passage of the stria terminalis, or the stria terminalis itself, resulted in reduced secretion of trophic hormones, except for that of LTH (i.e. prolactin), which, on the basis of positive decidual responses, may have been increased (Moore, Woehler and Tarry, 1965).

From what was known of the structure of the limbic system and its relationships to the hypothalamus, it had already seemed likely to us that the amygdaloid complex could well be involved in the modulation of pituitary function, and might play a role in the secretion or release of prolactin. Accordingly, we began a programme of work in which minute local oestrogen implants were placed bilaterally in the brains of pseudopregnant rabbits (*Figure 1*). It was found that lactogenesis ensued when the oestrogen lay in the basal amygdala (for a preliminary account *see* Tindal and Knaggs, 1966). In this, and more recent studies (Tindal and Knaggs, unpublished), responsive sites were found in the basomedial portion of the basal nucleus, the medial and central nuclei, and the stria terminalis. In addition, part of the underlying pyriform cortex and cortical amygdaloid nucleus may also be involved. We considered the possible criticism that the implanted oestrogen might be dissolved and carried by the bloodstream direct to the anterior pituitary, as suggested by Bogdanove (1963) for an analogous situation, but do not believe it to be valid because implants in closely adjacent sites yielded negative results.

In the rabbit, the medial, cortical, and medial portion of the basal amygdaloid nuclei project, via the supra-commissural component of the stria terminalis, to hypothalamic structures as far caudal as the rostral face of the ventromedial hypothalamic nucleus (Ban and Omukai, 1959). It is possible, therefore, that oestrogen-sensitive neurons in the amygdala exert their influence on prolactin secretion by synaptic means at the level of the basal hypothalamus.

Amongst its many functions the amygdala is concerned with olfaction (*see* Goddard, 1964) and it seems possible that it may be concerned with the olfactory block to pregnancy in the mouse, caused by the presence of strange males (Bruce, 1959, 1965). The block itself was shown to be due to the inhibition of prolactin secretion, since the block to pregnancy did not occur if exogenous prolactin was also given (Bruce and Parkes, 1960). In the rat, the olfactory bulbs project to the anterior amygdala area, the cortical, medial, and basomedial amygdaloid nuclei, and the surrounding pyriform cortex (White, 1965). Furthermore, the pyriform cortex has been shown to project, along with the amygdala, in the ventral amygdalofugal bundle to the lateral pre-optic and lateral hypothalamic regions (Powell, Cowan and Raisman, 1963).

Although, therefore, we have been discussing several different species, evidence is accumulating to suggest that the amygdala may be a modulator of prolactin secretion, through the mediation of the hypothalamo-hypophysial system. This view may be substantiated

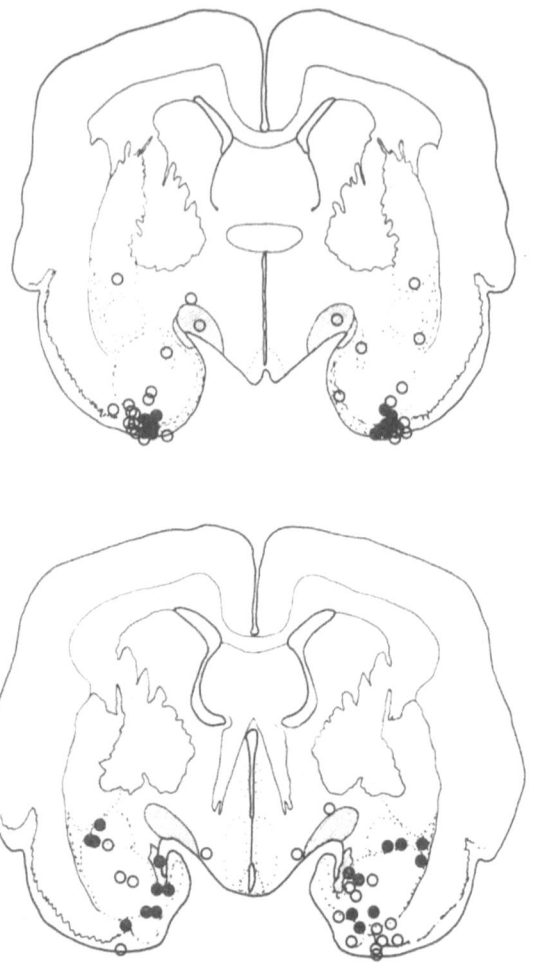

Figure 1.   Transverse sections through the amygdaloid region of rabbit brain showing sites of bilateral oestrogen implants.   Upper section is 1 mm anterior to the lower section.   Solid circles denote occurrence of lactogenesis after oestrogen implantation·in the pseudopregnant rabbit; open circles denote absence of lactogenesis (From J. S. Tindal and G. S. Knaggs, unpublished)

further by our recent preliminary finding that bilateral lesions of the amygdaloid complex caused lactogenesis in the pseudopregnant rabbit (Tindal and Knaggs, unpublished).

THE SUCKLING STIMULUS

Effects of suckling stimulus on anterior pituitary function

The suckling stimulus has a well-established role in triggering the release of oxytocin, as part of the milk-ejection reflex (*see* Denamur, 1965, for review). In addition, the suckling stimulus can also affect anterior pituitary function, and has been shown to cause discharge of prolactin in the rat (Grosvenor and Turner, 1957, 1958; Moon and Turner, 1959; Grosvenor, McCann and Nallar, 1965) and of ACTH in the rat (Grégoire, 1946), goat and sheep (Denamur, Stoliaroff and Desclin, 1965). Furthermore, there is now cytological evidence for activation of the anterior pituitary by the suckling stimulus (Pasteels, 1963). When lactating rats were separated from their litters for 10 h, the erythrosinophilic 'prolactin cells' became engorged with granules. A short period of nursing caused a severe degranulation without affecting the size or number of the cells. Under the electron microscope, the prolactin granules were seen to be excreted from the pole of the cell into the perisinusoidal spaces (Pasteels, 1963), and prolactin granules appeared to be excreted for about 1 h after the end of nursing (R. E. Smith, quoted as personal communication by Grosvenor *et al.*, 1965).

The suckling stimulus presumably exerts its effect by inactivating or preventing the synthesis or release of prolactin-inhibiting factor in the hypothalamus, and the *in vivo* injection of acid extracts of bovine stalk-median eminence has been shown to inhibit the fall in pituitary prolactin concentration which occurs after the combined stimuli of nursing and stress (Grosvenor *et al.*, 1965), thus indicating a physiological role for a prolactin inhibiting factor. Ratner and Meites (1964) failed to detect any prolactin-inhibiting activity in hypothalami obtained from suckled rats, and suggested that suckling depleted the hypothalamus of PIF. This has been denied by Grosvenor (1965b) since as little as one-third of a rat hypothalamus, obtained either from pre-nursed, post-nursed or stressed lactating rats, was able to inhibit nursing-induced or stress-induced discharge of prolactin in the lactating rat. It appears, therefore, that hypothalamic PIF is extremely potent and long-lasting in its effect, and it has been suggested that the acute stimuli of nursing and stress elicit prolactin release in the rat, not by destroying, but by inhibiting the release of PIF from the hypothalamus (Grosvenor, 1965b).

It should be mentioned that in addition to the suckling stimulus itself, other sensory modalities may also be involved in the nursing-induced discharge of pituitary hormones. The mere presence of a rat's litter, even though physically separated from the mother, can cause discharge of prolactin from the maternal pituitary (Grosvenor, 1965a), suggesting the involvement of visual, auditory or olfactory influences.

An understanding of anterior pituitary activation by suckling or milking may well depend on a broader interpretation than has been considered up till now, since, on the one hand, a variety of stressful procedures were shown to initiate mammary secretion in the rat (Nicoll, Talwalker and Meites, 1960), and stress was also found to be as effective as the suckling stimulus in causing discharge of prolactin in the rat (Grosvenor et al., 1965), while on the other hand, in the goat and sheep, the milking stimulus leads to a rapid release of ACTH from the pituitary. The quantity of ACTH released after the milking stimulus in the small ruminant is proportionally as great as that caused by stress in the rat (Denamur et al., 1965), which raises the question as to whether the suckling or milking stimulus itself should be considered to be stressful. It seems more probable that the sensory inflow resulting from these two phenomena may share a final common path to the hypothalamo-hypophysial system at brainstem levels, and to attempt a rigid distinction between the two seems unreasonable. The milking or suckling stimulus is known to trigger rumination in the goat (see Andersson, Kitchell and Persson, 1958) and to cause activation of the EEG in the rabbit (Holland and Cross, unpublished, quoted by Cross, 1961a), so that when more is known of the central nervous pathways involved, and the nature of central nervous activation caused by stress and by the suckling or milking stimulus, the jig-saw may fall into place.

The suckling stimulus not only evokes the secretion of those pituitary hormones concerned with the maintenance of lactation, but in the human, it may also play a vital, indirect, sociological role. It is well known that the menstrual cycle is normally suppressed during the course of lactation—lactational amenorrhoea—and this can be of crucial importance, especially in the developing countries, where the mother's milk may be the only high-quality food available to the baby during infancy. An early post-parturitional pregnancy would result in the premature cessation of lactation, possibly with tragic nutritional results for the infant. Further knowledge of the suppression of gonadotrophin secretion by the suckling stimulus, therefore, has quite obvious implications in the field of population control.

Importance of the suckling stimulus for the maintenance of lactation in different species

When considering the mechanisms responsible for the release of the lactation-maintaining trophic hormones, it is clear that the importance of the suckling stimulus varies from species to species. Spinal section in the cat results in the inhibition of lactation, even when oxytocin is administered to ensure milk-ejection (Beyer, Mena, Pacheco and Alcaraz, 1962a). In the lactating rat, spinal section was reported to cause death of the litter (Eayrs and Baddeley, 1956), although if the operation is performed in mid-lactation when the pups are more vigorous, lactation can continue at a much reduced level, provided oxytocin is administered (Grosvenor, 1964), and a similar result was achieved by anaesthetizing the lactating rat during nursing periods (Yokoyama and Ôta, 1965). Lactation was found to be arrested in the rat when the young were confined to deafferented nipples, but was restored after injection of prolactin and cortisol (Eayrs and Edwardson, 1965).

In contrast to the cat and rat, the sheep and goat present a completely different picture. It is little short of astonishing that in these small ruminants the suckling or milking stimulus is not essential for the maintenance of lactation. This has been demonstrated beyond all doubt by the studies of Tverskoy and of Denamur and Martinet in which the udder was deafferented either by cutting the nerve supply to the gland or by spinal-cord section (for review of this work *see* Folley, 1961) and more recently by transplantation of the mammary gland to the neck (Linzell, 1963) and by cyclopropane anaesthesia during milking (Yokoyama and Ôta, 1965). Furthermore, in the goat, not only is the milking stimulus unnecessary for lactation maintenance in terms of anterior pituitary activation, but, provided thorough hand milking is carried out, the milk-ejection reflex itself is not essential. Convincing evidence for this was obtained recently when oxytocin levels in jugular blood were measured in the goat during hand milking. In many cases, no detectable release of oxytocin occurred during the milking process (Folley and Knaggs, 1966). To the sheep and goat, one may perhaps add the cow to those species in which the milking stimulus is not essential for anterior pituitary activation, since normal milk production will continue in cows in the complete absence of the milking or suckling stimulus if the teats are permanently cannulated so that milk drains from the cisterns and oxytocin is injected intravenously twice daily to eject the alveolar milk (Mielke and Brabant, 1963).

The rabbit appears to lie in an intermediate position between the rat and cat on the one hand and the small ruminants on the other, since Tindal, Beyer and Sawyer (1963) showed that when the lactating rabbit was deeply anaesthetized with pentobarbitone during a once-daily nursing period, and injected with oxytocin, milk yield was unaffected, even when the regime continued for 12 days. It was also observed that when lactation was depressed due to infection and consequent loss of appetite, once the infection had responded to antibiotic therapy, the milk yield rose again, even under the regime of daily anaesthesia. In a related study in the rabbit (Mena and Beyer, 1963) lactation was inhibited by spinal section, and was only partially restored by oxytocin. However, if in addition either prolactin or ACTH were injected for a few days, milk yield was restored and was maintained even after withdrawal of the trophic hormone.

Since suckling evokes discharge of prolactin (Grosvenor and Turner, 1957) and ACTH (Grégoire, 1946) from the anterior pituitary, and also stimulates food intake in the rat (Cotes and Cross, 1954), it seems paradoxical that the small ruminant can lactate satisfactorily in the absence of the central effects of the suckling stimulus. In an attempt to provide an explanation, Cowie and Tindal (1965b) proposed that the utilization of milk precursors by the mammary gland was responsible for triggering appetite-stimulation, whose motivational mechanisms reside in the medial and lateral hypothalamus, as well as in such regions as globus pallidus and amygdala (*see* Morgane, 1961, 1962; Goddard, 1964). It seemed to us that if appetite mechanisms could be activated metabolically, that trophic hormone release mechanisms in the hypothalamus and limbic system might also be activated, either as a direct result of the metabolic changes, or as a secondary effect of appetite stimulation. Neurophysiological evidence has been claimed for the activation of the hypothalamic 'feeding' and 'satiety' centres by the circulating level of blood glucose (Anand, Chhina, Sharma, Dua and Baldev Singh, 1964) although Sutin (1963) failed to demonstrate any change in the rates of discharge of single neurones ('unit firing') in the ventromedial hypothalamic nucleus after either systemic injection or local microinjection of glucose, so that one may have to look elsewhere in the brain to find other metabolic sites of action.

When viewed in the light of metabolic activation, not only is there a sliding scale of species as far as the vital nature of the suckling stimulus is concerned, but there may also be an inverse ranking correlation, i.e. the less important the suckling stimulus the more important the metabolic stimulus. To express it in another way, in the lactating animal is appetite more under the control of the suckling stimulus in species such as the cat or rat, and more under the control of metabolic changes in the goat and sheep? Mena and Beyer's (1963) finding in the cord-sectioned rabbit adds support to our theory, since, once lactation was restored by either ACTH or prolactin, it became self-maintaining, suggesting that above a certain critical level metabolite utilization may be able to evoke release of prolactin and/or ACTH. However, such a hypothesis awaits experimental proof.

In recent years, evidence has accumulated to suggest that somatotrophin is dynamically involved in rapid adjustments to the metabolic needs of the body, at least in man, since the level of STH in plasma was found to rise during hypoglycaemia, fasting or after exercise (Roth, Glick, Yalow and Berson, 1963a, 1963b). The same may apply to lower animals, since a reduction in pituitary STH content was reported after insulin hypoglycaemia in the rat (Müller and Pecile, 1965b). The secretion of STH in the human, therefore, appears to respond to the moment-to-moment metabolic needs of the body, and if a similar mechanism operates in non-primates, it may have some bearing on the mechanisms responsible for lactation maintenance in the small ruminant. It is known that STH, together with prolactin, is an essential hormone for the maintenance of lactation in the hypophysectomized goat (Cowie and Tindal, 1960, 1961b; Cowie, Knaggs and Tindal, 1964), while in the hypophysectomized rat, ox STH is without beneficial effect on the maintenance of lactation (Cowie, 1957). The metabolic demands made on the body by the mammary glands of a high-yielding goat must surely be of a very high order, and, considering the evidence presented here, it seems reasonable to postulate that the metabolic demand could be responsible for releasing STH, and hence could at least partly explain the maintenance of lactation in the small ruminant in the absence of the central effect of the suckling or milking stimulus.

Although the inessential role of the milking stimulus in the maintenance of lactation in the goat is not disputed, we have recently obtained evidence that this stimulus is a powerful mammogenic and lactogenic agent in this species. Application of the milking stimulus to the virgin ovariectomized goat caused mammogenesis, lacto-genesis and a considerable daily milk yield (Cowie and Tindal, 1965c), an effect which was abolished by pituitary stalk section (Cowie, Knaggs and Tindal, unpublished). There are two points of interest here. First, that the stimulus exerts such a powerful action in this species, and second, that it can elicit mammogenesis in the absence of the ovaries. Although we are aware of the possibility that adrenal corticosteroids may be partly substituting for the ovarian steroids, the second point raises the question of whether, in some species at least, a major role of the ovarian steroids in causing mammogenesis is exerted via the central nervous system and pituitary, rather than directly on the mammary gland, and that mammary growth in normal pregnancy may result from the secre-tion of far larger quantities of trophic hormones than have been suspected so far.

Inhibition of the release of oxytocin

It is well known that emotional factors can inhibit the milk-ejection reflex. Two types of inhibition have been recognized, peri-pheral inhibition, in which vasoconstriction in the mammary glands prevents oxytocin gaining access to the myoepithelium or circulating adrenaline inhibits the response of the myoepithelium to oxytocin (Chan, 1965), and central inhibition, where the release of oxytocin itself is blocked (*see* Cross, 1961b, for review of earlier work). Recently, experimental evidence for cerebral cortical involvement in the central inhibition of the reflex has been reported. Taleisnik and Deis (1964) subjected lactating rats to painful procedures which inhibited the milk-ejection reflex in response to normal suckling and prevented the young from obtaining milk. The inhibition was overcome by direct application of KCl solution to the cerebral cortex, resulting in the well-known phenomenon of cortical spread-ing depression (Leão, 1944), which, in effect, is a temporary, reversible 'pharmacological decortication'. It was suggested that the focus of the cortical inhibition might be at the mesencephalic tegmental level.

The recent report in abstract that the release of oxytocin in the cord-sectioned rabbit elicited by stimulation of forebrain structures was inhibited by simultaneous stimulation of more caudal structures (Aulsebrook and Holland, 1965) must await confirmation. It was claimed that inhibition of release resulted from stimulation of a wide range of mesencephalic and thalamic structures. One of the sites claimed to be inhibitory was the mesencephalic central grey, but since both Cross (1961a, 1961b) in the rabbit, and ourselves (Tindal and Knaggs, unpublished) in the guinea-pig have elicited milk ejection by electrical stimulation of this structure, the data presented by Aulsebrook and Holland (1965) might possibly be explained in terms of sympathetico-adrenal activation and the consequent peripheral inhibition of milk ejection. It is apparent that much remains to be learnt about the nature and the neural substrate of the central inhibition of the milk-ejection reflex.

Mapping the ascending path of the suckling stimulus in the spinal cord and brain

Although the determination of the pathway taken by the suckling stimulus in the central nervous system might be thought of as fundamental research, the considerable practical implications should not be overlooked. Further knowledge should lead to a better understanding of the processes concerned in the removal of milk, not only in the field of dairy husbandry and the achievement of more efficient milking procedures, but also on the clinical side in breast feeding, where emotional factors might be expected to play a more important role than in non-primates. A greater knowledge of the central nervous pathways and structures traversed by the stimulus will be required if we are to have a better understanding of control mechanisms for lactation. The most feasible method of tackling this problem is the determination of the ascending path of the milk-ejection reflex, even though the assumption has to be made that, for the most part, the suckling stimulus will ascend by similar or even identical paths, whether it is activating the anterior or the posterior pituitary glands.

Studies involving lesions of the central nervous system*

Eayrs and Baddeley (1956) reported that the ascending path of the suckling stimulus lay deep in the lateral funiculus of the spinal cord, and, since Marchi degeneration could be traced only as far as the medulla, the pathway probably became multisynaptic further rostrally (*see* Cross, 1961b, for review). The path was claimed to be ipselateral in the goat (Tsakhaev, 1953) and to be mainly ipselateral, with a minor contralateral component, in the rat (Eayrs and Baddeley, 1956), although more recently it has been suggested that the path in the rat may be both crossed and uncrossed, or that there

---

* The reader who is not fully conversant with the neuroanatomical terms used here may find it helpful to consult a text such as *Craigie's Neuroanatomy of the Rat*, W. Zeman and J. R. M. Innes, New York and London; Academic Press, 1963.

may be a double decussation (Stutinsky and Terminn, 1964a). It may be pertinent to recall that a multisynaptic pathway has been described in the cat, extending from the ground bundles of Bechterew in the lateral funiculus of the cord, through the hind-brain and mid-brain reticular formations, from which projections arise to several structures, including the mesencephalic central grey matter (Nauta and Kuypers, 1958). From here, fibres of the ascending component of the dorsal longitudinal fasciculus distribute to the intralaminar thalamic nuclei, and to the hypothalamus (Nauta, 1960). There is, therefore, an anatomical substrate for the ascending path of the suckling stimulus as suggested by Beyer *et al.* (1962a) following the finding that discrete lesions at the meso-diencephalic junction which destroyed either rostral central grey or the dorsal part of the caudal hypothalamus blocked the milk-ejection reflex in the cat (Beyer *et al.*, 1962a, 1962b; Beyer, personal communication) and rat (Stutinsky and Terminn, 1964b). Furthermore, in the rabbit, the mesencephalic central grey projects, via the dorsal longitudinal fasciculus, to paramedial structures in the hypothalamus, including the supraoptic nuclei (Beyer, Tindal and Sawyer, 1962).

Studies involving electrical stimulation of the brain

Turning now to experiments involving electrical stimulation of discrete sites in the brain, milk ejection has been elicited in the goat by stimulation of the medial lemniscus (Andersson, 1951), and in the rabbit from widely separated regions in the brain stem, central grey, reticular formation, medial lemniscus, subthalamus, supramamillary area, mamillary peduncle, as well as from limbic structures such as septum and fimbria of the fornix (Cross, 1961a, 1961b). Cross (1961b) proposed that the milk-ejection reflex involves diffuse, afferent, non-thalamic pathways, passing directly to the hypo-thalamus, and that the limbic structures may be concerned with conditioning or modification of the reflex.

Oxytocin release has also been elicited in the rabbit by stimulation of pyriform cortex, amygdala, hippocampus, fornix, septum, lateral hypothalamus, subthalamus, and ventral-postero-lateral and reti-cular thalamic nuclei (Holland, Aulsebrook and Woods, 1963), and, on the basis of uterine contractions, from stimulation of the cingulate gyrus of the cortex in the cat (Beyer, Anguiano and Mena, 1961).

The tracing of the ascending path of the milk-ejection reflex by electrical stimulation of the brain has been carried out up till now almost exclusively in the rabbit. However, we considered the guinea-pig to be a more suitable subject, since it has the built-in advantage of two relatively large mammary glands with a discrete nervous and arterial supply, and teats which are extremely simple to cannulate. We have now begun a stereotaxic study in order to trace the ascending path of the reflex in this species, using the co-

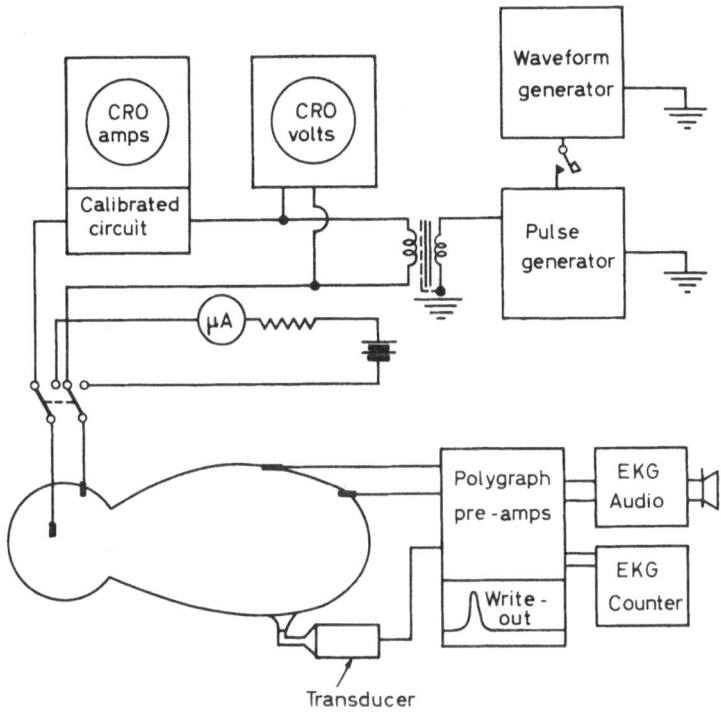

Figure 2. Diagram of circuit used for mapping ascending path of milk-ejection reflex in guinea-pig. Stimulation pulses are produced by waveform and pulse generators, voltage and current are monitored oscilloscopically, and pulses pass between monopolar electrode in brain and indifferent electrode in scalp.   The battery-resistor-meter circuit is used to place marker lesions in the brain (From J. S. Tindal and G. S. Knaggs, unpublished)

ordinates of Tindal (1965) (*Figure 2*).   A major problem in this type of work is that the electrode which is used to stimulate the brain may activate descending pathways in the brainstem which cause sympathetico-adrenal activation (*see* Cross, 1961b).   The resulting peripheral vasoconstriction will thus prevent oxytocin reaching the

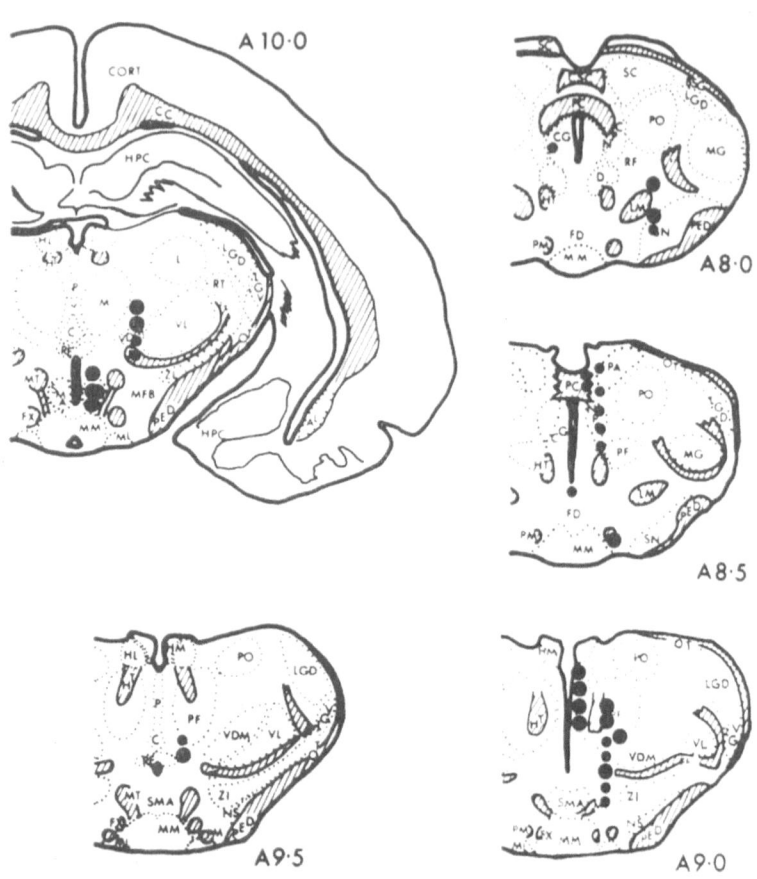

Figure 3. Milk-ejection responses obtained by electrical stimulation of guinea-pig brain stem. Small circles denote releases of less than 0·2 mU oxytocin, large circles greater than 0·2 mU oxytocin (From J. S. Tindal and G. S. Knaggs, unpublished)

96

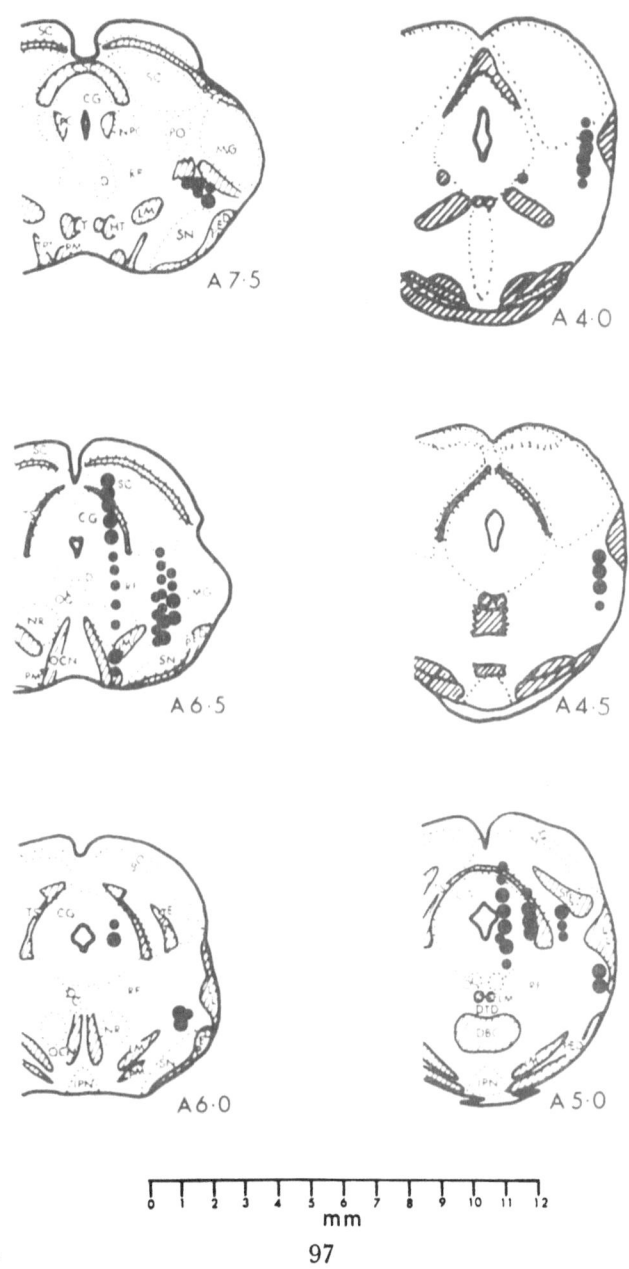

4+

mammary myoepithelium and circulating adrenaline itself may inhibit the action of oxytocin (Chan, 1965) thus masking any release of oxytocin which may have occurred consequent to the electrical stimulus. This has been overcome by making a surgical transection through the cerebellum and hindbrain. Milk-ejection responses were recorded after stimulation of a number of sites in the brainstem, including the tectal region, mesencephalic central grey, the periventricular region of the mesodiencephalic boundary and the parafascicular thalamic nucleus (*Figure 3*). In the mesencephalon, the greatest release of oxytocin was obtained from stimulation of a pathway in the lateral tegmentum which lay close to the lateral lemniscus at caudal levels, while more rostrally it passed the medial border of the medial geniculate nucleus and appeared to distribute to the tectum, central grey and parafascicular nucleus (Tindal and Knaggs, unpublished).

Indirect evidence for the ascending path of the reflex has come from studies carried out in other laboratories in which urine output was measured as an indication of antidiuretic hormone (ADH) release. In the monkey, ADH release occurred after stimulation of hypothalamus, mesencephalic reticular formation, ventral tegmental area of Tsai and central grey, but not after stimulation of dorsolateral hypothalamus, thalamus, red nucleus, tectum, pons and optic chiasma. It was suggested that three ascending pathways may be involved, the dorsal longitudinal fasciculus, the mamillary peduncle and the subthalamic limb of the central tegmental tract (Hayward and Smith, 1964). Mills and Wang (1964) found positive sites for ADH release in the dog in the lateral reticular formation, central grey, and central tegmental tract. In contrast, the latter workers suggested that ascending pathways involved in ADH release did not include the mamillary peduncle, and that impulses may reach the hypothalamus via the dorsal longitudinal fasciculus, as well as by relays in the thalamus.

There is a considerable amount of evidence from stimulation experiments to implicate the amygdala in the mechanisms governing the release of oxytocin. On the basis of uterine contractions, oxytocin has been detected after electrical stimulation of this complex in the cat (Gastaut, Vigouroux, Corriol and Badier, 1951; Koikegami, Yamada and Usui, 1954; Shealy and Peele, 1957) and by milk-ejection responses in the rabbit (Holland *et al.*, 1963). Direct evidence for the participation of the amygdaloid complex in the milk-ejection reflex came from the placement of bilateral lesions in the amygdalae which blocked milk ejection in the lactating rat

(Stutinsky and Terminn, 1965). The effective lesion sites included the lateral and central nuclei, the claustrum and the ventral putamen. Further studies are required to determine whether the amygdala lies on the ascending path of the suckling stimulus, whether it is a collateral branch of the pathway, or whether it facilitates transmission in more caudal brainstem pathways.

Neurological considerations: speculation on pathways of the suckling stimulus and possible mechanisms of action on the anterior pituitary

It would be useful to speculate a little on the light which recent studies on brain function may throw on the ascending path of the suckling stimulus, and on the possible mechanisms by which the suckling stimulus might regulate both anterior and posterior pituitary function. While this section might be of particular interest only to the specialist, no apology is made for including it here.

A feature of our preliminary studies in the guinea-pig (Tindal and Knaggs, unpublished) has been the distribution of the ascending path of the milk-ejection reflex in the caudal mesencephalon. Here, the path seems to be restricted to a discrete region in the lateral tegmentum, and from its position it seems likely that the spinothalamic and, possibly, the spinotectal tracts are involved. The possibility of spinothalamic involvement in the milk-ejection reflex was hinted at by Eayrs and Baddeley (1956), received indirect support from the finding that the pathway was probably bilateral (Stutinsky and Terminn, 1964a), but has lacked experimental evidence until now. Further rostrally, the pathway moves dorsomedially to reach the tectum, mesencephalic central grey and the thalamic parafascicular nucleus, which is of particular interest, since the spinothalamic tract is known to project to the parafascicular nucleus (Gerebtzoff, 1940) and central grey (Nauta, 1960).

As the pathway ascends from the lateral tegmentum and moves across the brainstem to tectum, central grey and parafascicular nucleus, it traverses an anatomically ill-defined region, the posterior thalamic complex, whose main afferent path is thought to be the spinothalamic tract (Poggio and Mountcastle, 1960; Whitlock and Perl, 1959, 1961). In this region, posterior to the ventro-basal thalamic complex and medial to the medial geniculate nucleus, the body was shown to be represented bilaterally in the cat (Whitlock and Perl, 1959; Poggio and Mountcastle, 1960) and rat (Emmers and Leeb, 1963; Davidson, 1965). Whereas the lemniscal system terminates contralaterally in the ventrobasal thalamic complex with a topographical organization, the spinothalamic system terminates both ipselaterally and contralaterally in the posterior thalamic complex (and also in the adjoining parafascicular nucleus, Gerebtzoff, 1940) with little or no such organization (Darian-Smith, 1964; Davidson, 1965). It is just such a system which would fulfil the requirements for the ascending path of the suckling stimulus, especially since Calma (1965a) considers that the function of the posterior complex is not related to transmission of detailed information. In addition, occlusion or inhibitory interaction can occur in this complex (Calma, 1965a) and there is evidence that corticofugal descending impulses may inhibit the response of the complex to peripheral stimulation (Calma, 1965b), which raises the question as to whether the central inhibition of the milk-ejection reflex could occur at this level.

Regarding the rostral pathway of the suckling stimulus, it may ascend via central grey, dorsal longitudinal fasciculus and hypothalamic periventricular system. This would seem a possible path for two reasons, since, according to Nauta (1960) the central grey receives spinothalamic afferents, and the arcuate nucleus of the hypothalamus is itself an extension of the periventricular system and is heavily implicated in the control of anterior pituitary function. The problem is complicated by the possibility of there being one or more fine-control mechanisms capable of modulating the action of incoming sensory information, either before it reaches, or as it reaches, the hypothalamic level. Since the elegant studies of Nauta (1960, 1963) there is at least a neuro-anatomical substrate on which such mechanisms could be based. Nauta envisages a paramedian region in the mesencephalon—the limbic midbrain area—and a group of forebrain limbic structures comprising amygdala, hippocampus, septum, striate body and cingulate cortex. There are reciprocal connections between the two poles of the circuit by means of the ascending mamillary peduncle and dorsal longitudinal fasciculus, and the descending fornix and stria medullaris, with two-way traffic in the medial forebrain bundle. In addition, there is a fine network extending outwards from the central grey to most parts of the mesencephalic tegmentum—Weisschedel's radiatio grisea tegmenti—which is in a strategic position to affect the conduction of sensory information ascending through this region (*see* Nauta, 1960). The circuit contains sufficient side branches to the basal hypothalamus to imply a role for this system in pituitary control.

Support for this concept of dualism and reciprocity in the central nervous mechanisms which may be concerned with pituitary control has come from recent neurophysiological studies (Tsubokawa and

Sutin, 1963). It was shown that high-frequency stimulation of the dorsomedial mesencephalic tegmentum increased the amplitude of the potential evoked in the hypothalamic ventromedial nucleus (VMH) by septal stimulation, and decreased the VMH response evoked by amygdaloid stimulation. Conversely, stimulation of the lateral mesencephalic reticular formation increased the amplitude of the amygdaloid-VMH response, and decreased the size of the septal-VMH response. If the assumption is made that the impulses of certain types of incoming sensory information can exert a similar effect as high-frequency (100/sec) electrical stimulation of the mid-brain, then one has the neurophysiological basis on which to propose a mechanism by which ascending impulses of the suckling stimulus could affect pituitary function. If, for the sake of argument, the suckling stimulus influenced the amygdala-VMH pathway, either facilitating or inhibiting a tonic barrage of impulses impinging on the basal hypothalamus, and at the same time affected the septal-VMH pathway in the reverse direction, then this could be a possible mode of action by which the stimulus affects the basal hypothalamic neurohumours and hence pituitary function.

As mentioned above, we have shown that the amygdala is involved in the tonic control of prolactin secretion, since the local implantation of oestrogen in certain parts of the complex will cause release of the hormone (Tindal and Knaggs, 1966). Furthermore, we have found that the optimal site in the midbrain for triggering oxytocin release by electrical stimulation lies in the far lateral tegmentum (Tindal and Knaggs, unpublished) close to the lateral lemniscus and medial geniculate nucleus, probably involving the spinothalamic and possibly the spinotectal tracts. This is a similar site to that found in the cat by Tsubokawa and Sutin (1963) in which 100/sec stimulation increased amygdala-VMH potentials and decreased septal-VMH responses. In a later evoked potential study, Wepsic and Sutin (1964) showed that potentials evoked in the basal VMH and tuberal region by stimulation of the medial magnocellular portion of the medial geniculate nucleus (MMG) apparently travelled via the amygdala, since destruction of this complex drastically reduced or even abolished the tuberal potentials. In addition, high-frequency stimulation of MMG facilitated unit firing in the globus pallidus (which is an efferent path of the putamen) and inhibited unit firing in the amygdala. Rose and Woolsey (1949) considered MMG to be more closely related to the posterior thalamic group of nuclei than to the medial geniculate body itself, and it is interesting that MMG is yet another structure which receives spinothalamic afferents (Wepsic

101

and Sutin, 1964). Lesions in the amygdala which blocked the milk-ejection reflex in the rat (Stutinsky and Terminn, 1965) included the lateral nucleus which is connected with the lateral hypothalamic and pre-optic areas by the longitudinal association bundle, the central nucleus which contains many fibres of passage of the stria terminalis *en route* to medial pre-optic area and hypothalamus, the claustrum (a cortical structure of unknown function) and the ventral part of the putamen. There are, therefore, several mechanisms or pathways which could have been destroyed by the lesions, and it will be interesting to see whether one or several were necessary for the proper functioning of the milk-ejection reflex.

The pathway of the suckling stimulus may ascend to the hypothalamus via the amygdala, but equally, the role of the amygdala in oxytocin release may be that of a modulator, exerted at mesencephalic levels. Descending influences from the rhinencephalic cortex and basal ganglia can influence unit activity in the mesencephalic reticular formation (Adey, Buchwald and Lindsley, 1960), and many of these descending influences have been shown to travel through the ventral thalamus, dorsal hypothalamus and subthalamus (Adey, Dunlop and Sunderland, 1958). Furthermore, tonic influences arising in or passing through the subthalamic region, in addition to those arising more caudally in the pons, exercise a tonic excitatory influence on the mesencephalic reticular formation, and may play a part in determining the responsiveness of the tegmentum to peripheral sensory inputs (Adey and Lindsley, 1959; Lindsley and Adey, 1961; Batini, Moruzzi, Palestini, Rossi and Zanchetti, 1958). It cannot necessarily be assumed, therefore, that elicitation of a milk-ejection response by electrical stimulation of a particular part of the brain implies that the electrode tip rests in the ascending path of the suckling stimulus. It could be lying in a facilitatory path which converges with and activates the pathway in which the mainstream of information is ascending. Although the ascending path of the suckling stimulus is now, perhaps, a little clearer, at least as far as the rostral mesencephalon, beyond this point there may be several pathways concerned with both anterior and posterior pituitary function.

In conclusion, the pathways and role of the suckling stimulus, like the wider problem of all the central nervous mechanisms controlling lactation, must remain for the present in the tantalizing state of being partly elucidated, with large gaps separating the many pieces of fragmentary information. Fortunate is the future reviewer who can expound these mechanisms in detail.

ACKNOWLEDGEMENT

Grateful acknowledgement is made to the Ford Foundation and the Rockefeller Foundation for generous financial support of the recent studies described in this paper.

REFERENCES

Adey, W. R., Buchwald, N. A. and Lindsley, D. F. (1960). *Electroenceph. clin. Neurophysiol.* **12,** 21–40
— Dunlop, C. W. and Sunderland, S. (1958). *J. comp. Neur.* **110,** 173–203
— and Lindsley, D. F. (1959). *Expl Neurol.* **1,** 407–426
Anand, B. K., Chhina, G. S., Sharma, K. N., Dua, S. and Baldev Singh (1964). *Am. J. Physiol.* **207,** 1146–1154
Andersson, B. (1951). *Acta physiol. scand.* **23,** 8–23
— Kitchell, R. and Persson, N. (1958). *Acta physiol. scand.* **44,** 92–102
Aulsebrook, L. H. and Holland, R. C. (1965). *Anat. Rec.* **151,** 319–320
Averill, R. L. W. and Purves, H. D. (1963). *J. Endocr.* **26,** 463–477
Ban, T. and Omukai, F. (1959). *J. comp. Neur.* **113,** 245–279
Batini, C., Moruzzi, G., Palestini, M., Rossi, G. F. and Zanchetti, A. (1958). *Science, N.Y.* **128,** 30–32
Ben-David, M., Dikstein, S. and Sulman, F. G. (1964). *Proc. Soc. exp. Biol. Med.* **117,** 511–513
Bercovici, B. and Ehrenfeld, E. N. (1963). *J. Obstet. Gynaec. Br. Commonw.* **70,** 295–300
Beyer, C., Anguiano, G. and Mena, F. (1961). *Am. J. Physiol.* **200,** 625–627
— and Mena, F. (1965). *Am. J. Physiol.* **208,** 289–292
— — Pacheco, P. and Alcaraz, M. (1962a). *Fedn Proc. Fedn Am. Socs exp. Biol.* **21,** 353
— — — — (1962b). *Am. J. Physiol.* **202,** 465–468
— Tindal, J. S. and Sawyer, C. H. (1962). *Expl Neurol.* **6,** 435–450
Bintarningsih, Lyons, W. R., Johnson, R. E. and Li, C. H. (1958). *Endocrinology* **63,** 540–548
Bogdanove, E. M. (1963). *Endocrinology* **73,** 696–712
Bruce, H. M. (1959). *Nature, Lond.* **184,** 105
— (1965). *Proc. 2nd Int. Congr. Endocr.* (London, 1964) *Excerpta Medica Int. Congr. Series* No. 83: pp. 193–197
— and Parkes, A. S. (1960). *J. Endocr.* **20,** xxix–xxx
Calma, I. (1965a). *J. Physiol., Lond.* **180,** 350–370
— (1965b). *Nature, Lond.* **205,** 394–396
Chan, W. Y. (1965). *J. Pharmac. exp. Ther.* **147,** 48–53
Cotes, P. M. and Cross, B. A. (1954). *J. Endocr.* **10,** 363–367
Cowie, A. T. (1957). *J. Endocr.* **16,** 135–147
— (1961). In: *Milk: The Mammary Gland and its Secretions,* Chap. 4. Ed. by S. K. Kon and A. T. Cowie. New York and London; Academic Press
— Daniel, P. M., Knaggs, G. S., Prichard, M. M. L. and Tindal, J. S. (1964). *J. Endocr.* **28,** 253–265

Cowie, A. T. and Folley, S. J. (1961). In: *Sex and Internal Secretions,*Chap. 10. Ed. by W. C. Young. Baltimore; Williams and Wilkins

— Knaggs, G. S. and Tindal, J. S. (1964). *J. Endocr.* **28,** 267–279

— and Tindal, J. S. (1960). *Acta endocr., Copenh.* **35,** Suppl. 51, 679–680

— — (1961a). *J. Endocr.* **22,** 403–408

— — (1961b). *J. Endocr.* **23,** 79–96

— — (1965a). *Rep. natn. Inst. Res. Dairy* (1965)

— — (1965b). *Proc. 2nd Int. Congr. Endocr.* (London, 1964) *Excerpta Medica Int. Congr. Series* No. 83: pp. 646–654

— — (1965c). *Proc. 2nd Int. Congr. Endocr.* (London, 1964) *Excerpta Medica Int. Congr. Series* No. 83, p. 1276

— — and Benson, G. K. (1960). *J. Endocr.* **21,** 115–123

Cross, B. A. (1961a). In: *Oxytocin,* pp. 24–47. Ed. by R. Caldeyro-Barcia and H. Heller. London; Pergamon Press

— (1961b). In: *Milk: The Mammary Gland and its Secretion,* Chap. 6. Ed. by S. K. Kon and A. T. Cowie. New York and London; Academic Press

Daniel, P. M. and Prichard, M. M. L. (1958). *Am. J. Path.* **34,** 433–469

Danon A. Dikstein, S. and Sulman, F. G. (1963). *Proc. Soc. exp. Biol. Med.* **114,** 366–368

Darian-Smith, I. (1964). *J. Physiol., Lond.* **171,** 339–360

Davidson, N. (1965). *J. comp. Neur.* **124,** 377–390

Denamur, R. (1965). *Dairy Sci. Abstr.* **27,** 193–224 and 263–280

— Stoliaroff, M. and Desclin, J. (1965). *C.r. hebd. Séanc. Acad. Sci., Paris* **260,** 3175–3178

Deuben, R. R. and Meites, J. (1964). *Endocrinology* **74,** 408–414

— — (1965). *Proc. Soc. exp. Biol. Med.* **118,** 409–412

DeVoe, W. F., Ramirez, V. D. and McCann, S. M. (1966). *Endocrinology* **78,** 158–164

Donovan, B. T. and van der Werff ten Bosch, J. J. (1957). *J. Physiol., Lond.* **137,** 410–420

Eayrs, J. T. and Baddeley, R. M. (1956). *J. Anat.* **90,** 161–171

— and Edwardson, J. A. (1965). *Acta endocr., Copenh.* **50,** Suppl. 100, 154

Ehni, G. and Eckles, N. E. (1959). *J. Neurosurg.* **16,** 628–652

Elwers, M. and Critchlow, V. (1960). *Am. J. Physiol.* **198,** 381–385

Emmers, R. and Leeb, I. J. (1963). *Fedn Proc. Fedn Am. Socs exp. Biol.* **22,** 394

Everett, J. W. (1964). *Physiol. Rev.* **44,** 373–431

Flament-Durand, J. and Desclin, L. (1964). *Endocrinology* **75,** 22–26

Folley, S. J. (1960). In: *Clinical Endocrinology I,* Chap. 8. Ed. by E. B. Astwood. New York; Grune and Stratton

— (1961). *Dairy Sci. Abstr.* **23,** 511–528

— and Knaggs, G. S. (1966). *J. Endocr.* **34,** 197–214

— Scott Watson, H. M. and Bottomley, A. C. (1940). *J. Physiol., Lond.* **98,** 15–16P

Gala, R. R. and Reece, R. P. (1964). *Proc. Soc. exp. Biol. Med.* **115,** 1030–1035

Gale, C. C. (1963). *Acta physiol. scand.* **59,** 269–283
— (1964). *Acta physiol. scand.* **61,** 228–237
— and Larsson, B. (1963). *Acta physiol. scand.* **59,** 299–318
— and McCann, S. M. (1961). *J. Endocr.* **22,** 107–117
— Taleisnik, S., Friedman, H. M. and McCann, S. M. (1961). *J. Endocr.* **23,** 303–316
Gastaut, H., Vigoroux, R., Corriol, J. and Badier, M. (1951). *J. Physiol. Path. gén.* **43,** 740–746
Gerebtzoff, M. A. (1940). *Cellule* **48,** 91–146
Goddard, G. V. (1964). *Psychol. Bull.* **62,** 89–109
Grégoire, C. (1946). *J. Endocr.* **5,** 68–87
Grosvenor, C. E. (1964). *Endocrinology* **74,** 548–553
— (1965a). *Endocrinology* **76,** 340–342
— (1965b). *Endocrinology* **77,** 1037–1042
— McCann, S. M. and Nallar, R. (1965). *Endocrinology* **76,** 883–889
— and Turner, C. W. (1957). *Proc. Soc. exp. Biol. Med.* **96,** 723–725
— — (1958). *Endocrinology* **63,** 535–539
Grosz, H. J. and Rothballer, A. B. (1961). *Nature, Lond.* **190,** 349–350
Guillemin, R. (1963). *J. Physiol., Paris* **55,** 7–44
Halász, B. (1965). *Proc. 2nd Int. Congr. Endocr.* (London, 1964) *Excerpta Medica Int. Congr. Series* No. 83: pp. 517–521
— Pupp, L. and Uhlarik, S. (1962). *J. Endocr.* **25,** 147–154
Haun, C. K. and Sawyer, C. H. (1960). *Endocrinology* **67,** 270–272
— and Sawyer, C. H. (1961). *Acta endocr., Copenh.* **38,** 99–106
Hayward, J. N. and Smith, W. K. (1964). *Am. J. Physiol.* **206,** 15–20
Holland, R. C., Aulsebrook, L. H. and Woods, W. H. (1963). *Fedn Proc. Fedn Am. Socs exp. Biol.* **22,** 571
Jaszmann, L. (1963). *J. Obstet. Gynaec. Br. Commonw.* **70,** 120–124
Kanematsu, S., Hilliard, J. and Sawyer, C. H. (1963a). *Endocrinology* **73,** 345–348
— — — (1963b). *Acta endocr., Copenh.* **44,** 467–474
— and Sawyer, C. H. (1963a). *Endocrinology* **72,** 243–252
— — (1963b). *Am. J. Physiol.* **205,** 1073–1076
— — (1963c). *Proc. Soc. exp. Biol. Med.* **113,** 967–969
Kawakami, M. and Sawyer, C. H. (1959). *Endocrinology* **65,** 631–643
Kehl, R., Audibert, A., Gage, C. and Amarger, J. (1956). *C.r. Séanc. Soc. Biol.* **150,** 981–983
Koikegami, H., Yamada, T. and Usui, K. (1954). *Folia psychiat. neurol. jap.* **8,** 7–31
Krulich, L., Dhariwal, A. P. S. and McCann, S. M. (1965). *Proc. Soc. exp. Biol. Med.* **120,** 180–184
Leão, A. A. P. (1944). *J. Neurophysiol.* **7,** 359–390
Lindsley, D. F. and Adey, W. R. (1961). *Expl Neurol.* **4,** 358–376
Linzell, J. L. (1963). *Q. Jl exp. Physiol.* **48,** 34–60
McCann, S. M. and Friedman, H. M. (1960). *Endocrinology* **67,** 597–608
— Mack, R. and Gale, C. (1959). *Endocrinology* **64,** 870–889

Meites, J. (1957). *Proc. Soc. exp. Biol. Med.* **96,** 728–730

— and Hopkins, T. F. (1960). *Proc. Soc. exp. Biol. Med.* **104,** 263–266

— Kahn, R. H. and Nicoll, C. S. (1961). *Proc. Soc. exp. Biol. Med.* **108,** 440–443

Mena, F. and Beyer, C. (1963). *Am. J. Physiol.* **205,** 313–316

Mielke, H. and Brabant, W. (1963). *Arch. exp. VetMed.* **16,** 909–919

Mills, E. and Wang, S. C. (1964). *Am. J. Physiol.* **207,** 1399–1404

Moon, R. C. and Turner, C. W. (1959). *Proc. Soc. exp. Biol. Med.* **101,** 332–335

Moore, W. W., Woehler, T. R. and Tarry, K. (1965). *Anat. Rec.* **151,** 390

Morgane, P. J. (1961). *J. comp. Neur.* **117,** 1–26

— (1962). *Proc. 22nd Int. Congr. Physiol. Sci.*, Leiden. *Excerpta Medica Int. Congr. Series* No. 47: pp. 670–676

Müller, E. and Pecile, A. (1965a). *Proc. Soc. exp. Biol. Med.* **119,** 1191–1194

— — (1965b). *Boll. Soc. ital. Biol. sper.* **41,** 581–583

— — (1966). *Experientia,* **22,** 108

Nauta, W. J. H. (1960). In: *Electrical Studies on the Unanesthetized Brain,* Chap. 1. Ed. by E. R. Ramey and D. S. O'Doherty. New York; Hoeber

— (1963). In: *Advances in Neuroendocrinology,* Chap. 2. Ed. by A. V. Nalbandov. Urbana; Univ. of Illinois Press

— and Kuypers, H. G. J. M. (1958). In: *Reticular Formation of the Brain,* pp. 3–30. Ed. by H. H. Jasper, L. D. Proctor, R. S. Knighton, W. C. Noshay and R. T. Costello. Boston; Little, Brown

Nicoll, C. S. and Meites, J. (1962). *Endocrinology* **70,** 272–277

— Talwalker, P. K. and Meites, J. (1960). *Am. J. Physiol.* **198,** 1103–1106

Pasteels, J. L. (1961a). *C.r. hebd. Séanc. Acad. Sci.*, Paris **253,** 2140–2142

— (1961b). *C.r. hebd. Séanc. Acad. Sci.*, Paris **253,** 3074–3075

— (1963). *Arch. Biol. (Liège)* **74,** 439–553

Pecile, A., Müller, E., Falconi, G. and Martini, L. (1965). *Endocrinology* **77,** 241–246

Poggio, G. F. and Mountcastle, V. B. (1960). *Johns Hopkins Hosp. Bull.* **106,** 266–316

Powell, T. P. S., Cowan, W. M. and Raisman, G. (1963). *Nature, Lond.* **199,** 710–712

Ramirez, V. D. and McCann, S. M. (1964). *Endocrinology* **75,** 206–214

Ratner, A. and Meites, J. (1964). *Endocrinology* **75,** 377–382

— Talwalker, P. K. and Meites, J. (1963). *Proc. Soc. exp. Biol. Med.* **112,** 12–15

— — — (1965). *Endocrinology* **77,** 315–319

Rose, J. E. and Woolsey, C. N. (1949). *J. comp. Neur.* **91,** 441–466

Roth, J., Glick, S. M., Yalow, R. S. and Berson, S. A. (1963a). *Science, N.Y.* **140,** 987–988

— — — — (1963b). *Metabolism* **12,** 577–579

Sawyer, C. H. (1957). *Anat. Rec.* **127,** 362

STUDIES ON THE NEUROENDOCRINE CONTROL OF LACTATION

Sawyer, C. H., Haun, C. K., Hilliard, J., Radford, H. M. and Kanematsu, S. (1963). *Endocrinology* **73,** 338–344

Schally, A. V., Steelman, S. L. and Bowers, C. Y. (1965). *Proc. Soc. exp. Biol. Med.* **119,** 208–212

Shealy, C. N. and Peele, T. L. (1957). *J. Neurophysiol.* **20,** 125–139

Stutinsky, F. and Terminn, Y. (1964a). *J. Physiol., Paris* **56,** 443–444

—— (1964b). *C.r. Séanc. Soc. Biol.* **158,** 833–835

—— (1965). *J. Physiol., Paris* **57,** 279–280

Sulman, F. G. and Winnik, H. Z. (1956). *Lancet* **1,** 161–162

Sutin, J. (1963). *Electroenceph. clin. Neurophysiol.* **15,** 786–795

Taleisnik, S. and Deis, R. P. (1964). *Am. J. Physiol.* **207,** 1394–1398

Talwalker, P. K., Ratner, A. and Meites, J. (1963). *Am. J. Physiol.* **205,** 213–218

Tindal, J. S. (1960). *J. Endocr.* **20,** 78–81

— (1965). *J. comp. Neur.* **124,** 259–266

— Beyer, C. and Sawyer, C. H. (1963). *Endocrinology* **72,** 720–724

— and Knaggs, G. S. (1966). *J. Endocr.* **34,** ii–iii

Tsakhaev, G. A. (1953). *Dokl. Akad. Nauk SSSR* **93,** 941–944

Tsubokawa, T. and Sutin, J. (1963). *Electroenceph. clin. Neurophysiol.* **15,** 804–810

Tverskoy, G. B. (1960). *Nature, Lond.* **186,** 782–784

Vernikos-Danellis, J. (1965). *Vitam. Horm.* **23,** 97–152

Wepsic, J. G. and Sutin, J. (1964). *Expl Neurol.* **10,** 67–80

White, L. E. (1965). *Anat. Rec.* **152,** 465–480

Whitlock, D. G. and Perl, E. R. (1959). *J. Neurophysiol.* **22,** 133–148

—— (1961). *Expl Neurol.* **3,** 240–255

Yamada, T. and Greer, M. A. (1960). *Endocrinology* **66,** 565–574

Yokoyama, A. and Ôta, K. (1965). *J. Endocr.* **33,** 341–351

DISCUSSION

S. J. FOLLEY (*Reading*)

It is claimed that the placement of oestrogens in the hypothalamus is an efficient way of getting oestrogens into the pituitary by way of the blood stream. Would that apply to any degree in your experiments where you are implanting oestrodes in the amygdala?

TINDAL

I don't think so. The distances involved here are much greater between the amygdaloid complex and the pituitary, and are of the order of several millimetres. We did find that in closely neighbouring sites to the positive ones there was no response at all. There appeared to be a definite area of sensitivity within the amygdala. In one region we obtained a positive response but in a very closely related region half a millimetre away there was no response. I think that it is a genuine local response in the amygdala and that the oestrogen is not being transported to the pituitary.

REPRODUCTION IN THE FEMALE MAMMAL

S. M. McCann (*Texas, U.S.A.*)

I was very interested in the question of oestrogen implants in the amygdala. I must say that I would hate to move the sites of the feed-back action over to the amygdala. Consequently, what I would like to ask is, have you done studies to determine the sensitivity of this effect as compared to those done in the hypothalamus? In other words, does it take, for example, 10 times more oestrogen implanted in the amygdala than it would to do the same thing in the hypothalamus? Have you done any studies of this kind?

Tindal

No, we have done no studies of this kind at all yet, since we are still in a very early stage of trying to convince ourselves that the response really does exist. This, however, must be done and we are about to measure the amounts of oestrogen left in these oestrodes at autopsy by gas chromatography so that we can get some idea of the absolute amount that has been absorbed by the brain during the period of implantation.

B. T. Donovan (*London*)

There are two questions I wish to ask. The first is, have you tried progesterone implants and the other question is, do you think the oestrogen applied to the amygdala is inhibiting or stimulating the nerve cells in that area?

Tindal

The answer to the first question is that we haven't tried progesterone yet. It would be very interesting to do this. With regard to the second question there is no evidence to indicate whether the oestrogen is facilitating or inhibiting. We have made very preliminary studies on the effect of lesions in the amygdala which seem to indicate that these led to lactogenesis. Now, if that is the case one would assume the oestrogen is inhibiting something which can also be inhibited by physically removing it. We are not disputing that the final control of the pituitary is the basal hypothalamus by means of neurohumours, what we are saying is that we think we have a mechanism here which can modulate the secretion of these neurohumours concerned with prolactin secretion.

I. Rothchild (*Cleveland, U.S.A.*)

It is interesting to hear that factors which increase appetite in the goat and sheep are also associated with prolactin secretion and that this may be one of the mechanisms involved in addition to the suckling stimulus in maintaining the secretion of prolactin. It was a pleasant surprise to have this piece of information which fits in with a proposition I suggested about the role of the appetite controlling centres in prolactin secretion as well as other modalities. Do you have any additional information on the way in which the other modalities I spoke about are affected? I know that ovulation tends to be in abeyance during lactation in the sheep and goat. Is there any further information on this?

Tindal

No, there is not. I rather deliberately trailed my coat this morning over the appetite mechanism. We were driven to this in desperation in thinking of some way of explaining how the mammary glands can continue functioning in the absence of the suckling stimulus, assuming that there was some regulatory mechanism. I must say that we have no definite proof to substantiate our hypothesis. This is just an idea thrown out, but it seems quite a possible mechanism in view of what happens to the levels of somatotrophin in the human, and particularly as ox somatotrophin in our experiments seemed important in maintaining lactation in the ruminant whereas it seems to have far less importance in the small laboratory animals.

A. L. FINDLAY (*Cambridge*)

I would like to know what Dr. Tindal thinks of the fact that there are two very different sorts of lactation from either of the mammary glands in the marsupial. There is presumably the same hormonal environment, but with two very different sorts of afferent stimulation.

TINDAL

The explanation I have heard is that it may be related to the different suckling habits of the baby kangaroos, i.e. whether they are small or large, the small ones applying an almost constant suckling stimulus whilst the older ones suckle at more infrequent intervals, and the differences in suckling behaviour have been correlated with differences in milk composition. I think you will find that you get a much higher fat content in the milk from those animals which have a longer interval between suckling, e.g. the rabbit, which normally suckles only once a day when the pups are a week old, compared with animals which have an almost constant suckling regime every few minutes.

II. SEPARATION, PURIFICATION AND IMMUNO-ASSAY OF GONADOTROPHINS

THE PURIFICATION AND CHEMICAL PROPERTIES OF HUMAN PITUITARY FOLLICLE STIMULATING HORMONE

W. R. BUTT

The United Birmingham Hospitals Department of Clinical Endocrinology, Birmingham and Midland Hospital for Women

FOLLICLE stimulating hormone (FSH) of fairly high potency is easily obtained from acetone-dried or fresh frozen human pituitary glands. The purification of the hormone, however, is not so easy and traces of luteinizing hormone (LH) and proteins such as albumin remain with the FSH through many systems of chromatography, electrophoresis and salt precipitation. Moreover, the stability of FSH appears to decrease as purification proceeds and therefore the true biological potency and chemical nature of the hormone are uncertain. The present work is concerned with attempts to prepare highly purified FSH and with details of its chemical properties and the effects of certain reagents on its biological activity.

INITIAL EXTRACTION AND PRELIMINARY PURIFICATION

Several methods are available for the initial extraction of gonadotrophins from human pituitary glands. Many of these are based on the early methods of Koenig and King (1950) who used mixtures of acetates and ethanol, and of Ellis (1958) who used aqueous salt solutions. An important point about these methods is that they are suitable for the recovery of other pituitary hormones besides the gonadotrophins.

Subsequent purification in all the modern methods involves chromatography on the ion-exchange celluloses, and gel filtration. Luteinizing hormone is preferentially adsorbed from weak salt solutions on to carboxymethyl cellulose (CM-cellulose) while FSH is adsorbed on diethylaminoethyl cellulose (DEAE-cellulose) and LH is easily eluted. In Table 1 the potencies of some preparations of FSH obtained by such methods are given: these preparations appear to be stable and have been obtained as dry powders.

In the method of Butt, Crooke and Cunningham (1961) the initial steps are similar to those of Steelman, Segaloff and Andersen (1959)

Table 1

Potencies of stable preparations of human pituitary FSH

Outline of method	FSH potency		LH potency		Reference
	As reported	in I.U./mg	As reported	in I.U./mg	
Acetone-dried pituitaries: Amm. sulph. extraction: DEAE-cellulose	28·9 × NIH-FSH-S1 or 5,820 × IRP-HMG	831	0·06 × NIH-LH-S1	92	Reichert and Parlow (1964a)
Acetone dried pituitaries: Amm. acet./ethanol: CM-cellulose: DEAE-cellulose: Calcium phosphate: ethoxy ethanol/urea/amm. sulph. partition	7,250 × IRP-HMG	1,036	360 × IRP-HMG	180	Butt, Crooke and Cunningham (1961), and Butt, Crooke and Wolf (1965)
Fresh pituitaries: amm. sulph. extraction and fraction-ation: zone electrophoresis on cellulose: starch gel electrophoresis	60 × NIH-FSH-S1	1,560	not given		Saxena (1966)

using the Koenig and King method of extraction followed by chromatography on CM-cellulose, DEAE-cellulose and calcium phosphate. It has now been found that the same high order of activity as originally described can be obtained in better yields by carrying out the later stages of the method by batch techniques at 0°C. After separation of most of the LH on CM-cellulose the FSH fraction is dissolved in mM-Na_2HPO_4 and a suspension of calcium phosphate is added. After allowing time for any adsorption to occur the unadsorbed material containing the FSH is mixed with a suspension of DEAE-cellulose equilibrated in 0·01 M-ammonium acetate. The FSH is adsorbed under these conditions and is eluted subsequently in several washings of 0·2 M-ammonium acetate. It is then precipitated by adding 5 volumes of ethanol and is washed with ethanol and ether and dried. This is the material we have called CP 1 and it contains on the average 500 I.U. FSH/mg (ovarian augmentation method in mice) and 150–180 I.U. LH/mg (ovarian ascorbic acid depletion method in rats, OAAD method). It also contains albumin and several other proteins which may be demonstrated by starch gel electrophoresis (Butt, Crooke, Cunningham and Wolf, 1963). It is stable indefinitely when kept dry at 4°C however, and is very suitable for clinical use (Crooke, Butt, Palmer, Morris, Edwards and Anson, 1963; Crooke, Butt, Palmer, Bertrand, Carrington, Edwards and Anson, 1964).

Further purification is achieved by partitioning in the system 6 M-urea : ammonium sulphate (40 g/100 ml) : 2-ethoxy-ethanol (50 : 100 : 100 by vol.). The CP 1 is dissolved in the solution of urea and kept at 4°C for 1 h. This time should not be extended or FSH activity may be lost; LH is destroyed more rapidly, however. The ammonium sulphate solution and ethoxy-ethanol are now added and after gentle shaking for a few minutes the mixture is centrifuged slowly (about 2,000 rev/min). Some denatured protein appears at the interface while the FSH enters the organic layer. This upper layer is separated, placed in Visking tubing and dialysed against water for about 24 h. The FSH is precipitated by the addition of 5 volumes of ethanol and it is finally washed and dried. The average potency of the product is about 1,000 I.U. FSH/mg (Table 1) and it will be observed that the proportion of LH is decreased but is still appreciable. The product is stable in the dry state at 4°C and the yield from CP 1 is good (Table 2).

FINAL PURIFICATION OF PITUITARY FSH

Extremely high potencies have been reported for some preparations of FSH in solution. Reichert and Parlow (1964a) submitted their preparation containing 830 I.U./mg (Table 1) to gel filtration on Sephadex G-100 or to gradient elution chromatography on a column of DEAE-cellulose and obtained fractions containing 86–94 units NIH-FSH-S1/mg (2,200–2,450 I.U./mg) judged by ultra-violet

Table 2

Recovery of FSH from human pituitaries

	Wt. per 100 pituitaries	Potency I.U./mg	Total yield I.U./100 pituitaries
Acetone-dried powder	12 g	2	24,000
↓			
Amm. acet./ethanol extract	240 mg	90	21,600
CM-cellulose fraction (CM 1)	120 mg	140	16,800
↓			
Calcium phosp : DEAE-cellulose fraction (CP 1)	20 mg	500	10,000
↓			
Urea : amm. sulph. : ethoxy ethanol	8·5 mg	1,000	8,500

absorption at 215–225 mμ. In a similar study Parlow, Condliffe, Reichert and Wilhelmi (1965) obtained a potency of 74·5 times NIH-FSH-S1 (1,940 I.U./mg). This potency was maintained if deep frozen but it dropped to 47 times NIH-FSH-S1 (1,220 I.U./mg) when freeze-dried and 5 weeks later had dropped still further to 38 times NIH-FSH-S1 (990 I.U./mg).

Roos and Gemzell (1965) used fresh frozen pituitaries and fractionated their material by precipitation with ammonium sulphate followed by chromatography on DEAE-cellulose, gel filtration on Sephadex G-100 and then electrophoresis on a column of polyacrylamide gel. The FSH fraction is reported to have the amazing potency of 428 times NIH-FSH-S2, but it must be remembered that this estimate is provisional since the protein concentration was based on the ultra-violet reading at 230 mμ. Full assay details are not available on the possible LH contamination; by the assay in rats depending upon the seminal vesicle response the contamination is less than 6 I.U. chorionic gonadotrophin per milligramme. Again freeze-drying was not attempted because of loss of biological activity.

Butt et al. (1963) described a method depending on starch gel electrophoresis for the removal of LH and an extremely high ratio of FSH to LH was obtained thereby (Table 3). Difficulties were encountered, however, in the preparation of a dry powder. Furthermore, it is difficult to be sure that all starch is removed from the final

Table 3

FSH : LH ratios in purified FSH

Method	FSH I.U./ml	LH I.U./ml	FSH/LH	Reference
CP 1 : partition : starch gel electrophoresis	134	< 12	> 11	Butt *et al.* (1965)
CP 1 : gel filtration : DEAE-Sephadex	143	< 10	> 14	Amir *et al.* (1966a)

Table 4

Composition of human FSH
(From Amir and Somers, 1966)

	μg/ml	Molar ratio to nearest integer
Aspartic acid	29·6	17
Threonine	21·5	14
Serine	39·2	29
Glutamic acid	41·3	19
Proline	10·5	7
Glycine	21·8	22
Alanine	16·6	14
Valine	24·2	16
Isoleucine	9·2	5
Leucine	16·8	10
Tyrosine	14·1	6
Phenylalanine	14·0	6
Lysine	16·0	8
Histidine	8·0	4
Arginine	11·5	5
Cysteine	2·0	1
Tryptophan	3·4	1
Methionine	0·0	0
Galactose	4·4	2
Mannose	40·6	17
N-acetyl glucosamine	21·0	7
6-Deoxyhexose	6·4	3
N-acetyl neuraminic acid	12·8	3

Note: The optimum times for hydrolysis were determined for each amino acid since the rates of liberation and of destruction varied considerably.

product so that the method is not very suitable when further chemical studies are contemplated. An alternative method has therefore been devised making use of gel filtrations on Sephadex G-100 or Biogel P-150 and chromatography on DEAE-Sephadex (Amir, Barker, Butt and Crooke, 1966a). Gel filtration of the CP 1 fraction described above did not effect a separation of FSH and LH or albumin but some inert material was removed. Next, chromatography on DEAE-Sephadex was performed in 0·05 M-phosphate buffer at pH 7·0 with a salt gradient to 0·8 M-NaCl. It was noted that the FSH activity was not uniformly distributed about the ultraviolet peak but was eluted slightly earlier. In order to achieve maximum separation of FSH from LH and albumin it was necessary to use two passages down DEAE-Sephadex. The elution of LH, which was slightly ahead of the main FSH region, and of albumin, which was slightly behind FSH, was detected immunologically by the haemagglutination-inhibition method. The final product was not freeze-dried because of the possible risk of loss of biological activity, but was concentrated in Visking tubing by dialysis under reduced pressure. No contamination with albumin was recognized by the haemagglutination-inhibition test and the ratio of FSH to LH was greater than 14 (each in I.U., *see* Table 3).

CHEMICAL PROPERTIES OF PURIFIED FSH

The amino acid (except for cysteine and methionine which are at present being estimated) and carbohydrate composition of the preparation of Amir, Barker, Butt and Crooke (1966a) is shown in Table 4 (Amir, 1966). The carbohydrates therefore make up about 23 per cent of the molecule. The hexoses have been identified as galactose and mannose and the terminal carbohydrate is N-acetyl neuraminic acid. No N-glycollyl neuraminic acid or uronic acid was detected.

Treatment of FSH with 0·05 M-NaNO$_2$ in acetate buffer at pH 5·0 for 48 h at 4°C caused 85 per cent loss of activity, but most of this was recovered after dialysis against 0·1 M-mercaptoethanol for 20 h. This raises the possibility of an oxidation reduction centre in the molecule, oxidation of —SH—SH— to —S—S— groups by nitrite causing loss of activity and reduction by mercaptoethanol back to —SH—SH— groups regenerating it. Alkylation of the —SH— groups after treatment with urea by iodoacetamide destroyed activity.

Performic acid caused complete loss of activity but formic acid alone at −10°C for 5 h caused some 50 per cent loss of activity.

Acidic media in general cause loss of activity of FSH. This is possibly associated with the cleavage of the terminal neuraminic acid and enzymic hydrolysis with *Vibrio cholerae* neuraminidase at pH 5·6 for 3 h at 37°C completely destroys activity.

Activity is retained after limited digestion with certain enzymes. Attack by trypsin appears to be relatively slow and using one part of trypsin to ten of protein no loss was encountered after 3–4 h at 37°C. After 16 h, however, only 16 per cent of the original activity remained. With only one part of trypsin to fifty of protein, 70 per cent of the activity remained after 7 h and 30 per cent after 48 h. No loss of activity was encountered on treatment of FSH with chymotrypsin at 37°C for 13 or 24 h, but pronase caused complete loss of activity after only 2 h at pH 7·0 and 34°C.

COMMENT

It is clear that one of the major difficulties in the final purification of human pituitary FSH is the question of its instability. The reason for this is not known, but it appears to develop at the stage when contaminating proteins such as albumin are removed. Albumin may therefore act as a protective protein during freeze-drying, etc., in the earlier stages of purification. Better results may be obtained by ethanol precipitation but at the final stages the amount of material available is usually too small for adequate recovery. Loss of activity may also be associated with the presence of undetected proteolytic enzymes which are known to be present in crude pituitary powders and have been recognized in some highly purified gonadotrophins (Reichert and Parlow, 1964b).

It is interesting that limited digestion with an enzyme such as trypsin does not destroy the activity of human FSH. Such treatment has been used as a method of reducing LH contamination of ovine NIH-FSH-S1 by Jutisz (1965), and a similar method was used in the case of porcine FSH by Steelman and Segaloff (1959). Recently it has been claimed that HCG retains activity after treatment with trypsin and that active peptides can be separated by gel filtration and ion-exchange chromatography (Kikutani and Tokuyasu, 1965). Further developments of this nature will be awaited with great interest.

The high carbohydrate content of human FSH is similar to that reported for purified HCG and PMS (Bourrillon and Got, 1960; Got and Bourrillon, 1960). Neuraminic acid accounts for 7·5 per cent and this contrasts with the extremely low amounts in LH of ovine origin (Adams-Mayne and Ward, 1964). It is interesting that the action of LH does not depend on the presence of even traces of neuraminic acid since treatment with neuraminidase has no effect, but it is essential for FSH activity.

The part played by neuraminic acid in the mode of action of FSH is not at present clear. The acid itself has no FSH activity in hypophysectomized mice but some interesting effects have been noted when it is given to mice in combination with human FSH. Thus, when mixed with the CP 1 fraction in the proportion of 1 mg to 1 μg CP 1, considerable augmentation of activity in the uterine weight and the ovarian augmentation assays is observed (Amir, Barker, Butt and Crooke, 1966b, *Figure 1*). In some assays the relative

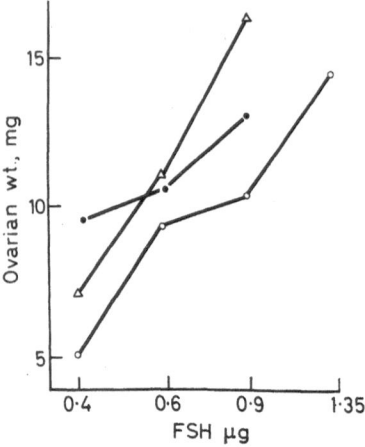

Figure 1. *The effect of N-acetyl neuraminic acid on the FSH potency of CP 1 in the ovarian augmentation assay*

O———O CP 1 alone
●———● ⎫ *Two experiments using mixtures*
△———△ ⎬ *of N-acetyl neuraminic acid (1 mg) and CP 1 (1 μg)*
Relative potencies in terms of control
●———● 156(96–246) *per cent*
△———△ 168(106–256) *per cent*

potency has been nearly doubled but with preparations of varying degrees of purity the effect may be less. It may be speculated that the neuraminic acid occupies some vacant site on a protein constituent of CP 1, or that it is concerned with the transport of the hormone or has some action in activating the receptor sites at the target organ.

Now that methods are available for the production of highly purified preparations of human FSH, although only in solution, it should be possible to study more satisfactorily its specific biological properties and its mode of action.

SUMMARY

1. Stable preparations of human FSH containing more than 1,000 I.U./mg have been described.

2. A common contaminant of pituitary FSH is albumin and when this is removed the stability decreases.

3. Preparations virtually free of albumin and LH may be obtained by methods of electrophoresis or gel filtration and ion-exchange chromatography; immunological tests are useful for identifying these impurities.

4. Purified human FSH contains a high proportion (about 23 per cent) of carbohydrate.

5. Activity is retained after limited digestion with trypsin and chymotrypsin but pronase rapidly destroys FSH.

6. *N*-acetyl neuraminic acid is essential for activity and when added to certain preparations of FSH activity is enhanced.

ACKNOWLEDGEMENTS

The original work described here was carried out with the encouragement and direction of Dr. A. C. Crooke. The author is grateful also to Dr. Barker, Dr. Amir and the staff of the Chemistry Department, University of Birmingham, for making available much information before publication. The work was supported by a grant from the Medical Research Council and by funds for technical assistance from Organon Laboratories Ltd. The pituitary fraction (CM 1) was kindly supplied by Dr. Anne Hartree, Biochemistry Department, University of Cambridge, from the collection organized by the Clinical Endocrinology Committee of the Medical Research Council.

REFERENCES

Adams-Mayne, M. and Ward, D. N. (1964). *Endocrinology* **75,** 333–340

Amir, S. M. and Somers, P. J. (1966). Personal communication

— Barker, S. A., Butt, W. R. and Crooke, A. C. (1966a). *Nature, Lond.* **209,** 1092–1093

— — — — (1966b). *J. Endocr.* **35,** 425–426

Bourrillon, R. and Got. R. (1960). *Acta endocr., Copenh.* Suppl. **51,** 683

Butt, W. R., Crooke, A. C. and Cunningham, F. J. (1961). *Biochem. J.* **81,** 596–605

Butt, W. R., Crooke, A. C., Cunningham, F. J. and Wolf, A. (1963). *J. Endocr.* **25,** 541–547

—— and Wolf, A. (1965). In: 'Gonadotropins: Physicochemical and Immunological Properties', *Ciba Fdn Study Grps*, No. 22, pp. 85–97. Ed. by G. E. W. Wolstenholme and Julie Knight

Crooke, A. C., Butt, W. R., Palmer, R. F., Bertrand, P. V., Carrington, S. P., Edwards, R. L. and Anson, C. J. (1964). *J. Obstet. Gynaec. Br. Commonw.* **71,** 571–585

——— Morris, R., Edwards, R. L. and Anson, C. J. (1963). *J. Obstet. Gynaec. Br. Commonw.* **70,** 604–635

Ellis, S. (1958). *J. biol. Chem.* **233,** 63–68

Got, R. and Bourrillon, R. (1960). *Acta endocr., Copenh.* Suppl. **51,** 1091

Jutisz, M. (1965). In: 'Gonadotropins: Physicochemical and Immunological Properties', *Ciba Fdn Study Grps* No. 22, pp. 113–114. Ed. by G. E. W. Wolstenholme and Julie Knight

Kikutani, M. and Tokuyasu, K. (1965). *J. Biochem., Tokyo* **57,** 598–603

Koenig, V. L. and King, E. (1950). *Archs Biochem.* **26,** 219–229

Parlow, A. F., Condliffe, P. G., Reichert, L. E. and Wilhelmi, A. E. (1965). *Endocrinology* **76,** 27–34

Reichert, L. E. and Parlow, A. F. (1964a). *Endocrinology* **74,** 236–243

—— (1964b). *Endocrinology* **74,** 809–810

Roos, P. and Gemzell, C. A. (1965). In: 'Gonadotropins: Physicochemical and Immunological Properties', *Ciba Fdn Study Grps* No. 22, pp. 11–25. Ed. by G. E. W. Wolstenholme and Julie Knight

Saxena, B. (1966). *J. Endocr.* In press

Steelman, S. L. and Segaloff, A. (1959). *Recent Prog. Horm. Res.* **15,** 115–125

—— and Andersen, R. N. (1959). *Proc. Soc. exp. Biol. Med.* **101,** 452–454

DISCUSSION

W. H. Hansel (*Cornell, U.S.A.*)

We have been examining the FSH content of pituitary glands by measuring the uptake of ^{32}P in the testes of the day-old chick. We had hoped that we could assay the glands and then inactivate the relatively low FSH component by incubation with neuraminidase and get a lower figure. Instead we often get a higher figure—sometimes significantly higher. Does this make sense in the light of your comments?

Butt

If you incubate the crude pituitary glands with neuraminidase, there are a lot of other mucoproteins present containing sialic acids and it is possible that these are being hydrolysed and the hormone is not affected.

L. E. Reichert (*Atlanta*)

Do you have any values for the extinction coefficient of your purified FSH in mg/ml solution at 280?

Butt

I can find the information for you (0·41 in 1 cm cell).

122

HUMAN PITUITARY FOLLICLE STIMULATING HORMONE

REICHERT

Is it less than 1 mg/ml?

BUTT

It is and I suppose you noticed from the amino acid composition that the tyrosine content is pretty low. It is very difficult to measure the purified preparation with the Folin reagent or to stain it on starch gel with ordinary protein stains for it gives a very weak stain. I presume that the high carbohydrate content also accounts for the low extinction coefficient.

H. A. ROBERTSON (*Aberdeen*)

Did I hear correctly that you said pepsin had no effect on the biological activity; at what temperature was this carried out?

BUTT

After 6 h at 37°C there is relatively no loss of biological activity.

R. A. WELCH (*Cambridge*)

Do I understand correctly that the action of neuraminic acid in potentiating FSH applies only to the mouse and that when you applied this to the rabbit you got no potentiating effect. Have you ever examined the mouse ovaries to determine what happens or do you assume that a weight increase means a growth of follicles?

BUTT

Could I correct you on one thing, we haven't tried this in the rabbit. We have tried it in the human and mixed with the CP 1 fraction there has been no potentiation of activity. You are quite correct that we have only demonstrated it in the mouse and with purer preparations of FSH the effect is even less.

We haven't examined the ovaries histologically. We realize we are giving an excess of actual gonadotrophin. As far as we can tell the mixture has the same dose/response effect as the gonadotrophin itself; the slopes are not significantly different. Thus we assumed that the mechanisms in the bio-assay are the same. Experiments have been done on hypophysectomized rats and they have all been completely negative; there is no evidence that the neuraminic acid itself has a follicle stimulating effect.

C. GEMZEL (*Uppsala, Sweden*)

Could you give the yield of your CP 1 fraction?

BUTT

From the acetone dried pituitary to the CP 1 stage the yield is around 40 per cent.

A. STOCKWELL-HARTREE (*Cambridge*)

Have you yet determined the dry weight of your high-potency FSH solution? Would not your amino acid analyses be more meaningful if they could be expressed on a weight basis, which you could do by drying a small aliquot of your solution?

BUTT

We have determined the dry weight and you are quite right that when we have completed the amino acid analyses we will present them per milligramme dry weight. Had you in mind what was the specific activity of the product? If so, I cannot claim that it is very much more than $60 \times$ NIH at the time we were assaying it.

PURIFICATION AND PROPERTIES OF PITUITARY AND URINARY LUTEINIZING HORMONE

LEO E. REICHERT, JR.

Department of Biochemistry, Division of Basic Health Sciences,
Emory University, Atlanta, Georgia 30322

WE HAVE for some time been interested in the development of methods for purification of luteinizing hormone (LH, ICSH) from a variety of sources, and have recently initiated studies on the comparative chemical and biological properties of this hormone. In the following sections, we will outline procedures which have proved effective for the purification of LH from human, ovine, bovine, porcine and equine pituitary glands and from human urine, and will summarize current information relative to its comparative physical and chemical properties.

BIOLOGICAL ASSAY

A necessary prerequisite to the development of effective purification protocols for LH or, indeed, for any biologically active protein, is the availability of a reliable, specific and sensitive method of assay. There are in current use a number of biological assays for LH, varying widely in their basic character. These are: (*a*) the hypophysectomized rat ventral prostate (VP) assay (Greep, Van Dyke and Chow, 1941); (*b*) the ovarian ascorbic acid depletion (OAAD) assay (Karg, 1957; Parlow, 1961); (*c*) the ovarian hyperaemia assay (Ellis, 1961a); and (*d*) the ovarian cholesterol depletion assay (Bell, Mukerji and Loraine, 1964). Unfortunately, none of the above satisfy all the requirements of the ideal bio-assay. Of those listed, the ones most widely utilized have been the VP and OAAD assays. An analysis of factors affecting the performance of these assays and the interpretation of the resulting data is clearly beyond the scope of the present discussion, but has been considered elsewhere (Rosemberg, Solod and Albert, 1964a; Albert, 1961; Reichert, 1966a). The measurements of LH reported here were obtained by use of the OAAD assay. This test, based on the observations of Claesson, Hillarp, Hogberg and Hakfelt (1949) and closely paralleling in concept the traditional adrenal ascorbic acid assay for ACTH, was

125

first applied to the quantitative measurement of LH by Parlow (1961), and has since undergone several operational modifications. The procedure as employed by us is described elsewhere (Reichert and Parlow, 1963), except that rats of the Sprague-Dawley strain were used, and the doses of the reference preparation NIH-LH-S1, were 1·0 and 4·0 μg.

REFERENCE PREPARATIONS

In Table 1 are summarized factors which permit conversion of LH relative potency estimates from NIH-LH-S1 units/mg (one unit = activity in one milligramme of this standard) into terms of other commonly employed reference preparations. For convenience, factors permitting conversion of FSH relative potency estimates from NIH-FSH-S1 units/mg (one unit = activity in one milligramme of this standard) into terms of similar reference preparations are also included. It is important to note that NIH-LH-S1 is a partially purified ovine pituitary LH, while the IRP-HMG preparations

Table 1

Potency of various gonadotrophin reference preparations in terms of NIH-LH-S1 and NIH-FSH-S1

| Reference preparation | LH potency μ/mg* | | I.D. VP/OAAD | FSH potency μ/mg* HCG-augmentation assay |
	OAAD assay	Ventral prostate assay		
HMG-24†	0·00032‡	0·009§	28·1	0·0052‡
Pergonal-23‡	0·0052‡	0·120§	23·1	0·30‡
NIH-HPG-UE‡	0·0021‡	—	—	0·46‡
227-80 (Armour)	0·94¶	1·31¶	1·4	0·018 ‖
264-151-X (Armour)	—	—	—	0·55‡
2nd Int. Std. HCG¶	900	—	—	—
2nd IRP-HMG**	0·00065	0·015	23·1	0·037

* One unit = activity in one milligramme of NIH-FSH-S1 or NIH-LH-S1
† HMG-24 = 1st-IRP-HMG
‡ Reichert and Parlow, 1964a
§ Rosemberg, Lewis and Solod, 1964b
‖ Parlow and Reichert, 1963
¶ One milligramme NIH-LH-S1 is equivalent to 900 I.U. HCG (Reichert, unpublished)
** The 2nd IRP-HMG is Pergonal-23 which has been diluted with an inert carrier. These specific activities are per milligramme of the diluted preparation. It should be noted that the 2nd IRP-HMG has been distributed in ampoules containing a total of 40 I.U. of LH and/or FSH. To convert NIH-LH-S1 units into International Units, multiply by 1,538 (OAAD assay) or 66·6 (VP assay). To convert NIH-FSH-S1 units into International Units, multiply by 26·6

represent crude human urinary gonadotrophin fractions. The conversion factors in Table 1 were obtained by assaying the listed preparations in terms of the ovine standard in the OAAD or VP assay. The differences in the value for the conversion factors between these assays (the index of discrimination, I.D.) has been explained on the basis of differing biological properties of LH from various sources (Parlow, 1963; Reichert and Parlow, 1964a; Rosemberg et al., 1964a). Human pituitary and urinary LH fractions give considerably greater relative potency estimates (up to 25-fold greater, Reichert and Parlow, 1964a; (Table 1)) when measured in terms of the ovine standard in the VP as compared to the OAAD assay. Somewhat similar results have been reported by Reichert (1966b) for the assay of human LH in terms of NIH-LH-S1 when compared in the ovarian hyperaemia (Ellis, 1961a) and OAAD assays. However, assay of human pituitary or urinary LH fractions in terms of an appropriate (human) LH reference preparation, such as IRP-HMG, would be expected to give similar relative potencies in each assay (Rosemberg et al., 1964a).

PURIFICATION OF OVINE LUTEINIZING HORMONE

A variety of methods have been proposed for purification of ovine LH (Ellis, 1958; Ward, McGregor and Griffin, 1959; Squire and Li, 1959; Woods and Simpson, 1961). The specific activities of LH prepared by the above procedures have been compared by Reichert and Parlow (1963) and found to be essentially equivalent, ranging from 1·6 to 2·4 NIH-LH-S1 units/mg. The approach we have employed in the preparation of ovine LH for the chemical studies reported here involves sequential extraction of fresh glands with water, pH 5·5, and 0·1 M ammonium sulphate, pH 4·0 (Ellis, 1961b), followed by precipitation with ammonium sulphate, fractionation with metaphosphoric acid and ethanol, chromatography on IRC-50 (Reichert and Parlow, 1963) and molecular sieving through the dextran gel, Sephadex G-100 (Reichert and Jiang, 1965). The specific activity of the final lyophilized product varies between 2·0 and 3·0 NIH-LH-S1 units/mg.

PURIFICATION OF BOVINE LUTEINIZING HORMONE

To our knowledge, the only procedure which has been described in detail for purification of bovine LH involves sequential extraction of fresh glands with water, pH 5·5 and 0·1 M ammonium sulphate, pH 4·0 (Ellis, 1961b), followed by ammonium sulphate precipitation, IRC-50 adsorption (Reichert, 1962) and gel filtration through the dextran gel Sephadex G-100 (Reichert and Jiang, 1965). The specific activity of the final lyophilized preparation varies between 2 and 3 NIH-LH-S1 units/mg (Table 2).

Table 2

Relative potencies of purified ovine, bovine, porcine and equine luteinizing hormones expressed in terms of NIH-LH-S1 (OAAD assay)

Reference	Species	Relative potency units/mg*	Contaminating activities	
			TSH (USP units/mg)	FSH*
Ellis (1958)	Ovine	2·2†	§	‖
Squire and Li (1959)	Ovine	2·4†	‖	‖
Woods and Simpson (1961)	Ovine	1·6†	‖	‖
Ward, Walborg and Adams-Mayne (1961)	Ovine	1·9	§	§
Reichert and Parlow (1963)	Ovine	2·8	< 0·1	< 0·012
		4·5 (in soln.)	§	§
Reichert (1962)	Bovine	1·7	0·05	< 0·0028‡
Reichert and Jiang (1965)	Bovine	2·4	§	§
Reichert (1964)	Porcine	0·80	0·05	0·06
Adams-Mayne and Ward (1964)	Porcine	0·43	§	§
Reichert and Wilhelmi (1965)	Equine	0·96	0·0025	0·25
Reichert and Jiang (1965)	Equine	1·40	§	§

* One unit = activity in one milligramme NIH-LH-S1 (OAAD assay) or NIH-FSH-S1
† Data taken from Reichert and Parlow (1963)
‡ Originally reported as showing no response at 8,000 µg. Subsequent studies reveal no positive response in Steelman-Pohley assay at 32,000 µg. (Reichert, unpublished)
§ Not given
‖ Quantitation not possible on basis of information provided

PURIFICATION OF PORCINE LUTEINIZING HORMONE

Except for a brief report in abstract form by Adams-Mayne and Ward (1962), there has been relatively little progress made in purification of porcine LH since the classical studies of Shedlovsky, Rothen, Greep, Van Dyke and Chow (1940). Using a protocol which parallels those described above for purification of ovine and bovine LH, we have obtained a lyophilized porcine LH fraction which is essentially equivalent in potency to NIH-LH-S1 (OAAD assay). The procedure involves sequential extraction of fresh glands with water pH 5·5 and 0·1 M ammonium sulphate, pH 4·0, followed by precipitation with ammonium sulphate, fractionation with metaphosphoric acid and ethanol, DEAE-cellulose and IRC-50 chromatography (Reichert, 1964) and gel filtration through Sephadex G-100 (Reichert and Jiang, 1965).

PURIFICATION OF EQUINE LUTEINIZING HORMONE

It has been possible to obtain a purified equine LH preparation by extraction of whole equine pituitaries with potassium chloride, fractionation of the resulting extract with ethanol and trichloroacetic acid, tandem chromatography on DEAE-cellulose and IRC-50 (Reichert and Wilhelmi, 1965) and gel filtration through Sephadex G-100 (Reichert and Jiang, 1965). The most potent preparations obtained by this sequence have a relative potency 1·4 times NIH-LH-S1 (OAAD assay). Using a different approach, Segaloff and Steelman (1959) have described the preparation of equine LH having a potency 500–600 per cent of the Armour 227-80 reference preparation. This potency estimate was determined by a modification of the ventral prostate assay of Greep *et al.* (1941). Because of this, and in light of studies already referred to (Parlow, 1963; Reichert, 1966b) it is not possible to compare the potency of their preparation with that of the equine LH fraction described above.

A summary of the relative potencies of currently available ovine, bovine, porcine and equine pituitary LH fractions, together with an estimate of their TSH and FSH contaminations is given in Table 2.

PURIFICATION OF HUMAN PITUITARY LUTEINIZING HORMONE

There have been numerous procedures described for purification of luteinizing hormone from human pituitary glands. Segaloff and Steelman (1959) were probably the first to prepare a purified human LH fraction. These workers, using an ammonium-acetate-ethanol extraction followed by CM-cellulose chromatography, reported a preparation having a relative potency 6–10 times greater than the Armour ovine reference preparation 227-80 (which is approximately equipotent to NIH-LH-S1, Table 1). FSH contamination was less than 1·0 times the Armour porcine reference preparation 264-151-X. (Armour 264-151-X has been shown to be 0·55 times as potent as NIH-FSH-S1, Table 1.) No estimate of TSH contamination was provided, and the LH potency estimate was obtained using the ventral prostate assay. Thus, a purified human LH fraction was evaluated in terms of an ovine reference

5+   129

preparation in an assay which has since been shown to be particularly sensitive to species differences in this hormone (Parlow, 1963). Therefore, the stated potency is probably over-estimated by a factor of from 12 (Rosemberg et al., 1964a) to 25 (Reichert and Parlow, 1964a). Squire, Li and Anderson (1962), using a calcium oxide extraction, followed by chromatography on IRC-50 and CM-cellulose, obtained a preparation which subsequently was reported as having a specific activity of 500 I.U./mg when assayed in terms of the International Standard for HCG, in the ventral prostate assay (Lostroh, Squire and Li, 1963). No quantitative estimate of FSH and/or TSH contamination was provided. Hartree, Butt and Kirkham (1964) using an ammonium acetate-ethanol extraction, followed by chromatography on IRC-50 and DEAE-cellulose obtained a preparation which was 6,400 times as potent as the first IRP-HMG/mg. The TSH potency was estimated by an *in vitro* technique to be less than 0·006 USP units/mg. FSH contamination was estimated as being less than once the first IRP-HMG. Using a primary ammonium sulphate extraction procedure developed by Wilhelmi (1961) (Parlow, Wilhelmi and Reichert, 1965), followed by DEAE-cellulose chromatography (Reichert and Parlow,

Table 3

Relative potency of purified human pituitary luteinizing hormone

Reference	NIH-LH-S1 units/mg	Armour 227-80 units/mg	1st IRP-HMG I.U./mg	2nd IRP-HMG I.U./mg	HCG I.U./mg
Segaloff and Steelman (1959)	6–10 ‖	—	—	—	—
Squire et al. (1962)	—	—	—	—	500*
Reichert and Parlow (1964)	3·2‡	3·4§	10,000§	614§	2,880§
Hartree et al. (1964)	2·04	2·2	6,400†	391§	1,836§

* Ventral prostate of Greep et al. (1941)
† OAAD assay, I.P. injection modification (Bell, Loraine, Mukerji and Visutakul, 1965)
‡ OAAD assay (Parlow, 1961)
§ Values calculated by application of conversion factors obtained with OAAD assay. (*See* Table 1)
‖ For discussion of this estimate, *see* text

1964a), IRC-50 chromatography (Reichert and Parlow, 1964b) and gel filtration through the dextran gel Sephadex G-100 (Reichert and Jiang, 1965) we have prepared a fraction assaying 3·2 NIH-LH-S1 units/mg. TSH contamination was 0·18 USP units/mg, and FSH contamination was estimated at 0·11 NIH-FSH-S1 units/mg. A summary of the relative potency of currently available human LH fractions is given in Table 3. Where possible, the published values have been converted into terms of a common reference preparation.

PURIFICATION OF HUMAN URINARY LUTEINIZING HORMONE

In an earlier report (Reichert and Parlow, 1964c) we described the use of DEAE-cellulose for partial purification of postmenopausal and eunuch urinary luteinizing hormone, and the separation to a significant degree of this gonadotrophin from its most tenacious contaminant, FSH. The most potent preparations from postmenopausal urine showed a specific activity of 0·022 times NIH-LH-S1/mg, with an FSH/LH ratio of less than 2·2. Since then,

Table 4

Summary of procedure for further purification of postmenopausal urinary LH

Fraction	Description	LH* units/mg	FSH* units/mg	FSH/LH ratio
F-1	Precursor fraction, see text	0·018	1·01	56
F-2	F-1 adsorbed on CM-cellulose in 0·005 M phosphate, pH 6·0. LH eluted with 0·5 M NaCl	0·034	0·24	7·1
F-3	F-2 after filtration through G-100 column precalibrated with purified pituitary LH	0·081†	< 0·25	< 3·1

* One unit = activity in one milligramme of NIH-LH-S1 or NIH-FSH-S1. To convert into terms of IRP-HMG, use conversion factors listed in Table 1.
† Mean value from three independent assays

Anderson and Albert (1965) have described experiments utilizing carboxymethylcellulose in ammonium acetate buffer, pH 6·0, to achieve a further separation of urinary LH and FSH. Working independently, we have utilized carboxymethylcellulose, followed by gel filtration through Sephadex G-100 to obtain a significant further purification of postmenopausal urinary LH. These experiments, which will be described in detail elsewhere, are summarized in Table 4. In essence, the procedure involves the following purification sequence: Adsorption of gonadotrophins from urine by the kaolin-acetone procedure of Albert (1956); concentration of the gonadotrophins by ammonium acetate-ethanol fractionation (Albert, Derner, Stellmacher, Leifman and Barnum, 1962); further concentration of the gonadotrophins by adsorption of impurities on to DEAE-cellulose in 0·05 M phosphate buffer, pH 7·0 (Albert *et al.*, 1962); partial separation and further purification of the LH by chromatography on DEAE-cellulose in 0·007 M phosphate–0·003 M borate buffer, pH 8·0 (Reichert and Parlow, 1964c). The fraction unadsorbed on DEAE-cellulose at pH 8·0 (F-1) was then chromatographed on CM-cellulose in 0·005 M phosphate buffer, pH 6·0, and a 'separable LH' was adsorbed and eluted with 0·5 M NaCl (F-2). As can be seen in Table 4, the CM-cellulose chromatography effected a simultaneous increase in urinary LH activity and improvement in the FSH/LH ratio for this preparation. Finally, the urinary LH was filtered through a column of Sephadex G-100 which had been 'precalibrated' by filtration of a highly purified human pituitary LH preparation. Fractions in that portion of the elution profile from the filtration of the urinary LH which corresponded to the pituitary 'LH-area' were combined and assayed (F-3). The relative potency of F-3 was 0·081 times NIH-LH-S1/mg (OAAD assay), or 124·8 I.U./mg (Table 1). Thus far, there has not been sufficient material on hand to permit a critical evaluation of FSH contamination. Preliminary results, however, show the FSH contamination of F-3 to be less than 0·25 NIH-FSH-S1 units/mg.

COMPARISON OF RELATIVE POTENCIES OF CURRENTLY AVAILABLE URINARY LH FRACTIONS

In Table 5 are summarized the relative potencies of currently available urinary LH preparations, as well as their FSH contaminations. For convenience, these values have been expressed in terms of the second IRP-HMG. Fraction D-NR (Albert, 1965) was reported as having a relative potency of 0·6 × NIH-LH-S1/mg obtained with the ventral prostate assay. The index of discrimination (I.D.) between the OAAD and VP assays, when measuring human urinary or pituitary LH in terms of the ovine NIH-LH-S1 reference preparation, ranges from 12 to 25 (references cited). Therefore, when expressed in terms of the OAAD assay Fraction D-NR may be expected to contain between 0·024 to 0·050 NIH-LH-S1 units/mg. Similarly, the FSH/LH ratio of this fraction would be expected to range from 6 to 12. We have arbitrarily used a value of 18·5 (the mean of the two proposed values for the I.D. OAAD/VP) in calculating the values expressed in Table 5. Donini, Puzzuoli and Montezemolo

Table 5

Relative potency of currently available postmenopausal urinary luteinizing hormone fractions expressed in terms of 2nd IRP-HMG

Author	Fraction	I.U. LH/mg	I.U. FSH/mg	FSH/LH ratio
Reichert and Parlow (1964c)	D-PM-2-1	33·8	< 1·33	< 0·03
Donini et al. (1964)	1–10	57	88·7	1·55
Albert (1965)	D-NR*	39·9	8	0·20
Reichert (This paper)	F-3	124·8	6·6	< 0·052

* Relative potency as reported = 0·6 NIH-LH-S1 units/mg. See text for discussion. Values obtained by application of conversion factors listed in Table 1

(1964) also used the ventral prostate assay to arrive at the relative potency estimate of fraction 1–10. Since they used an appropriate human LH reference preparation (Pergonal-23), however, it may be expected that their results can be compared directly with those obtained using the OAAD assay (Rosemberg et al., 1964a). The FSH/LH ratios are also expressed in terms of the second IRP-HMG. In those instances where the FSH activity was reported in terms of NIH-FSH-S1, the appropriate conversions into terms of the second IRP-HMG were made on the basis of factors provided in Table 1. The results suggest that F-3 probably represents the highest level of urinary LH activity yet reported.

A COMPARISON OF HUMAN PITUITARY AND HUMAN URINARY LUTEINIZING HORMONE

Although, as indicated, fractions equivalent to F-3 represent a high degree of biological activity for a urinary LH preparation, it should be noted that the order of potency attained is some 30 times less than that reported for LH purified from human pituitary glands (Reichert and Parlow, 1964a, b, c, d). Recently described gel filtration experiments (Reichert and Jiang, 1965), as well as those reported here, show that urinary LH fractions with high biological activity emerge from columns of G-100 with Ve/Vo ratios identical to those observed for purified human pituitary LH preparations. These results have led us to postulate that a relatively small peptide or glycopeptide, presumably required for full expression of LH activity in the OAAD assay, may be removed from the pituitary LH molecule during its metabolism, giving rise to a urinary LH metabolite which is structurally modified, but not to an extent detectable by the gel filtration techniques referred to above. Disc electrophoresis studies of fractions equivalent to F-3 have suggested heterogeneity, and also have indicated that the urinary LH may be considerably more acidic in character than its pituitary precursor molecule (Reichert, 1967).

GENERAL COMMENTS ON METHODS
FOR PURIFICATION
OF LUTEINIZING HORMONE

As has been described in the preceding sections, a high purification of luteinizing hormone may be achieved through use of ion-exchange chromatography, especially on CM-cellulose, DEAE-cellulose and IRC-50. Some general properties of LH's in this regard may be noted. Ovine, bovine, porcine, equine, human pituitary and human urinary LH are all *unadsorbed* by the high capacity anion-exchanger DEAE-cellulose in 0·007 M phosphate– 0·003 M borate buffer, pH 8·0 (References cited). Therefore, a convenient concentration of LH activity may usually be obtained by treatment of crude preparations in this manner. Also, the FSH activity of ovine, bovine, porcine, equine, human pituitary and human urinary fractions (Reichert, unpublished; Reichert and Parlow, 1964a, c) are all *adsorbed* by the exchanger under the conditions stated. Therefore, an important separation of LH from one of its most tenacious contaminants, FSH, is simultaneously accomplished. Further, all of the above listed LH's are *adsorbed* by the cation exchanger IRC-50 in the same buffer system (0·007 M phosphate– 0·003 M borate, pH 8·0) and may be easily eluted in stepwise fashion with 0·5 M NaCl. Thus, it has been possible to run both ion-exchange treatments concurrently, in tandem, to obtain a high purification of LH from a wide variety of sources. In addition, this sequence may be run in either column or batch operation and is, therefore, amenable to a virtually unlimited 'scaling up'. Porcine, ovine, bovine LH (Adams-Mayne and Ward, 1964) and human LH (Reichert, unpublished) may also be purified by adsorption on to CM-cellulose in 0·005 M phosphate buffer, pH 6·0. In this system the FSH is unadsorbed, but the LH must be differentially eluted from the column.

SEPARATION OF LUTEINIZING HORMONE FROM FOLLICLE STIMULATING HORMONE

As mentioned in the preceding section, a partial separation of LH and FSH activities may be achieved by ion-exchange chromatography on DEAE-cellulose and/or CM-cellulose. To our knowledge, however, no pituitary LH fraction has been found completely 'free' of FSH activity when attempts to detect this gonadotrophin were carried out in a rigorous fashion in a specific bio-assay system. The problem here is the relative insensitivity of currently available assays for FSH and the consequent requirement for large amounts of LH for the exclusion assay. Of all the LH's bovine LH (Reichert, 1962) appears to be most completely free of FSH contamination. Originally reported as being devoid of FSH activity when tested at the 8,000 µg level in the HCG-augmentation assay (Steelman and Pohley, 1953), it has since been shown not to elicit a positive response in this assay when tested at 32,000 µg (Reichert, unpublished).

SEPARATION OF LUTEINIZING HORMONE FROM THYROID-STIMULATING HORMONE

In addition to FSH, a major biological contaminant associated with most LH fractions is TSH, and the problem of complete separation of these activities has proven a difficult one. Condliffe, Bates and Fraps (1959) reported a separation of bovine LH and TSH by chromatography on DEAE-cellulose in 0·005 M glycine buffer, pH 9·5. Separation of human LH and TSH has also been accomplished by chromatography under these conditions. (Hartree et al., 1964; Parlow, Condliffe, Reichert and Wilhelmi, 1965). Reichert (1962) and Reichert and Parlow (1963) were able to separate bovine and ovine TSH and LH by differential elution of these activities from columns of IRC-50. Fractionation with metaphosphoric acid and ethanol has been successfully employed to separate ovine, bovine and porcine LH and TSH (Ellis, 1958; Reichert, 1962; Reichert, 1964). It should be emphasized, however, that while useful, none of the above procedures will effect a *complete* separation of LH and TSH. Usually a considerable percentage of LH remains associated with TSH in fractions other than those designated 'TSH-free'. Also, where the LH fractions have been tested at sufficiently high dose levels, it has invariably been possible to detect TSH contamination. A comparison of degrees of TSH contamination among currently available LH preparations is complicated by the use of a variety of *in vivo, in vitro* and immuno-assay systems, as well as by recently-recognized problems associated with species differences in TSH (*see* World Health Organization Bulletin: Human Thyrotrophin Research Std. A).

SOME PHYSICAL CONSTANTS
OF LUTEINIZING HORMONE

The molecular weight of *ovine LH* has been reported to be about 30,000 on the basis of ultracentrifugation studies carried out in phosphate buffer, pH 6·8 (Squire and Li, 1959). The sedimentation coefficient for the Squire and Li preparation was 2·70, with an isoelectric point of 7·3 and a carbohydrate composition of 11 per cent. Ward *et al.* (1959), on the basis of similar studies in glycine buffer

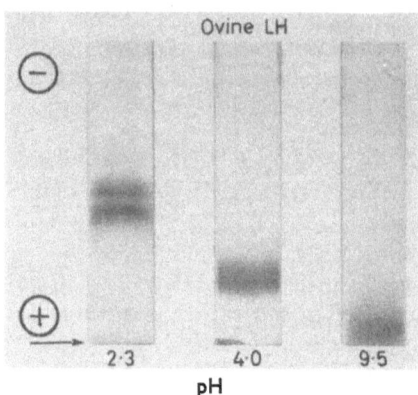

Figure 1. Electrophoresis of ovine LH fraction O-LH-2 (Reichert, 1966c) on polyacrylamide gel at various pHs as described by Reisfeld, Lewis and Williams (1962), and by Ornstein and Davis (1962). For the electrophoresis at pH 2·3 and 4·0, the anode is located at the bottom, and the cathode at the top of the gel. For electrophoresis at pH 9·5, the poles are reversed. 125 μg were applied to each column

pH 6·0, reported a molecular weight for *ovine LH* of 28,000, an isoelectric point of 7·7, a sedimentation coefficient of 2·32 and a carbohydrate composition of 25 per cent. Recently, a molecular weight of 16,000 was proposed for *ovine LH* on the basis of ultracentrifugation studies in a KCl-HCl buffer pH 1·3 (Li and Starman, 1964). It has been suggested that this lower molecular weight estimate may be due to an acid catalysed hydrolysis or dissociation of the native hormone molecule (Reichert and Jiang, 1965). Electrophoresis studies of purified ovine LH at pH 9·5, 4·0 and 2·3 (*Figure 1*) would seem to confirm that a dissociation of the ovine hormone is occurring at the acid pH.

Human LH has been reported to have a molecular weight of 26,000 (ultracentrifugation studies in formate buffer, pH 3·6), a sedimentation coefficient of 2·71, an isoelectric point of 5·4, and a carbohydrate content of 4·4 per cent. Noteworthy here is the pronounced acidic nature of the isoelectric point for *human LH* as compared to that reported for ovine LH, as well as the significantly lower carbohydrate content. The molecular weight most quoted for *porcine LH* is 90,000 (Shedlovsky *et al.*, 1940). We have recently studied the behaviour of porcine, as well as ovine, bovine, human pituitary, human urinary and equine LH fractions on calibrated columns of Sephadex G-100, as well as in sucrose density gradients (Reichert and Jiang, 1965). The purified LH's studied (equine LH excepted) *all* emerged from the G-100 column with essentially identical Ve/Vo ratios of 1·74. Based on calibration of the column with simple proteins of known molecular weight, a molecular weight estimate of 45,000 for the LH's was obtained. It is clearly recognized that, because of the glycoprotein character of luteinizing hormone, this estimate probably represents an upper limit. It is also appreciated that gel filtration behaviour is determined more by differences in effective molecular dimensions than weight. The glycoprotein anomaly noted by others (Whitaker, 1963; Andrews, 1964, 1965) and illustrated by the studies referred to above, underlines the point that Ve/Vo ratios are not simply equated to molecular weights. Whatever the absolute molecular weights of various LH's are, however, it is suggested that they are quite similar. Based on ultracentrifugation studies in a sucrose density gradient, the molecular weight for ovine, bovine, porcine and human LH was estimated to fall between 25,000 to 33,000 (Reichert and Jiang, 1965). This is closer to the values reported for ovine and human LH (cited above). In each case the molecular weight value for *porcine LH* fell considerably short of the 90,000 figure so often quoted. It would appear that porcine LH is similar to that of other species with respect to molecular weight and/or molecular dimensions. Of interest was the result obtained upon gel filtration of equine LH. Whilst ovine, bovine, porcine, human pituitary and human urinary LH activities were retarded by Sephadex G-100, the equine LH fraction of highest specific activity was not, and emerged with the Vo (outer volume), thus suggesting it to be considerably larger in molecular weight and/or dimensions than the other LH's studied.

N- AND C- TERMINAL AMINO ACID ANALYSIS OF LUTEINIZING HORMONE

Ward *et al.* (1959) and Ellis (1958) have reported the detection of serine and threonine following dinitrophenylation of ovine luteinizing hormone. Adams-Mayne and Ward (1964) proposed lysine as the probable C-terminal amino acid for this hormone, but have recently revised this position and now report serine and aspartic acid

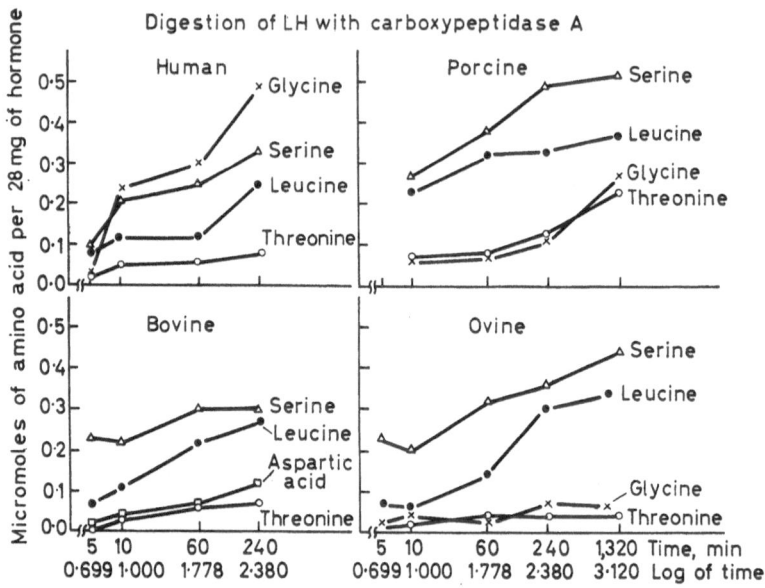

Figure 2. Digestion of purified human, bovine, porcine and ovine LH with carboxypeptidase (For conditions of digestion, see Reichert, 1966c)

to be the C-terminal residues for ovine LH (Ward, Firgino and Lamkin, 1966). We have conducted an extensive series of studies on the response of purified LH preparations to dinitrophenylation, digestions with carboxypeptidase A and hydrazinolysis (Reichert, 1966c). Dinitrophenylation of ovine, bovine and porcine LH showed serine, threonine and phenylalanine to be the major amino acids detected. Serine was also found in human LH, along with

high levels of valine and aspartic acid. Threonine and phenylalanine were relatively lacking in the human LH fractions studied.
All of the amino acid residues were recovered in yields considerably
less than 1 μ mole/μ mole of hormone, assuming a molecular weight
of 28,000 for LH (Reichert and Jiang, 1965). Carboxypeptidase A
digestion released serine and then leucine from ovine and bovine LH.
Serine and leucine, as well as glycine, were released rapidly from
human and porcine LH, although the sequence of release was not as
clear-cut as with the ovine and bovine hormones (*Figure 2*). Hydrazinolysis revealed serine as a major residue in each species LH,
together with surprisingly large amounts of phenylalanine. A summary of the major amino acid residues detected as well as the

Table 6

Characterization of luteinizing hormone by dinitrophenylation, hydrazinolysis,
and digestion with carboxypeptidase A*

Species	Dinitrophenylation		Hydrazinolysis		Carboxypeptidase A sequence of release
	amino acid	yield	amino acid	yield	
Ovine	Ser.	0·28	Ser.	0·70	serine-leucine
	Thr.	0·14	Phe.	0·49	
	Phe.	0·12	Gly.	0·22	
			Leu.	0·21	
			Pro.	0·16	
Bovine	Ser.	0·21	Ser.	0·79	serine-leucine
	Thr.	0·19	Phe.	0·59	
	Phe.	0·12	Gly.	0·32	
			Leu.	0·20	
Human	Ser.	0·36	Ser.	0·51	(serine-leucine)-glycine
	Val.	0·26	Glu.	0·46	
	Ala.	0·16	Phe.	0·43	
	Asp.	0·15	Gly.	0·19	
			Asp.	0·17	
Porcine	Ser.	0·25	Ser.	0·85	(serine-leucine)
	Phe.	0·24	Phe.	0·45	
	Thr.	0·22	Leu.	0·43	
	Ala.	0·12	Gly.	0·30	
			Asp.	0·23	

* Data are summarized from Reichert (1966c).   Yield of amino acids expressed as micromoles per
micromole of LH, assuming a molecular weight of 28,000.

apparent sequence of release of amino acids following enzymic digestion is shown in Table 6. These results, which are amenable to a variety of interpretations, have been discussed in detail elsewhere (reference cited). The multiplicity of amino acids detected with each type of analysis may well be considered to reflect heterogeneity in the preparations studied. An alternate possibility would be that LH is composed of more than one polypeptide chain, so that it may indeed have more than a single amino- or carboxy-terminus.

AMINO ACID COMPOSITION
OF LUTEINIZING HORMONE

The amino acid composition of ovine LH has been reported by Ward et al. (1961) and by Papkoff, Gaspodarowicz, Candiotti and Li (1965). A feature of the amino acid composition of this species LH is its apparent high content of proline and of cystine. The high proline content is thought to be unfavourable for an alpha helical structure and would support the suggestion of Jirgensons (1960), based on optical rotatory dispersion studies, that ovine LH is a non-helical protein. We have essentially confirmed the amino acid data previously cited for the ovine hormone (Kathan, Reichert and Ryan, 1967). In addition, we have found the composition of bovine LH to closely approximate that of the ovine hormone. Interesting differences in the amino acid and carbohydrate composition of human LH and of ovine and bovine LH, were also noted. The human LH was significantly lower in proline content, and showed a relatively reduced content of the basic amino acids. Further, human LH showed a markedly higher content of sialic acid (2 per cent) than did the ovine and bovine hormones (0·3 per cent). These latter two observations would tend to explain an earlier report of Squire and Li (1959) that human LH has a more acidic isoelectric point than does the ovine hormone.

IMMUNOLOGICAL STUDIES
OF LUTEINIZING HORMONE

The cross-reactivity of antibody to human and bovine LH with homologous and heterologous LH has been studied (Reichert and Treadwell, 1967). Antibodies to human LH were capable of inhibiting the OAAD depleting activity, as well as giving positive precipitin and complement fixation reactions, with human pituitary and urinary LH and HCG. Such antisera, however, did not inhibit the biological activity of bovine, porcine, equine or ovine pituitary LH, or of PMSG. Further, no in-vitro serologic cross-reactivity was observed with LH from these sources. Antibodies to bovine LH, however, inhibited the biological activity of LH from all the above sources, including those from the human. Interestingly, positive serological cross-reactivity was observed only with bovine, porcine and ovine LH. Further, it required eight times as much bovine LH antibody to neutralize human pituitary LH than an equivalent amount of the homologous hormone. It appears that antibodies to human LH contain reactive sites which are rather specific for the homologous hormone, whether of pituitary, urinary or chorionic origin, while those against bovine LH have a much less restricted specificity.

SUMMARY

Methods which have proved effective for purification of LH from ovine, bovine, porcine, equine and human pituitary glands have been reviewed. A procedure for preparation of a potent urinary LH through use of CM-cellulose chromatography and gel filtration through Sephadex G-100, has been described. The potency of LH from the above sources have been compared, and presented in terms of a number of widely employed reference preparations. Current information on the molecular weight, isoelectric point, sedimentation coefficient and carbohydrate content of LH has been summarized. Recent studies bearing on the nature of the N- and/or C-terminal amino acid of LH, as well as its amino acid composition have been discussed. Wherever possible, the above topics have been considered from a comparative point of view.

ACKNOWLEDGEMENTS

It is a pleasure to acknowledge the expert technical assistance of Miss Karin Westphal, Mrs. Lilly Chang, Miss Jean Batchelder, Mrs. Dianne Murphy and Mr. William Fugate in various phases of this work. The author is indebted to Professor P. Donini, Istituto Farmacologico Serono, Rome, Italy, for a generous supply of the postmenopausal urinary gonadotrophin fraction Pergonal-29.

The investigations reported here were supported in part by Grant AM-3589 USPHS and Grant M-3616, Population Council. This is Publication No. 849 from the Division of Basic Health Sciences, Emory University, Atlanta, Georgia.

REFERENCES

Adams-Mayne, M. E. and Ward, D. N. (1962). *Fedn Proc. Fedn Am. Socs exp. Biol.* **21,** 197
— — (1964). *Endocrinology* **75,** 333

REPRODUCTION IN THE FEMALE MAMMAL

Albert, A. (1956). *Recent Prog. Horm. Res.* **12,** 227
— (1961). *Human Pituitary Gonadotropins.* Springfield; Thomas.
— (1962). *J. clin. Endocr. Metab.* **22,** 996
— Derner, I., Stellmacher, V., Leifman, J. and Barnum, J. (1962). *J. clin. Endocr., Metab.* **22,** 1962
— (1965). *Endocrinology* **25,** 1119
Andrews, P. (1964). *Biochem. J.* **91,** 222
— (1965). *Biochem. J.* **96,** 595
Anderson, R. N. and Albert, A. (1965). *Endocrinology* **77,** 1085
Bell, E. T., Mukerji, S. and Loraine, J. A. (1964). *J. Endocr.* **28,** 321
— Loraine, J. A., Mukerji, S. and Visutakul, P. (1965). *J. Endocr.* **32,** 1
Claesson, L., Hillarp, N., Hogberg, B. and Hakfelt, B. (1949). *Acta endocr., Copenh.* **2,** 249
Condliffe, P. G., Bates, R. W. and Fraps, R. M. (1959). *Biochim. biophys. Acta* **34,** 430
Donini, P., Puzzuoli, D. and Montezemolo, R. (1964). *Acta endocr., Copenh.* **45,** 321
Ellis, S. (1958). *J. biol. Chem.* **233,** 63
— (1961a). *Endocrinology* **68,** 334
— (1961b). *Endocrinology* **69,** 554
Greep, R. O., Van Dyke, H. B. and Chow, B. F. (1941). *Proc. Soc. exp. Biol. Med.* **46,** 664
Hartree, A. S., Butt, W. R. and Kirkham, J. (1964). *J. Endocr.* **29,** 61
Jirgensons, B. (1960). *Arch Biochem. Biophys.* **91,** 123
Karg, H. (1957). *Klin. Wschr.* **35,** 643
Kathan, R., Reichert, L. E. jun. and Ryan, R. (1967). *Endocrinology.* In press
Li, C. H. and Starman, B. (1964). *Nature, Lond.* **202,** 291
Lostroh, A. J., Squire, P. G. and Li, C. H. (1963). *J. Endocr.* **29,** 61
Ornstein, L. and Davis, B. J. (1962). *Disc Electrophoresis.* Preprinted by Distillation Products Industries, Rochester, N.Y.
Papkoff, H., Gospodarowicz, D., Candiotti, A. and Li, C. H. (1965). *Archs Biochem. Biophys.* **111,** 431
Parlow, A. F. (1961). In: *Human Pituitary Gonadotropins*, p. 301. Ed. by A. Albert. Springfield; Thomas
— (1963). *Endocrinology* **73,** 509
— Condliffe, P. G., Reichert, L. E. (jun.) and Wilhelmi, A. E. (1965). *Endocrinology* **76,** 27
— and Reichert, L. E. (jun.) (1963). *Endocrinology* **73,** 377
— Wilhelmi, A. E. and Reichert, L. E. (jun.) (1965). *Endocrinology* **77,** 1126
Reichert, L. E. (jun.) (1962). *Endocrinology* **71,** 729
— (1964). *Endocrinology* **75,** 970
— (1966a). 'Drugs of Animal Origin.' Ed. by A. Leonardi. In: *Chemistry and Biology of Pituitary Gonadotropins.* Milan, in press
— (1966b). *Endocrinology* **78,** 815
— (1966c). *Endocrinology* **78,** 186

Reichert, L. E. (jun.) (1967). *Endocrinology* **80,** 319
— and Jiang, N. S. (1965). *Endocrinology* **77,** 78
— and Parlow, A. F. (1963). *Endocrinology* **73,** 285
— — (1964a). *Endocrinology* **74,** 236
— — (1964b). *Endocrinology* **75,** 815
— — (1964c). *J. clin. Endocr., Metab.* **24,** 1040
— — (1964d). *Endocrinology* **74,** 809
— and Treadwell, P. (1967). *Endocrinology.* In press
— and Wilhelmi, A. E. (1965). *Endocrinology* **76,** 762
Reisfeld, R. A., Lewis, U. J. and Williams, D. E. (1962). *Nature, Lond.* **195,** 281
Rosemberg, E., Lewis, W. B. and Solod, E. A. (1964b). *J. clin. Endocr., Metab.* **24,** 675
— Solod, E. A. and Albert, A. (1964a). *J. clin. Endocr., Metab.* **24,** 675
Segaloff, A. and Steelman, S. (1959). *Recent Prog. Horm. Res.* **15,** 127
Shedlovsky, T. A., Rothen, A., Greep, R. O., Van Dyke, H. B. and Chow, B. F. (1940). *Science, N.Y.* **92,** 178
Squire, P. G. and Li, C. H. (1959). *J. biol. Chem.* **234,** 520
— Li, C. H. and Anderson, R. N. (1962). *Biochemistry, N.Y.* **1,** 412
Steelman, S. L. and Pohley, F. M. (1953). *Endocrinology* **53,** 604
Ward, D. N., Firgino, M. and Lamkin, W. M. (1966). *Fedn Proc. Fedn Am. Socs exp. Biol.* **25,** 348
— McGregor, R. F. and Griffin, A. C. (1959). *Biochim. biophys. Acta* **32,** 305
— Walborg, E. F. (jun.) and Adams-Mayne, M. E. (1961). *Biochim. biophys. Acta* **50,** 224
Whitaker, J. R. (1963). *Analyt. Chem.* **25,** 1950
Wilhelmi, A. E. (1961). *Can. J. Biochem. Physiol.* **36,** 1659
Woods, M. C. and Simpson, M. E. (1961). *Endocrinology* **68,** 647

DISCUSSION

DR. LORRAINE (*Edinburgh*)

In what proportion of assays is parallelism not obtained?

REICHERT

Less than 20 per cent.

DR. HEALD (*London*)

The presence of proteinases in allegedly purified preparations of gonadotrophins seems to present a problem. Do you make any attempts to detect such activity at various stages of fractionation? Can one remove proteolytic activity before fractionation?

REICHERT

We do not attempt to do this. However, I believe Dr. Popkoff, of the West Coast Group, heats his primary extracts to destroy proteolytic activity.

REPRODUCTION IN THE FEMALE MAMMAL

DR. HEALD

Do their preparations differ from yours?

REICHERT

No. There is no evidence to date causally relating pituitary proteinases to hormone breakdown in the type of extraction procedures we use.

BETTERIDGE (*Reading*)

At Shinfield we have tried to prepare pure pig LH with a view to developing a radio immuno-assay. We had difficulties in assessing its homogenicity. Several preparations obtained from G 100 as a single peak corresponded to Dr. Reichert's preparation mol, wt. = 45,000. We obtained single bands at pH 4 by disc electrophoresis but were able to show multiple bands at pH 9·5 spreading as a smear from the origin. After ^{131}I iodination of these preparations we were interested to find that only a small proportion of the iodinated material entered the gels at pH 4, whereas at pH 9·5 disc electrophoresis accounted for all the radioactivity. We did not feel happy about equating a stained band on a disc electrophoretogram with LH unless we could find biological activity. It proved very difficult to get biological activity from the gels, and I would like Dr. Reichert to comment on whether pH 4 disc electrophoresis is a better test of homogenicity than at pH 9·5.

REICHERT

I believe the test at pH 9·5 would be more sensitive, particularly for gonadotrophins, which tend to be somewhat acidic. For basic proteins, electrophoresis at pH 4·0 would be better.

DR. BUTT (*Birmingham*)

I was interested to hear that you found so much carbohydrate in your human LH in view of the fact that we reported a high carbohydrate content in human FSH. I wondered if you have any information on the nature of this carbohydrate in view of the reports which I believe Ward and Adams made on ovine LH of an almost complete lack of sialic acid. This is strange in view of the fact that HCG contains a high proportion of sialic acid, and we believe that the chorionic and luteinizing hormones are similar. I wondered if you have any information on pituitary LH.

REICHERT

The carbohydrate component consists of hexose, hexosamine and sialic acid, and represents approximately 15 per cent of the hormone on a dry weight basis. The sialic acid content of bovine and ovine LH was 0·3 per cent, and of human LH, almost three times as much, or about 1·4 per cent. This might account for the relative acidity of the human hormone when compared to ovine and bovine LH.

DR. HARTREE (*Cambridge*)

I believe that comparison of the amino acid composition of carbohydrates from different species should be interpreted cautiously at this stage because I do not believe that these LH preparations are purified proteins. I noticed on your slides that they had different biological potencies and you found some evidence for heterogenicity in the amino-terminal and carboxy-terminal amino acids. We

144

have prepared LH which is considerably higher in potency than any of those you showed, and this has been assayed by two different laboratories by the ovarian ascorbic acid depletion method. I think your material could be purified further.

REICHERT

The LH fractions which we used showed no evidence of heterogeneity when examined by polyacrylamide gel electrophoresis, upon immunoelectrophoresis or in the ultracentrifuge. Therefore, I believe it reasonable to assume they are pure. Your most potent human preparation is claimed to be five times NIH-LH-S1/mg, while the fraction we use is three times NIH-LH-S1/mg. A rather large number of assays must be performed before this difference in potency is proved significant.

We have used ovarian ascorbic acid depletion assay exclusively in monitoring our fractions during purification.

DR. DONOVAN (*London*)

Have you noticed any changes in biological activity (for example in hypophysectomized animals) of your LH as you obtained further purification? Have you relied entirely on ascorbic acid depletion?

REICHERT

No, we haven't investigated biological activity other than ascorbic acid depletion.

IMMUNOLOGICAL STUDIES ON HUMAN GONADOTROPHINS (WITH SPECIAL REFERENCE TO FSH)

W. R. BUTT

The United Birmingham Hospitals, Department of Clinical Endocrinology

THERE are so many useful applications of immunological techniques to the study of gonadotrophins that a good deal of effort to overcome the technical difficulties is justified. A successful immunological assay for one of the gonadotrophins, human chorionic gonadotrophin (HCG), has already been developed and it has revolutionized the field of pregnancy diagnosis. Some of the potential applications of immunological techniques to the study of pituitary hormones will be considered here.

THE INTRACELLULAR DETECTION OF GONADOTROPHINS

The fluorescent antibody technique of Coons and Kaplan (1950) requires highly specific antisera and for this reason no conclusive results have so far been obtained with the pituitary gonadotrophins. More progress has been made using antisera to HCG and there is a good deal of evidence regarding its localization in the placenta. The major site of production has been considered to be the cytotrophoblast but fluorescent staining, both by the direct and the indirect techniques, has been noted only in the cytoplasm of the syncytotrophoblast (Leznoff and Davis, 1963; Sciarra, Kaplan and Grumbach, 1963; Pierce and Midgley, 1963).

Antisera to HCG have also been used to study the localization of gonadotrophin in the pituitary since it is now well recognized that anti-HCG cross-reacts with luteinizing hormone (LH). Koffler and Fogel (1964) purified rabbit HCG-antiserum by absorbing it with human menopausal gonadotrophin (HMG) which contains primarily follicle stimulating hormone (FSH). They also absorbed antiserum to HMG with HCG to remove possible antibodies to LH and with these two antisera they attempted to identify the cellular origin of FSH and LH. They found no consistent pattern other than a correlation of the fluorescein-stained cells with the basophils. Midgley (1964), using anti-HCG found positive reactions randomly distributed throughout the anterior lobe in cells that were lightly to moderately PAS-positive while Herlant (1964), found that anti-HMG was also localized in PAS-positive cells but they were distinct from those stained with anti-HCG, i.e., the LH cells.

The only study reported in which antiserum to pituitary FSH has been used was not very successful because it produced generalized staining in the pituitary (McGarry and Beck, 1963).

A different approach has been employed by Bourdel and Li (1963). They neutralized the production of LH in rats by treating the animals with antiserum to ovine LH. Histological examination of the pituitaries revealed certain important changes: the basophils increased in number, were enlarged and stained less intensely. The peripheral cells showed no further alteration but the central cells had undergone degenerative changes. This is in agreement with the concept that LH is located in the central cells. No similar experiments with the purified antiserum to FSH now available have been reported.

THE PREPARATION OF 'BIOLOGICALLY' PURE GONADOTROPHINS

The removal of traces of FSH from preparations of LH by chemical fractionation is not too difficult. The contamination of the best preparations of ovine and human LH with FSH is very slight indeed. In the case of FSH, however, contamination with LH is a more serious problem—indeed some authorities still doubt if complete removal of LH is possible since it may be an intrinsic part of the FSH molecule. Most of the LH in crude extracts of FSH can be removed by procedures described in the accompanying paper (*see* page 113). Removal of the remaining LH by a specific antiserum to LH however, is an attractive possibility and Li, Mougdal, Trenkle, Bourdel and Sadri (1962) have presented evidence that this is indeed possible in the case of ovine FSH. The calculated quantity of anti-LH required to neutralize the 0·2 per cent contamination was added to the FSH: after this treatment the preparation was still active in the ovarian augmentation method for FSH but it did not increase the uterine weight in hypophysectomized rats. In parallel experiments with male hypophysectomized rats, Lostroh, Johnson and Jordan (1963) showed that FSH with traces of LH repairs the seminiferous tubules and after 14–21 days spermatids are in evidence. When treated with anti-LH, however, the FSH stimulates the Sertoli cells but no spermatids are formed. Kaiser (1964) has used a similar technique to prepare FSH for the study of its specific action in *in vitro* experiments.

Supplies of anti-human LH serum have not been so plentiful and there do not appear to be any similar experiments reported on its use in purifying human FSH. However, anti-HCG serum acts as a substitute and Lunenfeld and Donini (1965) have reported that treatment of HMG with anti-HCG results in an increase in the FSH : LH ratio of 80-fold. They indicate that even better results may be expected by careful adjustment of the amount of antiserum used, so this method looks most promising.

EFFECTS OF ANTISERUM TO GONADOTROPHINS ON FERTILITY

In 1936 Parkes and Rowlands demonstrated that antiserum to bovine pituitary extract inhibited ovulation in rabbits. Later workers, using more specific antisera, have obtained similar effects. Administration of antiserum to ovine LH inhibits endogenous LH in experimental animals (Bourdel, 1961) and prevents oestrus and ovulation (Bourdel and Li, 1963; Young, Nasser and Hayashida, 1963). The time factor is important here and it has been shown that provided the anti-LH is given 36 h before expected oestrus, then oestrus is prevented: if given only 12 h before, oestrus is not prevented although Kelly, Robertson and Stansfield (1963) showed that ovulation was blocked. In similar experiments in male rats, inhibition of spermatogenesis has been demonstrated (Young *et al.*, 1963; Hayashida, 1963).

Less work of this nature has been reported with anti-FSH but both ovine and human anti-FSH inhibit follicular growth in experiments with hypophysectomized rats (Hayashida and Chino, 1961; Wolf, 1963).

No serious attempt appears to have been made to control fertility in the human by means of antisera to gonadotrophins although the idea has appealed to several authorities. The effect of antibodies to gonadotrophins is likely to be temporary since antibodies would be produced to the heterologous proteins of the antiserum. It would be interesting therefore, to know whether active immunity can be produced and, if so, whether it would have a temporary or prolonged effect. There is no clear evidence that antibodies to gonadotrophins are produced in patients undergoing prolonged treatment with human gonadotrophins (Crooke, Butt and Bertrand, 1966) although this has been suspected in some cases. Two patients out of about 20 so far treated appeared to lose sensitivity to exogenous gonadotrophins. The first originally responded to dosages of about 900 I.U. FSH but then failed to respond to higher dosages which within 11 months had been increased to over 4,000 I.U. FSH. Local skin reactions to injections of FSH were noted but HCG caused no reaction. Samples of serum (0·1 ml and 0·2 ml) from this patient however, injected intraperitoneally into immature mice, did not reduce the effect of exogenous FSH as assessed by the uterine weight response, so there was no evidence for the presence of circulating antibodies. The second patient originally responded to about 400 I.U. FSH but in the course of 3·5 years required progressively more until finally she was being given 1,500 I.U. FSH. In this instance there were no local skin reactions but her blood was not examined for antibodies.

CROSS-REACTIVITY BETWEEN GONADOTROPHINS AND BETWEEN GONADOTROPHINS OF DIFFERENT SPECIES

Experiments concerned with the inhibition of biological activity and with precipitin reactions in agar gel have shown that LH is less species specific than some of the other pituitary hormones. Thus rabbit antiserum to ovine LH inhibits the LH activity of pituitary extracts of the rat, pig, whale and human as well as that of pregnant mare serum gonadotrophin (PMS) and ovine LH itself. It does not, however, cross-react with pituitary extract of the chicken (Moudgal and Li, 1961) and only inhibits the effect of HCG at much higher doses (Flux and Li, 1965). Antiserum to HCG however, reacts with homologous antigen or human LH and not with LH of non-primates (Shahani and Rao, 1964). Several groups have demonstrated a spur formation in agar gel diffusion experiments between LH and HCG against anti-human LH or anti-HCG (Taymor, Goss and Buytendorp, 1963; Goss and Lewis, 1964; Moritz and Booth, 1965; Butt, 1964) and cross-reactions occur in the haemagglutination inhibition method too.

Antisera to human FSH have been obtained that show virtually no cross-reaction with HCG or ovine FSH (Saxena and Henneman, 1964; Butt, Crooke and Wolf, 1965). Some degree of cross-reaction with human LH has been observed but it is difficult to decide if this is due to the traces of FSH remaining in the preparation (Wolf, 1965).

DEVELOPMENT OF IMMUNO-ASSAYS FOR GONADOTROPHIN

The cross-reaction between anti-HCG and human LH makes possible the development of immuno-assays for LH using the readily available antisera to HCG. Haemagglutination assays have been described by Wide, Roos and Gemzell (1961); Butt, Crooke, Cunningham and Ingrassia (1964); and Sato, Greenblatt and Mahesh (1965) and the use of latex particles by Rizkallah, Taymor, Park and Batt (1965). Although the correlation between immunological and biological activities of highly potent pituitary extracts is good it is more difficult to demonstrate this with preparations obtained from urine. Fluctuations in the amount of LH excreted during the normal menstrual cycle determined immunologically appear to be similar to those demonstrated by biological assay. However, the quantities reported by some groups appear to be much higher by immuno-assay. It is necessary to concentrate urine first and the result appears to depend on the method of extraction (Butt, 1964); the results obtained after direct acetone precipitation of urine are three to five times higher than those obtained by extraction with kaolin or benzoic acid–tungstic acid. Interest therefore, centres on the possibility of developing radio immuno-assays which are much more sensitive and could be applied to serum as well as urine. Preliminary work has already been reported by Wilson and Hunter (1966) using ^{131}I-labelled human LH and antiserum to LH, and by Wilde, Orr and Bagshawe (1967) using ^{131}I-labelled HCG.

The first problem in the development of a reliable immuno-assay for human FSH is concerned with the purification of the antigen. Butt *et al.* (1965) used the method given on page 115. After partition in the urea; ammonium sulphate : ethoxy ethanol system, starch gel electrophoresis usually reveals four main protein bands (*Figure 1*). By careful selection of the portion of the gel free of LH and albumin, as judged by immuno-assay, it is possible to obtain a suitable antigen. Traces of antibodies to these contaminants in the antiserum are easily removed by absorption with HCG and albumin. A very similar method has been used by Saxena and Henneman (1964) and Saxena (1966); electrophoresis on cellulose columns, and repeated electrophoresis on starch gel or polyacrylamide gel produced satisfactory results.

In the haemagglutination-inhibition method the pyruvic aldehyde method of preparing red cells was found to be less satisfactory for FSH than for HCG (Butt, Crooke and Cunningham, 1961). There is no explanation for this since it appears from bio-assay that FSH is absorbed to the cells but haemagglutination is poor. Somewhat better results were obtained if the treated cells are also incubated with tannic acid before sensitization with FSH. When the method is applied to extracts containing varying amounts of FSH fairly good correlation with bio-assay is obtained, particularly with preparations 'of high potency (Butt *et al.*, 1965). Saxena and Henneman (1964) have reported similar findings.

The method is useful for following the elution patterns of FSH

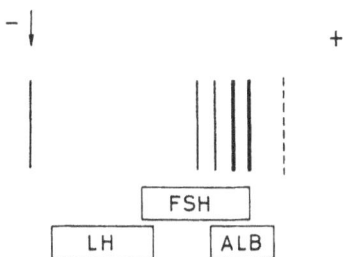

Figure 1. Starch gel electrophoresis of human FSH showing positions of protein bands (nigrosine staining), FSH, LH and albumin (the latter determined immunologically)

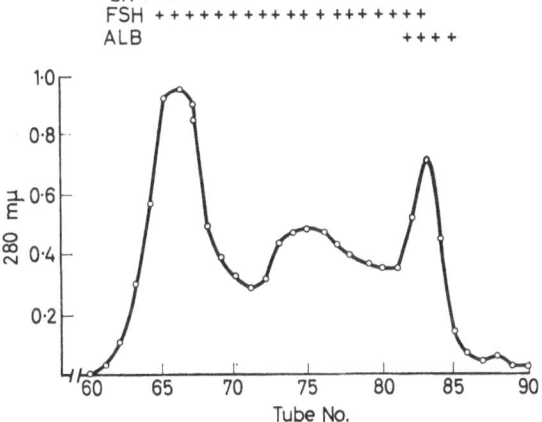

Figure 2. DEAE-Sephadex chromatography at pH 7·0 (phosphate) with gradient elution to 0·8 M-NaCl showing u.v. absorption at 280 mμ and positions of FSH, LH and albumin determined by haemagglutination-inhibition

during fractionation procedures. The presence of other proteins such as albumin, or of LH, can also be detected by a similar haemagglutination-inhibition method: an example is given in *Figure 2*.

Another technique of great interest, micro complement fixation

(Wasserman and Levine, 1961), has recently been shown to be of value in the investigation of growth hormone (Tashjian, Levine and Wilhelmi, 1965a, b). It is particularly useful for demonstrating slight differences in the spatial configurations of antigens. The method has now been applied to human FSH (Butt and Stacey, 1966).

In this investigation the rabbit anti-human FSH serum was first treated with complement (Taran, 1946) to remove anti-complementary factors. Serial dilutions of antigen were prepared in

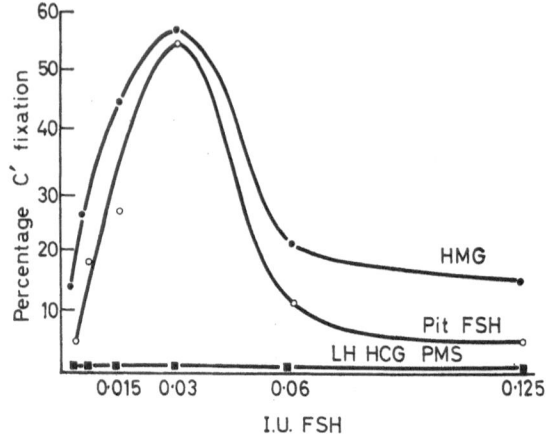

Figure 3. Complement fixation curves for gonadotrophins

O———O *Human pituitary FSH (CP 1 fraction: mean of 6 experiments)*
●———● HMG (*mean of 6 experiments*)
■———■ *Human pituitary LH, HCG, PMS and porcine FSH*
(From Butt, W. R. *The Chemisty of the Gonadotrophins*, 1967.
Courtesy of Charles C. Thomas, Publishers, Illinois)

volumes of 0·1 ml saline. Complement (0·2 ml) in a suitable dilution was added, followed by 0·1 ml of the antiserum. The mixtures were then incubated at 4°C for 18 h. Sensitized sheep red cells were then added, being prepared by incubating an equal volume of a 3 per cent red cell suspension with haemolytic serum for 1 h at 37°C immediately before the assay. A tris-gelatin buffer was used as diluent, all volumes being made up to 0·8 ml. After incubation at 37°C for 1 h the degree of haemolysis was obtained by reading the absorption of the solution at 415 mμ.

The degree of complement fixation by pituitary FSH and by FSH in urine was very similar (*Figure 3*). However, under the same con-

ditions no fixation occurred with human pituitary LH (kindly
supplied by Dr. A. S. Hartree, University of Cambridge), HCG,
porcine FSH, or PMS.

The quantity of pituitary FSH giving maximum fixation of
complement was 0·06 μg (0·03 I.U. FSH). The quantities of other
preparations of pituitary FSH and of HMG giving equivalent
fixation were related to their biological FSH activities (Table 1).
Preparations from the urine of younger women however, required
special treatment since at the concentrations required to fix complement, the presence of anti-complementary factors was still in evidence. This problem requires further investigation.

Table 1

The biological potencies of these preparations of FSH were assessed by the ovarian
augmentation method in immature mice. They are each expressed in terms of
I.U. FSH/mg. The amount of each, in I.U., giving maximum complement
fixation is given in the first column. The estimated potencies by the immuno-
assay are given in the last column using the first preparation as standard

(From Butt, W. R. *The Chemistry of the Gonadotrophins*, 1967.
Courtesy of Charles C. Thomas, Publishers, Illinois)

	Max C'-fixation	Potency I.U./mg	
	I.U.	Bio-assay	Immuno-assay
Pit. FSH	0·03	500	500 (adjusted)
	0·03	500	500
	0·08	530	200
	0·05	80	50
HMG	0·04	28	21
	0·125	10	3
	0·125	20	5
	0·075	28	12
	0·06	25	13
	0·06	28	14
	0·20	30	5

DISCUSSION AND SUMMARY

The preparation of human pituitary FSH sufficiently pure to use as
an antigen has been described. It should be possible, therefore, to
gain much more information by immunological techniques about

the cellular origin and the species specificity of FSH and about the biological effects of antisera to FSH.

The complement fixation method appears to be highly specific for the immuno-assay of FSH and suitable for application to the assay of FSH in pituitary extracts and in HMG prepared from urine. The sensitivity of the present method is probably insufficient for assays in serum or in urine without preliminary concentration and the specificity of the method when applied to such concentrates remains to be established.

Haemagglutination-inhibition methods appear to be less promising for FSH and the final solution to the problem of the immuno-assay of this hormone in blood and urine will probably come from radio immuno-assay. Labelling of FSH with ^{131}I by the method of Greenwood, Hunter and Glover (1963) is known to be possible (Butt and Wolf, 1965; Saxena, 1966) and once the technical details have been worked out a radio immuno-assay should be available.

ACKNOWLEDGEMENTS

The original work described here was carried out with the encouragement and direction of Dr. A. C. Crooke. The author is grateful also to Miss Janet Stacey for assistance with the immunological assays and Mr. D. Whyman for the biological assays. The work was supported by a grant from the Medical Research Council and by funds for technical assistance from Organon Laboratories Ltd.

REFERENCES

Bourdel, G. (1961). *Gen. comp. Endocr.* **1**, 375–380
— and Li, C. H. (1963). *Acta endocr., Copenh.* **42**, 473–479
Butt, W. R. (1964). Unpublished observations
— Crooke, A. C. and Cunningham, F. J. (1961). *Biochem. J.* **81**, 596–605
— — — and Ingrassia, F. (1964). *Proc. R. Soc. Med.* **57**, 851–854
— — and Wolf, A. (1965). *Ciba Fdn Study Grps* No. 22, 'Gonadotropins: Physicochemical and Immunological Properties'. Ed. by G. E. W. Wolstenholme and Julie Knight, pp. 85–97
— and Stacey, J. (1966). To be published
— and Wolf, A. (1965). Unpublished observations
Coons, A. H. and Kaplan, M. H. (1950). *J. exp. Med.* **91**, 1–13
Crooke, A. C., Butt, W. R. and Bertrand, P. V. (1966). *Acta endocr., Copenh.* Suppl. III, 22–23
Flux, D. S. and Li, C. H. (1965). *Acta endocr., Copenh.* **48**, 61–71
Goss, D. A. and Lewis, J. (1964). *Endocrinology* **74**, 83–86
Greenwood, F. C., Hunter, W. M. and Glover, J. S. (1963). *Biochem. J.* **89**, 114–123

Hayashida, T. (1963). *J. Endocr.* **26,** 75–83
— and Chino, S. (1961). *Anat. Rec.* **139,** 236
Herlant, M. (1964). *Proc. 2nd Int. Congr. Endocr.* 468–481
Kaiser, J. (1964). *Acta endocr., Copenh.* **47,** 676–688
Kelly, W. A., Robertson, H. A. and Stansfield, D. A. (1963). *J. Endocr.* **27,** 127–128
Koffler, D. and Fogel, M. (1964). *Proc. Soc. exp. Biol. Med.* **115,** 1080–1082
Leznoff, A. and Davis, B. A. (1963). *Can. J. Biochem. Physiol.* **41,** 2517–2521
Li, C. H., Moudgal, N. R., Trenkle, A., Bourdel, G. and Sadri, K. (1962). *Ciba Fdn Colloq. Endocr.* **14,** 20–32
Lostroh, A. J., Johnson, R. and Jordan, C. W. (1963). *Acta endocr., Copenh.* **44,** 536–544
Lunenfeld, B. and Donini, P. (1965). *Ciba Fdn Study Grps* No. 22, 'Gonadotropins: Physicochemical and Immunological Properties', pp. 85–97. Ed. by G. E. W. Wolstenholme and Julie Knight
McGarry, E. E. and Beck, J. C. (1963). *Fert. Steril.* **14,** 558–564
Midgley, A. R. (1964). *Expl Cell Res.* **32,** 606–609
Moritz, P. M. and Booth, W. D. (1965). *J. Endocr.* **33,** viii–ix
Moudgal, N. R. and Li, C. H. (1961). *Archs Biochem. Biophys.* **95,** 93–95
Parkes, A. S. and Rowlands, I. W. (1936). *J. Physiol.,* **88,** 305–308
Pierce, G. B. and Midgley, A. R. (1963). *Am. J. Path.* **43,** 153–173
Rizkallah, T., Taymor, M. L., Park, M. and Batt, R. (1965). *J. clin. Endocr. Metab.* **25,** 943–948
Sato, T., Greenblatt, R. B. and Mahesh, V. B. (1965). *Fert. Steril.* **16,** 223–228
Saxena, B. B. (1966). *J. Endocr.* 35. i.
— and Henneman, P. H. (1964). *J. clin. Endocr. Metab.* **24,** 1271–1276
Sciarra, J. J., Kaplan, S. L. and Grumbach, M. M. (1963). *Nature, Lond.* **199,** 1005–1006
Shahani, S. K. and Rao, S. S. (1964). *Acta endocr., Copenh.* **46,** 317–330
Taran, A. (1946). *J. Lab. clin. Med.* **31,** 1037–1038
Tashjian, A. H., Levine, L. and Wilhelmi, A. E. (1965a). *Endocrinology* **77,** 563–573
— — — (1965b). *Endocrinology* **77,** 1023–1036
Taymor, M. L., Goss, D. A. and Buytendorp, A. (1963). *Fert. Steril.* **14,** 603–609
Wasserman, E. and Levine, L. (1961). *J. Immun.* **87,** 290–295
Wide, L., Roos, P. and Gemzell, C. A. (1961). *Acta endocr., Copenh.* **37,** 445–449
Wilde, C. E., Orr, A. H. and Bagshawe, K. D. (1967). *J. Endocr.* **37,** 23–36
Wilson, P. and Hunter, W. M. (1966). *J. Endocr.* 35, i
Wolf, A. (1963). *Nature, Lond.* **198,** 1308–1309
— (1965). *Nature, Lond.* **205,** 504–505
Young, W. P., Nasser, R. and Hayashida, T. (1963). *Nature, Lond.* **197,** 1117

DISCUSSION

Professor L. E. Reichert (*Atlanta, Georgia*)

The specific activity of purified FSH is apparently about 400 units and the specific activity of LH is 3 units so the FSH : LH ratio is approximately 133. In a preparation which has 130 units FSH/mg and 1 unit LH/mg, the preparation probably has the same number of molecules of FSH and LH in it so that if you immunize an animal with a preparation which has 100 units FSH and 1/10 unit LH you are really immunizing the animal with only about 10 times more molecules of FSH than LH so you are generating antibodies to a considerable number of molecules of LH. When one tries to demonstrate that the antisera has inactivated the FSH by the Steelman-Polley assay then one must consider whether the antisera to LH is affecting the HCG used to prime the animal and whether a negative response is due to the anti-LH which has been produced. Have you demonstrated that your anti-serum to human FSH does have the ability to negate the effect of FSH, and do you feel that the question of LH contamination is important in your work?

Butt

Concerning the number of molecu'es of LH in FSH preparations, I think you directed this question to me about a year ago and I still cannot deny the validity of the point you make. The *in vivo* effect of anti-FSH is certainly directed towards FSH. If we use hypophysectomized rats in the assay and examine the ovaries histologically for lack of follicular growth, this shows the anti-FSH effect.

AN IMMUNOLOGICAL ESTIMATION
OF PROLACTIN IN SHEEP BLOOD

M. A. SAJI

University of Nottingham School of Agriculture,
Sutton Bonington, Loughborough, Leicestershire

DURING the last two decades several biological assays have been developed to estimate the activity of hypophyseal prolactin. The methods are based on the physiological activity of prolactin in different species, and in addition to their debatable specificity, they lack sensitivity and practicability and therefore could not be adequately applied to the estimation of the low levels of prolactin normally found in blood.

The availability in recent years of relatively pure protein hormones has permitted the use of immunological techniques in endocrinology. The production of antisera against these hormones has led to the development of *extremely* sensitive immuno-assays which can be used for the estimation of the minute amounts of hormone in biological fluids.

As early as 1937, Kabak and Stuloval demonstrated the antigenicity of prolactin and, subsequently, Young (1938), Strangeways (1938), Rowlands and Young (1939) and Bischoff and Lyons (1939) using more purified preparations of this hormone, demonstrated antibody production in rabbits. More recently, the availability of highly purified preparations of ovine prolactin has led to the production of potent antisera in rabbits (Levy and Sampliner, 1961 and 1962; Hayashida, 1962; Emmart, Bates, Condliffe and Turner, 1963). The capacity of such antisera to neutralize the biological effect of prolactin has been demonstrated by Hayashida (1962) using the pigeon crop-sac assay of Riddle, Bates and Dykshorn (1933). With regard to the development of an immuno-assay for prolactin, Levy and Sampliner (1961) observed that it was possible to detect 0·026 µg of purified prolactin in saline by using a haemagglutination inhibition technique. Hayashida (1962) detected 0·1–0·2 µg of purified ovine prolactin by using the precipitin ring test and 0·02 µg by a haemagglutination inhibition technique. No immunological estimation of prolactin in sheep blood has been reported to date.

157

This paper describes studies carried out to develop an immunological technique for the estimation of prolactin in sheep blood using a haemagglutination inhibition technique.

PREPARATION OF ANTISERA

Ovine hypophyseal prolactin (NIH-P-S4; NIH-P-S5; NIH-P-S6 and NIH-P-S7) was obtained from the Endocrinology Study Section of the National Institute of Health, Bethesda (U.S.A.). Antisera against these preparations were prepared in male and female rabbits. The use of bentonite as an adjuvant for the injection of prolactin was more effective than Freund's adjuvant which produced emaciated conditions in the animals. The results of immunization measured by the haemagglutination technique of Boyden (1951) using pyruvic aldehyde-treated sheep erythrocytes (Butt, Crooke and Cunningham, 1961) suggested that prolactin was not as effective in producing antibodies when injected intravenously compared with injections given intradermally or subcutaneously. This may be due to rapid elimination from circulation or rapid destruction before being fixed by the immunologically competent system.

During these studies it was observed that fresh preparations of prolactin were more antigenically active than preparations which had been stored for long periods. Not all rabbits produced antibodies which gave clear inhibition reactions and the antisera from such animals were discarded. Two animals proved satisfactory as a source of antibody production and they produced titres of antibodies reaching 1 : 10,000 to 1 : 20,000.

BIOLOGICAL ACTIVITY OF THE ANTISERUM

Following production of an antiserum it was necessary to demonstrate its biological activity. The local lactogenic response to prolactin in pseudopregnant rabbits (Lyons, 1942; Meites and Turner, 1947; Bradley and Clarke, 1956; Chadwick, 1962 and 1963) was used to study the biological activity of the antiserum to ovine prolactin.

The methods and preparation of the rabbits were as employed by Chadwick (1963) and differed only in detail. The mammary intraductal injections of prolactin and the antiserum were made on the eleventh day of pseudopregnancy. The thoracic pair of the mammary glands were treated as controls and were not injected. Of the remaining three pairs of the abdominal mammary glands, those glands on the left side of the rabbit were each injected with 0·15 ml antiserum to ovine prolactin (NIH-P-S4) while each gland on the right side was injected with the same amount of normal rabbit serum. In each case the injected ducts were marked with a permanent dye for identification. After a period of 45–60 min graded doses of ovine prolactin dissolved in isotonic saline solution were injected into the same ducts which had previously been injected with serum. The rabbits were examined on the seventh day after the intraductal injections, and samples of individual glands taken for reducing sugar and nitrogen determinations.

The study demonstrated that rabbit antiserum prepared against ovine prolactin inhibited the biological effects of purified ovine prolactin, i.e. lactogenesis (Plate 1). The results show that 0·15 ml of antiserum was capable of neutralizing 50 μg or more of ovine prolactin. Although visible milk secretion in response to ovine lactogenic hormone was inhibited completely at all three dose levels of lactogenic hormone used, the weight, nitrogen content and total reducing sugar content of mammary glands treated with the antiserum were higher at the 75 μg dose level than at lower levels (Table 1). This indicated a lactogenic effect either direct or indirect, of non-neutralized prolactin. In spite of the careful selection of rabbits to ensure uniformity with regard to age and body weight, considerable variation was encountered between rabbits in their response to lactogenic hormone. This variation was more marked when the criteria of mammary gland weight and nitrogen content of the mammary glands were considered than in the case of total reducing sugars. The lactogenic response to prolactin was so marked on the right side of rabbits (i.e. the side without antiserum) that there can have been no appreciable diffusion of antiserum across the epithelium of the alveoli to the adjacent tissues nor can it have been diffused to any degree in the general circulation.

There are several possibilities to explain this local effect of the antiserum. Firstly, the reaction between prolactin and its specific antiserum may have been rapid so that most of the antigen was quickly bound by the antibodies, leaving little to escape to adjacent tissues. Secondly, the impermeability of the epithelium of the alveoli to the antiserum, and the distension phenomena of the mammary glands by the injection techniques used might have been responsible for retaining the antibodies in the injected sectors. Thirdly, it is conceivable that distension of the alveoli and associated ducts could affect the circulation between blood vessels and the interstitial fluids of the connective tissues (Harkness and Harkness, 1956). Lastly, it is known that in colostrum the antibody titre exceeds the titre of similar antibodies in serum, and that this increase is restricted to a period at the end of pregnancy (Garner and

159

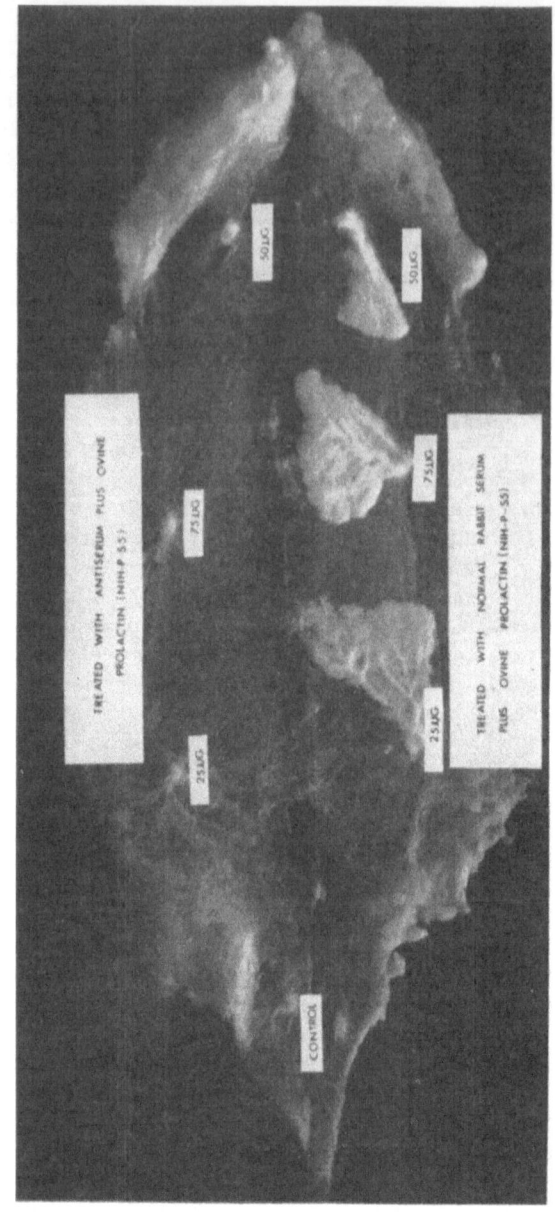

Plate 1. *Effect of rabbit antiserum to ovine prolactin on the lactogenic response of the rabbit*

Crawley, 1958). This suggests that the epithelial cells of the mammary gland are perhaps permeable to antibodies from the serum for a limited period after which the permeability may be altered or that the passage of the antibodies may occur in one way only.

Table 1

Mean effect of antiserum on lactogenic response of rabbit mammary gland to ovine prolactin

		Prolactin dose levels (μg)						Comparison with antiserum/ without antiserum	Effect of increasing dose
	Control	with antiserum			without antiserum			level of statistical significance	level of statistical significance
		25	50	75	25	50	75		
Mammary gland weight (g)	1·57	2·62	2·46	3·23 S.E. ± 0·024	3·20	3·87	5·18	P < 0·01	P < 0·05
Total nitrogen content (mg)	9·64	13·87	15·06	17·02 S.E. ± 1·4	20·18	25·06	29·80	P < 0·01	P < 0·05
Total reducing sugar content (mg)	1·90	3·09	3·59	4·63 S.E. ± 0·17	5·02	9·31	12·83	P < 0·01	P < 0·01

Whatever the mechanisms responsible, these results demonstrate that the antiserum to prolactin is capable of inhibiting the lactogenic effect of the hormone.

SPECIFICITY OF THE ANTISERUM

The double diffusion technique of Ouchterlony (1953) and immuno-electrophoresis tests (Scheidegger, 1955) were employed to determine the specificity of the antiserum and the homogeneity of the antigen.

The results obtained with these techniques demonstrated that the antiserum to all four preparations of ovine prolactin reacted by forming broad precipitin lines which fused completely with each other, thus demonstrating the immunological identity of the standard preparations.

A single precipitin line was also observed as a result of the reaction between the antiserum to ovine prolactin and a sheep pituitary extract. The precipitin lines of the standard prolactin and the pituitary extracts were confluent, demonstrating the immunological identity of the two reactants. No precipitin lines were observed between the antiserum to ovine prolactin and ovine follicle-stimulating hormone (FSH as NIH-FSH-S2), ovine luteinizing hormone (LH as NIH-LH-S3) and ovine growth hormone (GH as NIH-GH-S6 and NIH-GH-S7) at concentrations ranging from 1–200 µg.

Antiserum to ovine prolactin (NIH-P-S4) cross-reacted with sheep serum, forming a faint precipitin line at 1 : 32 and 1 : 64 dilutions. These lines did not fuse with the precipitin lines of the standard, and disappeared after absorbing the antiserum with sheep serum. No such reaction was observed with the antisera against ovine prolactin (NIH-P-S5; NIH-P-S6). From these results it was concluded that the antiserum to the ovine prolactin NIH-P-S4 also contained antibodies against serum proteins.

The specificity of the haemagglutination caused by the reaction of rabbit anti-ovine prolactin serum with purified ovine prolactin sensitized erythrocytes, was evaluated by determining the amount of an antigenic test solution which, when added to the antiserum, inhibited haemagglutination. It was observed that as little as 0·15 µg/ml of purified ovine prolactin completely inhibited the haemagglutination. Neither serial dilutions from 500 µg/ml of ovine FSH (NIH-FSH-S2), 1,000 µg/ml of ovine LH (NIH-LH-S7) and 0·2 ml of sheep serum caused any inhibition of haemagglutination. Only ovine GH (NIH-GH-S6) at a concentration of 100 µg/ml inhibited the haemagglutination reaction between antiserum to ovine prolactin and prolactin-sensitized cells, thus demonstrating the presence of reacting material in this growth hormone preparation.

Erythrocytes sensitized with ovine FSH, LH and GH failed to agglutinate when mixed with various dilutions of the antiserum to ovine prolactin.

IMMUNO-ASSAY OF PROLACTIN

The antiserum, after testing as described above, was used to develop an immuno-assay for the estimation of prolactin in sheep blood. The haemagglutination inhibition technique as used by Butt *et al.* (1961) was employed with some modifications.

During preliminary investigations a constant problem in applying the haemagglutination inhibition techniques to the assay of lactogenic hormone was its lack of reliability. An immuno-assay which fails is difficult to interpret in terms which indicate the causes of failure since a number of factors can be responsible. On occasion an assay which fails can be repeated with success without obvious change in conditions. The reasons can only be determined by a very careful and systematic check of all the factors involved. An investigation into some of the possible reasons for difficulties in application of the assay led to modifications which were incorporated into the method and led to improvement in its reliability. A general review of these modifications is given below.

DEVELOPMENT AND MODIFICATION OF THE HAEMAGGLUTINATION INHIBITION ASSAY

It was observed that the success of a haemagglutination inhibition assay depended on several factors, the main ones being the quality of sheep erythrocytes, the rate of dilution of the antiserum, the diluent used and the time and temperature of incubation during the assay.

The majority of early work on the haemagglutination inhibition reaction for the quantitative estimation of hormones in body fluids has involved the use of tannic acid-treated erythrocytes. These were found to be unstable in several instances, i.e. when sensitized with insulin (Arquilla and Stavitsky, 1956a, b) or with human growth hormone (HGH) (Reed and Stone, 1958). Later work showed that cells treated with formalin prior to tannic acid were less variable for HCG and HGH (Wide and Gemzell, 1960; Grumbach and Kaplan, 1962). Although few workers have used sheep blood cells stabilized with pyruvic aldehyde, similar results have been obtained with both tannic acid and pyruvic aldehyde treated cells (Wide, 1962). Sensitive assays for ovine prolactin have been developed by Hayashida (1962) using tannic acid-treated human 'O' type cells and by Levy and Sampliner (1961) using formalinized tannic acid-treated sheep erythrocytes.

In the present studies pyruvic aldehyde-treated cells were used and were found to be stable, but their sensitivity varied from one batch to another. This inconsistent behaviour of cells was found to affect not only the rate of dilution of the antiserum, but also the percentage of the diluent used, and consequently the haemagglutination inhibition reaction. Freshly treated cells were found to be less sensitive and more prone to agglutination than the cells stored for 6–8 weeks after treatment with pyruvic aldehyde. The aged cells were also less variable in their settling pattern. After coating the cells with the hormone, better results were obtained if the cells were washed, kept in the cold room at 5°C and used as soon as possible. No differences were observed when the sensitized cells were washed with sheep serum or with normal rabbit serum.

It is known that the haemagglutination reaction can be inhibited by pre-incubating the antiserum with excess corresponding hormone, thus making fewer antibodies available to react with hormone-coated cells. By dilution of the antisera and the use of a suitable range of hormone concentrations in the pre-incubation stage it is therefore possible to produce a variable reduction in antibody concentration. The degree of dilution of antiserum was found to be negatively correlated with the amount of hormone which causes inhibition of haemagglutination. It was demonstrated that each batch of anti-sera could be standardized prior to use in the assay with regard to its rate of dilution so that inhibition of haemagglutination could be obtained at 0·5 µg/ml of standard prolactin. This standardization procedure was repeated for every new batch of antiserum and new batch of erythrocytes. Once this dilution rate for a particular batch of antiserum was fixed for one batch of cells it was found to be stable and fluctuated little (3–4 per cent). A method similar to this has been used by Ehrlich and Randle (1962). Other workers have preferred to use a dilution rate about two-thirds of the original titre of the antiserum (Hartog and Fraser, 1961).

A clearer end-point in the assay was observed when the antiserum was diluted with large volumes of 100 to 500 ml (c. 1 : 1,000 to 1 : 5,000) of isotonic saline. The reason for this is not known. It is possible that dilution with a large volume of isotonic saline gave more thorough mixing and less wastage of antibodies. A further possibility is that by the use of large volumes of saline, the non-specific antibodies may be diluted sufficiently to prevent them affecting the assay. Bovine albumin, normal rabbit serum and sheep serum were tried in the diluent but sheep serum was preferred because ovine prolactin was used for raising the antiserum. It was thought that the impurities, if any, would most probably eliminate some of the non-specific antibodies. No difference in the haemag-glutination inhibition reaction was in fact observed when using sheep serum or normal rabbit serum. Horse, guinea-pig and rabbit serum have also been used in the diluent by other research workers.

During the course of haemagglutination inhibition studies it was observed that prolactin used for the sensitization of the sheep erythrocytes, tended to cause the agglutination of these cells with increased duration of storage of opened samples of the hormone at −15°C. Thus, wherever possible, a new bottle was opened as soon as difficulties arose with batches already sampled.

The time of incubation of haemagglutination trays at room temperature after adding the antiserum and before adding the prolactin-sensitized cells was also found to be related to the rate of dilution of the antiserum and to a certain extent to the sensitivity of the assay. An improved settling pattern of cells was observed if the trays were kept overnight at 3°C after the addition of cells.

The incorporation of these modifications gave an improved assay when applied to the measurement of sheep prolactin. However, even with rigid adherence to procedure, the reliability is not 100 per cent and assays still occasionally fail through unknown causes.

It is realized that although the assay can be used in its present state many of the modifications are empirical in nature and conditions are probably far from the best. A great deal of further study is undoubtedly necessary in order to determine the biological reasons for such failures before the optimum balance of all conditions can be derived.

INVESTIGATIONS INTO THE APPLICABILITY OF THE ASSAY AND THE RESULTS OBTAINED BY ITS USE

The first experiment carried out was to study the practicability of using the haemagglutination inhibition assay to measure amounts of prolactin added firstly to saline and secondly to lamb serum.

These studies demonstrated that the technique was specific and sensitive for the NIH ovine prolactin-anti-ovine prolactin system. The technique would detect levels of 0·15 µg/ml of this hormone in isotonic saline and a clear end-point was obtained. Similar results have been obtained by other workers (Levy and Sampliner, 1961; Hayashida, 1962).

However, distorted settling patterns of cells were obtained when the assay was applied to ovine prolactin added to lamb serum and the end-point was difficult to read. Similar results were obtained when attempts were made to assay blood serum from lactating sheep. It was considered that the distorted settling pattern could be due to a high concentration of serum protein or non-specific factors in the sera, since the distortion pattern and agglutination was more pronounced in wells containing high concentrations of serum than in wells with a low concentration. Non-specific effects have been observed in sera and have been shown to interfere with the measurement of other hormones, particularly HGH by the haemagglutination inhibition technique (Read and Stone, 1958; Read and Bryan, 1960a, b; Grumbach and Kaplan, 1962; Frazer and Hartog, 1962; Irie and Barrett, 1962; Dominguez and Pearson, 1962; Read, Eash and Najjor, 1962). Grumbach and Kaplan (1962) have claimed partial elimination of these factors by heating and absorption with sheep erythrocytes. In the present study, heating and absorption of sera did not significantly improve the settling pattern of the cells. It seemed desirable, therefore, to develop a method for removing any biologically active non-specific factors and an extraction procedure was used, based on the method of Sulman (1956). The main advantages of this procedure are that it inactivates hormone-decomposing enzymes in blood and it also precipitates the inert proteins. Since whole blood is used instead of serum the method avoids loss of hormone activity which may occur at the temperatures used to obtain serum. The main disadvantage is that on some occasions the final extracts contain haemolysed blood which interferes with the haemagglutination inhibition reaction. A similar procedure was used by Lyons and Page (1935) and Lyons (1937) for the extraction of prolactin from urine.

The extraction procedure gave samples which could be assayed for prolactin with a fair degree of success by the haemagglutination inhibition technique and the end-points were clear. Tests for recovery were satisfactory and the revised method involving extraction of the prolactin from sheep blood was applied to unknown samples.

The results suggested that prolactin was present at all times irrespective of the physiological state of the animal but that the levels show wide variations. In non-lactating, non-pregnant sheep, the blood levels of prolactin gradually decreased during the months of May, June and July and returned to the original level by September. The reason for this seasonal variation is not known but may be concerned with a general depression of the endocrine activity of the sheep hypophysis during the period. Three environmental factors known to affect the activity of the hypophysis in certain circumstances are the daylength (Yeates, 1949), environmental temperature (Dutt and Bush, 1955) and the nutritional status of the animal (Lamming, 1966). It is known that the release of pituitary gonadotrophins in sheep is depressed during anoestrus (Robinson, 1951) and that anoestrus is influenced by daylength. The low prolactin level coincides with the seasonal anoestrus and may be due to the general depression of pituitary activity during this period. However, the largest change of daylength is earlier than May and it could be expected that if daylength were the major factor influencing pituitary activity the fall in blood level would have taken place earlier than observed. Similarly, environmental temperature changes are not marked during the period when the changes in prolactin levels occurred.

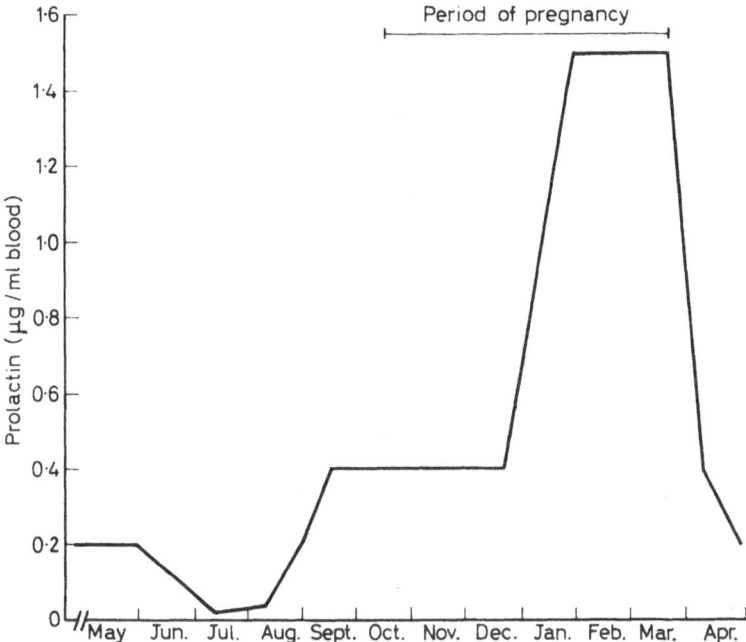

Figure 1.   Seasonal levels of prolactin in sheep blood.   Average levels from determinations completed in several groups of sheep

The nutritional status of the animals was altered at approximately the time of the fall in levels of prolactin since the animals were transferred from indoor feeding using a processed diet to feeding on pasture.   A change in plane of nutrition was not likely but the question remains of the effect of alteration of dietary regime on blood levels of prolactin.

The levels of prolactin in pregnant sheep blood extracts indicated that there was a progressive increase during the second half of pregnancy.   The increase is approximately fourfold and remains high until parturition. At parturition there is a progressive fall to the low level found in non-pregnant animals within 2–3 weeks and the level remains low throughout lactation. A summarized illustration of these changing levels of prolactin in sheep blood is given in *Figure 1*. The importance of prolactin as a mammotrophin and lactogenic hormone in many species and a luteotrophic hormone in a few species is well established but it has not been investigated to any extent in the sheep and its true biological function in this species remains obscure. There is evidence in this and other work that the observed rise in levels during pregnancy is of pituitary rather than placental origin, although the placenta is known to secrete very high levels of a lactogenic hormone in other species during pregnancy (Josimovich and MacLaren, 1962). The work of Friesen (1965) and preliminary experiments described here indicate that no cross-reaction exists between sheep placental tissue extracts and antiserum prepared to pituitary prolactin. Further, the levels decrease gradually after parturition and do not give the sudden drop which could be expected after expulsion of the placenta if the hormone was of placental origin. The fall in levels of placental lactogenic hormone in the human occurs within 4 h of parturition (Beck, Parker and Daughaday, 1965).

This evidence does not preclude the possibility, however, that the placenta is secreting high levels of a biologically active mammotrophin with chemical properties which render it immunologically dissimilar to pituitary prolactin.

COMPARISON OF RESULTS OBTAINED BY HAEMAGGLUTINATION INHIBITION WITH THOSE FROM A BIO-ASSAY

Although the subjective assessment of immuno-assay tends to limit the expression of biological variation the results obtained are unexpectedly constant between groups. This may be due to a singularly low level of biological variation in amounts of prolactin in sheep under similar conditions and it was decided to investigate this further. Also it has been demonstrated in other work that the results obtained for levels of hormone in sera by the haemagglutination inhibition reaction are higher than the results obtained by bio-assay on similar material, a difference that would be explained if the antisera reacts with both biologically active hormone and the hormone in inactive form. Hence, after having observed the presence of immunologically active prolactin in sheep blood extracts, it was considered essential to demonstrate the biological activity in a quantitative manner and compare the two. Unfortunately, as stated earlier, there are few reliable sensitive bio-assay methods available for the quantitative estimation of prolactin in blood. The local micro-method of pigeon crop-sac assay of Grosvenor and Turner (1958) is the most sensitive of all the known methods and was chosen in an attempt to correlate biological activity with immunological amounts of prolactin in sheep blood. Blood extracts from non-pregnant sheep and from sheep during the last few weeks of pregnancy were measured by immuno-assay and for biological activity by the pigeon crop-sac assay. Although these samples gave definite differences in epithelial proliferation between groups, with maximum biological activity being demonstrated in the samples from pregnant sheep when scored in a subjective manner from 0–3 (*Figure 2*), quantitative estimation of prolactin by the bio-assay was

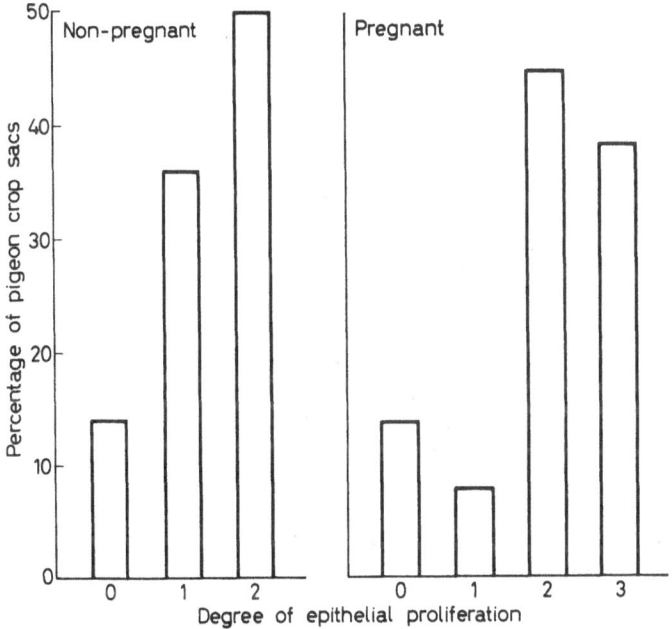

Figure 2. Histological assessment of pigeon crop-sac response to sheep blood extracts

not possible due to the subjective method of assessment and the large variation within all groups of pigeons, variation which was not removed despite uniform selection of birds on the basis of weight and age. However, histological comparison of the response to blood extracts and purified prolactin showed these to be similar and it appears that the proliferative reaction observed in the birds injected with the sheep blood extracts was probably due to the presence of prolactin. Other substances are known which can cause thickening of the epithelium of pigeon crop-sacs, e.g. non-specific factors present in human serum or urine, which are thought to be mucopolysaccharide in nature (Lahr, Bates and Riddle, 1943; Bahn and Bates, 1956). In order to check for non-specificity, sheep liver extracts were taken as controls using the same procedure adopted for extracting prolactin from sheep blood. The results of this experiment demonstrated that no visual or histological thickening of the epithelium of the pigeon crop-sac was observed following the administration of sheep liver extracts. While these results indirectly support the likelihood that the crop-sac stimulating effect of sheep blood extracts was not due to non-specific factors, the final confirmation of this could be obtained only by the administration of blood extracts of hypophysectomized sheep. Such samples were not available at the time of these investigations.

It appears that results obtained by application of the haemagglutination inhibition reaction to the determination of prolactin in sheep blood extracts are encouraging. However, at the present time these results cannot be verified by a direct comparison with a bio-assay on the same samples, the results of which could be subjected to analysis by acceptable statistical techniques.

SUMMARY

The results obtained and reported indicate that ovine prolactin preparations (NIH-P-S4; NIH-P-S5; NIH-P-S6 and NIH-P-S7) were antigenic. Antisera could be produced in the rabbit against these preparations and these were hormone specific but not species specific. The techniques of immuno-diffusion, immuno-electrophoresis and haemagglutination failed to demonstrate the presence of antibodies against other ovine hypophyseal gonadotrophins in the antisera.

The biological inhibition assay of the rabbit anti-ovine prolactin serum demonstrated the presence of specific antibodies against ovine prolactin which neutralized its lactogenic effect in pseudo-pregnant rabbits.

Attempts were made to develop a haemagglutination inhibition reaction for the estimation of prolactin in sheep blood serum. In spite of careful standardization with regard to factors such as dilution of antisera, sheep cells, age of the standard hormone and the diluents used for the antiserum, the technique did not prove reliable for the estimation of prolactin in untreated sheep serum due to the presence of non-specific factors. However, a procedure was developed to extract prolactin from sheep blood in an attempt to remove such factors. The haemagglutination inhibition assay could be successfully used for estimation of the hormone in these sheep blood extracts.

Preliminary studies using this modified technique showed that the hormone was present at low levels in sheep blood irrespective of their physiological state. A seasonal variation in the level was observed in non-pregnant and non-lactating sheep with the lowest levels occurring during anoestrus. The highest levels observed were found during the latter half of pregnancy and these levels decreased shortly after lambing to the levels found in non-pregnant animals. In spite of continued suckling of lambs the levels of prolactin remained low in lactating sheep.

The physiological meaning of the varying levels of the hormone is not understood and will be the subject of future studies.

It should be stressed that whilst these results are encouraging, more study is required to define the optimum conditions for the assay and of the true biological meaning of the measurements made including comparison with an acceptable bio-assay before the method can be put to routine use with complete confidence.

REFERENCES

Arquilla, E. R. and Stavitsky, A. B. (1956a). *J. clin. Invest.* **35,** 458–466
— — (1956b). *J. clin. Invest.* **35,** 467–474
Bahn, R. C. and Bates, R. W. (1956). *J. clin. Endocr. Metab.* **16,** 1333–1346
Beck, P., Parker, M. L. and Daughaday, W. H. (1965). *J. clin. Endocr. Metab.* **25,** 1457–1462
Bischoff, H. W. and Lyons, W. R. (1939). *Endocrinology* **25,** 17–27
Boyden, S. V. (1951). *J. exp. Med.* **93,** 107–120
Bradley, T. R. and Clarke, P. M. (1956). *J. Endocr.* **14,** 28–36
Butt, W. R., Crooke, A. C. and Cunningham, F. J. (1961). *Biochem. J.* **81,** 596–605
Chadwick, A. (1962). *Biochem. J.* **85,** 554–558
— (1963). *J. Endocr.* **27,** 253–263
Dominguez, J. M. and Pearson, O. H. (1962). *J. clin. Endocr. Metab.* **22,** 865–872
Dutt, H. E. and Bush, L. F. (1955). *J. Anim. Sci.* **14,** 885–896
Ehrlich, R. M. and Randle, P. J. (1962). *Ciba Fdn. Colloq. Endocr.* **14,** 117–132

Emmart, C. W., Bates, R. W., Condliffe, P. G. and Turner, W. A. (1963). *Proc. Soc. exp. Biol. Med.* **114,** 754–763
Fraser, R. and Hartog, M. (1962). *Ciba Fdn. Colloq. Endocr.* **14,** 105–114
Friesen, H. (1965). *Endocrinology* **76,** 369–381
Garner, R. J. and Crawley, W. (1958). *J. comp. Path. Ther.* **68,** 112–114
Grosvenor, C. E. and Turner, C. W. (1958). *Endocrinology* **63,** 530–534
Grumbach, M. M. and Kaplan, S. L. (1962). *Ciba Fdn. Colloq. Endocr.* **14,** 63–102
Harkness, M. L. R. and Harkness, R. D. (1956). *J. Physiol., Lond.* **132,** 476–481
Hartog, M. and Fraser, R. (1961). *J. Endocr.* **22,** 101–106
Hayashida, T. (1962). *Ciba Fdn. Colloq. Endocr.* **14,** 338–359
Irie, M. and Barrett, R. J. (1962). *Endocrinology* **71,** 277–287
Josimovich, J. B. and MacLaren, J. A. (1962). *Endocrinology* **71,** 209–220
Kabak, J. M. and Stuloval, O. P. (1937). *Bull. exp. Biol. Med., U.S.S.R.* **4,** 297–299
Lahr, E. L., Bates, R. W. and Riddle, O. (1943). *Endocrinology* **32,** 251–259
Lamming, G. E. (1966). *Nutr. Abstr. Rev.* **36,** 1–13
Levy, R. P. and Sampliner, J. (1961). *Proc. Soc. exp. Biol. Med.* **106,** 214–215
—— (1962). *Proc. Soc. exp. Biol. Med.* **109,** 672–673
Lyons, W. R. (1937). *Proc. Soc. exp. Biol. Med.* **35,** 645–648
— (1942). *Proc. Soc. exp. Biol. Med.* **51,** 308–311
— and Page, E. (1935). *Proc. Soc. exp. Biol. Med.* **32,** 1049–1050
Meites, J. and Turner, C. W. (1947). *Am. J. Physiol.* **150,** 394–399
Ouchterlony, O. (1953). *Acta path. microbiol. Scand.* **32,** 231–240
Read, C. H. and Bryan, G. T. (1960a). *Recent Prog. Horm. Res.* **16,** 187–213
—— (1960b). *Ciba Fdn. Colloq. Endocr.* **13,** 68–84
— Eash, S. A. and Najjar, S. (1962). *Ciba Fdn. Colloq. Endocr.* **14,** 45–61
— and Stone, D. B. (1958). *Am. J. Dis. Child.* **96,** 538
Riddle, O., Bates, R. W. and Dykshorn, S. W. (1933). *Am. J. Physiol.* **105,** 191–216
Robinson, T. J. (1951). *Biol. Rev.* **26,** 121–157
Rowlands, I. W. and Young, F. G. (1939). *J. Physiol. Lond.* **95,** 410–419
Scheidegger, J. J. (1955). *Int. Archs. Allergy appl. Immun.* **7,** 103–110
Strangeways, W. I. (1938). *J. Physiol. Lond.* **93,** 47–48
Sulman, F. G. (1956). *J. clin. Endocr. Metab.* **16,** 755–774
Wide, L. (1962). *Acta. Endocr. Copenh.* Suppl. 70
— and Gemzell, C. A. (1960). *Acta endocr. Copenh.* **35,** 261–267
Yeates, N. T. M. (1949). *J. Agric. Sci., Camb.* **39,** 1–43
Young, F. G. (1938). *Biochem. J.* **32,** 656–664

DISCUSSION

Dr. T. A. Lorraine (*Edinburgh*)

There has been some criticism of haemagglutination inhibition reactions when applied to other hormones particularly in relation to body fluids. I think the comparison of biological and immunological methods of assay is the key point in this particular investigation. I wondered whether Mr. Saji has been able to conduct this comparison in a rather more quantitative manner, for example, by estimating the index of discrimination when these two assays were conducted on the same sample of blood, a procedure similar to that shown by Professor Reichert in relation to his pituitary extracts.

SAJI

Unfortunately the pigeon crop-sac assay is only semi-quantitative so we could only compare results of blood samples from pregnant and non-pregnant sheep. We checked the validity of the haemagglutination inhibition method by repeatedly assaying blood to which known amounts of prolactin had been added. Quantitative measurements suitable for statistical evaluation have not been possible so far with either technique.

PROFESSOR S. J. FOLLEY (*Reading*)

I was interested in Mr. Saji's work on prolactin levels because some years ago Dr. Chadwick did similar work using a haemagglutination inhibition method using goats. He raised an antiserum to NIH-P-S3 in rabbits, using Freund's adjuvant rather than Bentonite, but the response was not as good as yours. Later, Dr. Forsythe found that the guinea-pig was a better animal for making antiserum to sheep prolactin. Chadwick, using his antiserum with titre 1 in 1,000, did some preliminary experiments in goats. He obtained what, at first, seemed to be meaningful results with levels similar to yours, 1 or 2 μg/ml in lactating goats. They were higher in lactating than non-lactating females and higher in the latter than in males. In hypophysectomized goats he could find no prolactin levels at first, but later experiments gave positive results. I wondered if you have done any experiments in hypophysectomized sheep. One thing that surprises me about your results is that the levels of prolactin are so much higher than the currently accepted levels for growth hormone in humans obtained by the radio immuno-assay, which are of the order of 20 mμg/ml. It was these high levels that Chadwick obtained which made us suspect that the haemagglutination inhibition method was not very satisfactory for the purpose. It was about that time that the method came under suspicion and the radio immuno-assay took its place. With regard to the local pigeon crop method, Dr. Forsythe developed a method whereby the stimulated area is weighed and measured so that you do not rely on a subjective measure but take into account both the area and the weight of the stimulated glandular tissue. Perhaps you would find this method more satisfactory. Dr. Forsythe, using Simpkin and Goodhart's technique for extracting prolactin from human blood has not been able to find any activity in women in the *post-partum* period, heavily lactating goats or in cows.

SAJI

I agree that the interpretation of the pigeon crop assay is difficult and needs expressing in a quantitative manner. Regarding the preparation of anti-serum we used several rabbits but in two only was the antiserum suitable for quantitative work.

PROFESSOR A. V. NALBANDOV (*Illinois*)

I would like to draw attention to the fact that in our hands a great many non-specific substances such as denatured plasma, liver extracts and muscle extracts

173

will give beautiful local crop responses which means such substances have to be used as controls. In some instances we have judged the local responses equivalent to 2–3 μg prolactin upon injection of a liver extract. I suggest therefore that in future assays you include a control system of this type.

SAJI

We were aware of this possible non-specific proliferation, hence the additional histological evaluation. We have not observed any non-specific effects so far, but accept that adequate controls are essential.

III. CYCLIC VARIATION OF GONADO-TROPHIN SECRETION AND RELEASE

EXPERIMENTALLY-INDUCED ALTERATIONS IN THE SECRETION OF GONADOTROPHINS IN THE RAT

G. P. VAN REES

Department of Pharmacology, University of Leiden,
The Netherlands

IT IS possible to affect the secretion of the gonadotrophic hormones in many ways. However, in the case of LH and FSH we may make a distinction between two kinds of experimentally-induced alterations in the secretion of these hormones, related to two essentially different regulatory mechanisms. Firstly, there are the mechanisms regulating basal, or tonic, secretion of LH and FSH; and secondly, those which in the female rat cause rhythmic peaks in the secretion of LH, and possibly also of FSH, thus leading to ovulation and the formation of corpora lutea.

Basal secretion is affected, for instance, by the chronic administration of oestrogens and androgens. Rhythmic secretion may be inhibited by the administration of androgens shortly after birth, by lesions in the anterior hypothalamus or by exposing the animals to constant light.

Favouring this assumption of dual control of the secretion of LH and possibly also of FSH is the fact that in circumstances when the central ovulation-inducing stimuli are absent, essentially normal relationships between oestrogen and the secretion of LH and FSH are still operative. It has been shown by Harris and Levine (1965), that when animals, in which no ovulations take place owing to an injection of testosterone shortly after birth, are gonadectomized, castration cells do develop in their pituitary glands and this development can be inhibited by the administration of oestrogen. Also pituitary LH levels in these animals appear to react normally to gonadectomy or to gonadectomy and subsequent administration of oestradiol, as shown in the following experiment (van Rees and Gans, 1966).

Female rats (*Figure 1*) were injected subcutaneously on the fourth day after birth with 2 mg of testosterone propionate (androgen-sterilized rats) or with an equal volume of solvent oil (controls).

From the thirteenth week after birth, daily vaginal smears were taken in order to check the effects of the post-natal injections. On the first day of the fourteenth week two-thirds of both androgen-sterilized rats and controls were gonadectomized, leaving the remaining one-third intact, after which another interval of 4 weeks was

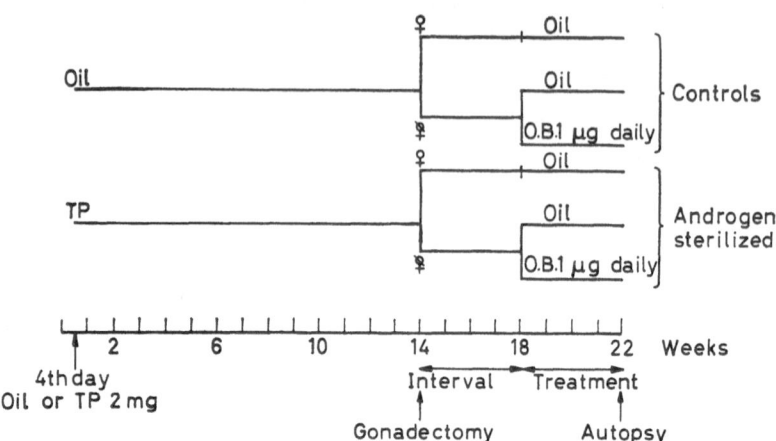

Figure 1. Design of experiment to study the reaction of pituitary LH to gonadectomy or to gonadectomy and subsequent administration of oestradiol in androgen-sterilized female rats. 12–14 animals per group

allowed to elapse. Starting on the eighteenth week after birth, the animals were treated for 4 weeks with daily injections of oil (intact and gonadectomized animals) or with the submaximally active dose of 1 μg of oestradiol benzoate (OB) daily (gonadectomized animals only). Autopsy was carried out 1 day after the last injection, the pituitary glands were removed, homogenized in saline and injected into freshly-hypophysectomized immature male rats for 6 days, starting the day after hypophysectomy. The weight of the ventral prostate on the day after the last injection was then taken as a criterion of the amount of LH administered. The pituitary homogenates were injected at only one dose level, but the six groups of donor rats were assayed simultaneously and no supramaximal amounts of homogenate were administered to the recipients.

The results are given in *Figure 2*, in which the ventral prostate weights of the recipient animals injected with the pituitary homogenates are presented. It can be seen that both in controls and in androgen-sterilized rats, pituitary LH levels increased to the same extent after gonadectomy, the amount of the increase being about fivefold as known from previous quantitative estimations of pituitary LH in intact and gonadectomized female rats. Administration of oestradiol had a similar effect in androgen-sterilized animals and

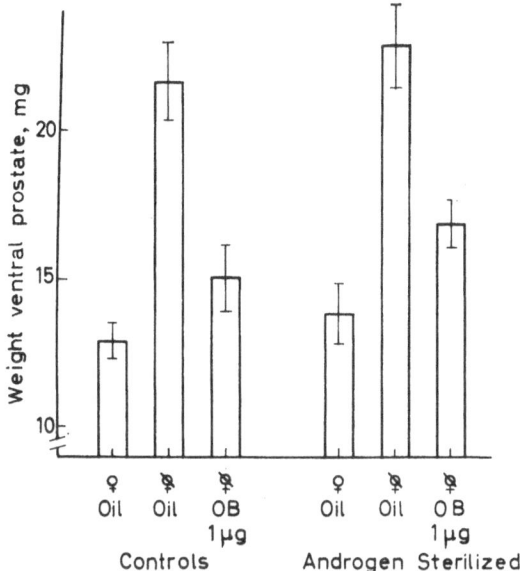

Figure 2. Mean *ventral-prostate weight of hypophysectomized immature rats injected once daily with* 0·5 ml *of pituitary homogenate for* 6 days (*total dose per recipient equivalent to* 0·5 *hypophyses*), *from groups depicted in Figure 1. Bars indicate standard error,* 10–14 *recipient animals per group*

controls. It would appear, therefore, that regarding the effects of oestrogens on basal secretion of LH, no differences exist between androgen-sterilized animals and normal female rats, thus confirming and extending the observations of Harris and Levine cited above.

A similar anovulatory state such as occurs in androgen-sterilized females is also seen when lesions are made in the anterior hypothalamus of normal rats. Therefore the question arises whether in these animals as well, an essentially normal regulation of basal secretion of gonadotrophins is operative. The available data are less clear in this case: Taleisnik and McCann (1961) observed that after gonadectomy in such rats pituitary LH rose, although blood LH levels failed to increase significantly in gonadectomized-lesioned animals. Hence they concluded that the lesions specifically blocked the release of LH by the pituitary gland. However, Flerkó and Bárdos (1961a, 1961b) found that removal of half or more of the ovarian tissue in anovulatory rats bearing anterior hypothalamic lesions resulted in the formation of corpora lutea in the remaining ovarian fragments. Removal of both ovaries and re-implantation of one half of an ovary into the spleen also resulted in luteinization of the transplants. Moreover, administration of oestrogen to such animals prevented the formation of corpora lutea. Therefore, it is believed that in animals rendered anovulatory by anterior hypo-thalamic lesions, a decrease of circulating levels of oestrogen leads to an increased secretion of gonadotrophins by the pituitary gland.

A final argument in favour of normal basal secretion of FSH and LH is given by the histological picture of the ovaries, since in both cases the ovaries contain an abundance of mature follicles and only the corpora lutea are absent, thus resembling ovaries transplanted into castrated adult male rats.

Opposed to the selective inhibition of central ovulation-inducing processes is the experimentally induced decrease of basal secretion of FSH and LH. In these circumstances ovulations do not occur either, but the anovulatory state could be due to several circum-stances. Firstly, lowered blood levels of gonadotrophins will result in a retardation of follicular maturation in the ovary, leading to a low sensitivity of the ovary to LH with regard to ovulation and corpus luteum formation. A second possibility is that the pituitary gland is rendered insensitive to impulses of hypothalamic origin and thirdly we have also to take into account the possibility that the changes in the hypothalamus have taken place blocking or inhibiting the central stimuli reaching the pituitary gland.

These different possibilities of blocking ovulation pertain especi-ally to the anovulatory state seen during pseudopregnancy (psp) in the rat or to that resulting from the administration of progesterone. In what follows, some experiments will be described which were carried out in order to investigate the mechanisms by which ovulation is inhibited during pseudopregnancy.

It is well known that during the normal oestrous cycle in the rat no evident progestational phase is present; the corpora lutea formed are relatively small and there are no pronounced progestational changes in the uterus or vagina. Such a progestational state, or pseudopregnancy, however, can be induced readily in the rat, for instance by mating with a vasectomized male, by electrical or mechanical stimulation of the cervix uteri, or by pituitary grafts. All these stimuli are thought to result in increased blood levels of luteotrophic hormone, leading to activation of the corpora lutea present in the ovaries and hence to an increased secretion of progesterone. Progesterone is then held responsible for the absence of ovulations during pseudopregnancy by interfering with the secretion of gonadotrophins by the anterior pituitary gland.

Subsequently we will discuss (1) the sensitivity of the ovary to ovulation-inducing amounts of gonadotrophic hormones, (2) fluctuations in the pituitary levels of FSH and LH, and (3) the effects of electrical or chemical stimulation of the anterior hypothalamus on ovulation in pseudopregnant animals.

All animals used were adult female rats with a bodyweight of 170–230 g. They were kept under standardized lighting conditions, i.e. the animal room was illuminated artificially from 5 a.m. until 7 p.m. Only animals, which had shown at least two consecutive regular vaginal cycles of 4 or 5 days' duration immediately before use, were selected. Pseudopregnancy was induced by electrical stimulation of the uterine cervix on the day of vaginal oestrus resulting in about 80 per cent of cases with an average duration of 12 days, as judged by daily vaginal smears. Pseudopregnancy is then followed either by a day of vaginal pro-oestrus, followed by a day of oestrus, or by a change overnight from dioestrus into oestrus: of 76 animals 26 per cent showed no pro-oestrus at the end of pseudopregnancy. Nevertheless, all animals which showed vaginal oestrus at the end of pseudopregnancy had ovulated the previous night as judged by the presence of a normal number of ova in the oviduct.

Let us first consider the sensitivity of the ovary during pseudopregnancy.

It is known that during pseudopregnancy follicles in various stages of development may be seen in the ovary, but it seemed that a certain retardation of follicular development during pseudopregnancy might be difficult to detect by simple histological examination of the ovaries. Since, however, it was thought that differences in follicular development would be reflected in the sensitivity of the ovary to LH-induced ovulation, it was decided to assess quantitatively the ED 50 of Human Chorionic Gonadotrophin (HCG), i.e. that dose which is effective and brings about ovulation in 50 per cent of the animals. Since differences in the amounts of circulating gonadotrophins of pituitary origin at the time of injection of HCG might affect the amount of HCG necessary to obtain ovulation, it was decided to hypophysectomize all animals 2 h before the intravenous injection of HCG. In view of the half-life of rat LH in the blood of the rat, which is about 15 min (Parlow and Ward, 1961) this interval was judged to be sufficient to mask any pre-existing differences in blood LH levels between the experimental groups. Thus, all animals were hypophysectomized between 12.00 noon and 12.30 p.m. and HCG was injected into the jugular vein between 2.00 and 2.30 p.m. The next morning the animals were killed and the presence of ova in the oviducts was checked. ED 50's were determined by means of a sequential design (Dixon and Mood, 1948; Brownlee, Hodges and Rosenblatt, 1953), which means that each dose of HCG injected into each single animal was dependent upon the result obtained with the previous one injected the day before, thus allowing a reasonably accurate calculation of the ED 50 using a relatively small number of animals.

The complete experiment was carried out with the following groups of animals. Normal animals taken on the first or second day of dioestrus, pro-oestrus or oestrus served as controls; the experimental groups consisted of animals taken on the fourth or seventh day of pseudopregnancy and on the day of pro-oestrus immediately following pseudopregnancy.

Firstly, from Table 1 it can be seen that during the normal cycle

Table 1

Sensitivity of the ovary to HCG in cyclic and pseudopregnant (PsP) rats

Groups used (15 animals/group)	HCG* ED 50 I.U./100 g body weight	Mean number of ova ± S.E.
Dioestrus I	> 32	
Dioestrus II	1·73 (1·35–2·20)	10·7 ± 2·1
Pro-oestrus	0·43 (0·34–0·55)	9·2 ± 1·6
Oestrus	—	9·8 ± 2·8
4th day PsP	—	4·2 ± 1·0
7th day PsP	4·80 (3·64–6·32)	3·7 ± 0·9
Pro-oestrus after PsP	2·83 (1·71–4·68)	8·6 ± 1·2

* Figures in parentheses give 95 per cent confidence limits.

the sensitivity of the ovary fluctuates markedly. Thus in animals taken on the first day of dioestrus, even 32 I.U. of HCG per 100 g bodyweight failed to induce a single ovulation. On the second day of dioestrus sensitivity was much higher, reaching a maximum one day later, during pro-oestrus, when the ED 50 was at least 80 times less than on the first day of dioestrus. This increase in sensitivity of the ovary during the normal cycle probably represents progressive follicular maturation. No ED 50 is presented at oestrus because the sensitivity of the ovary varied so widely during the experiment that calculation of the ED 50 was not possible. Our explanation is that at oestrus some animals still possessed ovaries with large follicles which just escaped ovulation the previous night and thus would need very little HCG to ovulate, while in others either such follicles were absent because of a previous 'complete' ovulation, or because these had already become atretic.

Also on the fourth day of pseudopregnancy a wide variation was found, possibly because some ovaries were still at a stage resembling the first day of dioestrus of the normal cycle (very low sensitivity) and some had already undergone further development to the stage of the seventh day of pseudopregnancy. On the seventh day of pseudopregnancy sensitivity of the ovary was still remarkably low, which in our opinion reflects a retarded follicular maturation during pseudopregnancy in comparison to that which occurs during the normal cycle, thus leading to the conclusion that during pseudopregnancy basal secretion of FSH and possibly also of LH is inhibited to some extent.

Most remarkable was the finding that on the day of pro-oestrus at the end of pseudopregnancy the sensitivity of the ovary, although somewhat higher than on the seventh day, was still remarkably low, and even lower than that found on the second day of normal dioestrus; yet, these post-pseudopregnant pro-oestrous animals would have ovulated spontaneously the following night. Apparently the peak in blood LH levels causing ovulation at the end of pseudopregnancy is large enough to cause these insensitive ovaries to ovulate.

That follicular maturation is only retarded but still continues during pseudopregnancy is indicated by the fact that the sensitivity of the ovary was somewhat higher at the post-pseudopregnant pro-oestrus than on the seventh day, and by the number of ova counted in the cases which ovulated in response to injected HCG: it was low at the fourth and seventh day of pseudopregnancy, but normal at pro-oestrus following pseudopregnancy. Thus, if the findings result from a decreased basal secretion of gonadotrophins during pseudo-pregnancy, this must be of a limited nature.

Since it is generally assumed that the anovulatory state during pseudopregnancy results from endogenous progesterone, it would be interesting to know whether the administration of exogenous pro-gesterone to non-pseudopregnant females would result in a similar decrease of the sensitivity of the ovary. However, this experiment has not yet been carried out. Also in progesterone-treated non-pseudopregnant animals Graafian follicles may be seen in the ovaries as is the case with pseudopregnant animals, but still this does not mean that a certain retardation of follicular development may not occur.

The next series of experiments deals with fluctuations of pituitary levels of FSH and LH during pseudopregnancy, and, in one instance, with serum FSH concentration. In the first of these the following groups of animals were investigated: normal oestrus; fourth, seventh and twelfth day of pseudopregnancy and oestrus following pseudo-pregnancy. Pituitary LH estimations were carried out using the ovarian ascorbic acid depletion technique as described by Parlow (1958), and pituitary and serum FSH were estimated with the testis weight augmentation method (Paesi, de Jough, Hoogstra and Engebregt, 1955), in which testis growth is measured in hypophy-sectomized immature rats treated simultaneously with an overdose of HCG. Serum LH estimations are not presented here because serum effects obtained with the ovarian ascorbic acid depletion tech-nique could not be ascribed unequivocally to LH (van Rees and de Groot, 1965).

The results are presented in *Figure 3*. It can be seen that pituitary LH levels during pseudopregnancy were higher than those either at control oestrus or oestrus after pseudopregnancy, a result which agrees with those of Schwartz and Rothchild (1964). Pituitary FSH levels showed the same pattern. As in this case no quantitative estimations were carried out but the assay animals were injected with one dose of pituitary homogenate, the ratio between pituitary FSH levels on control oestrus and the seventh day of pseudo-pregnancy was estimated using a four-point design, and proved to be of the same order of magnitude as found for the pituitary LH levels in these two groups. The results obtained with animals on the twelfth day of pseudopregnancy will not be discussed: this group is heterogeneous since some of the animals may have been on the last day of pseudopregnancy whereas others were not. Moreover, all animals were killed in the afternoon (i.e. between 4.30

and 5.30 p.m.), which means that some of the animals were taken at a time, only a few hours before ovulation would have occurred.

The estimations of serum FSH showed that during all stages of pseudopregnancy some activity was present. It can be seen that no

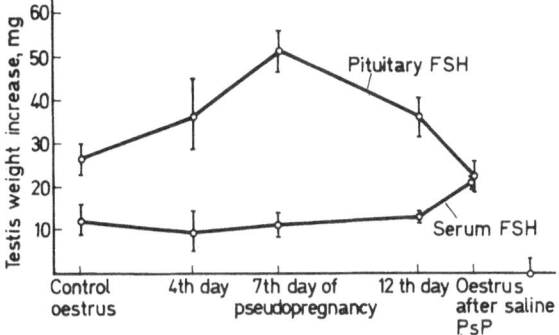

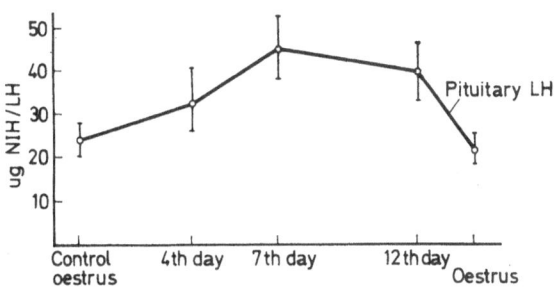

Figure 3. Pituitary LH and FSH content and serum FSH activity during pseudopregnancy. Bars indicate standard error of the mean response (FSH) or potency (LH), expressed as μg-equivalents of NIH-LH-S3 (From van Rees and de Groot, 1965, by courtesy of the Editor, *Acta Endocrinology*)

differences were apparent between sera from control oestrous animals or from pseudopregnant rats. On the basis of the decreased sensitivity of the ovary during pseudopregnancy one would have expected somewhat lowered blood levels during pseudopregnancy, but it must be stressed that at these barely detectable levels limited decreases are difficult to assess. Moreover, we do not know as yet whether serum FSH at normal oestrus is low when compared to the other stages of the oestrous cycle; if this were the case the results could be caused by a sustained relatively low secretion rate of FSH during pseudopregnancy, in contrast to the normal cycle. Surprising was the high serum FSH activity present in serum from animals at oestrus immediately following pseudopregnancy. It was thought that this might reflect an overshoot of FSH secretion brought about by the withdrawal of endogenous progesterone at the end of pseudopregnancy.

The changes in pituitary gonadotrophin levels were further studied in the following experiments (van Rees, to be published). Firstly, we were interested in the question of whether the relatively elevated levels of pituitary FSH and LH were still high on the morning of pro-oestrus following pseudopregnancy, and to this end these levels were compared in the following two groups: (a) animals sacrificed in the afternoon of the seventh day of pseudopregnancy (4.30–5.30) and (b) in the morning of pro-oestrus after pseudopregnancy (11.00–12.00).

Pituitary FSH levels were estimated using the testis weight augmentation technique already mentioned, but pituitary LH was

Table 2

Pituitary FSH during and following pseudopregnancy (PsP)

Status of donor	Dose administered to each recipient	Testis weight increase (mg) $\pm$ S.E.
7th day PsP (p.m.)	0·25 pit. eq.*	26·5 $\pm$ 4·1 (7)
Pro-oestrus after PsP (a.m.)	0·25 pit. eq.*	25·1 $\pm$ 5·2 (7)

* + 120 I.U. HCG.
Figures in parentheses indicate number of donors.

now studied using the ventral prostate weight method, after it had been ascertained that with this method the same quantitative difference was found between pituitary LH levels on the seventh day of pseudopregnancy and control oestrus as with the ovarian ascorbic acid depletion technique. The results of pituitary FSH estimations, carried out semiquantitatively at one dose level are presented in Table 2 and those relating to pituitary LH in Table 3. In both cases no differences were seen between these two groups, which indicates that at least on the morning of the day before the first ovulation these levels are still relatively elevated.

This experiment was extended by another in which possible

fluctuations of pituitary FSH and LH content in relation to ovulation were compared in normal cyclic females and in animals that had been pseudopregnant immediately before.

Thus, the following six groups were compared:

Normal cyclic females morning of pro-oestrus (11.00–12.00)

afternoon of pro-oestrus (4.30–5.30)

morning of oestrus (11.00–12.00)

Post-pseudopregnant females morning of pro-oestrus (11.00–12.00)

afternoon of pro-oestrus (4.30–5.30)

morning of oestrus (11.00–12.00)

Table 3

Pituitary LH during and following pseudopregnancy (PsP)

Status of donor	Dose administered to each recipient	Ventral prostate weight (mg) $\pm$ S.E.
7th day PsP (p.m.)	0·5 pit. eq.	$17·6 \pm 1·5$ (8)
Pro-oestrus after PsP (a.m.)	0·5 pit. eq.	$16·4 \pm 1·0$ (11)

Figures in parentheses indicate number of donors.

The results of pituitary FSH estimations (carried out at one dose level in which the six experimental groups were compared simultaneously) are given on *Figure 4*. It can be seen that during normal pro-oestrus a temporary decrease of pituitary FSH levels occurred. This result is in agreement with previous findings (Gans, de Jongh, van Rees, van der Werff ten Bosch and Wolthuis, 1964), although it should be mentioned that we have not always been able to find this decrease during normal pro-oestrus. On the morning of post-pseudopregnant pro-oestrus, pituitary FSH levels were still elevated, in agreement with the result obtained in the previous series. However, during pro-oestrus a marked drop of pituitary FSH content was observed, which continued until oestrus. At post-pseudopregnant oestrus, levels were equal to those at normal oestrus, a result which had already been obtained in the previous experiments.

187

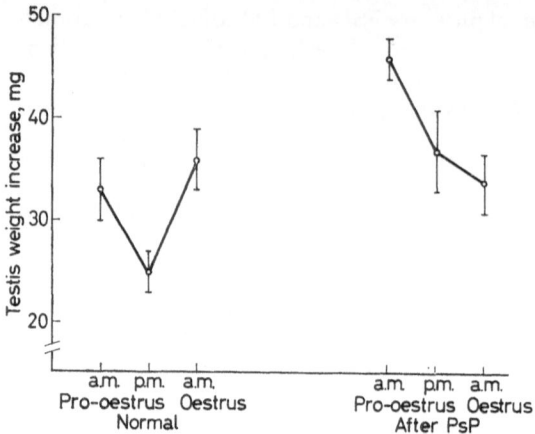

Figure 4. Pituitary FSH of pro-oestrous and oestrous animals taken either during the normal cycle or following pseudopregnancy. Given are the mean responses obtained in recipient animals (total dose per recipient equivalent to 0·5 hypophyses); bars indicate standard errors. Number of animals 10–15 per group

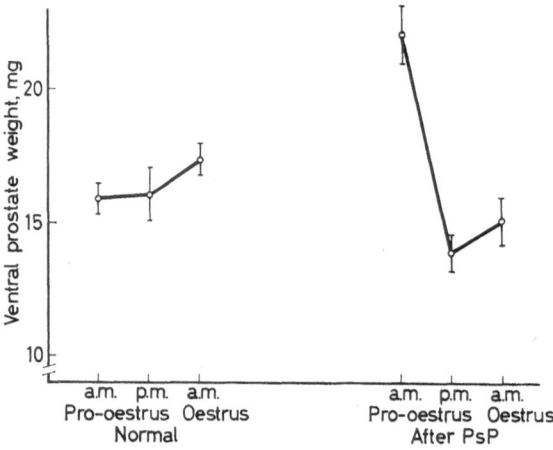

Figure 5. Pituitary LH. Description as for Figure 4

Pituitary LH estimations were again carried out at one dose level, the results of which are presented on *Figure 5*. During normal pro-oestrus no significant fluctuations were found. This is not in agreement with other reports. Thus Schwartz and Bartosik (1962), found maximal LH levels in the pituitary gland on the morning of pro-oestrus, which then declined until minimum levels were found at oestrus. However, in a previous experiment carried out in our laboratory (Gans and de Jongh, 1959) maximal levels in the pituitary gland were found on the last day of dioestrus, declining thereafter until the morning of pro-oestrus. At the same time, serum LH levels were observed to increase progressively from the afternoon of the last day of dioestrus onwards until maximal serum LH was found in the afternoon of pro-oestrus (Swelheim, 1965), which is in agreement with other reports (e.g. Ramirez and McCann, 1964). Therefore, it has been concluded that ovulation-inducing peaks of blood LH are caused both by increased synthesis and release of LH by the pituitary gland. Thus, a small shift in the balance between increases of synthesis and release of LH in different strains of animals might result in different fluctuations of pituitary LH content as found by various authors.

On the morning of post-pseudopregnant pro-oestrus pituitary LH levels were relatively high, as would be expected. During the day, however, such a marked decrease occurred that on the afternoon even subnormal levels were found. We have not yet carried out serum LH and FSH estimations but it seems probable that this acute marked drop of pituitary gonadotrophin levels results from a sudden release of these hormones. It is quite possible that pre-ovulatory peaks in blood-LH and FSH at post-pseudopregnant pro-oestrus are higher than during normal pro-oestrus.

These findings agree with those obtained when the effect of exogenous progesterone in non-pseudopregnant females was studied. Progesterone caused an increase both of pituitary FSH (van Rees, 1959) and LH (van Rees and de Groot, 1965; Hoffman and Schwartz, 1965). In this case it may be relevant that even 1 mg of progesterone daily is able to cause an elevation of pituitary FSH and LH content, although these animals continued to show normal ovarian cycles, suggesting that the block of ovulation and increase of pituitary FSH content were not directly related. However, an inhibition of release of these hormones is probably the cause of the relatively high pituitary gonadotrophin levels. In some instances, regardless of the inhibiting effect of progesterone on release, stimuli coming from the central nervous system should still be able to cause the release of ovulation-inducing amounts of gonadotrophins.

The last series of experiments to be discussed concerns the sensitivity of the hypothalamo-pituitary system. We investigated whether stimuli known to induce secretion of ovulation-inducing amounts of gonadotrophic hormones in normal rats, with spontaneous ovulations blocked by the appropriately-timed injection of pentobarbital (Everett and Sawyer, 1950) would have the same effect in pseudopregnant rats or in rats during post-pseudopregnant pro-oestrus.

Table 4

Effect of preoptic stimulatory lesions on ovulation in pro-oestrous and pseudo-pregnant (PsP) rats

Treatment	With ova	Without ova
Normal Pro-oestrous+		
Preoptic lesion	16	0
Med. Em. lesion	1	0
Lesion elsewhere*	0	6
Sham lesion	2	12
7th day PsP		
Preoptic lesion	1	12
Pro-oestrous after PsP		
Preoptic lesion	11	3
Sham lesion	3	10

* In posterior hypothalamus.

First it was found that the ovulation which occurs at the end of pseudopregnancy could be blocked by the intra-peritoneal injection of pentobarbital (30 mg/kg) shortly before 2.00 p.m., as is the case in normal pro-oestrous females. Then the procedure performed by Everett and Radford (1961) was followed. They showed that in normal pro-oestrous females in which spontaneous ovulations were blocked by pentobarbital, ovulations could be induced again by preoptic lesions made with an iron electrode and direct current, due to the deposit of stimulating amounts of iron from the electrode. The results of our experiment are shown in Table 4. Lesions were made using a unipolar stainless steel electrode and a direct current of 2 mA was passed for 30 sec. Sham lesions involved insertion of the electrode after injection of pentobarbital and withdrawal without a current being passed. It can be seen that, contrary to normal animals only one animal out of 13 could be induced to ovulate during pseudopregnancy, a result which agrees with that obtained

190

by Moll and Zeilmaker (personal communication) who also stimulated on other days during pseudopregnancy. Since we showed that during pseudopregnancy the sensitivity of the ovary was low, it might be supposed that in pseudopregnant animals the lesions induced secretion of subliminal amounts of gonadotrophins. However, at the post-pseudopregnant pro-oestrus the ovarian sensitivity is likewise low, yet ovulation was regularly induced in these animals. To test this hypothesis the following experiment was carried out. It was argued that when a lesion induced the pituitary gland to secrete an amount of LH, which was not sufficient to cause ovulation owing to low sensitivity of the ovary, less exogenous gonadotrophin would be necessary to induce ovulations in lesioned animals than when no lesions were placed. Thus in animals on the seventh day of pseudopregnancy, lesions were made as well as sham lesions. In both cases the ED 50 of HCG was determined and no difference was found. Thus it appears that during pseudopregnancy the stimulation of preoptic structures is unable to cause a significant secretion of gonadotrophins by the pituitary gland.*

This could be brought about either because the stimuli are unable to reach the pituitary gland or because the pituitary gland had become insensitive. Hence the next step will be to study whether hypothalamic extracts which are able to stimulate the pituitary directly, bring about ovulations in pseudopregnant animals.

In conclusion, the results may be summarized as follows. During pseudopregnancy the basal secretion of FSH and LH is decreased to a limited extent, since the ability of the ovary to ovulate is diminished. This results from an inhibited release of FSH and LH by the pituitary gland, causing pituitary levels to increase. At the same time the application of strong hypothalamic stimuli is unable to cause secretion of ovulation-inducing amounts of gonadotrophins. At the end of pseudopregnancy, i.e. on the morning of pro-oestrus, pituitary levels of FSH and LH are still elevated and the sensitivity of the ovary is still decreased, indicating that basal secretion of these hormones is still low. However, on this day spontaneous ovulations are initiated and when these have been blocked by the administration of pentobarbital, ovulations can be induced again by stimulation of the preoptic hypothalamus. Thus the results indicate that the

* *Note added in proof.* This conclusion should be corrected since in later experiments (van Rees, to be published) it was found that in pseudopregnant animals sham-lesions already caused a discharge of LH, of the same magnitude as that brought about by lesions. Thus, the absence of effects of lesions (and sham-lesions) during pseudopregnancy seems to be a primary result of the low sensitivity of the ovaries.

decrease of basal secretion of FSH and LH during pseudopregnancy may not be directly responsible for the absence of ovulations.

These effects have been ascribed to progesterone secreted during pseudopregnancy. This would mean that progesterone, when present, has at least two chronic effects, one causing a certain inhibition of release of FSH and LH from the pituitary gland, the other, having a higher threshold, blocking the effect of central ovulation-inducing stimuli. The divergence of basal secretion and ovulation seen during pro-oestrus following pseudopregnancy could thus result from the gradually decreasing blood levels of progesterone at the end of pseudopregnancy.

REFERENCES

Brownlee, K. A., Hodges, J. L. and Rosenblatt, M. (1953). *J. Am. statist. Ass.* **48,** 262

Dixon, W. J. and Mood, A. M. (1948). *J. Am. statist. Ass.* **43,** 109

Everett, J. W. and Sawyer, C. H. (1950). *Endocrinology* **47,** 198

— and Radford, H. M. (1961). *Proc. Soc. exp. Biol. Med.* **108,** 604

Flerkó, B. and Bárdos, V. (1961a). *Acta endocr. Copenh.* **36,** 180

— — (1961b). *Acta endocr. Copenh.* **37,** 418

Gans, E. and de Jongh, S. E. (1959). *Acta physiol. pharmac. néerl.* **8,** 447

— — van Rees, G. P., van der Werff ten Bosch, J. J. and Wolthuis, O. L. (1964). *Acta endocr. Copenh.* **45,** 335

Harris, G. W. and Levine, S. (1965). *J. Physiol., Lond.* **181,** 379

Hoffmann, J. C. and Schwartz, N. B. (1965). *Endocrinology* **76,** 626

Paesi, F. J. A., de Jongh, S. E., Hoogstra, M. J. and Engebregt, A. (1955). *Acta endocr. Copenh.* **19,** 49

Parlow, A. F. (1958). *Fedn Proc. Fedn Am. Socs exp. Biol.* **17,** 402

— and Ward, D. N. (1961). In: *Human Pituitary Gonadotrophins,* p. 204. Ed. by A. Albert. Springfield; C. Thomas

Ramirez, V. D. and McCann, S. M. (1964). *Endocrinology* **74,** 814

van Rees, G. P. (1959). *Acta physiol. pharmac. néerl.* **8,** 195

— and de Groot, C. A. (1965). *Acta endocr. Copenh.* **49,** 370

— and Gans, E. (1966). *Acta endocr. Copenh.* In press.

Schwartz, N. B. and Bartosik, D. (1962). *Endocrinology* **71,** 756

— and Rothchild, I. (1964). *Proc. Soc. exp. Biol. Med.* **116,** 107

Swelheim, T. (1965). *Acta endocr. Copenh.* **49,** 239

Taleisnik, S. and McCann, S. M. (1961). *Endocrinology* **68,** 263

DISCUSSION

I. ROTHCHILD (*Cleveland*)

This is a beautiful piece of work and I think it is the first to indicate the possibility that progesterone might have a direct effect on the pituitary in inhibiting ovulation. The closest approach to this is the paper by Kanematsu and Sawyer, who also used progesterone in the median eminence but it did not inhibit coitus-induced ovulation in the rabbit although nonethisterone did. There are so many indications of effects of progesterone on the CNS it would appear that the effect on ovulation-inhibition is via the CNS but we have never ruled out the possibility of a direct effect.

It is possible that the LH-RF or the hypothalamic extract may not be entirely from the median eminence. It may involve substances produced elsewhere which might work through the median eminence. Now I know McCann has shown that it will work in animals with lesions in the median eminence but I still think there may be objections to this evidence.

One point should be made with regard to other species since many have worked with the rat. Over 30 years' ago Dempsey showed that the removal of corpora lutea in the guinea-pig had no effect whatsoever on the rate of growth of the Graafian follicles in the ovary. Ovulation occurred earlier than expected, i.e. at about 11 days after the previous ovulation, rather than 16, and from smaller follicles than in a normal cycle. This would suggest that your data on the sensitivity of the follicle in the rat does not fit the picture in the guinea-pig because in this species progesterone has no effect on the growth of the follicle. This makes me wonder what you mean by sensitivity? What aspect of growth or maturation or hormone secretion does this involve? Facilitation of ovulation is a frequently-observed phenomena in many animals, including the rat, during the period following pseudopregnancy or equivalent conditions. In the constant oestrous-anovulatory type of rat, such as are induced by constant light or by anterior hypothalamic lesions, when a pseudopregnancy is induced in these animals they tend to ovulate spontaneously for at least several cycles afterwards before going back into the constant oestrous type of behaviour. Therefore progesterone is doing something, either to the ovary directly or to the pituitary directly, or to the CNS which is then expressed as facilitation of the ovulatory process. I do not know how this evidence compares with the data you show where it is very clear that there is a change in the ability to ovulate in response to an ovulation-inducing hormone.

VAN REES

You said my work indicated that progesterone inhibited the hypophysis directly or made it insensitive, though I did not claim to have shown this. All I did show is an indication that the pituitary gland has become insensitive during pseudo-pregnancy. I do not know whether or not this is a direct action. Concerning the insensitivity of the ovary, the point I wanted to make was that the absence of ovulation and the decreased rate of secretion due to pseudopregnancy are two different things. The fact that progesterone may decrease the secretion of FSH and possibly also LH, has nothing to do with its effects on inhibition of ovulation. The effect on the guinea-pig may be concerned with a limiting effect on ovulation but an absence of effect on basal secretion would explain the condition you described.

MR. WILLIAMS (*Imperial Cancer Research Fund*)

Two points of Van Rees' paper worry me. The first concerns hypophysectomizing test rats 2 h before carrying out a sensitivity test. I believe that when you take out the pituitary of the immature rat you cause a release of gonadotrophin. You can show this by weighing the ovaries at intervals after the operation and there is a slight increase on the following day, and there is also a slight increase in response to gonadotrophins. So, although gonadotrophin is stated to have a 15 min half-life, I am not convinced of this in relation to the responsiveness of the tissues concerned.

7+

Secondly, I would like to reinforce Dr. Lorraine's remarks about 'One Point' assays which I believe Dr. Van Rees has used. I thought we had outlived these techniques years ago and I am sorry to see them coming back.

VAN REES

In reply to the question about hypophysectomy, we were afraid to leave the pituitary glands in, because they could have affected the results. In comparing the sensitivity of the ovary, basic levels of LH could have differed and that would have affected the ED 50 of HCG. If there is already a lot of pituitary LH in the blood, then you will need only a small amount of LH to obtain ovulation. We will probably be able to repeat this experiment without hypophysectomy, since it appeared in a subsequent experiment on ED 50 of HCG that it made no difference whether we used a normal pro-oestrous animal which has been hypophysectomized, or a pro-oestrous animal in which the pituitary gland had been blocked by Nembutal. We did not know this at the beginning of the work and so I shall repeat some of these experiments using Nembutal-blocked animals.

Concerning the semiquantitative assays, I feel rather strongly about this and think that the use of quantitative assays is now more or less a question of fashion. It is not always necessary to use them and it can involve a waste of material when you know the exact amount of material to give to recipient animals. If you are not interested in exact quantitative differences, then I think there is nothing against using a single dose level. You will get results much faster and although you cannot give a figure to the difference between pituitary homogenates you can say that differences are significant.

GONADOTROPHIN SECRETION IN RELATION TO OESTRUS AND TO OVULATION

H. A. ROBERTSON

*Division of Agricultural Biochemistry, Department of Biological Chemistry, University of Aberdeen**

TIMING OF OVULATION

ONE OF the main prerequisites of any detailed study of the inter-related, mutually dependent factors controlling ovulation, is an accurate model for the timing of the various sequential events that occur. Further, in selecting a suitable female mammal to study, it is most desirable that she should have some readily observable or detectable, physiological event from which these sequential events can be accurately timed. In induced ovulators the time of coitus can be used as a marker for timing such events as ovulation, but in spontaneous ovulators the accurate timing of ovulation is sometimes very difficult, primates are a case in point. In some animals such as the rat, the timing of ovulation can be standardized by keeping the experimental animal in a controlled environment of which the dura-tion of light and of darkness are probably the most important single factors (Everett, Sawyer and Markee, 1949). Under such con-ditions the timing of certain of the immediate pre-ovulatory events in the rat can be predicted with some precision. One of the events which might merit more careful study in the rat is the more accurate timing of when copulation first occurs under the standard conditions of 14 h daylight and 10 h darkness.

The rat has, however, certain drawbacks, its normal oestrous cycle has a fairly condensed time scale and the amount of tissue and body fluids available are strictly limited, thus imposing severe restrictions on the possibility of studying the action of purified preparations on the same species of animal from which they have been derived.

Since farm animals are readily available and unlimited quantities of endocrine material are obtainable it would seem that they should warrant consideration as animals suitable for research in the basic aspects of reproductive physiology. The sheep has much to com-mend it as an experimental animal for such studies, particularly

* Present address: Animal Research Institute, Canada Department of Agri-culture, Ottawa, Canada.

when one considers the amount of background information relating to its reproductive cycle which has accumulated during the past 60 years.

THE ONSET AND DURATION
OF THE BREEDING SEASON

The early work on the onset and duration of the breeding season, on the onset and duration of oestrus and on the length of the oestrous cycle of the sheep has been summarized by McKenzie and Terrill (1937). Some of the main features of the oestrous cycle of the sheep are summarized below.

Marshall (1903) gave as the only reliable external indication of oestrus or approaching oestrus the behaviour of the ewe towards the ram. In our experience the behaviour of the ram towards the ewe is generally a more reliable index of impending oestrus than the behaviour of the ewe herself.

There have been varying reports as to the abruptness or otherwise of the beginning of oestrus, our own observations extend over many years and includes a continuous watch on a flock of 40 animals for 6 weeks. It would seem that the length of the period prior to active sexual receptivity during which approaching oestrus can be predicted, varies from one animal to another. In a few cases the ewe appears to be more actively interested in the ram than the ram is in her but generally it is the ram who constantly seeks out the ewe but at this stage she will not stand to be mounted. In this communication the term onset of oestrus will frequently be used. This is arbitrarily defined as the time at which the ewe will first stand to be mounted and is synonymous with the onset of sexual receptivity. This is a remarkably abrupt phenomenon and led us to believe that the timing of this is very precise in relation to subsequent events. Recent experiments have confirmed this view (Robertson and Rakha, 1965a, 1966).

The time of onset of oestrus in relation to the time of day has received little attention in relation to sheep although in some species there is well documented evidence for the onset of oestrus occurring at some particular time of the day. In the early reports on the sheep, by Quinlan, Maré and Roux (1932), Cole and Miller (1935), no evidence was presented to indicate that the onset of oestrus was associated with any particular time of the day. McKenzie and Terrill (1937) stated that a larger number of heat periods began between midnight and noon than between noon and midnight. Kelley (1937) thought that there was a tendency for ewes to come into oestrus more frequently during the day than at night. Hutchinson, O'Connor and Robertson (1964) found that during the months of October, November and December a high proportion of animals (Welsh Mountain) were mated between 06.30 and 07.30 h and this was attributed to an increase in the activity of the ram or to an increased number of ewes coming on oestrus during this time. In the above reports the exact conditions of observation and frequency of observation are not always clearly stated and so considerable errors may have arisen.

Perhaps the first detailed study was carried out by Robertson and Rakha (1965b) who kept a continuous watch on a flock of 40 Cheviot (North Country) ewes from 12th October to 28th November. Once an animal had accepted mating she was removed in order that the ram would not overlook the onset of sexual receptivity of another animal. The behavioural pattern of the ram was observed and it was noticed that he would go round the paddock in search of animals coming in oestrus every 20 to 40 min during the night. The frequency of this search increased when an animal was approaching oestrus and the authors considered that in the experiment, any delay in mating a receptive ewe which might be attributable to the ram was less than 10 min. The onset of oestrus was observed in 56 cases. The results indicated that at the beginning of the breeding season (mid October to the end of November) the time of onset of sexual receptivity is not uniformly distributed throughout the day. The time of onset of activity appeared to be bimodal with equal numbers coming into oestrus during the period 04.00 to 11.00 h and 13.00 to 22.00 h G.M.T. The periods of activity at the time of observation appeared to be centred around the mean time of sunrise and sunset (07.04 and 16.23 h G.M.T., respectively). It appeared that later in the breeding season, i.e. late in December, January and February, the time of onset of oestrus became more uniformly distributed throughout the 24 h.

The distribution of the time of day of onset of sexual receptivity may be a function of the period of the breeding season during which the observations are made. It is possible that the first ovulation of the breeding season, which is not accompanied by oestrus, occurs at a fixed time of day relative to the hours of light and hours of darkness at that time. The time of successive ovulations may then be determined by variations in the life span of the corpus luteum and, as a consequence, as the breeding season progresses the time of onset of oestrus tends to become more randomly distributed throughout the 24 h.

From a practical point of view it is easier to determine the onset of sexual receptivity than it is to determine the precise time at which the ewe will no longer accept mating. Considerable variability in the duration of oestrus has been observed within a single flock of ewes but most reports suggest a modal period of from 24–36 h. The end of sexual receptivity may well be related to the time of ovulation but this would be difficult to establish.

THE TIMING OF RELEASE OF GONADOTROPHINS

An extensive study of the time of ovulation in the ewe was made by Ivanow (1913) who concluded that in the majority of cases ovulation occurs 24 h after the onset of oestrus. The findings of Cole and Miller (1932 and 1935) and McKenzie and Terrill (1937) confirm that ovulation occurs rather uniformly between the twenty second and the thirtieth hour after the onset of oestrus although ovulation in the Merino may occur later (Quinlan and Maré, 1931).

The length of the oestrous cycle of the ewe, when defined as the time from the onset of sexual receptivity to the time of the next onset of sexual receptivity is remarkably constant even between different breeds of sheep. The mean values falling between 16–17·5 days. Many of the observations quoted in the literature were obtained from the study of flocks of animals where only intermittent observations were made throughout the day and seldom during the hours of darkness. Under these circumstances considerable errors in the oestrous cycle length of individual animals can arise. The mean oestrous cycle length of our present experimental animals (North Country Cheviots) is 16·5 days.

The classical experiments of Everett, Sawyer and Markee (1949) and Everett and Sawyer (1950) demonstrated that when rats are kept under controlled lighting conditions (illumination from 05·00–19.00 h) the neural stimulus which leads eventually to the release from the adenohypophysis of the gonadotrophins causing ovulation, occurs during pro-oestrus between 14.00 and 16.00 h, i.e. 10–12 h before ovulation. To release sufficient gonadotrophins for complete ovulation the stimulus in the rat has to last 60 min (Everett, 1965). The hormone release in this species being almost synchronous with the stimulus (Everett, 1956).

In the sheep, Santolucito, Clegg and Cole (1960); Hutchinson and Robertson (1960); Robertson and Hutchinson (1962) found that release of the gonadotrophins causing ovulation occurred sometime between the onset of oestrus and ovulation. The latter authors suggested that this release was likely to take place within 18 h after the onset of oestrus. If it is assumed that in the sheep, the stimulus and release of the gonadotrophins inducing ovulation are almost synchronous, then the determination of the timing of the stimulus would not only be of value itself but would be useful as a method for obtaining more precise information on the timing of the release of the gonadotrophins.

Hansel and Trimberger (1951) successfully blocked ovulation in the cow with atropine but the use of this blocking agent in large animals presents many problems. Chlorpromazine has been shown to be an effective inhibitor of ovulation in the rat (Barraclough and Sawyer, 1957) and in the mouse (Purshottam, Mason and Pincus, 1961). Apart from the possibility of local damage to tissues at the site of injection when large doses are given intramuscularly it is relatively free from side effects. Using chlorpromazine (single I.V. injection) in the ewe, Robertson and Rakha (1965a) obtained the results shown in *Figure 1*.

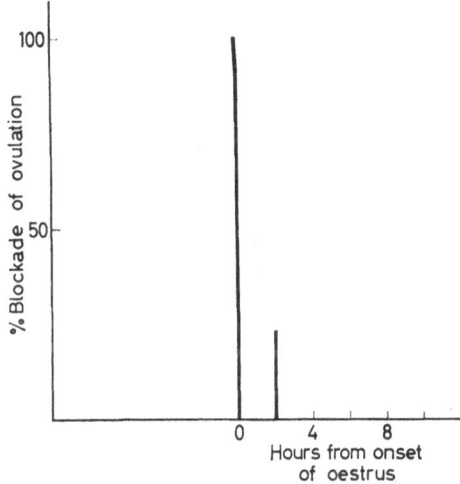

Figure 1. Time of injection of chlorpromazine (From H. A. Robertson and A. M. Rakha, 1965a, by courtesy of Cambridge University Press)

Ovulation was blocked in all 10 animals injected within 15 min of the onset of oestrus. Of 5 animals injected 2 h later, 4 ovulated and 1 was blocked. Ovulation was not blocked in any of the animals injected at 4 h, 6 h, or 10 h after the onset of oestrus.

Some support for the view that ovulation was only delayed in the blocked animals was obtained when a second laparotomy was performed on three of these animals 60 h after the onset of oestrus. All the animals had ovulated by 60 h. It is estimated that with the dose level used (12 mg/kg) ovulation occurred somewhere around 45 h, i.e. a delay of approximately 15–20 h.

It was concluded from these experiments that by the second hour after the onset of oestrus the hypothalamus had received a sufficient duration of stimulus to lead ultimately to ovulation. As a corollary to this, the release of the ovulating gonadotrophin(s) was likely to occur within the first few hours after the onset of oestrus. This conclusion left unanswered the question of whether the discharge of the gonadotrophins commenced before the onset of oestrus, and if so, were they responsible for the stimulation of the ovary to secrete the steroids necessary for behavioural oestrus. In the induction of behavioural oestrus the role of progesterone secreted by the corpus luteum must also be considered. This will be discussed later.

It is of considerable interest to note that the animals were sexually receptive during the period of sedation. Of 5 animals injected with chlorpromazine at the onset of oestrus 2 were still receptive 36 h later when the laparotomies were carried out; the other 3 were not, their sexual receptivity had ceased at approximately 24 h. The normal duration of oestrus in the ewe is 24–36 h. Since a general anaesthetic (sodium pentobarbitone) was used for the laparotomy at 36 h it was not possible to check for prolongation of the receptive period, although, since 3 animals had an apparently normal duration of receptivity it was concluded that under these experimental conditions where the period of sedation was short, no prolongation of oestrus had occurred. Recently, Radford (1966) using sodium pentobarbitone as the blocking agent has confirmed the findings of Robertson and Rakha (1965a) and has further suggested that there is a prolongation of oestrus in the blocked animals; the extension of the oestrus period being related to the duration of sedation. In the rat, Everett and Sawyer (1950) have shown that prolonged treatment with sodium pentobarbitone prolongs vaginal cornification.

To test the theory whether a release of gonadotrophins takes place before the onset of sexual receptivity, and if so, if this release might be responsible for inducing this state, we have maintained animals under continuous chlorpromazine sedation commencing on the fifteenth day of the cycle and terminating on the eighteenth day, i.e. approximately 1·5 days beyond the anticipated onset of oestrus. The injection schedule was adjusted to maintain each animal fairly heavily sedated but they were still capable of normal locomotion. Although the preliminary results obtained are not clear cut they suggest that under these conditions oestrus does not occur at the expected time but is in fact delayed until the chlorpromazine treatment is stopped. It should be noted that not all the animals came into oestrus on withdrawal of the chlorpromazine, some did not come into oestrus for another 16–17 days, i.e. at the next cycle.

The action of chlorpromazine in blocking oestrus may result either from a suppression of the release of an oestrus inducing gonadotrophin or by suppression of some behavioural centre in the brain. Although it has been shown (Robertson and Rakha, 1965a) that chlorpromazine will not suppress oestrous behaviour once it has commenced, it may, nevertheless, still be effective before the behavioural pattern has been initiated. The more likely effect, however, is in delaying the release of some gonadotrophins. It is possible to speculate that in these animals which missed their expected oestrus the chlorpromazine blockade had been carried on too long. This would result in the declining progesterone level falling to such a low level that either the oestrus inducing gonadotrophin was not released when the blockade was lifted or that the required steroid balance for initiating oestrous behaviour could not be achieved. Although a check was not made on these animals to see whether ovulation had occurred it is assumed that this was the case, since the next oestrus occurred at approximately a normal cycle length later, suggesting that a functional corpus luteum had been present.

The inference can be drawn from this experiment that a discharge of gonadotrophin(s) occurs before the onset of oestrus and is responsible for initiating sexual receptivity. Since ovulation can be blocked after this proposed release, i.e. by blockading with chlorpromazine after the onset of oestrus, it seems acceptable to postulate that a release of gonadotrophin(s) initiating sexual behaviour occurs before the ovulation-inducing discharge. The experiments of Kelly, Robertson and Stansfield (1963) on the suppression of ovulation in the rat by rabbit anti-ovine LH serum indicate that in the rat an oestrus-inducing gonadotrophin release may likewise precede an ovulation inducing release.

This anti-serum was subsequently shown by Robertson and Kelly (1963 unpublished) to neutralize not only rat LH but rat FSH. When it was administered on the day of pro-oestrus as an intraperitoneal injection at 12.30 h followed by a subcutaneous injection at 16.30 h ovulation was blocked in 10 out of 11 treated animals, all

7*   201

rats had the expected oestrous smear at autopsy the next morning. Bourdel and Li (1963) made a similar observation. In this type of experiment there is no possibility of a pharmacological interference with psychic oestrus as might exist with pharmacological ovulation blocking agents.

The relationship between suppressed oestrus (cornification) in the rat and suppressed ovulation is not clear cut in the report of Everett *et al.* (1949) but in the subsequent paper by Everett and Sawyer (1950) of 55 animals receiving a blocking dose of sodium pento-barbitone at 14.00 h, 58 per cent of these animals had an oestrous smear the following morning suggesting that the mechanism initiating vaginal changes had not been blocked in a high percentage of

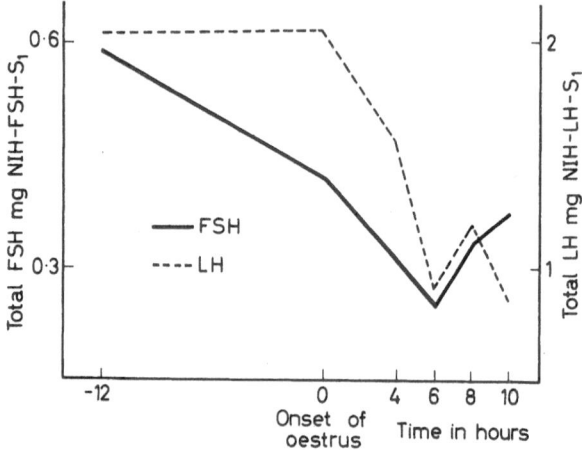

Figure 2. Changes in pituitary gonadotrophin levels of sheep in relation to oestrus and to ovulation (From H. A. Robertson and A. M. Rakha, 1966, by courtesy of Cambridge University Press)

these animals although ovulation had been. This is regarded as further evidence for a release of oestrus-inducing gonadotrophin(s) prior to the release of ovulatory gonadotrophin(s).

It could be argued, however, that all these effects may be due to quantitative differences in the amounts of gonadotrophin required to induce first oestrus and then ovulation; more gonadotrophins being required to induce ovulation than to induce oestrus.

These conclusions regarding the timing of the release of the gonadotrophins in relation to oestrus and to ovulation have been tested (Robertson and Rakha, 1966) in an experiment in which groups of ewes (North Country Cheviots, 6 animals per group) were killed at −12, 0, 4, 6, 8 and 10 h; these times being measured from the onset of oestrus. The −12 h group were all slaughtered at the beginning of the sixteenth day after the previous onset of oestrus on the assumption that they would all have a 16·5 day cycle. This group was included to see whether any release of gonadotrophins from the pituitary could be detected before the onset of oestrus.

Quantitative estimations of the FSH and of the LH concentration and total content of individual pituitaries were obtained. The results are shown in *Figures 2 and 3.*

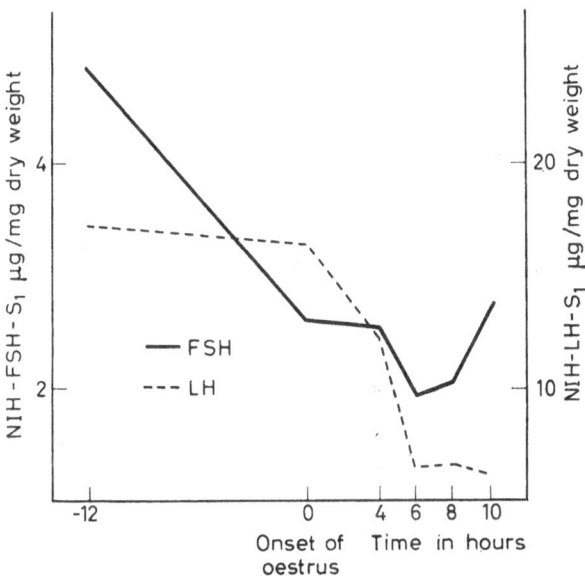

Figure 3. Changes in pituitary gonadotrophin concentrations of sheep in relation to oestrus and to ovulation (From data of H. A. Robertson and R. M. Rakha, 1966, by courtesy of Cambridge University Press)

Although there is, as might be expected, a considerable scatter of the individual values within each group, the variance is small enough to imply that all the animals were responding fairly uniformly with time. Over the period −12 h to 6 h there is a remarkable similarity in the overall changes of FSH and of LH. The overall depletion amounting to some 52 per cent of the total content of the pituitary in each case or, 61 per cent when expressed as a change in concentration. The difference between these values (52 per cent and 61 per cent) may be accounted for by a difference (non-significant) in the mean pituitary weights for the two groups of animals (122 mg and 160 mg respectively).

Perhaps the most significant conclusion from the above results is that the release of the two gonadotrophins from the pituitary is not synchronous because the release of FSH commences before that of LH. The release of FSH begins (*Figure 2*) approximately 8 h before the onset of oestrus, and continues for 14 h, i.e. until 6 h after the onset of sexual receptivity. The release of LH does not begin until just after the onset of oestrus, i.e. about 8 h after the release of FSH has started.

There have been many reports in the past on the constancy of the qualitative nature of the gonadotrophins in the pituitary or in urine (human) and this has led many workers to speculate on the possibility of there being only 'one gonadotrophin' having two activities. The chemistry of the isolated gonadotrophins was such as to preclude this as a tenable hypothesis. Robertson and Hutchinson (1962) found a remarkable constancy in the ratio of LH:FSH in the pituitary of ewes at different stages of their oestrous cycle. Rakha and Robertson (1965) showed, that in the cow, this ratio did alter around the time of the ovulatory discharge and this seemed to be due to differences in the percentages of the gonadotrophins secreted.

THE RATIO OF LH:FSH

The experiments described here show how easily any change in ratio could be missed for although the overall percentage of the two gonadotrophins appear to be identical (52 per cent of the total content in each case) the ratio of LH:FSH changes as a consequence of a differential release with time. It would be impossible to detect such changes when urinary gonadotrophins assayed from bulk urine samples were taken as the only parameter of changing gonadotrophin levels.

The rapid changes that occur in the levels of FSH and of LH are suggestive that either (*a*) a dramatic release of these gonadotrophins from the pituitary occurs during a precise period of a few hours, this release outstripping the rate of synthesis, or (*b*) assuming that the rate of release remains constant that a dramatic change in the rate of synthesis occurs over the same period. The evidence including the pharmacological blockade of ovulation in the ewe (Robertson and

Rakha, 1965a) already discussed is heavily weighted in support of the first of these two possibilities, namely, that a massive release of these two gonadotrophins takes place. Subsequent interpretation of these results does not preclude the possibility that these sudden changes are superimposed on a background of a steady release of one or both of these gonadotrophins during the cycle.

The early release of FSH, followed a few hours later by sexual receptivity, suggests that this gonadotrophin is responsible for stimulating production and secretion of those steroid hormones which induce behavioural oestrus. In this connection, it has been well documented that oestrous behaviour in the sheep can be induced by oestrogen treatment superimposed upon a prior period of treatment with progesterone (Cole and Miller, 1933; Hammond, 1945; Robinson, 1950, 1954, 1955a and b; Robinson, Moore and Binet, 1956). The observations that a period of progesterone treatment prior to oestrogen treatment is necessary before the full oestrous behaviour pattern can be developed can explain the 'silent' oestrus observed at the beginning of the breeding season where ovulation occurs without oestrous behaviour.

It is now necessary to speculate on the factors which may initiate the release of the gonadotrophin(s) which induce (a) sexual receptivity, (b) ovulation. The role of progesterone in inducing or in suppressing ovulation has been admirably reviewed by Rothchild (1965). In the cycling sheep with a functional corpus luteum it is likely that during its period of activity the corpus luteum suppresses the release of the ovulatory discharge of gonadotrophin. McKenzie and Terrill (1937) demonstrated many years ago that removal of the corpus luteum early in the cycle of the ewe resulted in ovulation occurring within 2–3 days, and conversely Robertson (1966, unpublished) has shown that when the level of progesterone is maintained by its administration commencing on the fifteenth day of the cycle, oestrus and ovulation are delayed until the progesterone administration is stopped.

THE GROWTH AND REGRESSION
OF THE CORPUS LUTEUM

The growth and regression of the corpus luteum of the ewe is well documented: Quinlan and Maré (1931); Grant (1934); Cole and Miller (1935); Hutchinson and Robertson (1966). The significant decrease in size (*Figure 4*) which occurs sometime between the tenth and the fifteenth day after the onset of oestrus suggests that the progesterone output may be dropping by this time. A more rapid rate of regression commences around the fifteenth day. A more functional approach has been made by Edgar and Ronaldson (1958) and by Short (1964) who determined the progesterone levels in ovarian venous blood. Their work suggests that the corpus luteum attains full secretory activity at about the sixth–eighth day of the cycle and continues to secrete progesterone at a fairly constant rate until the fifteenth day. The values of Short indicate that a slight

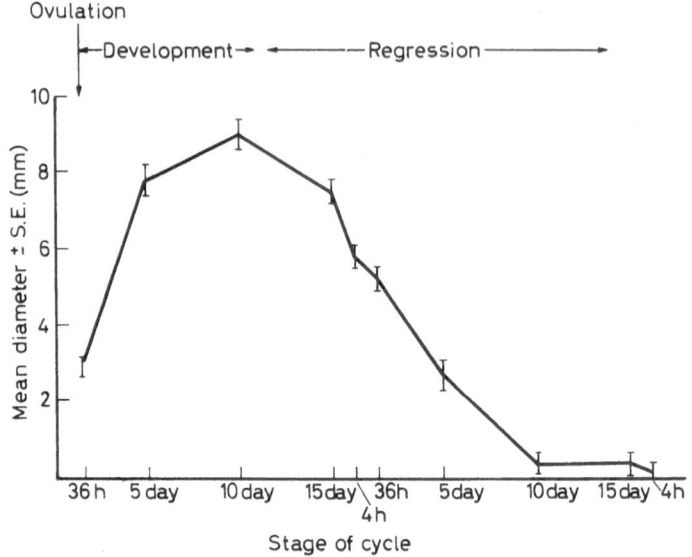

Figure 4. Changes in the mean diameter of corpora lutea in the ovaries of Welsh Mountain ewes during the oestrous cycle (From J. S. M. Hutchinson and H. A. Robertson, 1966, by courtesy of Blackwell)

decline in progesterone output may occur between the tenth and fifteenth day. Both reports state that the secretion of progesterone by the ovary stops quite abruptly within 24 h of the next expected oestrus. The most recent report of Deane, Hay, Moor, Rowson and Short (1966) on the relationship between the morphology and function of the corpus luteum of the sheep during the oestrous cycle is of interest in that evidence for the commencement of the regression of the corpus luteum was found on day 12 or 13 of the cycle by electron microscopy.

The present evidence suggests that the rapid degenerative phase of the corpus luteum associated with a rapid lowering of the level of progesterone secreted by the corpus luteum occurs at about the fifteenth day of the cycle. This precedes the sudden decrease in FSH and LH content of the pituitary by approximately 24–36 h. Evidence suggesting that LH and or FSH may be luteolytic, i.e. lead to the rapid regression of the corpus luteum, has been reviewed by Rothchild (1965) but if the rapid lowering of the LH and FSH content of the pituitary is interpreted as indicating a sudden release of these gonadotrophins occurring after the degeneration of the corpus luteum has been initiated, it would appear that neither LH nor FSH can be described as being luteolytic in the sheep. It is not intended to discuss here the relationship between the uterus and the activity of the corpus luteum.

A fairly precise model for the sequence and timing of the events leading to ovulation in the sheep is beginning to emerge. The weak link at the moment lies in the interpretation of the changes that occur in the levels of the gonadotrophins in the pituitary.

A true assessment of these changes must await the development of valid methods for the determination of the levels of these gonadotrophins in peripheral blood.

REFERENCES

Barraclough, C. A. and Sawyer, C. H. (1957). *Endocrinology* **61,** 341–351

Bourdel, G. and Li, C. H. (1963). *Acta endocr., Copenh.* **42,** 473–479

Cole, H. H. and Miller, R. F. (1932). *Anat. Rec.* **52,** Suppl. 1, 50

—— (1933). *Am. J. Physiol.* **104,** 165–171

—— (1935). *Am. J. Anat.* **57,** 39–97

Deane, H. W., Hay, M. F., Moor, R. M., Rowson, L. E. A. and Short, R. V. (1966). *Acta endocr., Copenh.* **51,** 245–263

Edgar, D. G. and Ronaldson, J. W. (1958). *J. Endocr.* **16,** 378–384

Everett, J. W. (1965). *Endocrinology* **76,** 1195–1201

Everett, J. W., Sawyer, C. H. and Markee, J. E. (1949). *Endocrinology* **44,** 234–250

—— (1950). *Endocrinology* **47,** 198–218

— (1956). *Endocrinology* **59,** 580–585

Grant, R. (1934). *Trans. R. Soc. Edinb.* **58,** 1–47

Hammond, J. (jun.) (1945). *J. Endocr.* **4,** 169–180

Hansel, W. and Trimberger, G. W. (1951). *J. Anim. Sci.* **10,** 719–725

Hutchinson, J. S. M. and Robertson, H. A. (1960). *Nature, Lond.* **188,** 585–586

— O'Connor, P. J. and Robertson, H. A. (1964). *J. agric. Sci., Camb.* **63,** 59–60

— and Robertson, H. A. (1966). *Res. vet. Sci.* **7,** 17–24

Ivanow, E. I. (1913). *Prelim. Commun. Zootech. Station, Askania-Nova*

Kelley, R. B. (1937). *Bull. Commonw. scient. ind. Res. Org.* 112

Kelly, W. A., Robertson, H. A. and Stansfield, D. A. (1963). *J. Endocr.* **27,** 127–128

McKenzie, F. F. and Terrill, C. E. (1937). *Bull. Mo. agric. Exp. Res. Stn.* No. 264

Marshall, F. H. A. (1903). *Phil. Trans. R. Soc. B* **196,** 47–99

Purshottam, N., Mason, M. M. and Pincus, G. (1961). *Fert. Steril.* **12,** 346–352

Quinlan, J. and Maré, G. S. (1931). 17th *Rep. Dir. vet. Sci. Anim. Ind. Un. S. Afr.* 663–703

—— and Roux, L. L. (1932). 18th *Rep. Dir. vet. Sci. Anim. Ind. Un. S. Afr.* 831–870

Radford, H. M. (1966). *J. Endocr.* **34,** 135–136

Rakha, A. M. and Robertson, H. A. (1965). *J. Endocr.* **31,** 245–249

Robertson, H. A. and Hutchinson, J. S. M. (1962). *J. Endocr.* **24,** 143–151

— and Rakha, A. M. (1965a). *J. Endocr.* **22,** 383–386

—— (1965b). *J. Reprod. Fert.* **10,** 271–272

—— (1966). *J. Endocr.* **35,** 177–184

Robinson, T. J. (1950). *J. agric. Sci.* **40,** 275–307

— (1954). *Endocrinology* **55,** 403–408

— (1955a). *J. Endocr.* **12,** 163–173

— (1955b). *J. agric. Sci.* **46,** 37–43

— Moore, N. W. and Binet, F. E. (1956). *J. Endocr.* **14,** 1–7

Rothchild, I. (1965). In: *Vitam. Horm.* **23,** 209–327

Santolucito, H. A., Clegg, M. J. and Cole, H. H. (1960). *Endocrinology* **66,** 273–279

Short, R. V. (1964). *Recent Prog. Horm. Res.* **20,** 303–340

DISCUSSION

Professor T. J. Robinson (*Sydney, Australia*)

There is similarity between the work reported here and that of my colleague Dr. Moore who is working with Dr. Brown in a joint programme between us and the University of Melbourne. Moore and Brown have been re-examining the levels of progesterone and oestrogen in ovarian vein blood using modern techniques and gas liquid chromatography (GLC). They have been re-examining the results of Short and Rowson in which Moore himself was originally involved.

The results for progesterone confirm previous evidence—it is produced over 12 days or so at a relatively uniform level, and falls sharply sometime before the onset of the next oestrus. The original concept of Short and Rowson concerning oestrogen is that it is produced at a fairly uniform level throughout the cycle and that oestrus occurs when the progesterone level falls. Using GLC this does not appear to be the case for there is a sudden increase in oestrogen some hours before the onset of oestrus, followed by an equally dramatic fall, so that at the onset of sexual receptivity, there is little if any oestrogen in circulation. This fits with my earlier work on the induction of oestrous behaviour in spayed sheep where, after a single injection of oestrogen there is normally a time lag of some 24 to 30 h before oestrus occurs, a time when one would expect the oestrogen level to have fallen practically to nothing.

This unpublished work of Moore and Brown fits in very nicely with the results of Dr. Robertson. For example, he points out that when chlorpromazine is given at 0 h it will block ovulation, but it is too late to stop oestrus. In other words the oestrogen that has been in circulation will stimulate the nervous system, and the animal will come into oestrus regardless of whether or not it will ovulate. Also, chlorpromazine given before the onset of oestrus will act as an antioestrogen and will block oestrus. It is an important concept that oestrous behaviour can take place in the absence of oestrogen. The oestrogen available to set the phenomenon in motion may not be needed for its manifestation.

I wholeheartedly agree that the onset and manifestation of oestrus by sheep is a very precise phenomena in terms of endocrine requirements to induce it and the time at which it can be expected.

PROFESSOR I. ROTHCHILD (*Cleveland, U.S.A.*)

I would like to add to Dr. Robertson's remarks that in other animals, particularly in the rat, the effects we ascribe to oestrogen during the pro-oestrus and oestrus phase of the cycle can also be shown to be due to a hormone secreted at an interval of some 48 h before that effect is actually observed. The sheep therefore is not unique in showing a particular type of behaviour at a time when the oestrogen is not acting. Secondly, it is of interest to know that the behavioural response to oestrogen is not inhibited by chlorpromazine although the neural mechanisms involved in ovulation are. This fits with other observations, such that in certain anovulatory conditions although animals will mate they do not ovulate spontaneously. Thirdly, concerning the interpretation of Dr. Robertson's graphs depicting the concentration over a short period, we can safely say that this represents release, and when the concentration of LH in the pituitary is falling rapidly, that this is an expression of release at a very rapid rate. However, when the concentration remains level between -12 h and 0 h I do not think this necessarily means that release is zero. It could represent an increased rate of synthesis of LH in the gland accompanied by a gradual increase in release. At the moment oestrus appears, the release rate associated with it may exceed synthesis and so the concentration in the pituitary falls precipitously.

ROBERTSON

In reply there was a point I forgot to mention, that these dramatic changes might be superimposed on a steady state of release, or on some other state. When using antigonadotrophic substances in the rat there is a distinction to be made between blocking oestrus and blocking ovulation. If on the day of pro-oestrus a rat is given a single injection of rabbit antiserum to ovine LH which, by the way, neutralizes both rat LH and FSH, at 12.30 p.m. ovulation was blocked but oestrus was not. In the rat likewise, there is some event occurring prior to the neural stimulus which institutes ovulation, and it might be that, in this species also there is release of FSH prior to the ovulatory release of LH.

DR. FALCONER (*Nottingham*)

With respect to the effects of chlorpromazine, the data you showed fitted well with the idea that it blocks the release of pituitary gonadotrophins but it could equally be possible that it acts at the ovarian level. Have you any evidence from bio-assay of pituitaries of animals blocked by chlorpromazine to indicate this type of activity?

REPRODUCTION IN THE FEMALE MAMMAL

ROBERTSON

I have not personally but I know that P. S. Brown has tried this technique of blocking with chlorpromazine to increase the sensitivity of animals. From the levels he used in his mice and the levels we use in our sheep we are reasonably certain that the action of chlorpromazine is not at the ovarian level.

PROFESSOR VAN REES (*Leiden, Holland*)

I would like to return to a discussion on the rat. Everett has said that injection of nembutal on the day of pro-oestrus will block ovulation but then ovulation occurs the next day. If you block ovulation in the sheep by injections of chlorpromazine, do you retard ovulation and, if so, when does it occur?

ROBERTSON

I think I mentioned this. It is difficult because we do the laparotomy under a general anaesthetic, and we are rather against giving a second general anaesthetic too soon, but three animals examined were blocked. These animals were examined again after 60 h and all had ovulated. We concluded that with the particular level of chlorpromazine we were using, the animals had probably ovulated at 45 h—that is ovulation was delayed by 20 h.

DR. HANCOCK (*Edinburgh*)

Did Dr. Robertson say that there were no differences in the duration of oestrus between sheep? Did he mean no differences between sheep of different breeds?

ROBERTSON

One can only survey the literature on this. With sheep of the same breed there is little variation and most breeds have an oestrous period of 24–36 h. I think the exception to this may be the Merino. It may well be that in sheep with longer oestrous periods the actual time of ovulation may be shifted slightly.

HANCOCK

From my experience of the literature, most workers have not been as critical as Dr. Robertson. There are two problems in determining the time of ovulation relative to the time and onset of heat. Firstly, the timing of heat and secondly, the timing of ovulation, which can only be done by laparotomy at repeated intervals. There is a dearth of this type of information and I place little reliance on literature about the subject. We have good evidence that the Finnish Landrace breed has a longer oestrous period than any of our native breeds. It can be as long as 3 days, which is distinctly different from the Scottish Blackface.

T. J. ROBINSON

In our experience, the duration of oestrus is quite variable, and is influenced by age and breed. The maiden Merino ewe may have an oestrous period of 10 min only and may be served once only. Here I am referring to the normal ewe, not a hormone-treated ewe. Dr. Lindsay has found a seasonal variation within the Merino which is normally considered to be a poly-oestrous animal breeding all the year round. He has found a seasonal variation not only in the duration of oestrus but also in its intensity. During the autumn months, oestrous behaviour is well marked. However, in the spring months, the ewe is less receptive and has to be sought out and courted by the ram.

Dr. Munro (*Reading*)

Have you any idea of the mechanism of sudden release of oestrogen prior to oestrus in the sheep?

Robertson

I am afraid I have not. I can only guess that the actual timing of this oestrogen activity occurs in relation to the postulated release of FSH. This FSH may have an influence on the ovarian synthesis of oestrogens, and this may then be followed by a release of LH. One point should be borne in mind, if Short's work is correct, the progesterone level seems to drop dramatically at about the fifteenth day of the cycle. This still leaves quite a long gap of about 36 h before FSH release occurs. Considering the short half-life of progesterone, if it is this drop which triggers off the release of FSH, one might have expected the release of FSH earlier.

CONTROL OF OVULATION WITH HUMAN GONADOTROPHINS

CARL GEMZELL

Department of Obstetrics and Gynaecology,
University of Uppsala, Uppsala, Sweden

THE CONTROL AND TIMING OF OVULATION

It is through the two gonadotrophins, the follicle stimulating hormone (FSH) and the luteinizing hormone (LH), that the anterior pituitary controls the timing of ovulation and the number of follicles that rupture. Under the influence of FSH, the follicular growth occurs at a constant rate for any given species (Hisaw, 1947). This rate cannot be accelerated by increasing the doses of exogenous FSH (Davis and Hellbaum, 1944). The number of follicles developing to maturity is also constant for each species but can be increased by the addition of exogenous FSH (Brambell, 1956). As each follicle seems to require a certain amount of FSH for maturation, the release of FSH from the anterior pituitary of each species may also be constant (Evans and Simpson, 1940). Thus, the amount of exogenous FSH needed for a normal ovarian response varies with different species. Too much or too little will cause too many or too few follicles to develop and mature.

Under the influence of LH, the FSH-primed follicle ruptures with discharge of an ovum. The ruptured follicle is then transformed into a corpus luteum in which progesterone and oestrogen are formed. LH has no effect on unripe follicles and consequently cannot influence the follicles which are not ready for rupture at the time of its administration.

In the human female, during the early part of each menstrual cycle, a group of follicles starts to grow. When they have reached a certain size, only one goes further to full maturation and rupture, while the others become atretic. The mechanism behind this selection of a single follicle is not understood. It may be that the amount of FSH released from the pituitary is only enough to evoke oestrogen production in the most receptive follicle. This endocrine response

212

of the follicle might depress the release of FSH below the reactive threshold of the other follicles. An alternative hypothesis is that the appreciable amounts of oestrogen and progesterone produced by the dominant follicle may desensitize the other follicles to FSH.

In clinical practice, gonadotrophins are almost exclusively used for induction of controlled ovulation by imitating the sequential release of FSH and LH from the anterior pituitary. The ideal result of treatment is pregnancy obtained from fertilization of a single ovum released at a predictable time. FSH followed by LH is the usual therapeutic pattern.

Pregnant mare serum (PMS), bovine, ovine and porcine pituitaries constitute the main animal sources of extracts containing follicle stimulating activity. These substances alone, or in combination with human chorionic gonadotrophin (HCG), which has a luteinizing effect, have received extensive clinical trial. Even if the results obtained clearly indicate that they were effective, the treatment with animal hormones has now been largely abandoned because of allergic reactions and antibody formation. Furthermore, evidence is available that chemical and biological differences exist between gonadotrophins from human and animal sources. To obtain a normal response in humans therefore, gonadotrophins from human sources should be used.

In 1958, an FSH preparation obtained from human pituitaries was tested on 7 patients with primary or secondary amenorrhoea (Gemzell, Diczfalusy and Tillinger, 1958). There was little doubt about its ability to stimulate the development of Graafian follicles and, with HCG treatment, to lead to ovulation and corpus luteum formation. Later, Lunenfeld, Menzi and Volet (1960) reported that gonadotrophins extracted from postmenopausal urine could stimulate human ovaries and, together with HCG, induce ovulation. The urinary preparation in common use today is Pergonal[R] prepared by Donini, Puzzuoli and Montezemolo (1964).

Gemzell *et al.* (1958) used lyophilized pituitaries as their starting material. After homogenization, the glands were extracted in $Ca(OH)_2$ solution at pH 9·2 and the gonadotrophin-containing material was obtained by ammonium sulphate fractionation. The yield per gland was 3–5 mg of a product which had an FSH potency of 12–24 I.U./mg by the ovarian augmentation assay and an LH potency of 25 I.U./mg, HCG by an immunological assay. During the last few years the preparation described above has been replaced by a product obtained from fresh frozen pituitaries. The gonadotrophins were extracted according to the first step of a procedure

213

developed by Roos for the isolation of pure FSH (Roos and Gemzell, 1965). The yield of the clinical grade material was about 18 mg/gland. The FSH activity was 24–35 I.U./mg and the LH activity 20–30 I.U./mg, HCG.

THE HUMAN PITUITARY EXTRACT

The human pituitary extract, which contains both FSH and LH activities, is called HPFSH. It should only be used on women who have an infertility problem and the ideal subject is under 35 years of age with normal non-functioning ovaries, primary or long-lasting secondary amenorrhoea, normally developed sex organs, and lack of urinary gonadotrophins as evidence of pituitary failure. In doubtful cases an 'FSH-test' is of great value as a method of exploring ovarian potential. There should also be no detectable barriers or contra-indications to conception and the husband should be normally fertile.

Extensive use of HPFSH during a 7-year period has led to a number of observations which seem to throw some light on the control of ovulation, and these observations will now be described and discussed.

The only absolute proof of induced ovulation is pregnancy. Short of this, various indirect methods of detecting ovulation and corpus luteum formation exist, such as sex hormone excretion patterns, examination of vaginal smears or cervical secretions, basal body temperature recordings, ovarian palpation and inspection, and endometrial histology. As none of these indirect methods are entirely reliable, the more that are used, the better.

The response a patient shows will depend on a number of factors: the metabolism and excretion rates of the gonadotrophins, the level of endogenous gonadotrophin production, the FSH/LH ratio of the preparation used, the total dosages of FSH and HCG, the timing of their administration and the state of the ovaries.

The metabolism and excretion rates of the administered FSH and HCG may vary with their states of purification. The clinical grade material of FSH was far from pure and it is likely that the absorption rate is slower than that of purer preparations. In hypophysectomized individuals about 15 per cent of a single injection of HPFSH could be recovered in the urine during a period of 4 days. This would indicate that the biological half-life of FSH is rather long. The activity of the patient during treatment might also be of some importance as it has been found that a woman at rest in hospital shows a greater response to a given dose of HPFSH than does a woman attending her usual work.

The ideal ovary for stimulation is the resting one. From one time to another its response to the same dose of HPFSH is uniform and usually the dose for a normal response can be determined. If elevated levels of urinary oestrogen or proliferative endometrial activity indicate that the ovary is functioning, it can be brought back to a suitable state for treatment by a course of FSH and HCG injections or by cyclical administration of oral oestrogen and gestagen combinations for 2–3 months.

Only the ovaries of individuals completely lacking gonadotrophins were at complete rest. This state of affairs was uncommon and only occurred in women who had been hypophysectomized or born without a pituitary. In most other cases, small amounts of FSH could usually be detected in urine. The ovarian response to subnormal levels of endogenous gonadotrophin seems to be the formation of follicles which never mature but are fairly sensitive to exogenous FSH. Larger amounts of endogenous FSH might produce follicles which mature sufficiently to become cystic and only require minute amounts of exogenous FSH to make them rupture.

Diczfalusy, Johannison, Tillinger and Bettendorf (1964) found that a pituitary extract with an FSH/LH ratio of 2 : 3 was effective at lower doses than a urinary one with a ratio of 0 : 37. Crooke, Butt, Palmer, Morris, Logan, Edwards and Anson (1963) similarly found that preparations with the least amount of LH gave the best results. In a limited trial, Gemzell, Roos and Loeffler (1966) noted that a preparation of FSH with a ratio of 2 : 1 gave no better results than extracts with a ratio of 1 : 2. There might well be an optimal FSH/LH ratio above which no increase in effect is achieved and below which there is a loss of potency.

HPFSH appears to display the 'all or none' phenomenon, with little difference between the dose that overstimulates and the dose that has no effect. It is desirable to keep the FSH doses as low as possible. Using the data obtained from the literature it can be calculated that the effective total dosages given over periods of up to 10 days have ranged from 450 to 2,000 I.U. of FSH activity. In our hands a single injection of 110 I.U. of HPFSH per day during a period of 7–10 days gave the best results as far as controlled ovulation is concerned. A daily dose of less than 70 I.U. has never proved to be effective while in some women a daily dose of 200 I.U. was necessary for response. A period of treatment longer than 10 days has not been used. If no ovarian response is noticed after

215

6 days of treatment the remaining 4 dosages were sometimes doubled.

A preliminary 'FSH-test' with careful assessment of resulting ovarian activity might help to achieve the aim of accurate dosage. It is also useful as a method of determining whether the responsiveness of the ovaries is sufficient to warrant attempts at induced ovulation. The fact that the ovary seemed less prone to overstimulation during a second course of treatment, constitutes an additional reason for advocating an FSH-test as a routine preliminary to treatment.

HCG was usually given after the follicles were primed with FSH in amounts between 9,000 and 45,000 I.U. over 1–6 days. An amount of 3,000 I.U./day for 3 days seemed in most cases to be adequate. In cases where the ovaries were overstimulated with HPFSH, careful administration of HCG with daily control of effect, was advisable in order to avoid the formation of large luteinized cysts.

TIMING OF TREATMENT

Timing of FSH and HCG administration has been extensively investigated by Crooke *et al.* (1963) using experiments of statistical designs on a series of 9 patients. With patterns of steroid excretion as their criteria, they considered that the most normal response was likely to occur when FSH was divided into three equal doses over 8 days and HCG was given as a single injection after an interval of under 96 h after the third dose.

In our hands the timing of treatment was assessed with the number of treatments required before conception as the measure of success. A daily dose of HPFSH for 10 days followed 24–48 h later by HCG for 3 days gave the best results as far as the number of pregnancies was concerned. The treatment/pregnancy rate was about 2·0. However, the ovarian response as judged by the urinary excretion of oestrogen was more than three times the normal response occurring at the time of a spontaneous ovulation. Brown's (1955) data indicated that following a spontaneous ovulation, the total amount of oestrogen excreted during a 24-h period was at the most 100 μg. The average excretion following the administration of HPFSH for 10 days was about 300 μg. Furthermore, the pattern of administration led to a number of multiple births of over 50 per cent. If HPFSH in the same total dose was administered for 5–7 days the total oestrogen excretion was more normal but the pregnancy rate decreased and the time of ovulation was more difficult to predict.

A normal ovarian response in terms of sex steroid excretion patterns might be ideal, as far as preventing multiple births is concerned, but also gave fewer pregnancies and a higher percentage of early abortions. It seemed likely therefore, that many of the women treated needed the high levels of oestrogen and progesterone that followed overstimulation in order to negotiate the first crucial weeks of gestation.

A prerequisite for response to HPFSH seemed to be the presence of Graafian follicles in the ovary. It was found that on several occasions when courses of gonadotrophin were given in rapid succession, the ovary only responded to alternate courses of treatment. If 6 weeks were allowed to elapse between treatments the occurrence of negative responses could be eliminated. In cases with polycystic ovaries, the amount of HPFSH necessary to bring about follicular maturation was rather small in comparison to the dose required for hypophysectomized individuals or women with primary amenorrhoea. Furthermore, it was difficult to induce superovulation in normally menstruating women. It seemed likely that these differences in ovarian response reflect the state of the ovaries and particularly the number of receptive Graafian follicles available at the time of HPFSH administration.

SUMMARY

During a 5-year period about 150 women with primary or secondary amenorrhoea were treated for sterility with HPFSH and HCG in order to allow fertilization after controlled ovulation had been induced. About 90 per cent ovulated on one or more occasions at the expected times and 62 became pregnant. The frequency of multiple births was equal to single births and the chances of conceiving triplets, or more, was one in ten. In about 10 cases, a pregnancy following administration of human gonadotrophins led to a request for further therapy. Almost all of these patients became pregnant again after one or more further courses of treatment, and one woman conceived four times. The abortion rate was about 25 per cent, most of them late due to multiple births. Of the 16 abortions noticed, 3 women were responsible for 7 abortions. In about 5 cases the treatment led to spontaneous ovulation and conception. These successful cases are not included in the 62 pregnancies following the induced ovulation.

All infants born at full term and almost all foetuses from abortions were carefully examined for malformations and none has been observed so far. The first two deliveries, both twins, took place in January 1960. These 4 children are now more than 6 years old, perfectly healthy and show no signs of either mental or physical deficiency.

The two main complications following treatment with human gonadotrophins are polycystic enlargement of the ovaries and super-ovulation leading to multiple births. These two undesirable manifestations of ovarian stimulation could be due to the size of the FSH dose, or could equally well be caused by some other phenomenon which is not yet understood.

ACKNOWLEDGEMENTS

This investigation was completed in collaboration with Dr. Paul Roos, Institute of Biochemistry, Uppsala and was supported by research grants AM 06060 from the National Institute of Health, United States Public Health Service and from the Swedish Medical Research Council.

REFERENCES

Brambell, F. W. R. (1956). *Marshall's Physiology of Reproduction*, 3rd edn., vol. I. Ed. by Parkes. London; Longmans, Green

Brown, J. B. (1955). *Biochem. J.* **60,** 185

Crooke, A. C., Butt, W. R., Palmer, R. F., Morris, R. Logan, Edwards, R. and Anson, C. J. (1963). *J. Obstet. Gynaec. Br. Commonw.* **70,** 604

Davis, M. E. and Hellbaum, A. A. (1944). *J. clin. Endocr.* **4,** 400

Diczfalusy, E., Johannison, E., Tillinger, K. G. and Bettendorf, G. (1964). *Acta endocr., Copenh.* Suppl. **90,** 35

Donini, P., Puzzuoli, D. and Montezemolo, R. (1964). *Acta endocr., Copenh.* **45,** 321

Evans, H. M. and Simpson, M. E. (1940). *Endocrinology* **27,** 305

Gemzell, C. A., Diczfalusy, E. and Tillinger, K.-G. (1958). *J. Endocr.* **18,** 1333

— Roos, P. and Loeffler, F. E. (1966). *J. Reprod. Fert.* **12,** 49

Hisaw, F. L. (1947). *Physiol. Rev.* **27,** 95

Lunenfeld, B., Menzi, A. and Volet, B. (1960). *Acta endocr., Copenh.* Suppl. **51,** 587

Roos, P. and Gemzell, C. A. (1965). *Ciba Fdn Study Grps* No. 22, 11

DISCUSSION

Dr. J. Lorraine (*Edinburgh*)

Concerning the prediction of the response to gonadotrophin therapy by the use of endogenous hormone levels, have you looked at the possibility of making such a prediction by estimating oestrogens in the urine? I ask this question because in our experience with the ovarian stimulant chlomaphine we have been able to show in patients of rather similar nature, that a satisfactory response is more likely to occur in the presence of reasonably high endogenous oestrogen levels prior to treatment— the sort of levels one would find in the follicular phase of the menstrual cycle. On the other hand, in patients with very low oestrogen levels characteristic of the menopause or the post-menopause, a satisfactory response to chlomaphine is not generally encountered. With regard to the endogenous gonadotrophin level in relation to chlomaphine, our evidence would suggest that a satisfactory ovulation response is more likely to occur if the pre-treatment gonadotrophin levels are low and much less likely to occur if the pre-treatment gonadotrophin levels are high and I wonder if a study of this type would give some indication whether or not a good response to gonadotrophin therapy might be expected and possibly might have some relationship to multiple pregnancy.

GEMZELL

I believe that the ideal case shows no oestrogen excretion; in other words, has a completely resting ovary and no or very low levels of total gonadotrophins in the urine. I am not convinced that one should treat women with high oestrogen excretion or those who produce gonadotrophins at normal levels. If the ovaries are functioning, then it is likely that they contain a certain number of follicles at different stages of maturation, and they may require very small amounts of LH for rupture. The time of ovulation is usually more difficult to predict in these women. There is one thing that can be done and that is to treat them with oestrogen or perhaps better with a combination of oestrogen and progesterone for 2–3 months to let the ovary rest and come to some sort of zero position and then start the treatment. I think chlomaphine is ideal for those patients who have an ovarian and pituitary function. There is probably a steady state between the ovaries and the pituitary and consequently they are not cycling. What chlomaphine does is to bring on cycling and ovulation.

DR. W. R. BUTT (*Birmingham*)

I would like to comment on the ratio of FSH to LH in the preparation used in treatment. We made a preparation containing a ratio of FSH to LH of 125 : 1, both expressed in terms of the first IRP standard. We then added to the preparation different amounts of HCG to make the ratio 25 : 1, 5 : 1 and 1 : 1 and these were given to a group of patients, maintaining the same FSH dose in each case and we could find no statistical difference in the effectiveness of these four preparations although they had widely different FSH : LH ratios. By effectiveness, I mean the response as judged by the excretion of oestriol and pregnanediol. I wonder if you have any comments on whether the ratio might in fact have an effect on the outcome, that is the incidence of multiple pregnancies; and also to ask you at the same time what units you were using when you quoted the ratios of your preparation?

GEMZELL

I think the crucial point here is the preparation and how to administer it. The preparation we have been using is much cruder than your preparation, which is about twenty times as active as ours. Our ratio between FSH and LH is 1 : 1. Of course there have been some differences in the ratio during the years especially when we have been using lyophilized pituitaries. In those preparations there is generally a higher ratio of FSH : LH than in preparations from fresh pituitaries. I am not convinced that the ratio FSH : LH is very important; large doses of HCG can be given from the first day of treatment with FSH and it still takes 7 or 8 days before ovulation takes place. There is less increase in oestrogen excretion following such treatment as compared with FSH given alone followed by HCG. The reason why we first give FSH, then LH, is that we are more likely to predict the exact time of ovulation and this has been shown to be valuable. One can tell the patient that ovulation will take place within 24 h.

ANON

Am I right in saying that the temperature curves that you showed seemed to go up before you injected HCG, in fact in the last 2 or 3 treatments with FSH? If so what is the explanation for that.

GEMZELL

Ovulation takes place 24 h before the temperature rise. Temperature is only an indication of progesterone production in the newly-formed corpus luteum. In these two cases they actually ovulated before HCG was administered. The women in question had some ovarian and pituitary function before treatment, and then it is always more difficult to predict when ovulation will take place.

MR. P. G. HIGNETT (Glasgow)

Referring to cases of women who are amenorrhoeaic presumably because of hypophyseal insufficiency and are given replacement therapy, I gather that, in common with the Birmingham group, you do not find it necessary to give any LTH to maintain pregnancy. Can we take this as evidence that in the human there is no such thing as a luteotrophin?

GEMZELL

I think there is no evidence for luteotrophic hormone in the human pituitary so far. In hypophysectomized patients or in those who lack pituitary function it was not necessary to add anything after ovulation was induced. Pregnancy occurred and went to full term without the support of any other hormone. In those cases where conception did not occur the functional period of the corpus luteum was normal, i.e. 2 weeks. In the case of the woman without a pituitary who was treated, she had some bleeding 2–3 weeks after conception. We gave high doses of human growth hormone for 2 weeks and measured the excretion of pregnanediol, but there was no change in the pregnanediol excretion and her bleeding did not change. I think the LTH activity in the human comes from the placenta.

DR. L. MARTIN (Imperial Cancer Research Fund, London)

Do you monitor oestrogen excretion for each of these patients throughout the period of gonadotrophin treatment? Do you find any difference in the occurrence of the oestrogen peak and does this correlate with the state of the ovaries after treatment?

GEMZELL

In approximately the first 50 patients treated with HPFSH we measured the three oestrogens, pregnanediol, corticosteroid and 17-ketosteroid excretion daily. The general pattern showed a slight increase in oestrone and oestradiol appearing at about the fifth or sixth day, and on the seventh or eighth day there was a similar rise of oestriol. Latterly we have not made these determinations daily because of the work involved but in each case we have collected a 24 h urine sample after the FSH treatment and after the whole treatment and have examined it for oestrogens, pregnanediol, corticosteroid and 17-ketosteroids. After ovulation there is in almost all cases a drop in oestrogen excretion but it is very difficult to say whether or not a certain pattern of oestrogen excretion indicates a particular state of the ovaries.

MARTIN

Have you not found cases in which you have the peak corresponding to the ovulatory peak which occurs in the normal cycle on say 2 days treatment?

GEMZELL

No.

DR. W. R. BUTT

Could I comment on some of our recent work where women were given a single injection of gonadotrophin. We have been rather surprised to find that in a group of 12 patients the maximum increase of oestriol did not occur until 10 days after the injection. I believe your patients were given daily injections, but in these instances there was only a single injection on day 1 and we were rather surprised that the maximum oestriol level was as late as 10 days afterwards.

GEMZELL

We have injected a group of 12 women with a single injection of 50 mg of FSH (5,000 I.U.) and we found the maximum level of total oestrogens in urine on the fifth and sixth day, then it dropped. If HCG was given to these women 48 h after the single injection, we usually found a tremendous increase in oestrogen on the fifth or sixth day reaching a maximum on day 10 or 12.

ANON

Could I ask if the human ovary responds to either ovine or bovine FSH and LH?

GEMZELL

Bovine and ovine FSH have been used in the clinic but I think the danger lies in the fact that they arouse antibody production and allergic reactions which can be quite serious. The general experience is that you can treat a woman once or twice and she may respond. If you then wait 6–8 months and try again, you may run into serious trouble with allergic reactions.

DR. J. P. BENNETT (*Godalming*)

I am working with primates and with monkeys it is customary to dose with PMS through the LH dosing period as well, while you dose separately, PMS first then HCG. Can you tell me why you need to separate this in the human when in other primates you get a much more satisfactory ovulation by continuing your FSH right through?

GEMZELL

Well, this is the result of 9 years' experience. The first year we tried different kinds of patterns including the one you suggested, i.e. both together; we gave the first hormone for 5 days and then the two together for 5 days. In fact, we tried all kinds of patterns and we have decided on this one because we think it gives the best results as far as pregnancy is concerned and the best results as far as the timing of ovulation, but I cannot say why.

MARTIN

In the normal human menstrual cycle, if I remember Brown's figures correctly, oestrogens have a second big peak in the luteal phase of the cycle. In some rodents you do not get implantation unless you get a surge of oestrogen at the right time. I think it has been suggested that this second peak in the human may be associated with implantation. Have you looked at oestrogen excretion after ovulation occurs, and, if there is no secondary peak, might this not explain the ovulations where you do not get pregnancy following? Could this be a lack of implantation?

GEMZELL

We have found this secondary rise of oestrogens but I cannot say anything about whether or not it occurs only in cases where the woman becomes pregnant. We have measured not only oestrogens but pregnanediol in those who ovulate. Their increase in pregnanediol excretion lasts for about 2 weeks and then declines. They also show the second rise and decrease of oestrogens. The interval from ovulation to the following menstrual period is astonishingly constant at 2 weeks.

EFFECTS OF LACTATION ON THE PITUITARY GONADOTROPHINS OF THE SOW

D. B. CRIGHTON

University of Nottingham School of Agriculture,
Sutton Bonington, Loughborough, Leicestershire

MECHANISMS OF LACTATION

THE REGULAR oestrous cycles of the sow are interrupted by two situations during its normal reproductive life, namely pregnancy and lactation. It is with the mechanisms underlying the second of these two anoestrous periods that this paper is concerned.

It is apparent that in the sow, the corpora lutea of pregnancy start to regress just before parturition. At this time, Kimura and Cornwell (1938) detected, by means of biological assay in adult rabbits, a decline in the progesterone content of the corpora lutea. Lauderdale, Kirkpatrick, First, Hauser and Casida (1965) showed that the total weight of luteal tissue in the ovaries and the progesterone content of the corpora lutea, measured by means of a chemical assay, were significantly lower on the day of parturition than 2 days earlier. Short (1960), using a chemical assay, was able to demonstrate in addition that the systemic blood progesterone level showed a decline at this time. The process of regression continues rapidly after parturition (Corner, 1919; Warnick, Casida and Grummer, 1950; Burger, 1952; Palmer, Teague and Venzke, 1965a, b). By the fifty-third to sixty-third day of lactation, the corpora lutea of pregnancy were represented only by dark brown spots in the ovarian tissue (Palmer *et al.*, 1965a).

There is agreement that post-partum oestrus occurs in the sow but the percentage of sows exhibiting oestrus has varied widely in different studies (Warnick *et al.*, 1950; Burger, 1952; Heitman and Cole, 1956; Self and Grummer, 1958). The variation in the expression of post-partum oestrus may be due in part to the widely differing breeds and strains studied and the different climatic and environmental conditions prevailing.

Post-partum ovulation does not normally occur in the sow, regardless of whether or not there is a post-partum oestrus. This has been established by examining the ovaries of sows slaughtered shortly after the post-partum oestrus (Warnick *et al.*, 1950; Burger, 1952) and about 10 days after parturition where post-partum oestrus had not occurred (Warnick *et al.*, 1950). Post-partum oestrus occurs at a time when large follicles are absent from the ovaries of the sow (Warnick *et al.*, 1950; Burger, 1952; Palmer *et al.*, 1965a). In addition, Palmer *et al.* (1965b), found that during the first week of lactation approximately 50 per cent of the ovarian follicles showed signs of atresia. The percentage of atretic follicles was much higher during this period than in later lactation.

Throughout the remainder of lactation subsequent to the immediate post-partum period, the sow does not normally exhibit oestrus or ovulation (Marshall and Hammond, 1937; Burger, 1952; Heitman and Cole, 1956; Allen, Lasley and Uren, 1957; Self and Grummer, 1958; Palmer *et al.*, 1965a). When lactation is terminated by removal of the young some 8 weeks after parturition, oestrus and ovulation occur within about 4–7 days (Marshall and Hammond, 1937; Burger, 1952; Self and Grummer, 1958; Smidt, Scheven and Steinbach, 1965). This is, in fact, a matter of common observation, being the basis of the traditional husbandry pattern whereby conception occurs at the post-weaning ovulation after an 8 week lactation. Thus lactation in the sow is accompanied by a state of follicular quiescence dating from parturition.

Weaning was found by Palmer *et al.* (1965a) to produce a marked increase in mean follicular diameter, which was most noticeable 3–4 days post-weaning. A similar pattern was seen when the number of ovarian follicles equal to or greater than 5·0 mm was recorded. This effect of weaning on follicular growth has also been observed in early lactation. Warnick *et al.* (1950) found that the total follicular volumes in sows killed about the sixth or tenth days of lactation were significantly less in animals which had been suckled up to slaughter than in those where the litters had been removed at birth. This finding is confirmed by the work of Lauderdale *et al.* (1965), who studied the effect of weaning 5 days before slaughter on the total follicular fluid weight and mean diameter of the four largest follicles in the ovaries of sows killed on the sixth, eleventh and sixteenth days post-partum. These workers found that the values for both criteria were significantly greater in the weaned than in the suckled groups of sows. There is evidence that the follicular quiescence during lactation demonstrated by these investigations is accompanied by atresia in about 30 per cent of the ovarian follicles (Palmer *et al.*, 1965b).

A number of these suggestions as to the underlying causes of the lactational anoestrus of the sow have been found in the literature. Warnick et al. (1950) suggested that suckling might operate via the pituitary gland to inhibit follicular development but did not contribute any evidence of altered levels of pituitary hormones during lactation to support this theory. Self and Grummer (1958) postulated a two-stage process responsible for the inhibition of ovulatory oestrus during lactation. The first stage envisaged so-called 'residual endogenous elements of the initiation process of lactation associated with parturition' combined with an unspecified effect of suckling. It was suggested that weaning during the period of dual responsibility for the suppression of ovulatory oestrus would result in the 'endogenous elements' operating for variable periods prior to their exhaustion, whereas weaning during the suckling-dependent anoestrous period would result in a more uniform return to oestrus. This hypothesis was used to explain results obtained from trials with 10-, 21- and 56-day weaning but was unsupported by experimental evidence. Smidt et al. (1965) stated that the lactational anoestrus of the sow is probably due to blockage of the FSH-releasing factor of the hypothalamus but again no evidence was provided for this hypothesis. Palmer et al. (1965a) suggested, on the basis of data collected on ovarian follicular and uterine changes during lactation and the early post-weaning period, that lactation, or the suckling stimulus, or both, may bring about a decrease in the synthesis or release of the pituitary gonadotrophic hormones which are responsible for follicular development and that this effect may be most marked during early lactation. These workers pointed out that two other pituitary hormones, prolactin and oxytocin, are known to be physiologically active during lactation and suggested that perhaps one or both of these may counteract the action of the gonadotrophic hormones or prevent their release from the pituitary gland. While these conclusions from ovarian and uterine data were reasonable, although unproven in their implication of a deficiency of the pituitary gonadotrophins, the hypotheses concerning prolactin and oxytocin in the sow are without supporting evidence.

There is, in fact, very little sound evidence to implicate the sort of gonadotrophic insufficiency postulated directly or indirectly by the above authors. One study has been made by Lauderdale et al. (1965) on changes in pituitary gonadotrophin levels coincident with early lactation. No significant differences in anterior pituitary FSH or LH levels were found during the stages of pregnancy or lactation studied. The anterior pituitary content of FSH during lactation up to the sixteenth day was significantly greater in suckled than in weaned sows but the LH content was unaffected by weaning. The observed increase in anterior pituitary FSH, coupled with the inhibition of follicular development over the same period was held by the authors to demonstrate an effect of suckling blocking the release of FSH from the anterior pituitary gland.

LACTATIONAL ANOESTRUS

We have become interested in the situation of lactational anoestrus because it is a factor restricting the annual production per sow and we have been examining the possibility of inducing oestrus and ovulation in the lactating sow and thus obtaining pregnancy concurrent with lactation. The results of experiments in the rat described by Rothchild (1960); McCann, Graves and Taleisnik (1961); McCann and Ramirez (1964) and others, suggested that the use of similar approaches in the sow might provide information on the underlying causes of lactational anoestrus. Accordingly, the ovarian and pituitary gland changes associated with lactation in the sow were investigated, partly in the hope that the results might provide a sound basis for any treatment for the induction of oestrus and ovulation in the lactating sow.

Two such experiments are described here. The first experiment was designed to provide information on the period between late lactation and the post-weaning oestrus and ovulation by examining any changes in the ovarian follicles and anterior pituitary gland content of FSH and LH which might take place over that period.

Thirteen Large White sows from the School of Agriculture herd were used in this experiment. These sows entered the experiment after approximately 45 days of lactation at which time they were nursing litters ranging from four to eleven piglets. All had normal reproductive histories, having reared at least two previous litters. They had been subjected to normal management during the existing lactation. The sows were divided into four groups on the basis of time of slaughter relative to a 56-day lactation, the presence or absence of oestrous behaviour immediately prior to slaughter and the presence or absence of ovulations on the ovaries obtained at slaughter. Details of the four groups are given in Table 1.

The second experiment was carried out in order to determine whether ovariectomy had any effect on the anterior pituitary LH potency of lactating sows and to compare any response to ovariectomy with that of sows ovariectomized during the oestrous cycle. A definite stage (day 5) of the oestrous cycle was chosen in order to allow a comparison of this with a definite stage of lactation. In addition, it was hoped that ovariectomy might raise serum levels of LH to a point where they might be detected by biological assay, thus providing an index of the relative abilities of lactating and cycling sows to release LH.

Fourteen Large White sows from the School of Agriculture Herd were used. Six entered the experiment after 10 days of lactation and the remaining eight during various stages of oestrous cycles initiated as a result of weaning.

The lactating sows were nursing litters of from 6–11 piglets at the start of the experiment and were divided at random into two groups

Table 1

Allocation of sows into groups in the first experiment

Group	No. of sows	Reproductive status at slaughter	Time of slaughter
1	4	Lactating, anoestrous	52–53 days of lactation
2	3	Post-weaning, anoestrous	3–4 days post-weaning after 56-day lactation
3	3	Post-weaning, oestrous, no ovulation	4 days post-weaning after 56-day lactation
4	3	Post-weaning, oestrous, recent ovulation	4–6 days post-weaning after 56-day lactation

on the basis of whether a laparotomy or ovariectomy was to be carried out on day 20 to 21 of lactation. Each laparotomized sow was tested daily with a boar from day 10 until slaughter, 25 days post-operation on day 45 or 46 of lactation. Each ovariectomized sow was tested daily with a boar from day 10 until ovariectomy on day 20 to 21 of lactation.

The cycling sows were at various stages of the oestrous cycle when they entered the experiment. Each sow was observed through two successive oestrous periods prior to surgery, which was carried out on day 5 of the cycle (the first day being designated day 1). The sows were divided on the basis of whether a laparotomy or ovariectomy was to be carried out on day 5.

In both experiments, quantitative, differential bio-assays were used to assess the anterior pituitary levels of FSH and LH. The

227

assay used for FSH was the Augmentation Assay of Steelman and Pohley (1953) as modified by Brown (1955) for use in mice. For LH the Ovarian Ascorbic Acid Depletion assay of Parlow (1958) was used with two modifications, a 4-h test as described by Schmidt-Elmendorff and Lorraine (1962) but using intraperitoneal injection of the test material, and a 1 h test as described by McCann and Ramirez (1964). The former was used for anterior pituitary material and the latter when serum was being injected. All anterior pituitary assays were carried out against standard preparations given by the National Institute of Health, U.S.A. Over 90 per cent of

Table 2

Allocation of sows into groups in the second experiment

Group	No. of sows	Reproductive status at operation	Operation performed	Time of slaughter
5	3	Lactating, 20–21 days, anoestrus	Laparotomy	25 days after operation
6	3	Lactating, 20–21 days, anoestrus	Ovariectomy	25 days after operation
7	4	Cycling, day 5 of cycle	Laparotomy	Day 5 of oestrous cycle subsequent to that of operation
8	4	Cycling, day 5 of cycle	Ovariectomy	25 days after operation

the anterior pituitary determinations were four-point assays, the remainder being of the three-point design.

The sows in both experiments were slaughtered in the following manner: Firstly they were stunned with a captive bolt pistol and then bled immediately, at which time a sample of blood was taken from the jugular venous effluent for serum collection. The pituitary glands were removed as soon as possible after slaughter and the anterior lobes were stored at $-20°C$. The reproductive tracts of these animals were also examined.

The mean diameter of the largest follicle present in either ovary showed an increase after weaning, reaching a maximum at oestrus, followed by a fall again when the follicles ovulated (*Figure 1*). These changes approach significance at $P = 0.05$ but do not reach it.

The number of large follicles (over 5·0 mm in diameter) increased significantly after weaning ($P < 0.001$) and decreased significantly ($P < 0.001$) after ovulation.

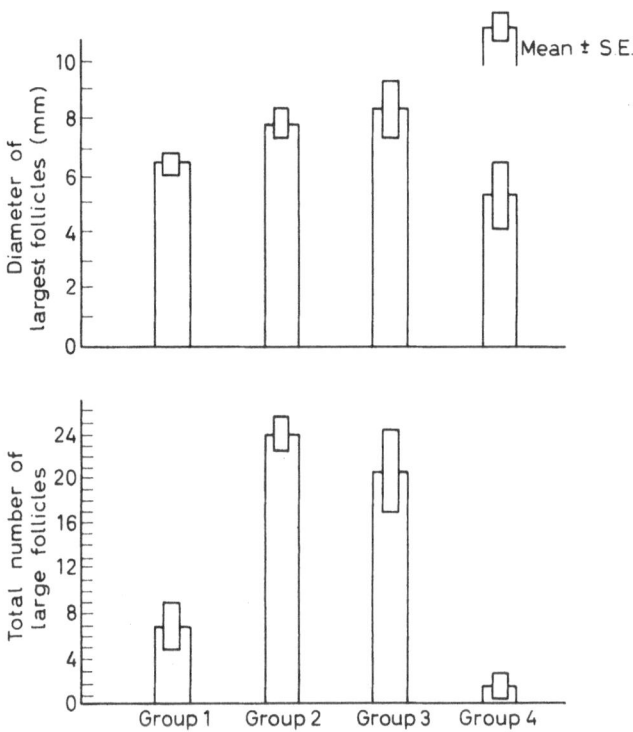

Figure 1.   Changes in the ovarian follicles of sows between late lactation and the post-weaning ovulation

From the second experiment a comparison can be made between the four groups for the weight and length of the two uterine horns of each animal taken together (*Figure 2*).   The mean weight for the normal lactating sows on days 45–46 of lactation (Group 5) is not significantly different from that of sows at the same stage of lactation but which had at that time been ovariectomized for 25 days (Group 6) or from cycling sows which had been ovariectomized for 25 days (Group 8).   The value for cycling sows on day 5 of the cycle is significantly higher than any of the other three values ($P < 0.001$ in each case).   When uterine development is expressed in terms of total length of the uterine horns, the picture is essentially similar to that seen for weight. Thus there is evidence that follicular growth is markedly suppressed during lactation, resulting in a substantial decrease in oestrogen secretion.

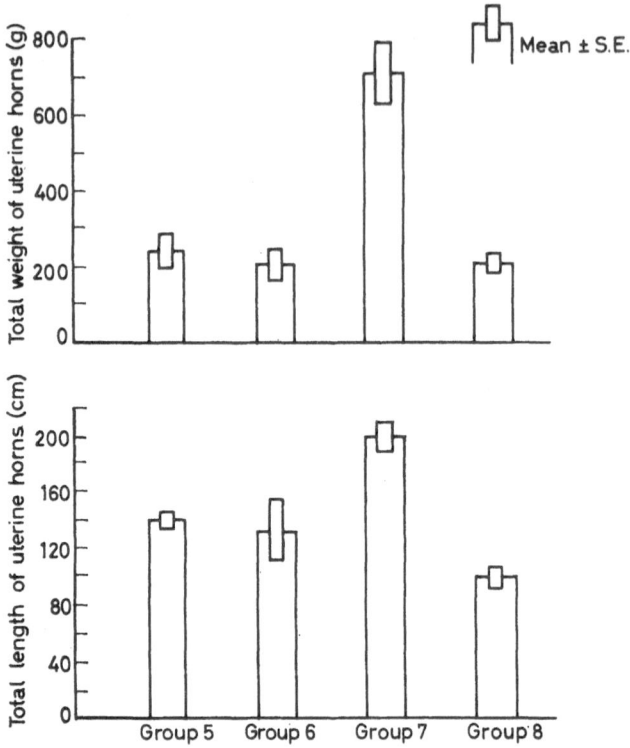

Figure 2. Total weight and total lengths of uterine horns of intact and ovariectomized lactating and cycling sows

With regard to the anterior pituitary gland assays from the first experiment, the results calculated for FSH are shown in *Figure 3*. The data presented in the form of the mean relative potencies per milligramme for each group show a slight and non-significant rise from late lactation to the post-weaning period before oestrus followed by a significant fall with the onset of oestrus. When these figures are converted to compare the total anterior pituitary contents of FSH the same picture is seen, with no significant change from lactation to the period immediately after weaning (in fact a slight fall in content is indicated).

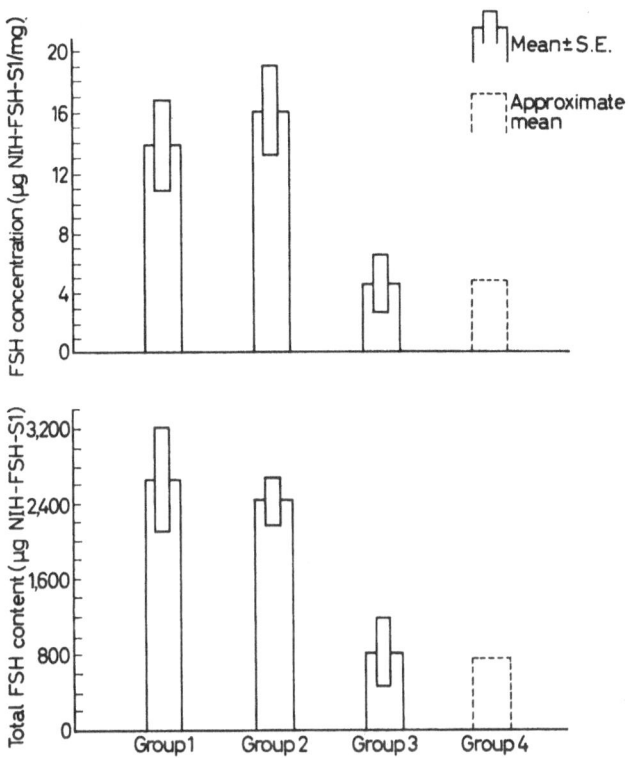

Figure 3. Changes in the anterior pituitary FSH content of sows between late lactation and the post-weaning ovulation

The mean results calculated for LH in the first experiment are shown in *Figure 4*. The data presented in the form of the mean relative potencies per milligramme for each group show a significant rise $(P < 0.01)$ between late lactation and the post-weaning period prior to oestrus, followed by a significant fall $(P < 0.01)$ with the onset of oestrus. When the results are converted to compare the total anterior pituitary gland contents of LH, the same picture is seen.

Thus the follicular changes occurring in the ovaries of sows over the period covered by the first experiment are accompanied by changes in the anterior pituitary gland content of FSH and LH. With regard to FSH, no effect of weaning on the pituitary could be demonstrated but in the case of LH there was a marked rise initially after weaning. This increase was interpreted as representing a

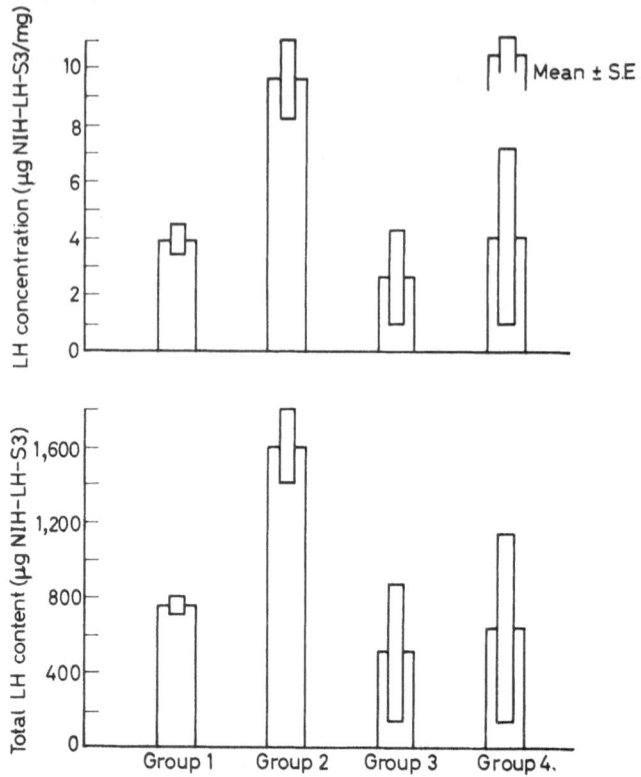

Figure 4. Changes in the anterior pituitary LH content of sows between late lactation and the post-weaning ovulation

blockage of the synthesis of LH during lactation which was relieved by weaning. The subsequent fall in both FSH and LH in the anterior pituitary gland with the onset of oestrus was interpreted, as it has been by many workers in several species, as representing a massive release of these hormones associated with rapid follicular development and ovulation. The results suggest that FSH release rather than synthesis may be inhibited during lactation, since the level was uniform and quite high before and after weaning. Alternatively, it is possible that a post-weaning increase in FSH synthesis comparable to that of LH occurred but was not demonstrable because it was balanced by a greater release of FSH at this time, resulting in rapid growth of follicles.

It was decided to test the hypothesis that LH synthesis and release are inhibited during lactation and the second experiment was designed for this purpose. Ovariectomy in the rat is known to cause a marked rise in the synthesis and release of LH as judged by increases in pituitary and blood levels of this hormone (Taleisnik and McCann, 1961; Ramirez and McCann, 1963; McCann and Ramirez, 1964). Increased pituitary levels after ovariectomy in the sow were found by Parlow, Anderson and Melampy (1964). Thus the second experiment was carried out and anterior pituitary and serum LH levels were examined.

The mean results calculated for anterior pituitary LH in the second experiment are shown in *Figure 5*. The data presented in the form of the mean relative potencies per milligramme for each group demonstrate that the level in lactating sows showed no significant change as a result of ovariectomy. There was, however, a significant rise $(P < 0.001)$ in anterior pituitary LH potency after ovariectomy during the cycle. The value for sows ovariectomized during the cycle was also significantly higher $(P < 0.01)$ than that for sows ovariectomized during lactation. When the results are converted to compare total anterior pituitary gland contents of LH, the same picture is seen.

The finding that the low anterior pituitary content characteristic of lactation failed to increase during the 25 day period after ovariectomy whereas the low level characteristic of day 5 of the oestrous cycle increased approximately fivefold is striking. It demonstrates an inability of the pituitary gland of the lactating sow to increase the synthesis of LH under circumstances where that of the cycling animal is capable of doing so. It also shows that the ovaries play no part in the maintenance of the low level of pituitary LH in the lactating sow.

In order to determine whether the serum samples might provide information supplementing the pituitary assay work, the 1 h modification of the Ovarian Ascorbic Acid Depletion assay was employed. Aliquots of 2·0 ml of serum were injected intravenously into the assay rats, the serum of each individual sow being injected into 6 rats. In addition, 10 rats were injected with 2·0 ml isotonic saline. The injection of saline produced a negative depletion (increase) in ascorbic acid ($-1·1$ per cent) whereas the mean depletions produced by the sow treatment groups were at least 44 per cent (Table 3). The differences between the means of the groups were not significant.

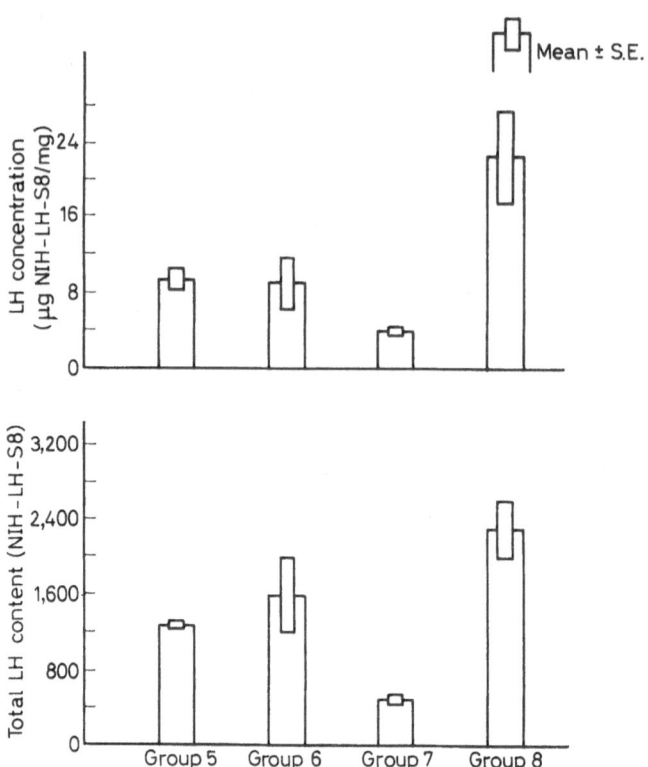

Figure 5. Anterior pituitary LH contents of intact and ovariectomized lactating and cycling sows

The consistency of the large ovarian ascorbic acid depletions produced by samples from sows of such widely differing reproductive status and anterior pituitary LH content is surprising. It suggests strongly that the depletions observed do not represent the levels of LH circulating in the blood during life. It is therefore pertinent to examine the possible reasons for the effect.

Firstly, it is possible that the depletions may be due to an artificial elevation of serum LH levels, by some feature in the treatment at slaughter, after the sows had been stunned with a captive bolt pistol.

This treatment may be responsible for releasing LH into the blood from the pituitary glands of the sows although the apparent contents of LH indicated by the depletions would seem to be so massive as to make this hypothesis unlikely. Such a situation would seriously undermine the value of the pituitary LH assays carried out in both our experiments, as measures of anterior pituitary LH during life, and would therefore render suspect the interpretation placed upon these results.

Table 3

Depletions of rat ovarian ascorbic acid produced by the serum of lactating and cycling sows

Group	Reproductive status at slaughter	Mean depletion of rat ovarian ascorbic acid ($\% \pm S.E.$)
5	Lactating, 45–46 days, anoestrous	$44 \cdot 8 \pm 3 \cdot 6$
6	Lactating, 45–46 days, ovariectomized 25 days previously	$48 \cdot 9 \pm 10 \cdot 4$
7	Cycling, day 5 of cycle	$48 \cdot 1 \pm 2 \cdot 0$
8	Ovariectomized 25 days previously on day 5 of cycle	$50 \cdot 8 \pm 5 \cdot 2$

Secondly, the depletions may be due to some agent in serum other than LH. Other hormones are capable of depleting ovarian ascorbic acid when present in very large quantities. Also a LH releasing factor is known to be present in the hypothalamus of the rat and may perhaps exist in the sow. This factor may be responsible for releasing LH from the pituitary glands of the assay rats resulting in ovarian ascorbic acid depletion.

Finally, it is possible that a non-specific effect may occur when large quantities of serum from the sow are injected into the rat, resulting in the release of the pituitary LH of the latter or acting directly on the ovaries to produce ascorbic acid depletion.

Thus at present no interpretation is placed upon the results of the serum assays. Further experiments are planned to investigate the nature of the ovarian ascorbic acid depleting agent(s) present in the serum of the sow.

SUMMARY

These results confirm the existence of a state of ovarian follicular quiescence during lactation in the sow, resulting in a low level of circulating ovarian hormones as indicated by the lack of uterine development. Weaning resulted in a rapid growth of ovarian follicles followed by oestrus and ovulation.

The level of pituitary FSH was high during lactation and was unaffected by weaning suggesting that the lack of follicular growth was due to a failure of FSH release rather than lack of FSH synthesis.

The level of pituitary LH rose significantly as a result of weaning and declined prior to ovulation. The LH content of the lactating sow failed to rise after ovariectomy. These findings suggest that lactation inhibits LH synthesis and FSH release.

ACKNOWLEDGEMENTS

Gifts of hormone preparations were received from Burroughs Wellcome and Co. (PMS and HCG) and from the Endocrinology Study Section, National Institute of Health, U.S.A. (NIH-LH-S3 and NIH-FSH-S1). The work was supported by a grant from the Pig Industry Development Authority.

REFERENCES

Allen, A. D., Lasley, J. F. and Uren, A. W. (1957). *J. Anim. Sci.* **16,** 1097. (Abstracts)

Brown, P. S. (1955). *J. Endocr.* **13,** 59–64

Burger, J. F. (1952). *Onderstepoort J. vet. Res.* Suppl. No. 2, p. 218

Corner, G. W. (1919). *Am. J. Anat.* **26,** 117–183

Heitman, H. (jun.) and Cole, H. H. (1956). *J. Anim. Sci.* **15,** 970–977

Kimura, G. and Cornwell, W. S. (1938). *Am. J. Physiol.* **123,** 471–476

Lauderdale, J. W., Kirkpatrick, R. L., First, N. L., Hauser, E. R. and Casida, L. E. (1965). *J. Anim. Sci.* **24,** 1100–1103

Marshall, F. H. A. and Hammond, J. (1937). *Bull. Minist. Agric. Fish. Fd., Lond.* No. 32

McCann, S. M., Graves, T. and Taleisnik, S. (1961). *Endocrinology* **68,** 873–874

— and Ramirez, V. D. (1964). *Recent Prog. Horm. Res.* **20,** 131–170

Palmer, W. M., Teague, H. S. and Venzke, W. G. (1965a). *J. Anim. Sci.* **24,** 541–545

— — — (1965b). *J. Anim. Sci.* **24,** 1117–1125

Parlow, A. F. (1958). *Fedn Proc. Fedn Am. Socs exp. Biol.* **17,** 402

— Anderson, L. L. and Melampy, R. M. (1964). *Endocrinology* **75,** 365–376

Ramirez, V. D. and McCann, S. M. (1963). *Endocrinology* **72,** 452–464

Rothchild, I. (1960). *Endocrinology* **67,** 9–41

LACTATION AND PITUITARY GONADOTROPHINS

Schmidt-Elmendorff, H. and Lorraine, J. A. (1962). *J. Endocr.* **23,** 413–421
Self, H. L. and Grummer, E. H. (1958). *J. Anim. Sci.* **17,** 862–868
Short, R. V. (1960). *J. Reprod. Fert.* **1,** 61–70
Smidt, von D., Scheven, B. and Steinbach, J. (1965). *Züchtungskunde* **37,** 23–36
Steelman, S. L. and Pohley, F. M. (1953). *Endocrinology* **53,** 604–616
Taleisnik, S. and McCann, S. M. (1961). *Endocrinology* **68,** 263–272
Warnick, A. C., Casida, L. E. and Grummer, R. H. (1950). *J. Anim. Sci.* **9,** 66–72

DISCUSSION

B. T. Donovan (*London*)

In view of the large litter size in the pig, would you imagine that this stimulation of lactation is the cause of the delay in the onset of oestrus? Have experiments been done where the litters have been reduced to half or less of the original size and the delay in oestrus measured?

Crighton

As far as I know no variation of the suckling stimulus has been studied in the pig but it is commonly observed that pigs suckling 2–4 piglets return to oestrus early in lactation.

I. Rothchild (*Cleveland, Ohio, U.S.A.*)

The early return to oestrus in sows suckling small litters is similar to the observations that I made in rats and in the mouse, the interval to post-partum oestrus is in inverse proportion to litter size. The castration changes in pituitary LH content conforms with work of McCann in the rat and the work I did using mouse uterine weight as an end-point of gonadotrophin potency in the pituitary. There appears to be an inverse relationship between the size of litter and response of the pituitary to castration in the rat and, from your results, the same appears true of the pig.

The content of LH in the blood should be measurable but as you have shown, it is not possible to interpret results obtained with the OAAD method. Could you examine this in another way by treating an animal during lactation with known amounts of LH or FSH or combinations and observe the response of the ovary? If you could stimulate the ovary directly and get oestrogen secretion then you would know that it is not an inability of the ovary to respond but a lack of release of gonadotrophins from the pituitary.

Crighton

One situation does have a bearing on the reduction of the suckling stimulus under certain conditions, for the separation of a sow from its litter for 12 h/day for 4–5 days will cause the sow to return to oestrus. Additional factors in the expression of oestrus involves the presence or absence of a boar and the use of PMS injections during the period of partial weaning. Several years ago Cole and his colleagues injected up to 1,600 I.U. PMS into sows in early lactation and obtained a comparatively poor response. However, after about 40 days of lactation the sows responded well with oestrus, ovulation and pregnancy. We have examined, in a series of small scale experiments the effects of injecting PMS on day 21 of lactation compared with the effect of separating sows for 2 days and then injecting with PMS on day 23. The sows responded poorly to direct injections but well to PMS given in combination with partial weaning.

H. KARG (*Munich, Germany*)

To prove these doubts of the specificity of LH in the serum, did you not check the serum removed from the animals before slaughter for comparison?

CRIGHTON

We plan to withdraw blood samples from sows before slaughter to make these comparisons.

H. A. ROBERTSON (*Aberdeen*)

We have killed animals both by stunning and by exsanguination and with neither have we observed an LH response.

J. A. LORRAINE (*Edinburgh*)

I would like to make a comment on the specificity of the OAAD test. There is a feeling now that the method was not as specific as the originator believed and that the work of Albert and Rosemberg has tended to show that where large amounts of FSH are given this may be capable of producing ascorbic acid depletion.

In the work we did in relation to the cholesterol method we found that very large amounts of Growth Hormone which we regarded as being unphysiological, as far as the application of the assay is concerned, might produce a depletion in ovarian cholesterol. So I think there are two hormones which may have affected your results in relation to serum. In studies on human subjects we have looked at the effect of operative stress on gonadotrophic activity in the urine, and have found that in a proportion of these subjects there is a marked rise in urinary gonadotrophin excretion associated with operation.

R. PINOT (*France*)

I have frequently found that the injection of plasma from hypophysectomized ewes depresses ovarian ascorbic acid. This is a non-specific effect. I have also found that egg albumin gave a depression of ascorbic acid. With low doses, egg albumin gave a depletion of ascorbic acid in normal rats, but not in the hypophysectomized rats. In these experiments it would have been difficult to distinguish between egg albumin and LH.

K. YOSHINAGA (*Cambridge*)

I have found that the effect of lactation on concurrent pregnancy in the rat was the same as you report in the sow. If the litter size is small there is no delay in implantation, but if the litter size is six or above then implantation is delayed. If oestrogen or LH is given to these rats, implantation occurs at the normal time. Thus I believe that the endocrine background is the same as for the pig.

STUDIES ON THE LH-CONTENT IN THE PITUITARY AND THE ANDROGENS IN THE TESTES OF THE BOVINE FOETUS

H. KARG

Institut für Tierphysiologie der Universität München, Munich, Germany

A DISCREPANCY exists between evidence of the functional importance of the testicular hormones in foetal or early post-natal life and the lack of analytical data of the pituitary-gonadal axis during these periods. We started to look at the LH-content of pituitaries (Karg, 1966) and at testicular androgens in the bovine foetus (Karg and Struck, 1966). The glands were obtained from slaughter house material and a few samples were taken from unborn calves by caesarian section. The ages of the foetuses were estimated by measuring the crown-rump length (Rüsse and Rüsse, 1963). The LH-content was determined by a modification (Karg, 1957) of the ovarian ascorbic acid depletion test according to Parlow (Parlow, 1961).

The detailed results of the estimation of foetal pituitary LH-content will be published elsewhere (Karg, 1962). Gonadotrophic activity was determined as soon as pituitary samples were available, i.e. from the end of the third month of gestation (crown-rump length approximately 14 cm). The LH-concentration showed a tendency to increase until 6–8 weeks *post partum*, when values were obtained which were in the range measured in adult bulls. In estimating the total LH-capacity of the individual, the increase in size of the gland development must be considered. Differences in the LH-concentrations between the foetal sexes were not statistically significant.

A remarkable finding is the depletion of the LH-concentration in calves during parturition. Five out of six estimations showed values lower than those obtained during the second half of the foetal life or after birth. We assume that this is due to the influence of hormones of maternal origin at the end of pregnancy.

The presence of androgens in the foetal testes was recently demonstrated by Lipsett and Tullner (1965) in the rabbit. Lindner and

239

Mann (1960) studied the relationship between testosterone and androstendione in the bull. They found that in the adult, testosterone was the main testicular androgen but in the calf there was an equal or slightly dominating occurrence of androstendione. We have used two-dimensional thin layer chromatography (Schink and Struck, 1964) to investigate the androgen content of purified foetal testicular extracts. Details of the method and the results will be presented elsewhere (Karg and Struck, 1966). We were, however, able to demonstrate that testosterone was the main androgen in the bovine foetal testes but from the late foetal period the concentration of androstendione increases.

The data suggests that hormones of the foetal pituitary-gonadal axis should be considered when discussing foetal-maternal hormonal relationships.

REFERENCES

Karg, H. (1957). *Klin. Wschr.* **35,** 643
— (1966). *Die Naturwissenschaften* **2,** 41
— (1967) *Zuchthygiene* **2,** 12
— and Struck, H. J. (1966). *Bericht. des 2. Int. Congr. of Hormonale Steroids,* Mailand; *Excerpta med.* No. 111, p. 288
Lindner, H. L. and Mann, T. (1960). *J. Endocr.* **21,** 341
Lipsett, M. B. and Tullner, W. W. (1965). *Endocrinology* **77,** 273
Parlow, A. (1961). In: *Human Pituitary Gonadotropins,* p. 300. Ed. by A. Albert. Springfield, Illinois; Charles C. Thomas
Rüsse, I. and Rüsse, M. (1963). *Tierärztl.* **18,** 309. Umschau
Schink, W. and Struck, H. (1964). *Med. Welt* **29,** 1525

DISCUSSION

DR. K. J. BETTERIDGE (*Birmingham*)

Professor Karg, do you consider that the fall in foetal pituitary LH content at parturition represents a release of LH from the foetus at that time? If so, have you any views on its possible significance in view of the fact that absence of the pituitary from foetal calves is associated with prolonged gestation?

KARG

I think it is a rather interesting point which should be considered in the future. At this stage I would not like to speculate further.

IV. COMPARATIVE ASPECTS OF REPRODUCTION IN POULTRY

RELEASING FACTORS AND LH IN THE PLASMA OF INTACT AND HYPOPHYSECTOMIZED CHICKENS

A. V. NALBANDOV and D. W. BULLOCK

Department of Animal Science—Genetics, University of Illinois, Urbana, Illinois

THE OVULATORY CYCLE OF BIRDS

ONE OF the more interesting problems of reproductive endocrinology lies in the mechanism of control of the ovulatory cycle of birds. An indeterminate layer like the domestic hen must have a mechanism which assures her that not more than one ovum is matured and ovulated on each successive day while the polytocous mammal must be able to mature groups of eggs at genetically predetermined intervals which are then ready for simultaneous ovulation. The daily ovulation of a single ovum clearly calls for a hypothalamo-hypophysial control mechanism which is distinctly different from the one which must prevent such an event and assure a much wider chronological spacing between single or multiple ovulations. It is now clear that in mammals a neurohumoral hypothalamo-hypophysial feed-back system operates in controlling the long interval ovulatory sequences. In the chicken it appears probable but not yet certain that a hypothalamic releasing factor system exists similar to the one in mammals. It is also well established that in mammals the gonadal steroids form an important link in the hypothalamo-hypophysial-gonadal (HHG) feed-back system. In some mammals, such as the reflex ovulators, a neural component also participates in the HHG system. Whether one should look for a steroid-controlled feed-back system, a neural one, or a neuro-humoral system in laying hens is at present not at all clear. The present discussion will present some data which will point out certain peculiarities of the endocrinology of the laying hen as well as of castrated and of normal males and which will attempt to direct attention to those aspects of the HHG feed-back system which will need to be studied further.

LH-RELEASE PEAKS

At a previous Nottingham Symposium as well as in our published reports (Nelson, Norton and Nalbandov, 1965a, b), we have amply documented our conviction that it is actually LH which we are measuring in plasma and in hypophyses using the ovarian ascorbic acid depletion method of Parlow (1961). I should like again to refer to data which concerns the time of LH release and the peaks of its synthesis during one complete interval in the laying cycle of a hen. Much to our amazement it turned out that the hen has not one LH-release peak as does the mammal, but three LH-release peaks (Nelson, 1965a). We have recently confirmed that the three peaks are really there and these new results are shown in *Figure 1*.

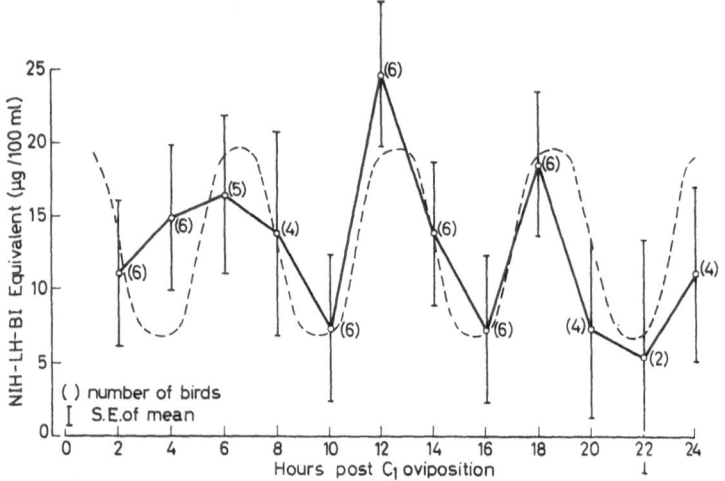

Figure 1. Mean plasma concentration of LH (solid line) during the interval preceding C_3 ovulation in laying hens and a fitted sine curve (broken line)

It turns out (Nelson *et al.*, 1965a) that both the second and third peaks shown are required for ovulation. This statement is based on the observation by van Tienhoven, Nalbandov and Norton (1954) that if the second peak is blocked by Dibenamine, a neural blocking agent, the majority of hens do not ovulate. Similarly, Rothchild and Fraps (1949) have found that in hens hypophysectomized at a time when the second LH peak had already occurred but the third release of that hormone had not yet taken place, ovulation was absent. Whether the first peak of LH release, the one occurring 20 h before ovulation, is involved in causing ovulation is at present not clear. Of itself the first peak is insufficient to cause ovulation since ovulation is prevented in all hens hypophysectomized 18 h prior to expected ovulation.

Inspection of Bullock's data (he has confirmed the three peaks found by Nelson et al., 1965a), shows that the three peaks appear to recur at 6 h intervals as if we were dealing with a rhythmic repetition of the same event. An attempt to test this possibility was made by fitting a sine curve with a period of 6 h to Bullock's data. Although the actual and the fitted curves are visually similar, the goodness of fit was not statistically significant. The question now arises whether we are dealing here with a pre-set recurrent system or with one which depends on a signal for each release. A neural signal is known to exist for the first peak, which occurs immediately after the forming egg passes from the albumin-secreting portion of the oviduct into the shell gland (Huston and Nalbandov, 1953). Conceivably this event or the first peak of LH release, could trigger subsequent cycles of synthesis and release of LH peaks which, in sum, could form an adequate stimulus to trigger ovulation. This might imply, then, that oscillations set up by the first neurally-triggered release would be followed 6 h and 12 h later by two other releases. Equally plausible is the assumption that the second and third LH releases are not rhythmic and are not consequences of the first but that each of them requires a separate neural or humoral signal.

There is no significant change in the size of the follicle destined to ovulate nor in the size of the remaining follicles in the hierarchy during the interval from the first to the second LH release and prior to the next ovulation. It is, of course, possible that during this time a significant qualitative or quantitative change in ovarian steroids may take place which may trigger LH releases 2 and 3. Unfortunately, no information on such changes in ovarian steroid hormones is available and until it is available no final interpretation of the events can be made. It appears improbable that neural signals emanate from either the ovary or the duct system during the period of LH releases 2 and 3 although here again information is wanting.

At present it appears probable that there may be a hypothalamic LRF in birds although this possibility is not at all definite. One of the difficulties lies in the fact that in our studies we have been using the usual rat systems for testing for avian LRF which may turn out to be unsuitable for demonstrating chicken LRF. Even though the evidence for, or against, the existence of an avian LRF is equivocal,

245

we must certainly include it in any scheme designed to explain the events of the ovulatory cycle. At the moment there is no real choice possible just as in the case of LH release, between a pre-set, rhythmic system which calls for LRF discharge at 6 h intervals (with possible exception of the LRF discharge responsible for LH-peak 1), and a signalling system which is controlled either by the ovary or the oviduct. In short, our data are not yet adequate to interpret our findings and much additional work will be required before we have an understanding of events equal to that of scientists working with rats.

That the avian hypothalamo-hypophysial-gonadal control system differs from that of mammals is also seen from experiments in which the rate of LH synthesis and release following castration was examined. According to the literature, castration of mammals and of chickens causes an increase in the total gonadotrophic hormone content in the pituitary gland. This is usually explained on the basis that in castrates the gonadal steroids are missing and that in their absence the gonadal-hypothalamo-hypophysial feed-back relationship has been upset. If we restrict our attention to the LH component only, we find that in the rat, for instance, there is an increase in the hypophysial LH content after ovariectomy. Taleisnik and McCann (1961), have also followed the changes in plasma LH of ovariectomized rats and have found an immediate and highly significant increase in plasma LH activity within 1 week after ovariectomy but from then on and for the next 16 weeks there is no further increase in plasma LH.

When male chickens are castrated and hypophysial and plasma LH activities are determined, one finds a picture which is distinctly different from that found in mammals. One sees (*Figure 2*) that, as in mammals, the plasma LH is significantly increased within 10 days after castration but that, unlike the mammal, the chicken continues to show a great increase in LH activity for the duration of the experiment. Unfortunately, the experiment was terminated 80 days after castration and we do not yet know how much longer this increase would have lasted and if it would have ever plateaued. Note that the hypophysial LH concentration in castrated male chickens actually *decreases*. This fact, together with the enormous increase in plasma LH suggests that the adenohypophysis is synthesizing LH at a very rapid rate and that this hormone is released just as fast as it is made and that none can thus be stored. In spite of the significant increase in hypophysial weight of castrates over intact controls, total anterior pituitary LH content of castrates is lower than that of controls.

These findings raise several problems for which there are no ready answers. If it is assumed that normally the steroid feed-back mechanism controls the RF and through it the rate of synthesis and release of LH, then in the absence of the feed-back it appears that

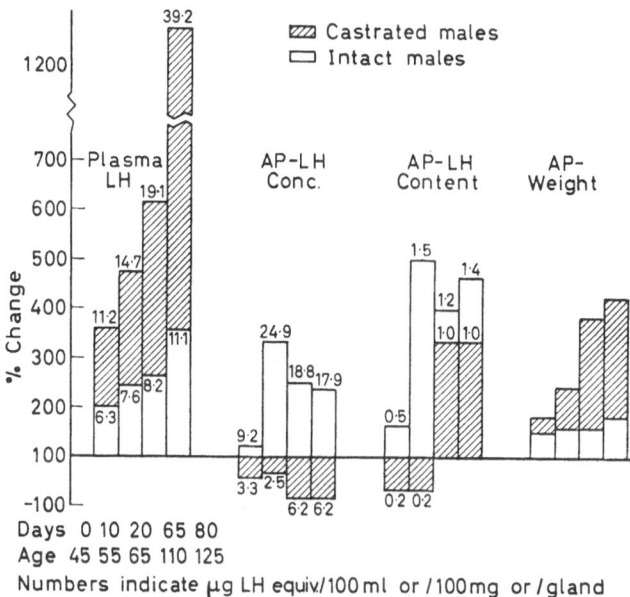

Figure 2. Comparison of changes in hypophysial and in plasma LH in intact and castrated cockerels

both the LRF and LH discharges become uncontrolled and, what is more surprising, one or both of them do not plateau for at least as long as 80 days after interruption of the feed-back system. The finding that the hypophysial LH concentration actually decreases when compared to the pre-castration LH concentration confirms the finding of others that the primary feed-back effect of the gonadal steroid is at the hypothalamic level (control of RF) rather than at the hypophysial level. The observation that in castrates there appears to be a limitless increase in LRF synthesis or secretion suggests that the primary control mechanism is humoral and not neural. Whether the control system in male chickens is distinctly different from the one in females remains, for the time being, unknown.

247

CHANGES IN LUTEINIZING HORMONE

Finally, I would like to discuss the appearance of a mystery factor in the plasma of male chickens after hypophysectomy. In the course of our experiments (Nelson *et al.*, 1965a, b), in which we wanted to be certain that we were actually measuring changes in LH in the plasma of castrated males and of laying hens, it became desirable to assay the plasma of hypophysectomized animals on the assumption that it would be completely devoid of LH activity. Much to our surprise we found that plasma of hypophysectomized males had a considerable ascorbic acid depleting potency when assayed in the pseudopregnant rat. The obvious interpretation that hypophysectomy had been incomplete was negated by serially sectioning the sella turcicae of hypophysectomized animals and by verifying that adenohypophysial tissue was absent. That the plasma substance giving ascorbic acid depletion was not due to LH was further supported by the findings that the combs and testes of hypophysectomized males rich in this substance were atrophic and that such plasma injected into hypophysectomized male rats gave no increase in prostate size and that no ovarian hyperemia response was obtained in female rats whereas the plasma of castrated cocks did give such a response. A comparison of the effects of plasma from intact and hypophysectomized cocks in the OAAD test is shown in Table 1. It should be noted that immediately after hypophysectomy, and for about 2 or 3 weeks thereafter, the plasma has no ascorbic acid-depleting factor (Frankel, Gibson, Graber, Nelson, Reichert and Nalbandov, 1965). These authors also found that this factor can be obtained after passing the plasma over Sephadex-G 100 (Table 2).

Table 1

OAAD activity of peripheral plasma from intact and hypophysectomized cockerels

No. of ♂ treatment	Blood obtained (days)		Mean AA depl. % ± S.E.	Lambda
	of age	After hypoph.		
6 Intact	65–73	—	21·4 ± 3·1*	0·11
3 Intact	365	—	31·9 ± 5·0*	0·18
5 Hypox	73–85	< 14	4·6 ± 6·1	0·11–0·21
7 Hypox	92–105	25–34	23·1 ± 5·1*	0·19–0·32
12 Hypox	150–211	> 105	34·0 ± 3·6*	0·11–0·32

* $P \leq 0.05$ or better; 3 ml of plasma/100 g rat.

Table 2

OAAD activity of Sephadex-G-100 fractions of plasma of intact and hypophysectomized cockerels

No. of ♂ treatment	Blood obtained (days)		Fraction No.*	Mean AA depl. % ± S.E.	Lambda
	of Age	After hypoph.			
4 Intact	188–228	—	1	54·9 ± 7·2†	
			2 + 3	46·8 ± 5·5†	0·15
			4	7·4 ± 8·1	
7 Hypox	188–202	127–133	1	22·3 ± 6·7	
			2	13·3 ± 7·5	
			3	23·3 ± 7·4†	0·15
			4	8·5 ± 11·2	

* Equivalent of 3/8 ml of plasma/100 g rat.
† $P \leq 0.05$ or better.

In view of the findings of Nallar and McCann (1965) and of
Schwartz and Caldarelli (1965) that an LRF can be demonstrated
in the plasma of hypophysectomized rats, the assumption seemed
logical that the OAAD factor in plasma of hypophysectomized
chickens was also an LRF. It was thought that the injection of such
plasma into a rat prepared for the OAAD assay caused release of the
rat's own LH which in turn acted on the luteinized ovary (Frankel
et al., 1965). However, on further experimentation the situation
seems not quite so simple. For instance, Stevens (1966) has found
that the plasma factor does not disappear, as would be expected,
after placing lesions in the hypothalamus of hypophysectomized
chickens or after injecting them with large doses of progesterone.
Using the in vivo rat systems which are specifically designed to
measure LRF activity, the hypophysectomized chicken plasma factor
gives equivocal results in that in some experiments it acts like a
typical LRF while in others these results are not reproducible. The
difficulties encountered in these assay systems, including the fact that
the plasma factor has caused ascorbic acid depletion even in some
experiments with hypophysectomized rats used as plasma donors, or
assay animals, find no ready explanation.

The assay systems used are all designed for testing mammalian
LRF and they depend on the ability of the injected RF to release the
animal's own LH within minutes after injection. It appears possible that when these systems are used for a species-alien RF they do

249

not perform reliably. The chicken plasma factor may be bound to a large protein molecule which may have to be broken before the LRF can have its releasing effect. We are now in the process of establishing whether chicken hypothalami contain an LRF, whether it can be tested in systems based on the rat or whether we must use the chicken as a test animal. Meanwhile, all we can say is that there is a factor in the plasma of hypophysectomized male chickens which causes ascorbic acid depletion in the Parlow rat. Whether this depletion is due to an LRF, or a non-specific factor which happens to have ascorbic acid depleting ability or whether it is a mixture of both, remains to be determined.

ACKNOWLEDGEMENT

This study was in part supported by NIH Grant AM 6976. We wish to acknowledge the generous gifts of purified hypophysial hormones by the Endocrinology Study Section, NIH.

REFERENCES

Frankel, A. I., Gibson, W. R., Graber, J. W., Nelson, D. M., Reichert, L. E. and Nalbandov, A. V. (1965). *Endocrinology* **77**, 651–657

Huston, T. M. and Nalbandov, A. V. (1953). *Endocrinology* **52**, 149–156

Nallar, R. and McCann, S. M. (1965). *Endocrinology* **76**, 272–275

Nelson, D. M., Norton, H. W. and Nalbandov, A. V. (1965a). *Endocrinology* **77**, 731–734

— — — (1965b). *Endocrinology* **77**, 889–896

Parlow, A. F. (1961). In: *Human Pituitary Gonadotropins*, pp. 300–310. Ed. by A. Albert. Springfield, Illinois; Charles C. Thomas

Rothchild, I. and Fraps, R. M. (1949). *Endocrinology* **44**, 134–149

Schwartz, N. B. and Caldarelli, D. (1965). *Proc. Soc. exp. Biol. Med.* **119**, 16–20

Stevens, K. R. (1966). Research in progress

Taleisnik, S. and McCann, S. M. (1961). *Endocrinology* **68**, 263–272

van Tienhoven, A., Nalbandov, A. V. and Norton, H. W. (1954). *Endocrinology* **54**, 605–611

DISCUSSION

Dr. HEALD (*London*)

I share Professor Nalbandov's frustration at the variability he gets with assays. We have attempted to assay LH in the plasma and serum of laying birds by the Parlow method and have been totally unsuccessful in getting any depletions at all even when using volumes of serum as high as 4 ml/rat. We have been able to recover LH added to the plasma, either as HCG or pituitary extracts. Perhaps your variability may be due to the strain of rat.

NALBANDOV

If this is the explanation then it is not the most obvious one because we were using the same strain of rat used for the LH bio-assays.

PROFESSOR H. KARG (Munich)

Is this mysterious factor heat stable?

NALBANDOV

No. It will pass through a dialysis bag and through the Sephadex column which shows that it must be a small molecular weight protein.

PROFESSOR R. M. MELAMPY (Iowa)

Have you tried any day-old chick tests with these preparations?

NALBANDOV

With LH but not with this unknown factor. This adds to our feeling that we are measuring LH as these results were comparable by the P32 assay and by the Parlow assay except that the former is far more sensitive.

PROFESSOR I. ROTHCHILD (Cleveland)

To test the reality of this releasing factor in the hypophysectomized chicken, it seems a very logical experiment to transplant the pituitary to an ectopic site. If the material really is a 'Releasing Factor', and is circulating, would not the transplanted pituitary be able to secrete LH and maintain spermatogenesis and testis weight?

Concerning the cycles of LH discharge in the hen in relation to ovulation, as you know there are variations in ovulation frequency from 24 h to 36 h or even longer. Would it not be interesting to see if there is maintenance of the 6 h rhythm in LH release, regardless of the ovulation frequency, or whether LH release is tied to ovulation frequency?

NALBANDOV

It is tied to ovulation; for there are no peaks in blood taken during the non-laying period. I do not think that a transplanted pituitary would necessarily prove that there is a releasing factor, because in the Nikitovitch-Winer system, she has to infuse the releasing factor into the transplanted pituitary itself in order to get synthesis and release of pituitary hormones. It does not necessarily follow that the RF would reach a transplanted pituitary.

PROFESSOR G. P. VAN REES (Leiden)

I have obtained similar results using rat material. Depletion occurred in blood from hypophysectomized rats and from hypophysectomized prolactin-maintained parallel rats. Median eminence extracts from these hypophysectomized animals also caused depletion. It is possible that our strain of animals may be reacting abnormally but, on the other hand, our results coincide with yours.

NALBANDOV

I am glad to hear this, because depletion in hypophysectomized rats bothers me very much, and I am sure it will bother Dr. Parlow more.

251

REPRODUCTION IN THE FEMALE MAMMAL

PROFESSOR S. M. MCCANN (*Texas*)

In view of your variability and the demonstration of what appears to be a releasing factor, I was wondering if there was any variability in your operative procedure as we have seen in the hypophysectomized rat. Are you damaging the hypothalamus in some experiments and not in others?

NALBANDOV

We take histological sections to confirm hypophysectomy and examine the hypothalamus. Damage of the hypothalamus does occur occasionally.

MR. P. C. WILLIAMS (*London*)

Concerning your suggestion that oviduct stimulation may trigger LH release, what happens if you remove the oviduct or anaesthetize it? Does it affect the next ovulation?

NALBANDOV

If you remove the oviduct, ovulation will continue into the body cavity. This is evidence against oviduct stimulation as the triggering mechanism.

DR. I. R. FALCONER (*Nottingham*)

Have you looked at the possibility that ovarian ascorbic acid depletion may be due to an excess of catecholamines in the serum of the animals which could be removed by mild oxidizing conditions.

NALBANDOV

Yes. We have looked at this and it is not due to catecholamines. You require enormous quantities to cause depletion.

DR. B. J. DONOVAN (*London*)

I am intrigued by the genesis of this 6 h rhythm. Is there any information on further rhythms in the chicken or is this a unique phenomenon for LH release? It seems difficult to explain this on the basis of a feed-back action of gonadal hormones for example. It seems far too rapid.

NALBANDOV

There are no other 6 h rhythms as far as I know, but the thyroid has a 12 h rhythm. I agree with your second point.

PROFESSOR I. ROTHCHILD

Is the variability in detection of this material between birds or between assays?

NALBANDOV

Between assays.

ROTHCHILD

In the Department of Biology, Dr. R. N. Foreman has developed an idea that the assay depends on a hyperaemic effect in the ovary. What we are measuring may not be a depletion, but a lowering of concentration because of an increase in weight of the ovary. It is conceivable that, in some circumstances, a change in concentration of ascorbic acid could occur because of the change in ovarian weight due to some substances which can increase it. With regard to Dr. Falconer's remark this is something worth considering.

H. KARG

In our 1959 paper on ovarian ascorbic acid depletion in rats, we calculated the time/dose response to HCG and calculated what was due to the weight of the ovaries and what was due to depletion and the work showed that both contribute to the effect, with confirmation of definite depletion.

THE MAINTENANCE OF SPERMATOZOA IN THE OVIDUCT OF THE DOMESTIC FOWL

P. E. LAKE

*Agricultural Research Council Poultry Research Centre,
The King's Buildings, West Mains Road, Edinburgh, 9*

INTRODUCTION

THE PHENOMENON of fertilization in animals accomplished by spermatozoa which have been stored within the oviduct for extremely long periods has been discussed recently by Parkes (1960). It occurs in the invertebrates, particularly insects, where the necessary storage of spermatozoa is usually carried out in special diverticulae of the oviduct, the spermathecae. The survival time in the female insect varies with the species; the desert locust (*Schistocerca gregaria* Forsk) can lay fertile eggs for 10 weeks after the last copulation (Hunter-Jones, 1960) whereas the queen bee has been known to retain viable spermatozoa for several years (Courrier, 1921).

Prolonged survival of spermatozoa within the female body is comparatively less common among the vertebrate animals. Viviparous fish have been known to retain spermatozoa for several weeks, and some reptiles for several months or years (*see* review by Parkes, 1960). The site of storage of spermatozoa in the viviparous fish is either folds in the wall of the oviduct or crypts in the ovary (Stolk, 1950). Kohlbrugge (1912) observed accumulations of spermatozoa just below the shell (jelly) gland of the oviduct of some elasmobranchs.

In reptiles, the spermatozoa are stored in epithelial glands of the oviduct either at the base of the infundibulum or in the anterior vagina (Fox, 1963).

Domestic poultry and other birds retain spermatozoa in the oviduct for prolonged periods, the mean duration of which depending upon the species, e.g., domestic fowl, about 12 days (Polge, 1951); geese, about 9 days (Johnson, 1954); ducks, about 7 days (Ash, 1962); Japanese quail, 6·3 days (Sittman and Abplanalp, 1965) and turkeys, about 30 days (Parker, 1949). Ring-doves have been known to lay fertile eggs 8 days after the removal of the male (Riddle and Behre, 1921). Several investigations have been made to discover how spermatozoa survive for these long periods within the oviduct of the domestic fowl. Iwanow (1924) suggested that spermatozoa fertilized immature oöcytes in the ovary, as well as mature ova, because he failed to terminate the laying of fertile eggs by hens after irrigating the oviduct and the peritoneal cavity with spermicidal fluids. Walton and Whetman (1933) confirmed the findings with spermicidal solutions but postulated that the spermatozoa must reside in crevices of the oviduct mucosa where they were protected from the fluids.

Van Drimmelen (1946) discovered glands, containing spermatozoa, in the chalaziferous region (posterior infundibulum) of the fowl oviduct and believed they were the residence sites for spermatozoa during the prolonged period that a hen lays fertile eggs following a single insemination. However, renewed interest in this phenomenon was aroused recently when Bobr (1962), Bobr, Lorenz and Ogasawara (1964a) and Fujii and Tamura (1963) observed aggregations of spermatozoa in epithelial glands at the junction of the uterus and vagina in the fowl. Bobr *et al.* (1964a) suggested that the residence of spermatozoa in the glands of the infundibulum, described by Van Drimmelen (1946), was an artefact of abnormal, experimental insemination techniques and that the utero-vaginal glands were the natural residence sites for spermatozoa following copulation or intra-vaginal artificial insemination. Accumulations of spermatozoa can be found in the utero-vaginal glands in progressively decreasing numbers for as long as fertile eggs are laid by fowl or turkey hens after single inseminations (Bobr *et al.*, 1964a; Verma and Cherms, 1965).

The fertilizing life of spermatozoa in the oviduct of mammals, and in that of the human, is generally not more than 2 days, with the exception of the horse (6 days), ferret (126 h) and bat (156 days) (*see* Chang, 1965). Ovulation and sexual receptivity in the majority of female mammals would appear to be closely synchronized and conditions for the sustenance of spermatozoa in the oviduct may only be favourable for relatively short periods. Thus, the short fertilizing life of mammalian spermatozoa in the oviduct may only be apparent, and limited by these circumstances rather than by inherent properties of the spermatozoa. This is an interesting possibility for future consideration.

This communication will deal with the phenomenon of the prolonged storage of spermatozoa in the oviduct of the domestic fowl and the investigations that were begun recently to discover the role

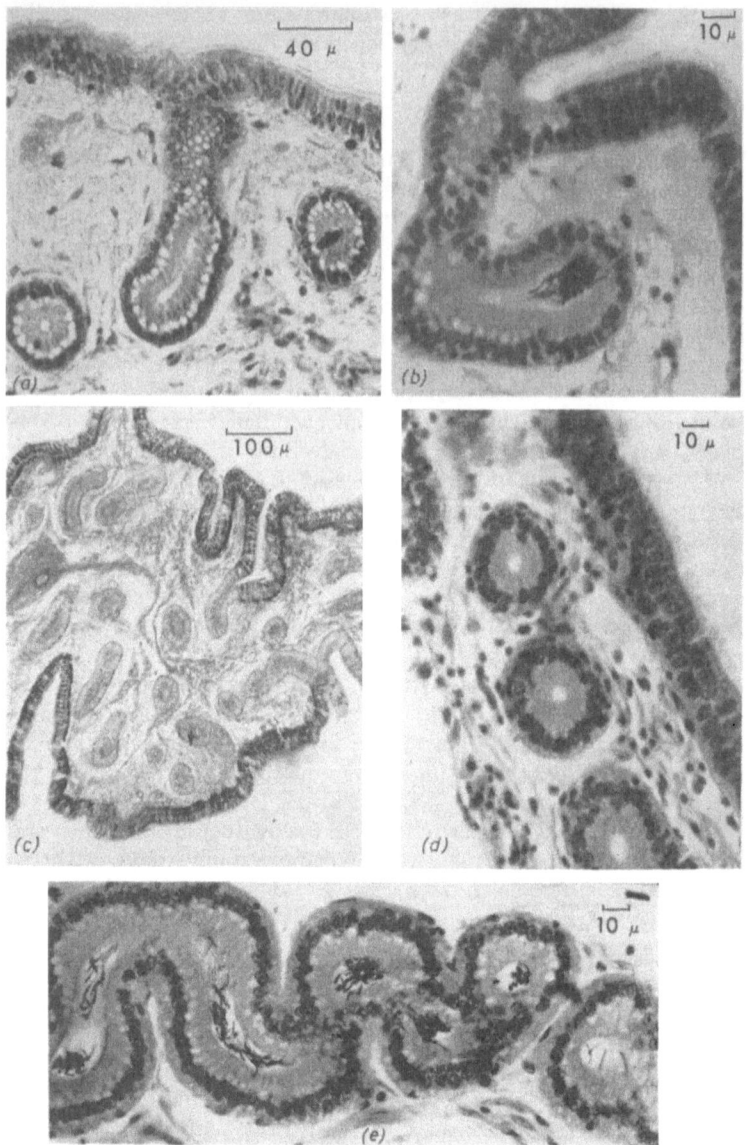

Plate 1

of the storage sites for spermatozoa in the oviduct in the process of reproduction in the bird. Preliminary work, pertinent to the discovery of possible mechanisms by which the spermatozoa are periodically activated to move to the upper parts of the oviduct for fertilization, are also discussed.

RESIDENCE SITES OF SPERMATOZOA IN THE OVIDUCT OF THE DOMESTIC FOWL

Large numbers of spermatozoa enter and reside in glands in a distinct region of the oviduct at the junction of the uterus (shell gland) and vagina after natural or artificial intravaginal insemination. The general mucosa in this area is folded and the structure of the glands in these folds has been described by Fujii and Tamura (1963) and Bobr *et al.* (1964a). Fujii and Tamura called them *vaginal glands*. Typical pictures obtained during the current work are shown in *Plate I*. The *glands* are of the simple, tubular type and are lined by columnar cells which display secretory activity; large supra-nuclear droplets are present (*Plate I, a* and *b*) which are believed to contain cholesterol esters. Phospholipids and mucopolysaccharides were not detected (Fujii, 1963). The general stratified, columnar epithelium, lining the folds of the mucosa (*Plate I, a*) contain cells which secrete abundant mucopolysaccharide but they are not found beyond the neck of the *vaginal glands* (*Plate I, c*). However, the secretion of mucopolysaccharide can often be found in the lumen of these glands, and it is feasible that this mucinous secretion provides protection to the spermatozoa and may account for the failure of Iwanow (1924) and Walton and Whetman (1933) to kill spermatozoa by irrigating the oviduct.

Further work is required to determine the precise composition of the secretions of the *vaginal glands* and establish whether any component(s) is utilized for the sustenance of the stored spermatozoa.

PLATE I (Reduced ⅔ on reproduction)

a. *Section of an active vaginal gland showing its simple tubular structure. Note the supra-nuclear secretory vacuoles in the columnar cells of the gland and a bunch of spermatozoa in the lumen of a section of a gland. Modified Carnoy-Lebrun fixation*

b. *Section of an active vaginal gland showing spermatozoa in the lumen. Note stereocilia on cells of general mucosa. Bouin fixation*

c. *Low-power section of oviduct at the utero-vaginal junction showing the secretion of mucopolysaccharide by the cells of the epithelial folds. Note little or no secretion of such substances by the cells of the vaginal glands. P.A.S. staining*

d. *Section of a vaginal gland from an infertile turkey. Note the absence of spermatozoa and supra-nuclear secretory vacuoles*

e. *Longitudinal section of a vaginal gland from a fertile turkey (control to that shown in Figure d). Note the abundance of spermatozoa and the supra-nuclear secretory vacuoles*

Recently, this region of the oviduct of the Brown Leghorn hen was shown to display intense red fluorescence when irradiated with ultra-violet light, which is indicative of the presence of porphyrin. Whether this compound is of any significance to the survival of spermatozoa or is concerned solely with the formation of the outer cuticle of the egg shell has yet to be investigated.

With regard to the prolonged survival of spermatozoa in the oviduct of the fowl, it may be significant that leucocytes, which phagocytose the spermatozoa of some mammals in the lumen of the oviduct (Austin, 1957; Chang, 1956), have never been observed in the bird. Compared with the cyclic mammal, a relatively constant production of hormones (oestrogen and progesterone) occurs in a hen as it ovulates on successive days (in sequences with a day's pause), often for a long period. This maintains the oviduct in an active state for the whole period, which may extend to several months. This relatively constant state of activity may be associated with the absence of phagocytosis of spermatozoa in the oviduct, as it is generally believed that a change in hormone balance during the oestrous cycle of the female mammal favours the phagocytic activity.

AN ASSOCIATION BETWEEN ACTIVE VAGINAL GLANDS AND THE MAINTENANCE OF THE FERTILIZING CAPACITY OF STORED SPERMATOZOA

A few observations have been made which indicate that the survival of spermatozoa in the oviduct, and, therefore, the production of fertile eggs by a hen, could depend upon the active functioning of the *vaginal glands*. For example, it is known that spermatozoa are less able to remain in atrophied glands as occur in moulting hens (Fujii and Tamura, 1963). Under these conditions there is a definite hypo-secretion of gonadal hormones and the normal functions of the oviduct are not maintained.

An infertility syndrome has recently been recognized in turkey hens which is characterized by the premature cessation of the production of fertile eggs midway through the breeding season. The number of ova ovulated and laid by the affected hens is not altered, but the majority of *vaginal glands* in the oviduct at *post mortem* examination are seen to be devoid of spermatozoa and the characteristic supra-nuclear, secretory droplets in the columnar cells are absent. Control, fertile hens mated to the same males show *vaginal glands* actively secreting and laden with spermatozoa (*Plate I, d* and *e*). The factor(s) causing the infertility is still obscure and may not be

258

due to changes in the secretion of gonadal hormones. Clearly in these cases the *vaginal glands* are adversely affected and spermatozoa are not stored in them.

The question of the control of the function of the *vaginal glands* under normal circumstances in the fowl is an interesting one for future study. Lamoreux (1940) showed that hens, laying eggs most intensively, laid fewer infertile eggs and had a consistently longer duration of fertility than poor layers. He postulated that the level of oestrogens in circulation in high-producing hens might play a part in maintaining a favourable environment for the support of spermatozoa in the oviduct. This is an attractive hypothesis which has yet to be substantiated.

THE SELECTION OF SPERMATOZOA IN THE OVIDUCT FOR FERTILIZATION AND THE RESTRICTION OF THE NUMBER REACHING THE SITE OF FERTILIZATION—A POSSIBLE ROLE OF THE VAGINAL GLANDS

Austin and Braden (1952), Braden (1953) and Mattner (1963) showed that the utero-tubal junction or cervix or both in the rat, rabbit and ewe exerted a restrictive action on the number of spermatozoa passing through. Allen and Grigg (1957) suggested that the utero-vaginal junction might restrict the number of spermatozoa reaching the site of fertilization in the hen, and, since the discovery of the storage of spermatozoa in the *vaginal glands*, more attention has been paid to the possible existence of natural mechanisms in the oviduct of the fowl for the selection of spermatozoa for fertilization.

Whilst it has not yet been proved conclusively that selection of spermatozoa occurs in the *vaginal glands*, certain observations have been made from which it can be inferred that the area of the oviduct containing them exerts a restrictive action on the transport of spermatozoa and that some property of the *vaginal glands* is most likely involved. Ogasawara, Lorenz and Bobr (1966) studied the activity in the oviduct of the spermatozoa from two males of proven low-fecundity. They produced semen of poor quality, as judged by certain *in vitro* tests. Few fertile eggs were obtained following natural mating or by intra-vaginal artificial insemination and few of their spermatozoa were found to enter the *vaginal glands*. An appreciable number of fertile eggs could be obtained if the semen was placed directly into the uterus when, concomitantly, large numbers of spermatozoa were found to enter the glands of the infundibulum, which were previously thought to be the natural

259

residence sites for spermatozoa in the upper part of the oviduct (Van Drimmelen, 1946). However, whereas the few fertile eggs obtained by intra-vaginal insemination produced mostly normal embryos, a high percentage of pre-oviposital, dead embryos were obtained after the intra-uterine insemination. This suggested to Ogasawara *et al.* (1966) the possibility that spermatozoa deposited above the region of the *vaginal glands* were not subject to selection and thus a number of abnormal spermatozoa were able to fertilize eggs. This result recalls the interesting observations made by Leonard and Perlman (1949), which showed that when rat spermatozoa were mixed with those of the bull, mouse or guinea-pig, and placed in the uterus of the female rat, very few of the foreign spermatozoa passed into the Fallopian tube. In this case there was a selection of different types of spermatozoa for transport to the Fallopian tube.

Apart from the possibility of the *vaginal glands* in the oviduct acting as a barrier to the progress of abnormal spermatozoa, they may be associated with the regulation of the numbers of spermatozoa reaching the site of fertilization at any given time. For instance, Van Krey, Ogasawara and Lorenz (1966) showed that there was an enormous increase in the numbers of spermatozoa entering the glands of the infundibulum, and also an increase in the length of the period during which fertile eggs were laid after a single insemination, if spermatozoa from normal males were allowed to enter directly into the magnum region of the oviduct, i.e. above the uterus. Again a very large number of pre-oviposital, dead embryos resulted. Van Krey *et al.* (1966) suggested, as a result of these experiments, that if unusually large numbers of spermatozoa are allowed to enter the upper parts of the oviduct *en masse*, high pre-oviposital, embryonic mortality may occur due either to an increased incidence of polyspermy or to an increased number of degenerating spermatozoa fertilizing the ova. Both hypotheses need further consideration, but the latter may be more pertinent because the length of the period during which fertile eggs were produced, was increased with intramagnal insemination, and Nalbandov and Card (1943) and Dharmarajan (1950) have shown that if spermatozoa remain in the oviduct beyond a certain limit of time, and then fertilize eggs, the chances of embryonic death are increased.

THE REGULATION OF TRANSPORT OF SPERMATOZOA IN THE OVIDUCT OF THE FOWL

If the spermatozoa are stored in the *vaginal glands* then a number of them must be released daily to ascend the oviduct to fertilize eggs.

At present we have no knowledge of the precise mechanism controlling this event. However, certain preliminary investigations have been made which could have a bearing on this problem. Bobr, Ogasawara and Lorenz (1964b) observed that spermatozoa were more frequently found freely motile in the lumen of the oviduct about

Table 1

The content of the principal inorganic ions (mEq/L; Mean ± S.E. of the mean) in the 'plumping' and 'oviposition' fluids from the uterus of Brown Leghorn hens (From M. H. El Jack and P. E. Lake, 1967, by courtesy of Blackwell)

	'Plumping' fluid	'Oviposition' fluid
Cations		
Sodium	139·1 ± 4·58	42·84 ± 2·11
Potassium	15·9 ± 0·79	75·0 ± 3·0
Calcium	28·3 ± 1·35	51·6 ± 6·85
Magnesium	2·74 ± 0·3	20·35 ± 2·45
Anions		
Chloride	79·9 ± 1·73	63·42 ± 3·89
Carbon dioxide	82·47 ± 1·75	91·26 ± 2·79

the time of ovulation or oviposition, suggesting that a stimulus for the mobilization of spermatozoa is provided at this time. With regard to the nature of this stimulus, El Jack and Lake (1967) examined the chemical composition of the secretions of the uterus (shell-gland) and obtained information on changes in its ionic composition at about the time of oviposition.

Fluid was taken from the uterus at about 15 h ('plumping' fluid), and less than 2 h ('oviposition' fluid), before oviposition and the ionic compositions were compared. The fluid present around the time of oviposition contained increased amounts of potassium, calcium, magnesium and carbon dioxide (Table 1). This investigation is to be continued in relation to the mechanism that mobilizes spermatozoa to ascend the oviduct for fertilization. Observations made so far could be pertinent to this problem for several reasons. Spermatozoa can ascend the oviduct of the hen within a few minutes (Mimura, 1939; Saeki, Tanabe, Katsuragi and Miyaga, 1963) and the peristaltic activity of the organ must be assumed to play a large part in the process. It is well known that potassium and calcium ions are implicated in the regulation of smooth muscle contractility (Burnstock, Holman and Prosser, 1963) and thus the great increase in

261

the concentration of potassium in the uterine fluid at about the time of oviposition might well be involved in the stimulation of muscular activity of the vagina and uterus and the expulsion of spermatozoa from their residence sites either just before, during or after oviposition (Bobr *et al.*, 1964b). Changes in the composition of the fluid environment of the uterus could also influence directly the activity of spermatozoa. In this respect it is interesting that a large amount of bicarbonate and carbon dioxide has been found in uterine and tubal fluids of rabbits (Vishwakarma, 1962; Hamner and Williams, 1964; Williams, Weinman and Hamner, 1964) as well as in the uterus of the hen (El Jack and Lake, 1967). These compounds are known to have a stimulatory effect on the respiration of rabbit, cock, bull and human spermatozoa when they are tested *in vitro*. Thus, the increase in carbon dioxide found in the uterine fluid of the fowl at the time of oviposition might feasibly be implicated in the mobilization of spermatozoa. Also, the rise in potassium could be a significant factor since it is known that this cation can stimulate the motility of fowl spermatozoa under certain conditions (Wales and White, 1958). These possibilities are to be the subject of future studies.

CONCLUSIONS

It has been the purpose of this communication to outline some of the intriguing problems confronting the physiologist who studies the storage of spermatozoa in the oviduct of the domestic fowl and the regulation of the transport of spermatozoa in the oviduct.

Work in this field is in its infancy and many questions remain to be answered. For instance, if spermatozoa are stored in the *vaginal glands*, at the junction of the uterus and vagina, they must be released periodically for transfer to the infundibulum, where fertilization occurs, during some part of the daily movement of ova down the oviduct. An interesting aspect of this phenomenon is that the spermatozoa must be released in groups and not *en masse*, since the period during which fertile eggs are produced after a single insemination extends over many days or weeks.

The phenomenon of 'capacitation', which is a type of maturation process undergone by the spermatozoa of mammals after they arrive in the oviduct and before they are capable of fertilizing eggs, is another aspect of the reproductive physiology yet to be considered in the fowl.

Lastly, extended studies of the biochemistry of oviduct secretions and of spermatozoa in the female fowl could probably provide information useful to the storage of fowl spermatozoa *in vitro*. Although they survive for long periods in the female, they are much more difficult to store *in vitro* than those of the bull.

ACKNOWLEDGEMENTS

The author gratefully acknowledges the collaboration of Professor F. W. Lorenz and Dr. F. X. Ogasawara of the Departments of Animal Physiology and Poultry Husbandry, Davis, California, U.S.A. in the preparation of this communication.

REFERENCES

Allen, T. E. and Grigg, G. W. (1957). *Aust. J. agric. Res.* **8,** 788–799
Ash, W. J. (1962). *Poult. Sci.* **41,** 1123–1126
Austin, C. R. (1957). *J. Endocr.* **14,** 335–342
— and Braden, A. W. H. (1952). *Nature, Lond.* **170,** 919–921
Bobr, L. W. (1962). 'Oviducal Distribution of Spermatozoa and Fertility of the Domestic Fowl.' *Ph.D. Thesis.* Univ. of California
— Lorenz, F. W. and Ogasawara, F. X. (1964a). *J. Reprod. Fert.* **8,** 39–47
— Ogasawara, F. X. and Lorenz, F. W. (1964b). *J. Reprod. Fert.* **8,** 49–58
Braden, A. W. H. (1953). *Austr. J. biol. Sci.* **6,** 693–705
Burnstock, G., Holman, M. E. and Prosser, C. L. (1963). *Physiol. Rev.* 482–527
Chang, M. C. (1956). *Annali Ostet. Ginec.* **4,** 74–86
— (1965). *J. exp. Zool.* **158,** 87–100
Courrier, R. (1921). *C.r. Séanc. Soc. Biol.* **85,** 941–943
Dharmarajan, M. (1950). *Nature, Lond.* **165,** 398
El Jack, M. H. and Lake, P. E. (1967). *J. Reprod. Fert.* **13,** 127–132
Fox, W. (1963). *Nature, Lond.* **198,** 500–501
Fujii, S. (1963). *Archvm. Histol. jap.* **23,** 447–459
— and Tamura, T. (1963). *J. Fac. Fish. Anim. Husb. Hiroshima Univ.* **5,** 145–163
Hamner, C. E. and Williams, W. L. (1964). *Fedn Proc. Fedn Am. Socs exp. Biol.* **23,** 430 (Abstr.)
Hunter-Jones, P. (1960). *Nature, Lond.* **185,** 336
Iwanow, E. (1924). *C.r. Séanc. Soc. Biol.* **91,** 54–56
Johnson, A. S. (1954). *Poult. Sci.* **33,** 638–640
Kohlbrugge, J. H. F. (1912). *Arch. EntwMech. Org.* **35,** 165–188
Lamoreux, W. F. (1940). *J. agric. Res.* **61,** 191–206
Leonard, S. L. and Perlman, P. L. (1949). *Anat. Rec.* **104,** 89–102
Mattner, P. E. (1963). *Aust. J. biol. Sci.* **16,** 877–884
Mimura, H. (1939). *Okajimas Folia anat. jap.* **17,** 459–476
Nalbandov, A. and Card, L. E. (1943). *Poult. Sci.* **22,** 218–226
Ogasawara, F. X., Lorenz, F. W. and Bobr, L. W. (1966). *J. Reprod. Fert.* **11,** 33–41

Parker, J. E. (1949). In: *Fertility and Hatchability of Chicken and Turkey Eggs,* Chap. 3. Ed. by L. W. Taylor. London; Chapman & Hall

Parkes, A. S. (1960). In: *Marshall's Physiology of Reproduction,* Chap. 9, Vol. 1, Part 2, 3rd edn. Ed. by A. S. Parkes. London; Longman's Green

Polge, C. (1951). *Proc. Soc. Study Fert.* **2,** 16–22

Riddle, O. and Behre, E. J. (1921). *Am. J. Physiol.* **57,** 228–249

Saeki, Y., Tanabe, Y., Katsuragi, T. and Miyaga, M. (1963). *Bull. Nat. Inst. Anim. Indy.* **3,** 91–98. Tokyo

Sittman, K. and Abplanalp, H. (1965). *Br. Poult. Sci.* **6,** 245–250

Stolk, A. (1950). Histo-endocrinological Analysis of the Gestation Phenomena in the cyprinodent, *Lebistes reticulatus* (Trans.) *Thesis.* University of Utrecht

Van Drimmelen, G. C. (1946). *Jl S. Afr. vet. med. Ass.* **17,** 42–52

Van Krey, H. P., Ogasawara, F. X. and Lorenz, F. W. (1966). *J. Reprod. Fert.* **11,** 257–262

Verma, O. P. and Cherms, F. L. (1965). *Poult. Sci.* **44,** 609–613

Vishwakarma, P. (1962). *Fert. Steril.* **13,** 481–485

Wales, R. G. and White, I. G. (1958). *Aust. J. biol. Sci.* **11,** 589–597

Walton, A. and Whetman, E. O. (1933). *J. exp. Biol.* **10,** 204–211

Williams, W. L., Weinman, D. E. and Hamner, C. E. (1964). *Proc. 5th Int. Congr. Anim. Reprod. and A.I.* Trento. Sec. 2, 367–370

DISCUSSION

Dr. M. C. Chang (*Massachusetts, U.S.A.*)

Do you think that the length of time that sperm remain viable is a factor of the environment in the female tract or of the sperm itself? Why is it longer in some species than in others?

Lake

I think that the life of the sperm in mammals is governed by conditions in the oviduct which are only favourable for very short periods. I do not think anybody has done experiments to find if mammalian sperm would live for a long period if conditions were favourable. If one could keep the animal in constant oestrus for long periods and then put spermatozoa in and test the fertilizing life, this could be very interesting.

Dr. J. R. Clark (*Oxford*)

I would like to ask whether there is any smooth muscle around the vaginal glands and what happens if you inject smooth muscle stimulants?

Lake

Work is in progress at both Edinburgh and Davis, California. The first thing we looked for was evidence of myoepithelial cells such as are present in the mammary gland of the mammal, and so far, we have not found any cells resembling them. In experiments with oxytocin and, in the case of the chicken, arginine vasotocin, the specific oxytocic principle in the bird, we have no encouraging results.

264

SPERMATOZOA IN THE OVIDUCT

PROFESSOR A. W. NALBANDOV (*Illinois, U.S.A.*)

Chicken sperm do not survive well *in vitro*. It is difficult to keep them alive for more than a few hours in contrast to most mammalian sperms. I believe the information about the mare and the bat is not now correct. You quoted 5 days' survival for the mare but it now seems to be as short lived as in other mammals. The bat apparently re-mates in the spring rather than use the sperm remaining from the fall mating.

PROFESSOR W. M. HANSEL (*Ithaca, U.S.A.*)

What do you know about the motility of the sperm when they are in the glands?

LAKE

With regard to the motility of the sperm, we have no direct evidence because direct observations are difficult to make. From the way in which the sperm are found in histological sections, I doubt whether they are freely motile within these glands during the period of survival. They are motile for a very short period after insemination and they must be stimulated to become motile under certain circumstances which we are trying to discover. Whilst they reside in the gland during the non-stimulated period, I doubt whether they are freely motile.

DR. J. M. BEDFORD (*London*)

I would like to ask you why you favour the utero-vaginal junction as a site of storage in the fowl rather than this upper infundibular region as this would seem more suitable for storing sperm. Secondly, I would like to make a comparison with the dogfish which seems to store sperm in the oviducal gland. As the egg distends the tube, the gland is distended and sperm are squeezed out physically. I wondered whether anything like this could happen in the chicken?

LAKE

During the moment of fertilization I think this happens. Fertilization definitely occurs in the infundibulum. With regard to your first question—I think sperm reside in the infundibular glands for very short periods and they can be regarded as the forward trenches. They are mobilized from the utero-vaginal glands and go up the oviduct and stay in the infundibular glands for a short period before fertilization and the actual mobilization from these sites is by distension of the oviduct. We favour these utero-vaginal glands because when glands were first reported in the infundibulum by Van Drimmelen, people who repeated the work could only find sperm in these glands if an excessively large number of sperm were put into the hen, either by intraperitoneal, intrauterine, or even by intravaginal insemination. Only in these circumstances could sperm always be found in these upper glands. These are the types of observation which at the moment, we feel, indicate that the actual storage site is the utero-vaginal glands and that periodically, batches are released to ascend the oviduct. If you examine the vitelline membranes of eggs after they have been laid you will find 200–300 sperm on each one. All the quantitative work which has been done on the number of sperm residing in the infundibular glands show that there are sufficient only to fertilize one or two eggs.

PROFESSOR T. J. ROBINSON (*Sydney, Australia*)

One point of great importance is the fundamental difference between birds and mammals regarding temperature regulation and the effect of temperature on spermatozoa. Earlier, Professor Nalbandov likened his birds to rats with feathers. I would liken them more to lizards with feathers. The evolutionary process has been such that in the mammal, nature has gone to an enormous amount of trouble to ensure effective cooling of the testes and maintenance of temperature below that of the body cavity. We know the extraordinary sensitivity of mammalian semen to temperature change, therefore I do not share Dr. Lake's optimism that one can manipulate the chemical environment within the reproductive tract of mammals so as to ensure longevity of spermatozoa. It would seem that there is a fundamental difference between mammals on the one hand which have delayed implantation and reptiles and birds on the other, where spermatozoa survive for long periods within the female tract, which is not known in any of the mammals.

Dr. J. M. Bedford

We know that male mammals will store semen in the epididymis and they have the ability to copulate daily or more than once per day. What is the situation with the bird? If the male bird in the natural environment has a large flock about him and is not competent to copulate more than once or twice a day I wonder if this is a factor why sperm are stored in the female rather than in the male?

Lake

Bird sperm are stored in the vas deferens and not the epididymis which is vestigial. There is some work which shows that sperm can survive in the vas deferens of the bird for at least 20 days. With regard to copulation frequency, it has been reported that the male fowl will mate forty or fifty times a day when running with a flock of hens but it has been shown that not all these copulations result in ejaculation and about 14 per cent are aspermic.

Dr. P. J. Heald (*London*)

We have conducted an extensive series of investigations and have isolated from the utero-vaginal junction and from the infundibular region a specific protein, a poly-L α-glutamic acid of molecular weight 80,000. This material, although well known as a synthetic substance has never been isolated from a natural source to our knowledge. When incubated *in vitro* this material immobilizes sperm at the body temperature of the bird (41·3°C), but will permit motility to be regained if the temperature is dropped to a lower level of about 37°C. This substance is found only in the utero-vaginal junction and in the infundibular region where the sperm are stored. This phenomenon of immobilization is absolutely specific to poly-L-glutamate as far as we can determine. We have tried many other compounds and tissue extracts but they were inactive in this respect. We have not had a great deal of success in preserving fertility *in vitro* which is disappointing but we feel that because of its localization and its apparent uniqueness of action, it could have some connection with the preservation of the sperm *in vivo* and I was very interested to hear that Dr. Lake too is considering materials of this type.

THE CIRCADIAN RHYTHM
IN FEMALE REPRODUCTION

WOLFGANG JÖCHLE

Syntex Research, AV. Insurgentes Sur 1457, Mexico 19, D.F. Mexico

RHYTHMS IN FEMALE REPRODUCTIVE FUNCTIONS

FOR A long time the rhythms in female reproductive functions and their seasonal periodicity have been well recognized, but it is only recently that the endogenous mechanisms responsible for these rhythmic changes have been fairly well understood (Jöchle, 1962 and 1963; Asdell, 1964; Nalbandov, 1964b). Within the last 20 years some of the exogenous factors have been identified which cause the seasonal variations in reproductive functions in horses (Nishikawa, 1959; Ortavant, Mauleon and Thibault, 1964), sheep, and goat (Yeates, 1949; Lamond, 1962; Ortavant *et al.*, 1964), cattle (Mercier, 1947; van Loen, 1962; Kelly and Hurst, 1963; Bane, 1964; El-Baschary and Fewson, 1964), pigs (Corteel, Signoret and du Mensil du Buisson, 1964; Münch, 1964), and wildlife (Ortavant *et al.*, 1964; Hollwich, 1964) and also the pathways by which these factors effect their remarkable influences (Tromp, 1963; Benoit, 1964; Gergen and MacLean, 1964; Hollwich, 1964). Comparatively recent is the concept that these rhythms are inborn and that environmental factors only modify their representation by inhibition or stimulation (Clegg, Cole and Ganong, 1964) thus ensuring optimal conditions for the maintenance of the species involved. It is mostly the seasonal variations in temperature and daylength which are responsible for these variations in the reproductive activity of male and female mammals (Tromp, 1963; Ortavant *et al.*, 1964). The eyes mediate the light influence, here working only as photoreceptors, and are connected directly with the so-called 'hypothalamic sex centre' (Blumcke, 1958; Benoit, 1964; Feldman, 1964; Gergen *et al.*, 1964; Hollwich, 1964; Knoche, 1957).

It has only been recently recognized that the most important event in female reproductive functions, namely ovulation, is in some species which ovulate spontaneously, linked closely to circadian rhythms (Everett, 1961). The daily change of light and dark seems to be the trigger for ovulation in rats (Everett, 1961), mice (Everett, 1961), and possibly sheep (Robertson and Rakha, 1965a and b). Important is the fact that even in prepubertal rodents, ovulation induced by gonadotrophins does not occur at a fixed time-interval, but after an interval related to environmental factors, as in the adult animal (Everett, 1961; Strauss and Meyer, 1962).

PMS administered to infantile mice in the evening produces significantly more ovulations than when injected in the morning (Lang and Lamond, 1966) and uterine and ovarian growth starts earlier (Lang and Lamond, 1966). In the well known HCG-FSH augmentation test, the response after hypophysectomy in the morning differs significantly from that following hypophysectomy in the afternoon. Animals hypophysectomized in the afternoon show decreased sensitivity to FSH, while their sensitivity to LH is increased (Lamond, 1964).

This situation gives rise to the following questions:

1. There is in rodents a diurnally recurrent hour of LH-release from the hypothalamus, making ovulation possible. Is this circadian rhythm of hypothalamic activity inborn or acquired from the environment, or has the neuroendocrine system been conditioned by rhythmic external factors?

2. Are these or similar mechanisms responsible for the timing of ovulation in other species which ovulate spontaneously?

3. Which environmental factors might be responsible for triggering ovulation in spontaneously ovulating species?

Our experiments using rodents may help to answer the first question and to understand some of the factors involved. Albino rats kept permanently in the light, show, sooner or later, a loss of the normal oestrous cycle and enter a condition of permanent oestrus (*Figure 1*) (Jöchle, 1956 and 1963). Urethane-narcosis, by blocking hypothalamic response to environmental influences, causes in those animals in permanent heat an immediate onset of dioestrus followed later again by permanent oestrus (*Figure 2*) (Jöchle, 1963). *Young female rats* exposed to constant light quickly shift from cyclic to permanent oestrus, with periods of continuous oestrus up to 100 days. *Adult animals* kept for the first 200 days of life in an environment alternating 12 h of light with 12 h of darkness, resist permanent oestrus-promoting activity of continuous light for 100 days, and their periods of permanent oestrus are shorter (Jöchle, 1963). Animals kept in constant darkness *maintain* their cyclic reproductive functions, showing only prolonged dioestrous phases (*Figure 1*) (Jöchle, 1956). Permanently illuminated C_3H-mice, however (a strain with normal agouti pigmentation) maintain their normal oestrous cycle, only the oestrous phase being somewhat prolonged (Jöchle, 1964a). However, in another C_3H substrain (C_3H-Jax) with a hereditary degeneration of the retina early in life, permanent illumination rapidly induces persistent oestrus (Jöchle, 1964a and b).

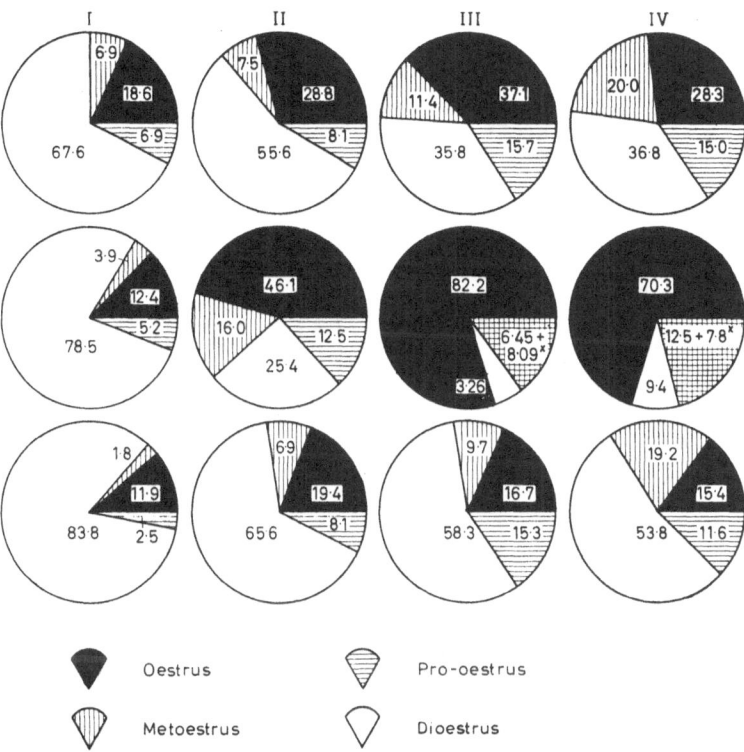

Figure 1. Percentage of heat cycle phases in young rats, kept alternately in 12 h *light and darkness (first line), in permanent illumination (second line) or in permanent darkness (line on the bottom). I to IV: each a* 20 *day period of recording, I starting immediately after puberty*

Persistent oestrus is always accompanied by ovaries which show only follicle growth and follicular cysts (Jöchle, 1963) instead of ovaries showing follicles and corpora lutea as observed in those animals with normal cyclic function, even if permanently illuminated (Jöchle, 1963).

Summarizing these experiences, one must assume that in rats and mice a circadian rhythm controlling ovulation is inborn, but maintained only during permanent illumination if the eye is normally pigmented. If albino animals have been conditioned by alternating light and dark, before exposure to permanent light, a similar circa-

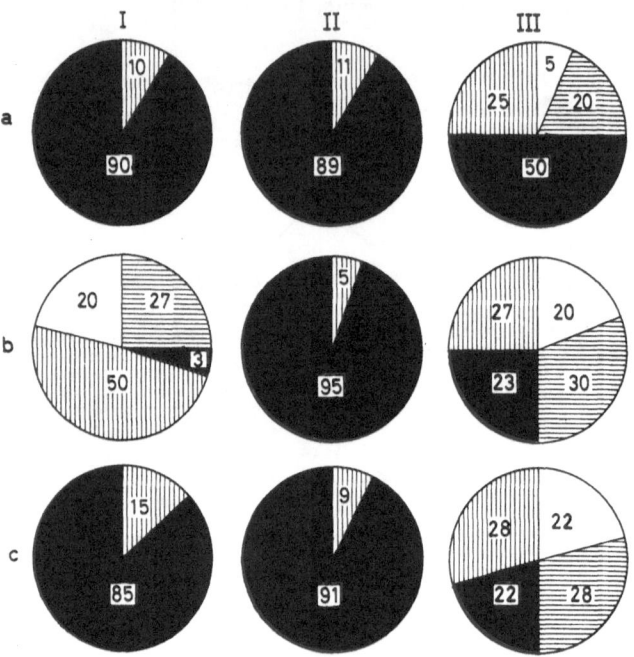

Figure 2. Percentage of heat cycle phases in adult rats, kept for 6 months permanently illuminated (I and II) or in 12 h light and darkness alternately (III)
a = situation in untreated animals during 5 days before treatment
b = 3 day reaction of group I after a 10 h urethane-narcosis; II and III serve as controls during the same days
c = days 4–8 after treatment: situation after treatment (group I) and in controls (II and III)

dian mechanism controlling ovulation is seen, but the normal 'Zeitgeber', the timing situation ensuring precise circadian ovulation seems to be the normal alternation of light and darkness.

Permanently illuminated rats with continuous persistent oestrus are not sterile; copulation causes ovulation, conception, pregnancy, undisturbed delivery and lactation (Maekawa, 1959; Jöchle, 1963). The rat, well known as a spontaneously ovulating species, is able to maintain its reproductive function in a strange environment only by induced ovulation. The question immediately arises whether this curious situation is a special way out of a trap, or if it is a mechanism always present, but not until now observed.

Perhaps there are, in other species, observations indicative of similar adaptive mechanisms?

INDUCED AND SPONTANEOUS OVULATION

The basic difference between species in which ovulation is induced and those spontaneously ovulating has been described in the following way:

The former produce follicles only at rhythmic intervals; ovulation is strictly dependent on cohabitation (or artificial cervical stimulation) which causes ovulation by LH release through neurohormonal pathways (Hansel, 1959). The latter ovulate (without copulation or stimulation) as a result of increasing oestrogen and progesterone blood levels, which induce during a circadian 'hour of special sensitivity' a sudden discharge of LH-releasers from hypothalamic nuclei (Everett, 1961).

In spontaneously ovulating domesticated animals the following observations are well documented:

Ovulation occurs late in heat or even hours after symptoms have subsided (Asdell, 1964; Nalbandov, 1964b). Copulation (or irrigation of the uterus) during oestrus hastens occurrence of ovulation in cattle (Marion, Smith, Willey and Barrett, 1950; Prandžev, Elezov and Bogdanov, 1964), sheep (Želtobrjuh and Rak, 1964) and pig (Radford, 1964) and shortens the oestrous period, and administration of oxytocin during oestrus has the same effect (Hansel, 1959; Sergeev, 1964). Copulation is known to induce oxytocin release (Van Demark and Hays, 1963; Harris, 1955). These results indicate clearly that there is in the spontaneously ovulating animal a period of hours in which the animals respond to copulation by induced ovulation (prior to the start of the mechanisms which result in spontaneous ovulation). The data Robertson (*see* page 195) has presented here seem to show the time sequence of events after induced ovulation (which the ram checking for oestrus has induced) and not after spontaneous ovulation.

If environmental conditions inhibit spontaneous ovulation, fertility can be maintained by induced ovulations. A good model for this situation is the vole (*Microtus agrestis*), which occupies an intermediate position with regard to spontaneity of ovulation: London moles need coitus to finish their oestrus and to ovulate, while Oxford moles have a short cycle around spontaneous ovulation (Chitty and

Austin, 1957). By 1922, strains of mice had been described which exhibited continuous oestrus and required copulation to ovulate (Allen, 1922). Spontaneous ovulation now can be defined as an inborn mechanism, acting only if ovulation induced by coitus does not occur. Its occurrence is determined by environmental factors. Normally the circadian change of light and darkness seems to be the governing influence.

One of the most important factors in zootechniques and the modern livestock industry is artificial insemination which seems to be based on the precise occurrence of spontaneous ovulation in cattle, horse, sheep, goat and pigs. It would be extremely useful to know whether in these species spontaneous ovulation can be expected around a circadian maximum of ovulations. One may ask if the light and dark dependent schedule of rodents is also the conditioning factor in other species. Are there other environmental influences changing or overrunning this mechanism?

The laying hen has been used in experiments to show interference between environmental influences and ovulation.

The laying hen is a spontaneously ovulating animal. Its optimal ovulatory capacity is claimed to be dependent on sufficient intensity and duration of light, and also on the circadian rhythm of darkness and light (Fraps, 1962, 1965; Wilson, 1964). A carefully arranged alternation of 16 h of artificial light (220 lux) and 8 h of darkness was therefore thought to ensure an equal distribution of ovipositions in approximately 100 single-caged hybrid hens during their first year of production. Complete air conditioning, feeding twice a week and watering three times a week were considered to have avoided most disturbances of the animals' environment.

During 3 years of experimentation, each year starting with 100 to 125 young hens replacing the old, the same observation was made: During the first 3 to 5 weeks ovipositions accorded to the predicted 'ideal-distribution of oviposition' during the 16 h of artificial illumination (*Figure 3*, Jan.–Feb.). Later the picture changed significantly; ovipositions culminated between 10 a.m. and 2 p.m. (*Figure 3*, Feb.–March).

In order to check whether the working hours of technicians could affect the animals' behaviour the relationship between hours of light and work was changed stepwise as each month light started an hour earlier. The results demonstrated that the influence of people's activity on the animals is diminished as the two stimulating factors, light and work, are separated (*Figure 3*, Aug.–Nov.). The necessity of re-establishing the original situation (*Figure 3*, Nov.–Dec.) shows the predominance of the possibly more attractive 'Zeitgeber' working hours.

There are hours each day in which the hypothalamus seems to discharge gonadotrophin releaser, inducing ovulation (Everett,

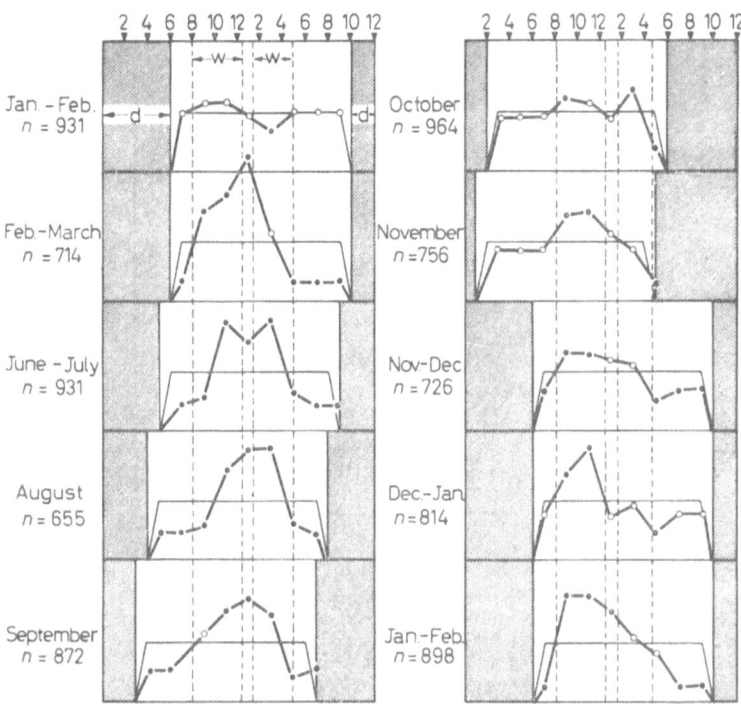

Figure 3. *Circadian-distribution of oviposition in single-caged laying hens during the first year of production, kept 16 h in light and 8 h in darkness alternately*

● = *two-hour-samples significantly different from the 'ideal distribution' (see page 272)*
○ = *two-hour-samples not significantly different from the 'ideal distribution'*
n = *total number of ovipositions recorded within the given period*
w = *working hours*
d = *dark period*

1961; Fraps, 1962). How these hours (8 to 9 according to Fraps, 1962, 1965) are synchronized with the hours of light or darkness can be explained by retino-hypothalamic pathways in the hen (Benoit, 1964).

How the many different stimuli of the technician's work affect this sensitive circadian periodicity is far more complex.

Another long-term experiment in hens has confirmed these results. Groups of one-day chicks have been raised and their laying controlled in the first production year by being kept in constant light, constant

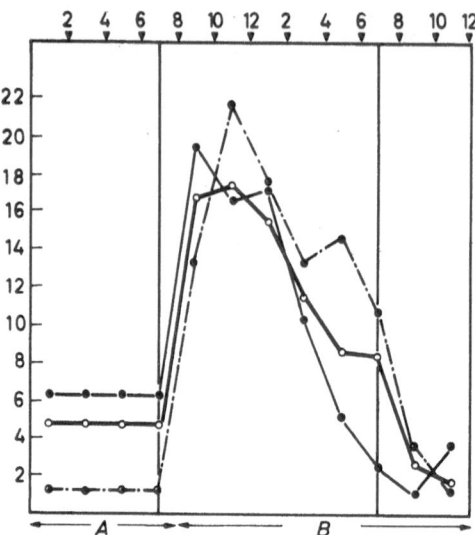

Figure 4. Circadian distribution of oviposition
●—● = animals kept in permanent darkness
○—○ = animals permanently illuminated
◑—·—◑ = animals kept in 12 h light–12 h darkness
(light: 7 a.m.–7 p.m.)

A = undisturbed period

B = period of egg collection every 2 h

darkness, or in 12 h light and darkness alternately. Daily from 8 a.m. to midnight every 2 h the eggs were collected from the singly-caged birds. In all three groups the major event synchronizing daily onset of ovulatory stimulation, was egg-collection itself (*Figure 4*). The only significant difference as far as the circadian distribution of egg laying is concerned, is the number of eggs laid during the 'undisturbed hours'; hens in constant environment without the light-darkness-cycle laid 19–24·4 per cent of their eggs during this period, normally kept animals not more than 5 per cent.

OVULATION RATE AND OESTRUS ACTIVITY
IN ANIMALS

Although similar observations in other animals are still awaited, some observations are available which indicate circadian periods of increased oestrus activity and ovulation rate. In cattle (*Bos taurus*) some authors report diurnal variations in onset of oestrous cycles and success rate of artificial insemination. In one paper 67 per cent of 613 dairy cows started oestrus during the night and 33 per cent late in the afternoon (Fallon, 1962). In another, 71·4 per cent of first inseminations in the morning resulted in pregnancy, but only 6 per cent of those in the afternoon (Prandžev, 1964). According to Trimberger (1948) more cows tended to go out of heat between 6 p.m. and midnight than at other times. It has also been reported that in cattle 60 per cent of heats start before noon, compared with 40 per cent in the period following (Roark and Herman, 1950). In a very careful study of follicle growth and ovulation it was stated that the hours generally preceding ovulation are early on day 2 of the cycle (Rajakoski, 1960). In Indian cattle (*Bos indicus*) the very short oestrous period is normally observed during the night (Howes, 1960). In goats, the temperature changes on the pro-oestrus and oestrus days are observed only in the mornings (Parer, 1963). In sheep, early in the season a circadian bimodial onset of heat may be observed, but not in the second half of the same season (Robertson, 1965a, b). Ovulation in sheep seems to be induced by the same trigger evoking heat symptoms, and therefore can be inhibited by central blocking substances administered immediately after onset of heat symptoms, but it has not been possible, up to now, to demonstrate a diurnal cyclic process (Robertson, 1965; Radford, 1966). Ewes kept with rams only from 5.00 p.m. to 8.00 a.m. have born significantly more twins, compared with ewes running with rams continually (Whiteman, 1965). Ewes kept for one cycle in permanent illumination show a significantly higher number of triplets without changes in cycle and oestrus length (Dutt, 1965).

In the sow, the standing reflex and conception rate is observed randomly without a diurnal maximum (Radford, 1964; Willemse and Boender, 1965).

A rise in temperature and undisturbed weather conditions (Brezowsky and Haeger, 1959; Kiesel and Grimm, 1961) clearly increase the percentage of dairy cows coming on heat, while changing weather has the reverse effect, especially if the temperature is declining. It has been shown that in women the temperature rise indicating ovulation occurs on a strict circadian schedule, being most often observed in the night. Between 0 and 6 a.m. a maximum of pregnandiol excretion in women is observed, independent on endogenous sources of progesterone (Corpus luteum or placenta) or administration of progesterone (Málek, 1962). Menstruation, on the other hand, starts significantly more often in the morning (Málek, 1962).

Olfactory influences in rodents and possibly in pigs, promote and synchronize ovarian functions in grouped animals by the so-called Lee-Boot and Whitten-effect (Parkes, 1963). Severe stress has been seen to start follicle growth and ovulation in ewes (Braden and Moule, 1964) and heifers (Hafez and Sugie, 1963). In the human, severe psychic and physical stress has been proved partly or totally to inhibit ovarian functions (Stieve, 1952), and induced ovulation by coitus in the human is well documented (Linzenmeier, 1947; Schaefer, 1949; Ibrügger, 1951; Phillipp, 1955; Herrligkoffer, 1956; Föllmer, 1961 and 1962).

High temperatures seem to interfere with seasonally stimulated reproductive functions in ewes (Shelton and Morrow, 1965) and cows (Fallon, 1962).

Possibly much more evidence for or against circadian influences on the oestrous cycle and ovulation is hidden in the literature. To explore these situations and to learn more about factors modifying or replacing the spontaneous reactions seems to be extremely important. Modern trends in live-stock production are to deal with and manipulate reproductive functions most effectively. To know the hour of highest efficacy of artificial insemination precisely has considerable economic value.

To know how to deal with environmental factors influencing the neurohormonal system will allow us to predict the hour of ovulation in the cycle, or after synchronized induced heat.

According to experience with seasonal differences in response to oestrogens and progesterone in heifers (Lamond, 1965) and ewes (Lamond, 1962), one may even learn about the circadian 'hours of maximal response' ensuring optimal reaction by a minimum of hormones administered in men and animals (Krähenbühl, 1965).

In conclusion, a well known circadian-affected function in female reproduction should be mentioned; the onset of delivery during the night and in the early morning hours (Málek, 1962). Cholinergic and oxytocic activities (Hough, Bearden and Hansel, 1955; Rüsse, 1965) are the most important parts of a complicated process resulting in expulsion of the egg cell or the foetus respectively.

The present state of knowledge of the circadian rhythm in female reproductive functions is inadequate, but the idea has been to focus attention on the problem, which does seem important and it has involved the discussion of some basic mechanisms of female reproduction. Intensified research in this area might help us to understand how exogenous factors and endogenous functions are linked together to secure reproduction.

REFERENCES

Allen, E. (1922). *Am. J. Anat.* **30,** 297–371
Asdell, S. A. (1964). *Patterns of Mammalian Reproduction,* 2nd edn. Ithaca, N.Y.; Cornell Univ. Press
Bane, A. (1964). *Br. vet. J.* **120,** 431–441
Benoit, J. (1964). *Ann. N.Y. Acad. Sci.* **117,** Art. 1, 23–34
Blumcke, S. (1958). *Z. mikrosk.-anat.* **48,** 261–282
Braden, A. W. H. and Moule, G. R. (1964). *Aust. J. agric. Res.* **15,** 937–949
Brezowsky, H. and Haeger, O. (1959). *Zuchthyg. FortpflStör. Besam. Haustiere* **3,** 272
Chitty, H. and Austin, C. R. (1957). *Nature, Lond.* **179,** 592–593
Clegg, M. T., Cole, H. H. and Ganong, W. F. (1964). *Proc. Conf. on Estrous Cycle Control in Domestic Animals.* Coop. State Res. Serv. and Agric. Res. Serv. U.S. Dept. of Agric. in Coop. with Univ. of Nebraska Misc. Dubl. **1005,** 96–103
Corteel, J. M., Signoret, J. P. and Du Mensil du Buisson, F. (1964). *5th Int. Congr. Anim. Prod. A.I.* (*Trento*), Vol. III, 536–540
Dutt, R. H. (1965). *J. Anim. Sci.* **24,** 916
El-Baschary, A. and Fewson, D. (1964). *Fortpfl. d. Haust.* **1,** 23–34
Everett, J. W. (1961). 'The Mammalian Female Reproductive Cycle and its Controlling Mechanisms'. In: *Sex and Internal Secretion,* Vol. I, 497–555. Ed. by W. C. Young. Baltimore; The Williams and Wilkins Comp.
Fallon, G. R. (1962). *J. Reprod. Fert.* **3,** 116
Feldman, S. (1964). *Ann. N.Y. Acad. Sci.* **117,** Art. 1, 53–68
Föllmer, W. (1961). *Arch. Gynaek.* **194,** 355
— (1962). *Geburtsh. Frauenheilk.* **88,** 1391
Fraps, R. M. (1962). 'Effects of External Factors on the Activity of the Ovary.' In: *The Ovary,* Vol. II, Chap. 19, 317–379. Ed. by Sir Solly Zuckerman. New York and London; Academic Press
— (1965). *Endocrinology* **77,** 5–18
Gergen, J. A. and MacLean, P. D. (1964). *Ann. N.Y. Acad. Sci.* **117,** Art. 1, 69–87
Hafez, E. S. E. and Sugie, T. (1963). *Acta Zool., Stockh.* **44,** 57–71
Hansel, W. (1959). In: *Reproduction in Domestic Animals,* Vol. I, Chap. 7, pp. 223–265. Ed. by H. H. Cole and P. T. Cupps. New York and London; Academic Press

Harris, G. W. (1955). *Neural Control of the Pituitary Gland.* Baltimore, Maryland; Williams and Wilkins

Herrligkoffer, K. M. (1956). *Münch. med. Wschr.* **98,** 1066

Hollwich, F. (1964). *Ann. N.Y. Acad. Sci.* **117,** Art. 1, 105–128

Hough, W. H., Bearden, H. J. and Hansel, W. (1955). *J. Anim. Sci.* **14,** 739–745

Howes, J. R., Warnick, A. C. and Hentges, J. F. (1960). *Fert. Steril.* **11,** 5–508

Ibrügger, A. (1951). *Zentbl. Gynäk.* **73,** 42

Jöchle, W. (1956). *Endokrinologie* **33,** 129–138

— (1962). *Angewandte Chemie, Internat.* **1,** 537–549

— (1963). *Zentbl. VetMed., Reihe, A.,* **10,** 653–706

— (1964a). *Ann. N.Y. Acad. Sci.* **117,** Art. 1, 88–104

— (1964b). *10. Symp. Dt. Ges. Endokrinologie* 305–308. Berlin–Göttingen–Heidelberg; Springer Verlag

Kelly, J. W. and Hurst, V. (1963). *J. Am. vet. med. Ass.* **143,** 40–43

Kiesel, K. and Grimm, H. (1961). *Zuchthyg.* **5,** 41

Knoche, H. (1957). *Z. Mikrosk. anat. Forsch.* **63,** 461–486

Krähenbühl, C. (1965). *5th Acta Endocrinologica Congr. Abstr. No. 45, Acta endocr., Copenh,* Suppl. 100, 77

Lang, D. R. and Lamond, D. R. (1966). *J. Endocr.* **34,** 41–50

Lamond, D. R. (1962). *J. Reprod. Fert.* **4,** 111–120

— (1964). *Proc. Conf. on Estrous Cycle Control in Domestic Animals.* Coop. State Res. Serv. and Agric. Res. Serv. U.S. Dept. of Agric. in Coop with Univ. of Nebraska Misc. Dubl. (Disse.) **1005,** 108–111

— (1965). *J. Reprod. Fert.* **9,** 41–46

Linzenmeier, G. (1947). *Zentbl. Gynäk.* **69,** No. 11

Loen, A. van (1962). *Zuchthyg.* **6,** 343–352

Maekawa, K. (1959). *Annotes. Zool. Jap.* **32,** 185

Málek, J. (1962). *Rep. 7th Conf. of the Soc. for Biol. Rhythm,* Siena, 1960, pp. 97–103. Edizione Panminerva Medica Turin

Marion, G. B., Smith, V. R., Willey, T. E. and Barrett, G. R. (1950). *J. Dairy Sci.* **33,** 855–859

Mercier, E. (1947). *J. Dairy Sci.* **30,** 747

Münch, H. J. (1964). *Giessen Schr. Reihe Tierz. Haustiergenet.* **12,** 74

Nalbandov, A. V. (1964a). *Proc. Conf. on Estrous Cycle Control in Domestic Animals.* Coop. State Res. Serv. and Agric. Res. Serv. U.S. Dept. of Agric. in Coop. with Univ. of Nebraska Misc. Dubl., **1005,** 92–95

— (1964b). *Reproductive Physiology,* 2nd edn. San Francisco and London; W. H. Freeman & Co.

Nishikawa, Y. (1959). *Studies in Reproduction in Horses,* Tokyo, Japan

Ortavant, R., Mauleon, P. and Thilbaut, C. (1964). *Ann. N.Y. Acad. Sci.* **117,** Art. 1, 157–1963

Parer, J. T. (1963). *Am. J. vet. Res.* **24,** 1223–1226

Parkes, A. S. (1963). *Proc. R. Soc. B.* **56,** 47–52

Philipp, E. (1955). *Dt. Med. Wschr.* **80,** 947

Prandzev, J., Elezov, G. and Bogdanov, M. (1964). *Vet. Sbir., Sof.* **61,** 24–29

Radford, H. M. (1966). *J. Endocr.* **34,** 135–136

Radford, P. (1964). *Vet. Rec.* **76,** 1013–1017

Rajakoski, E. (1960). *Acta Endocr. Suppl.* 52

Roark, D. B. and Herman, H. A. (1950). *Mo. Univ. agric. exp. Stn. Res. Bull.* 455

Robertson, H. A. and Rakha, A. M. (1965a). *J. Reprod. Fert.* **10,** 271–272

— (1965b). *J. Endocr.* **32,** 383–386

Rüsse, M. (1965). *Arch. exp. VetMed.* **19,** 763–870

Schaefer, G. (1949). *Zentbl. Gynäk.* **71,** 969

Sergeev, N. J. (1964). *5th Int. Congr. Anim. Prod. A.I.* (*Trento*), Vol. VI, 177–181

Shelton, M. and Morrow, J. T. (1965). *J. Anim. Sci.* **24,** 795–799

Stieve, H. (1952). *Der Einfluss des Nervensystems Aufbau und Tätigkeit der Geschlechtsorgane des Menschen.* Stuttgart; Georg. Thieme Verlag

Strauss, W. F. and Meyer, R. K. (1962). *Science, N.Y.* **137,** 860–861

Trimberger, G. W. (1948). *Bull. Neb. agric. Exp. Stn. Res.* 153

Tromp, S. W. (1963). *Medical Biometeorology: Weather, Climate and the Living Organism.* Amsterdam, London, New York; Elsevier Publ. Comp.

Van Demark, N. L. and Hays, R. L. (1953). *Iowa St. Coll. J. Sci.* **28,** 107

Whiteman, J. V. (1965). *J. Anim. Sci.* **24,** 932

Wilson, W. O. (1964). *Ann. N.Y. Acad. Sci.* **117,** Art. 1, 194–203

Willemse, A. H. and Boender, J. (1965). *Vet. Rec.* **77,** 659

Yeates, N. T. M. (1949). *J. agric. Sci.* **39,** 1–43

Zeltobrjuh, N. A. and Rak, L. P. (1964). *Ovtsevodstvo* **10,** 8–11

DISCUSSION

Dr. J. R. Clark (*Oxford*)

I don't think many people will dispute that the oestrous cycle is not quite such an internal clock as was originally thought and I would like to add a comment about the vole. In 1957, a paper was published suggesting that voles in the Bureau of Animal Population colony could switch from one type of ovarian function to the other, supporting a prior suggestion of Eckstein and Zuckermann that the distinction between spontaneous and induced ovulation is not clearcut. Mr. Breed and I, however, working with an Oxford colony of voles have been quite unable to reproduce the type of animal as found in 1957. All our voles are induced ovulators as in Dr. Austin's work at Mill Hill, despite manipulation of the social environment.

Professor A. W. Nalbandov (*Illinois, U.S.A.*)

I would not wish to let this assertion about induced ovulators pass without question. The impression that Dr. Robertson's sheep were induced ovulators is misleading because had he not allowed them to mate, they would have ovulated anyway. I would also like to ask you what evidence you have that induced ovulation occurs in women. You said it was well documented but I am not aware of this. My impression is that animals can use any trigger they wish for the synchronization of events, e.g. temperature in rats can be substituted for light.

279

REPRODUCTION IN THE FEMALE MAMMAL

JÖCHLE

Evidence for induced ovulation in women has been lacking until recently, but German papers have given details of pregnancy following rape at different stages of the cycle which demonstrated that ovulation could occur even during the menstrual phase. A second report concerns Arabic nations where girls are married at puberty at about 12 years of age. Fifty per cent have not menstruated and about one-third conceive soon afterwards without having menstruated. Others conceive after infrequent menstruation typical of puberty and only one-third conceive during mid-cycle. A further report gives details which show that these women may conceive during lactation (Föllmer, 1961, 1962).

DR. H. A. ROBERTSON (*Aberdeen*)

We have been thinking similarly about the possibility of induced ovulation in the sheep. If this is induced ovulation, it is true of course that the difference in time interval between induced and spontaneous ovulation will only be a matter of hours. The oestrogen peak referred to by Professor Robinson may determine when LH would be released in a spontaneous ovulation if copulation had not occurred. Our sheep were not allowed much sexual activity with copulation so any stimulus from this was fairly light. Once we saw any kind of mating behaviour we removed the ewe and so if this was induced ovulation, the stimulus required to produce it was small.

PROFESSOR C. GEMZELL (*Uppsala, Sweden*)

I think that one has to be very suspicious about induced ovulation in women. I think it can happen but ovulation is more likely to be spontaneous. Ovulation can occur more than once in the cycle. We have followed daily LH secretion in normal cycling women and have found indications of ovulation at least twice during the cycle. For example, ovulation and conception can take place two days before menstruation. With young girls at the time of puberty we know that ovulation does not usually occur for at least 2 or 3 years. Occasionally it can occur earlier but the usual pattern is 2 years of regular menstrual periods without ovulation, and when the girl reaches her fifteenth year, ovulation occurs more or less regularly.

CLARK

Do prepubertal rats given continuous light treatment have normal oestrous cycles after the cessation of treatment or is there permanent alteration of mechanisms.

JÖCHLE

If you treat rats from birth with permanent light you hasten the onset of puberty significantly by 5–8 days. They do not immediately show permanent oestrus but after some normal cycles they switch to permanent oestrus. I have not returned these animals to a normal 12 h light regime to see what happens.

DR. I. ROTHCHILD (*Cleveland, U.S.A.*)

Everett has shown with one particular strain of rats that the onset of the anovulatory pattern is hastened by subjecting them to constant light but it can also occur under a cyclic light pattern. Once the anovulatory pattern has developed they maintain this even when given alternate light and dark, but if the hours of light are reduced from fourteen to ten the animals begin to ovulate spontaneously. Ascheim has shown that the tendency towards anovulation in old rats, which can be accelerated by light, can also be accelerated by putting the animals into darkness.

Author's remarks added in galley proof

Further evidence for the fact that rats are able to ovulate spontaneously as well as induced by coitus has been obtained in oestrogenized and untreated female animals placed with males overnight, 24 h before spontaneous ovulation would have occurred. Ovulation and conception occurred in 30–50 per cent of these animals (Aron, Asch and Asch, 1961; Aron, Asch, Asch, Roos and Luxembourger, 1965; Aron, Asch and Roos, 1966; Roos, Asch and Aron, 1964). The correlation between the frequency of coitus and that of advanced luteinization was 0·995 (Aron, Asch, Roos and Luxembourger, 1964). Hemicastration at day 2 of the cycle resulted in a premature spontaneous ovulation during the following night (Roos and Roos, 1966). Both phenomena (advanced ovulation induced by coitus or by hemicastration) are mediated by the hypothalamus (Aron and Asch, 1962) and can be blocked by atropine treatment (Aron *et al.*, 1965; Roos and Roos, 1966). The reactivity to the copulatory stimulus differs between strains of rats (Roos, Asch and Aron, 1965).

REFERENCES

Aron, C., Asch, G. and Asch, L. (1961). *C.r. Séanc. Soc. Biol.* **155,** 2173–2176
Aron, C. and Asch, G. (1962). *C.r. hebd. Séanc. Acad. Sci.*, Paris **255,** 3056–3058
Aron, C., Asch, G., Roos, J. and Luxembourger, M. M. (1964). *C.r. Séanc. Soc. Biol.* **58,** 1426–1429
Aron, C., Asch, G., Asch, L., Roos, J. and Luxembourger, M. M. (1965). *Path. Biol.*, *Paris* **13,** 603–614
Aron, C., Asch, G. and Roos, J. (1966). *Int. Rev. Cytol.* **20,** 139–172
Roos, J., Asch, G. and Aron, C. (1964). *C.r. Séane. Soc. Biol.* **158,** 2467–2469
Roos, J., Asch, G. and Aron, C. (1965). *C.r. Séanc. Soc. Biol.* **159,** 2072–2076
Roos, M. and Roos, J. (1966). *C.r. hebd. Séanc. Acad. Sci.*, Paris **262,** 300–302

V. INTERRELATIONSHIPS BETWEEN THE PITUITARY GLAND AND THE CORPUS LUTEUM

HYPOPHYSIAL AND UTERINE INFLUENCES ON PIG LUTEAL FUNCTION

L. L. ANDERSON and R. M. MELAMPY

Department of Animal Science, Iowa State University, Ames

It is evident that the elucidation of the diverse patterns of mammalian reproduction is dependent upon research from a comparative viewpoint. The results of such studies will provide a basis for understanding the ways different kinds of animals meet their reproductive requirements and will be useful also in evaluating evolutionary as well as ecological aspects of reproduction. The discussion presented here deals with results of experimental work on the reproductive physiology of the female domestic pig (*Sus scrofa* L.).

OVARIAN MORPHOLOGY AND HISTOLOGY

The ovaries of sexually mature pigs may be described as berry-shaped organs because of the Graafian follicles and corpora lutea which protrude from their surfaces. Each pig ovary usually weighs from 3–7 g depending upon the number of follicles and corpora lutea present. The usual number of eggs shed by sows is 10–25, whereas non-parous females (gilts) ovulate 4 or 5 fewer ova. Oestrous cycles in the pig last about 21 days and the duration of pregnancy is approximately 114 days. The pig remains in oestrus (male acceptance) from 1–3 days and ovulation usually occurs during the latter part of this period. Morphological aspects of follicular growth during the oestrous cycle have been described by Robinson and Nalbandov (1951) and Parlow, Anderson and Melampy (1964). An increase in size of the ovarian follicles occurs at about day 16 of the cycle with a marked growth at the onset and during the period of oestrus. Graafian follicles are 8–12 mm in diameter at ovulation. Following ovulation and during the formation of corpora lutea, only small follicles, usually less than 4–6 mm in diameter, remain until about the fifteenth day. During gestation the majority of follicles are 4–6 mm in diameter. Follicles are small, usually 4 mm or less at day 14 of lactation.

Following rupture of the Graafian follicle and escape of the ovum, the walls of the follicle collapse usually around a central blood clot and reduce the diameter to 4–6 mm. The granulosa layer is retained intact after the rupture of the Graafian follicle, except for the loss of the cumulus oöphorus. The granulosa cells hypertrophy to become the lutein cells of the corpus luteum. The granulosa layer is invaded by capillaries from the theca interna forming a vascular plexus throughout the organ. The theca interna cells increase in number and migrate into the corpus luteum. They become lodged between the lutein cells and possibly persist throughout the functional life span of the organ. Within a week the corpora lutea have increased from a diameter of about 4–6 mm to 8–9 mm. If pregnancy follows, there is further growth to 10 mm or more. By day 7 the corpora lutea are usually solid and the cells have been fully differentiated. During the first 2 weeks of development corpora lutea of pregnancy are morphologically indistinguishable from those of the oestrous cycle. In the pig, Corner (1915, 1919) observed three principal types of luteal cells in the corpora lutea of pregnancy: (1) true lutein cells originating from the granulosa; (2) cells with smaller round or oval and more chromatic nuclei which appear on the periphery of the gland and along the connective tissue septa; and (3) cells with a spindle shape and a cytoplasm which stains dark brown or purple with Mallory's stain. There were transitional stages among the three types. In the pig, corpora lutea are necessary throughout gestation and show functional regression just prior to parturition.

Beginning on day 14 or day 15, a change takes place in the corpus luteum of animals with unfertilized ova. In 2 or 3 days, the corpus luteum has decreased to about 6 mm in diameter, its colour has changed from the pink of active capillary circulation to the white of scar tissue and its texture has become tougher and firmer. Microscopic examination shows that a very rapid degeneration has taken place with complete breakdown of the lutein cells and collapse of the capillaries. By the time of the next ovulation the regressed corpora lutea have diminished in diameter to 6 mm, by the mid-oestrual period to 4 mm and by the next ovulation to 2 mm, after which they disappear slowly. Eventually all that remains of the site of a Graafian follicle, and subsequently a corpus luteum of the oestrous cycle or pregnancy, is a small mass of scar tissue, a corpus albicans. Growth and regression of corpora lutea during the oestrous cycle are shown in *Figure 1* (Masuda, Anderson, Henricks and Melampy, 1967). This represents the total weight of corpora lutea from one ovary of cycling pigs.

The ovary is supplied with blood by the ovarian (utero-ovarian) artery which is a branch of the abdominal aorta (Oxenreider, McClure and Day, 1965). Several branches enter the ovary through the hilus and traverse the medulla as spiral vessels. These spread out at the bases of follicles and surround them in arterial

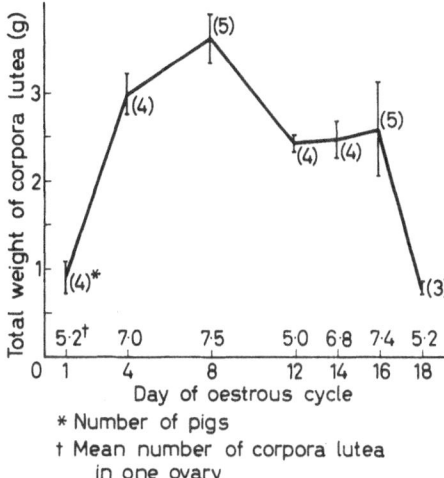

Figure 1. Total weight (Mean ± SE) of corpora lutea in one ovary of pigs at different stages of the oestrous cycle

wreaths. Arterioles and capillaries grow out from these wreaths during formation of corpora lutea and provide a vascular network. Follicles and corpora lutea are drained by corresponding venous and lymphatic wreaths.

Venules draining the wreaths unite in the stroma and mesovarium to form one or two short ovarian veins which either enter the large utero-ovarian vein or a plexus which is drained by it. The utero-ovarian vein receives most of the uterine blood and follows the course of the ovarian artery to empty into the caudal vena cava.

Lymph vessels draining the peri-follicular and peri-luteal wreaths enter into an abundantly anastomosing network of stromal lymphatics. Throughout the ovary lymphatics are found peri-arterially and as they emerge at the hilus of the ovary, leave the blood vessels and run in two parallel networks on either side of the mesovarium. These two networks coalesce 5–7 cm distal to the hilus and enter the sub-ovarian plexus where they are joined here by lymphatics from the oviduct and uterus. The resulting lymph vessels follow the ovarian artery and utero-ovarian vein to the midline where they drain into aortic lymph nodes. It is possible that a significant uterine–ovarian relationship in the pig may be associated with the intimate proximity of vessels carrying uterine lymph and venous blood and those carrying ovarian arterial blood.

PROGESTERONE IN OVARIAN VENOUS PLASMA AND CORPORA LUTEA

Ovarian venous plasma progesterone concentration increased until day 8 of the oestrous cycle and declined thereafter (*Figure 2*; Masuda

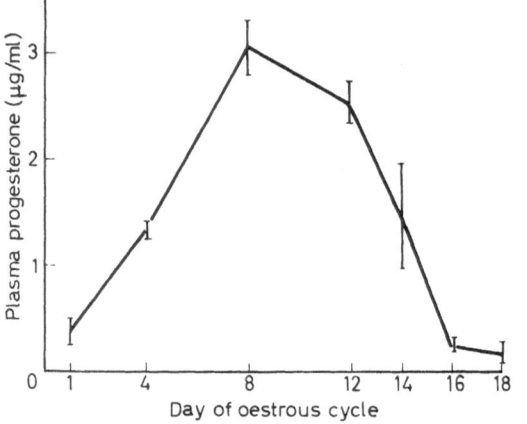

Figure 2. Ovarian venous plasma progesterone concentrations (Mean ± SE) from one ovary of pigs at different stages of the oestrous cycle

et al., 1967). By day 16 the concentration was one-twelfth of that found at day 8, even though the corpora lutea at the later stage were vascular and of considerable size. Plasma progesterone concentrations in this study were over twice those reported by Gomes, Herschler and Erb (1965); however, the rise and fall in hormone level occurred at the same stages of the oestrous cycle. The progesterone concentration was similar at day 14 after mating and at the same day of the cycle (*Figure 2*, Table 1; Masuda *et al.*, 1967). At day 18 the progesterone concentration was 10 times greater in pregnant pigs than in cycling animals. At day 25 of pregnancy, the level of this hormone was similar to that observed at days 14 and 18.

Even though corpora lutea were larger during late pregnancy than at previous stages, the progesterone concentration was reduced about one-half by day 102 (Table 1). Plasma level at day 110 was similar to that at day 102 and a further decline was observed on day 112.

Table 1

Progesterone concentrations in ovarian venous plasma and luteal tissue during pregnancy and following hysterectomy in the pig

(From H. Masuda *et al.*, 1967, by courtesy of the Editor, *Endocrinology*)

Reproductive stage (day)	Number of pigs	Progesterone*		
		Blood plasma	Luteal tissue	
			Total	
		(μg/ml) †	(μg) †	(μg/g) †
Pregnancy				
14	4	2·3 ± 0·2	246 ± 36	65 ± 3
18	3	2·6 ± 0·3	183 ± 20	87 ± 8
25	3	2·3 ± 0·4	178 ± 31	63 ± 6
102	4	1·3 ± 0·3	189 ± 35	54 ± 8
110	3	1·4 ± 0·5	115 ± 17	37 ± 2
112	3	0·9 ± 0·2	119 ± 7	37 ± 3
Following hysterectomy				
14	3	3·4 ± 0·4	205 ± 15	70 ± 2
18	3	2·5 ± 0·3	190 ± 20	65 ± 2
25	3	3·4 ± 0·4	165 ± 10	63 ± 4

* The blood and luteal tissue obtained from one ovary were used for progesterone analyses. Values estimate the amounts of the steroid in the blood or luteal tissue from one ovary.
† Mean ± SE.

Total luteal tissue progesterone content increased until day 8 of the cycle and declined thereafter to day 18 (*Figure 3*; Masuda *et al.*, 1967). Luteal progesterone concentration (μg/g) was at a maximum at day 12 and by day 16 was low. Rombauts, Pupin and Terqui (1965) reported that the highest luteal progesterone concentration was at day 13, and by day 15 to day 19 the hormone level was very low. Duncan, Bowerman, Hearn and Melampy (1960) observed that the highest luteal concentration was at the fifteenth day of the cycle, and 2 days later the hormone was undetectable. According to Masuda *et al.* (1967) the total luteal progesterone content 14 days after mating was twice that found at the same day of the cycle (*Figure 3*, Table 1). By day 18 this difference had increased at least six times. Total luteal progesterone at day 25 of pregnancy was similar to that at days 14 and 18 after mating. Progesterone concentrations in luteal tissue in this investigation are comparable with those reported by Duncan *et al.* (1960) and Rombauts *et al.* (1965); however, they are nearly twice those found

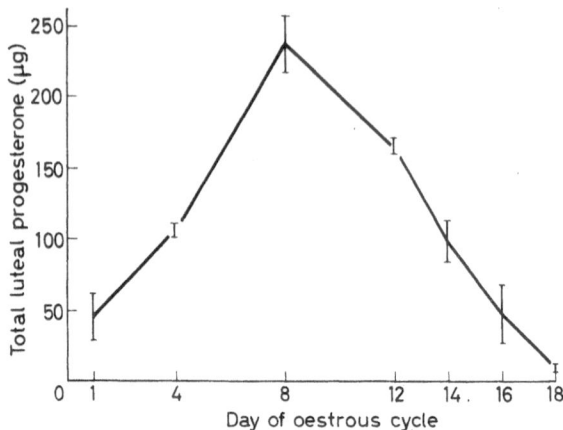

Figure 3. Total progesterone content (Mean ± SE) of luteal tissue from one ovary of pigs at different stages of the oestrous cycle

by Erb, Nofziger, Stormshak and Johnson (1962). By day 102 luteal progesterone content had declined to the level of early pregnancy (Table 1). A further decline in hormone content occurred by day 112. Progesterone levels in blood and luteal tissue followed a similar pattern during the oestrous cycle and early gestation. Morphology of the corpus luteum was not a satisfactory indicator of functional activity of the gland and this was particularly apparent during the period of luteal regression in the cycle and in late pregnancy.

Urinary metabolites of progesterone in non-pregnant and pregnant pigs were determined by Mayer, Glasgow and Gawienowski (1961). During the first 6 weeks of gestation the levels of conjugated and non-conjugated metabolites were about 2–5 mg/l of urine. From weeks 6 to 13 there was an increased level of conjugated metabolites (10–15 mg/l), and thereafter they declined rapidly to the fifteenth week. Levels of urinary metabolites may be correlated with luteal progesterone and ovarian venous plasma progesterone levels during early and late pregnancy (Masuda *et al.*, 1967). It is not known if pig placenta produces sufficient quantities of progestins to alter urinary excretion patterns of the metabolites. It may be that the placenta produces little, if any, progestins since the ovary is essential for maintenance of pregnancy in the species.

ANTERIOR PITUITARY GONADOTROPHINS, SOMATOTROPHIN AND THYROTROPHIN DURING DIFFERENT REPRODUCTIVE STAGES

Hypophysial gonadotrophic function may be evaluated by measuring follicle-stimulating hormone activity (FSH) and luteinizing hormone activity (LH, ICSH) in the pituitary. Pituitary content may reflect its secretory activity when gonadotrophic content is correlated with ovarian morphology at different reproductive stages.

Dry anterior pituitary glands from pigs at different reproductive stages were assayed for follicle-stimulating hormone activity by the human chorionic gonadotrophin augmentation assay of Steelman and Pohley (1953) and for luteinizing hormone activity by the ovarian ascorbic acid depletion assay of Parlow (1961). The anterior pituitary contents of FSH and LH, as well as the dry weights of the anterior pituitary, are shown in *Figure 4* for different stages in cycling, pregnant and lactating pigs. The pituitaries were assayed individually for FSH and LH from 5 pigs at each stage of the oestrous cycle and from 8 pigs at each stage of pregnancy or lactation. Mean values and their 95 per cent confidence limits are shown. The dry weight of the anterior pituitary did not change during the oestrous cycle or during the first half of pregnancy. By day 80 of pregnancy there was a marked increase in weight. At the fourteenth day of lactation the anterior pituitary gland weighed twice that found during the cycle or early pregnancy.

According to Parlow *et al.* (1964), pituitary content of FSH and LH during the oestrous cycle was low until day 4, increased between days 4 and 10 and was maintained at this level through day 18. Between the eighteenth day and oestrus a marked reduction occurred in pituitary FSH and LH. Ovarian follicular growth was evident between days 4 and 10 and again at days 18 to oestrus. None of the oestrous animals had ovulated when pituitaries were obtained for assay. A correlation of ovarian follicular development with pituitary gonadotrophin content suggested that increased release of FSH and LH had occurred as the pituitary synthesized gonadotrophins (days 4–10), and when there was a reduction in pituitary gonadotrophin content (day 18 to oestrus).

At day 13 of pregnancy, pituitary content of FSH and LH was

291

nearly twice that of the tenth day of the oestrous cycle. Furthermore, at day 18 of pregnancy the gland contents of these gonadotrophins remained twice those for the same stage of the cycle.

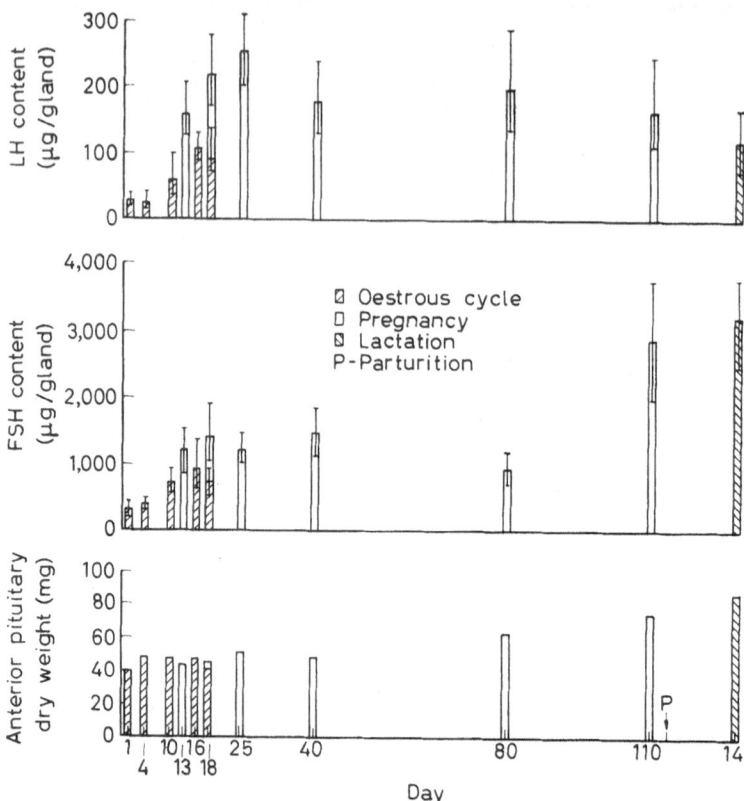

*Figure 4. Pituitary follicle-stimulating hormone activity and luteinizing hormone activity at
different stages in cycling, pregnant and lactating pigs. Anterior pituitary contents of FSH
and LH are expressed in equivalents of* NIH-FSH-S1 *or* NIH-LH-S1/mg *dry weight of the
gland. Mean values and their* 95 *per cent confidence limits are shown for each reproductive
stage* (Adapted from data of R. M. Melampy *et al.*, 1966, by courtesy of the Editor
of *Endocrinology*)

During pregnancy, pituitary content of FSH remained constant from
days 13–40. By day 80 it was lower than at previous stages and by
day 110 FSH content was four times the activity found at the
eightieth day. This high FSH level remained at day 14 of lactation.
After the initial increase in pituitary LH content between day 13

292

and day 25 of pregnancy, a gradual decline occurred between day 40 and day 110. By day 14 of lactation pituitary LH was reduced considerably compared with day 110 of pregnancy. The extent of ovarian follicular development remained relatively constant throughout pregnancy. After parturition, follicles and corpora lutea regressed during lactation, even though pituitary FSH content was at the highest level observed.

These results indicate that in mated pigs pituitary gonadotrophic content quickly adjusted with increased FSH and LH activities as compared with similar stages of the cycle. The relationship between this increased pituitary gonadotrophin content and pregnancy is not clear. Exogenous LH (bovine or equine) maintains corpora lutea in hypophysectomized-hysterectomized pigs (Anderson, Léglise, du Mesnil du Buisson and Rombauts, 1965), but it is not established that endogenous LH activity is luteotrophic in the pig. Elevation in FSH may have been caused by inhibition of FSH release by progesterone from the persisting corpora lutea. Pituitary FSH and LH levels at day 25 following hysterectomy, or day 25 following bilateral ovariectomy, were equal to those found at day 25 of pregnancy. There was no change in the ratio of pituitary FSH and LH content in cycling and pregnant pigs until the latter half of gestation. By day 110, the FSH : LH ratio increased three times, and at day 14 of lactation it was five times greater than during the oestrous cycle and early pregnancy (Melampy, Henricks, Anderson, Chen and Schultz, 1966).

Luteal enucleation during an early stage of the oestrous cycle altered pituitary gonadotrophic activities, subsequent ovarian function and oestrous behaviour in the pig (Anderson, Dyck and Rathmacher, 1966). When corpora lutea were enucleated at day 6 of the cycle, the animals returned to oestrus in 6 or 7 days. Pituitary FSH content was low 24 h after luteal enucleation and no consistent decrease was observed at oestrus. LH activity remained constant except at oestrus when it was reduced. Progesterone (4·4 mg/kg body weight), injected daily for 5 days beginning a day prior to luteal enucleation delayed oestrus, though pituitary FSH and LH did not decrease significantly after hormone treatment. When a similar dosage of progesterone was administered for 20 days in intact pigs, oestrus occurred 5 days after the end of the treatment period. Pituitary FSH content was significantly elevated a day following treatment, as compared with that found a day after progesterone treatment in -pigs with enucleated corpora lutea. Within 3 days after the last progesterone injection, FSH content was significantly reduced, whereas LH content remained constant. Pigs returned to oestrus within 6 days after the treatment period. Even at oestrus the LH content was significantly higher in these long-term progesterone-treated pigs than in the stages of oestrus of animals following luteal enucleation.

Pituitary FSH and LH increased 25 days after hysterectomy. After enucleation of the persisting corpora lutea in these hysterectomized pigs, LH content decreased until oestrus, 6 and 7 days later and FSH content remained unchanged during this time. Ovarian follicles increased in size during the quiescent period of the progesterone-treated hysterectomized pigs.

When the oestrous interval was reduced by enucleation of corpora lutea in an early stage of the cycle, the pituitary gonadotrophic content did not decrease until oestrus. Relatively long-term progesterone treatment or hysterectomy increased pituitary FSH and LH content to a higher level than observed in cycling pigs.

Growth promoting activity (somatotrophic hormone) in anterior pituitaries from pigs at different ages indicated that unit potency (hormone content in relation to body size) was significantly lower in animals less than 20 days of age than it was in older pigs (Baker, Hollandbeck, Norton and Nalbandov, 1956). The amount of available hormone per unit body weight increased up to 116 days of age, declined steadily to 300 days of age, and thereafter remained constant to 1,400 days. It has been stated by Nalbandov (1963) that the amount of somatotrophin entering each cell rapidly increases in young growing pigs and then decreases as the growth plateau is approached. It was suggested that the reduction in growth rate was the result of a decrease in the amount of available hormone. In this investigation, however, no significant changes were observed in unit potency or total potency of somatotrophic hormone activity in pituitaries from animals 56–98 days pregnant.

Somatotrophic hormone activity in pituitary glands during different reproductive stages in the pig is shown in Table 2. Increase in width of the tibial epiphysial cartilage of female rats hypophysectomized at 28 days of age was used to measure response to a standard growth hormone preparation and acetone-dried porcine anterior pituitary. This dried pituitary tissue was injected intraperitoneally once daily for 4 days into assay rats at total dosages of 200, 400 and 800 µg. The results were expressed in µg-equivalents of NIH-GH-B7 per milligramme dry anterior pituitary. Concentrations of somatotrophic hormone activity were similar in cycling, ovariectomized and hysterectomized pigs. There was a

Table 2

Anterior pituitary somatotrophic hormone activity during different reproductive stages in the pig

Reproductive stage (days)	Number of pigs	Mean body weight (kg)	Anterior pituitary dry weight (mg) *	Somatotrophic hormone activity		Total somatotrophic hormone activity per dry anterior pituitary (μg)
				Concentration (μg/mg) †	95% limits ‡	
8–10 oestrous cycle	5	131	49 ± 4	50·4	13–69	2,470
30 after ovariectomy§	5	135	54 ± 4	47·2	24–66	2,549
30 after hysterectomy§	5	135	59 ± 3	57·7	36–76	3,404
25 pregnancy	5	121	48 ± 2	34·7	10–55	1,666
40 pregnancy	5	122	50 ± 3	45·3	25–69	2,265
80 pregnancy	4	157	62 ± 1	37·7	20–47	2,337
110 pregnancy	4	179	80 ± 6	50·3	23–68	4,024
14 lactation	5	153	87 ± 6	72·5	59–81	6,308

* Mean ± SE.
† Expressed as μg-equivalents of NIH-GH-B7 per milligramme dry weight of anterior pituitary.
‡ The error between values was incorporated into the calculation of the confidence limits of the mean value of the group.
§ Animals were either ovariectomized or hysterectomized at days 8–10 of the oestrous cycle.

tendency towards increased concentration of somatotrophic activity between days 25 and 110 of pregnancy. At day 14 of lactation, hormone activity was significantly greater ($P < 0.05$) than observed at day 80 of gestation. An almost twofold increase in anterior pituitary weight occurred during the period from early pregnancy to day 14 after parturition. The increased weight of the anterior pituitary gland in pregnant and lactating animals was reflected by the progressive increase in total somatotrophic activity per dry anterior pituitary. Furthermore, a corresponding increase in body weight did not occur with the increased weight of the anterior pituitary during pregnancy and lactation. Increased body weight during the latter stages of gestation resulted from the growth of the conceptus. From these results it is concluded that pituitary somatotrophic hormone may provide an essential trophic stimulus for growth of organs associated with reproduction and lactation. Further increase in pituitary somatotrophic activity 2 weeks after parturition may indicate its role in lactation in the pig. The significance of a possible somatotrophic hormone-lactogenic hormone complex has not been defined for this species.

Relationship between pituitary thyrotrophic hormone activity and reproductive function in the pig is not clear. Robinson and Nalbandov (1951) reported that pituitary thyrotrophic hormone increased during the oestrous cycle. There appeared to be no association of thyrotrophic hormone activity with either ovarian activity or gonadotrophic potency of the pituitary. Later, however, Hollandbeck, Baker, Norton and Nalbandov (1956) observed a positive correlation between total pituitary potencies of the gonadotrophic and thyrotrophic hormones.

Table 3

Thyrotrophic hormone activity in anterior pituitary glands during different reproductive stages in the pig

Reproductive stage (days)	Number of pigs	Thyrotrophic hormone activity		Total thyrotrophic hormone activity per dry anterior pituitary	
		Concentration (USP units/mg) *	95% limits*	(USP units)	95% limits
8–10 oestrous cycle	5	0·077	0·005–0·148	3·77	0·25– 7·25
30 after ovariectomy	5	0·026	0·003–0·089	1·40	0·16– 4·81
30 after hysterectomy	5	0·023	0·001–0·129	1·35	0·06– 7·61
25 pregnancy	5	0·083	0·008–0·200	3·98	0·38– 9·60
40 pregnancy	5	0·133	0·109–0·266	6·65	5·45–13·30
80 pregnancy	4	0·128	0·030–0·236	7·94	1·86–14·63
110 pregnancy	4	0·059	0·017–0·066	4·72	1·36– 5·28
14 lactation	5	0·109	0·055–0·274	9·48	4·79–23·84

* Expressed as USP-equivalents per milligramme dry anterior pituitary.

Thyrotrophic hormone activity was determined in dry anterior pituitary glands from different reproductive stages in the pig. Results are presented in Table 3. Acetone-dried anterior pituitaries were assayed for thyrotrophic hormone activity by determining the increase in weight of the thyroid gland in chicks. Day-old cockerels were injected once daily for 5 days with either a standard thyrotrophic hormone preparation or porcine anterior pituitary tissue. Total dosages of the dry anterior pituitary ranged from 0·5–5·0 mg. Results from these assays indicated that concentrations of thyrotrophic hormone activity were similar in cycling, ovariectomized, and hysterectomized pigs. At day 40 of pregnancy, concentration of hormone activity was significantly greater ($P < 0·05$) than observed 30 days after ovariectomy. The concentrations of pituitary thyrotrophic activity remained the same during the four stages of pregnancy and at the fourteenth day of lactation. There was, however, a trend toward increased total activity during pregnancy and lactation. This apparent increase in hormone content was related primarily to increased weight of the pituitary gland.

THE UTERUS AND OVARIAN FUNCTION DURING THE OESTROUS CYCLE AND PREGNANCY

The influence of the uterus on ovarian and pituitary function in the pig is well established. In the cycling animal corpora lutea develop and regress during a 15–16 day period. In mated pigs corpora lutea persist throughout gestation, and for the maintenance of pregnancy they are essential for at least 106 days (du Mesnil du Buisson and Dauzier, 1957). Embryonic survival was maintained in bilaterally ovariectomized pigs between day 15 and day 25 with daily injections of 200 mg progesterone and 0·1 mg oestradiol benzoate (Day, Anderson, Emmerson, Hazel and Melampy, 1959). After day 40 of pregnancy, at least 5 corpora lutea were necessary for maintenance of normal gestation (du Mesnil du Buisson and Dauzier, 1959). Gestation was maintained by daily injections of 8 mg of progesterone in pigs that had only 3 corpora lutea remaining.

It is well known that hysterectomy prolongs the life span of corpora lutea in this species. Du Mesnil du Buisson and Dauzier (1959) reported that corpora lutea persisted until day 117 in pigs that were hysterectomized at day 8 of the oestrous cycle or day 30 of pregnancy. Furthermore, they observed large follicles after this period and by day 200 ovulation occurred. Spies, Zimmerman, Self and Casida (1960) also found that corpora lutea were maintained 32 to 119 days in pigs hysterectomized at the seventh day of the cycle.

The inhibition of cycling and maintenance of corpora lutea in the pig are affected by the amount of uterus retained following partial hysterectomy. Furthermore, the quantity of uterus retained determines whether bilateral or unilateral regression of corpora lutea result. The effect of partial hysterectomy on the maintenance of corpora lutea is shown in Table 4 (Anderson, Butcher and Melampy, 1961). Animals were partially hysterectomized during the luteal

Table 4

Oestrous cycle lengths following hysterectomy in the pig

(Adapted from data of L. L. Anderson *et al.*, 1961, by courtesy of the Editor, *Endocrinology*)

Group	Portion of uterus retained, indicated by shaded area	Number of pigs	Oestrous cycle interval following operation	
			First	Second
1		5	days none	
2		4 1	none 26	31
3		3 2	none 28 ± 1*	none
4		5	32 ± 6	23 ± 0
5		5	25 ± 2	25 ± 2
6		5	23 ± 1	24 ± 2
7		5	21 ± 0	22 ± 0

* Mean ± S.E.

phase, days 8–11, of the cycle. Pigs with only the posterior half of the cervix (group 1) or the uterine body and cervix (group 2) did not return to oestrus during a 120-day period after the oestrus prior to surgery. One-fourth of one uterine horn (group 3) approached the minimal quantity necessary for continuation of oestrous cycles; 2 of the 5 pigs returned to oestrus. Oestrous cycles continued in all pigs with more than one-fourth of a uterine horn (groups 4, 5, 6); however, they were longer than observed in sham-operated controls (group 7). Corpora lutea were maintained in pigs hysterectomized before day 16 of the cycle (Anderson, Butcher and Melampy, 1963). Removal of the uterus after this time resulted in oestrus, ovulation and persistence of the new corpora lutea. Du Mesnil du Buisson (1961a) partially hysterectomized pigs between days 7 and 9 of the cycle and found that when the amount of the remaining part of one uterine horn exceeded 26 cm (about one-fourth of a horn), oestrous cycles continued. When the amount of uterine horn was less than 26 cm, oestrus was inhibited. In those animals in which the uterus was less than 26 cm, corpora lutea appeared functional in the ovary contralateral to the remaining uterine fragment 50 to 55 days after the pre-operative oestrus. It was of particular significance that in these animals the corpora lutea regressed in the homolateral ovary. This unilateral luteal regression in the pig suggests an active role of a non-gravid uterus on the life span of corpora lutea, and furthermore, it appears to be a local luteolytic action. The nature of the uterine luteolytic effect in the pig is unknown. The uterus of the unmated sow is the source of luteolytic activity during the progestational phase of the cycle. This action is prevented by the conceptus and is absent following hysterectomy; hence, the persistence of corpora lutea and the absence of cycles.

The luteolytic effect of a non-gravid portion of the uterus is present also during pregnancy. Du Mesnil du Buisson (1961b) reported the interference of a non-gravid horn with maintenance of early pregnancy in the opposite intact uterine horn of the pig. When the non-gravid horn was removed before day 14, pregnancy was maintained; however, if the non-gravid horn was removed after day 16, the pregnancy failed. The luteolytic effect of the non-gravid uterine horn was eliminated by daily injections of 200 mg progesterone. Rathmacher and Anderson (1963) investigated the effects of different portions of the non-gravid horn upon the maintenance of pregnancy in the opposite intact uterine horn in the pig. Results are presented in Table 5. An almost intact non-gravid horn interfered with maintenance of early unilateral pregnancy (group 1). Only 4 of 27 (15 per cent) of the pigs maintained pregnancy for 35 days. Unilateral luteal regression resulted in 3 of the 4 animals that remained pregnant. These results are similar to those reported by du Mesnil

Table 5

Pregnancy and unilateral luteal regression in pigs with a non-gravid horn

(Adapted from data of L. L. Anderson, 1966, by courtesy of the Editor, *J. Reprod. Fert.* Suppl. 1, p. 21, and the Society of Fertility)

Group	Portion of reproductive tract retained, indicated by shaded area	Animals pregnant		Animals with uni-lateral luteal regression	
		No.	%	No.	%
1		4/27	15	3/4	75
2		7/12	58	5/7	71
3		4/6	67	4/4	100
4		2/8	25	2/2	100
5		7/10	70	3/7	43
6		8/11	73	0/8	0
7		2/2	100	0/2	0

du Buisson (1961b). When only an anterior half or one-fourth of one non-gravid horn remained, pregnancy was maintained in the intact horn in 11 of 18 (61 per cent) animals (groups 2 and 3). Unilateral regression of corpora lutea in the ovary homolateral to the non-gravid horn occurred in 12 of 15 (80 per cent) animals by day 35 of gestation when either seven-eighths (group 1), one-half (group 2) or one-fourth (group 3) of the horn remained. Pregnancy usually failed when the homolateral ovary was removed (group 4). The local luteolytic effect was not dependent upon the proximity of the non-gravid uterine fragment to the ovary. In unilaterally hysterectomized animals, pregnancy was maintained (8 of 11, 73 per cent) and the corpora lutea in both ovaries persisted (group 6). The luteolytic effect of the non-gravid horn was prevented by daily injections of progesterone. It appeared that the termination of pregnancy was caused by insufficient luteal progesterone.

Pregnancy was maintained in pigs in which, at the twelfth day, all of the embryos and the uterus were removed except for one embryo and its corresponding portion of uterine horn (du Mesnil du Buisson and Rombauts, 1963a). Pregnancy was maintained in the majority of animals when a fixed number of embryos and associated portions of the uterine horns were removed about day 40 and day 80. Removal of some embryos without removal of the corresponding uterine horn resulted in termination of pregnancy. It was suggested that the empty portion of uterine horn had a negative effect upon pregnancy after day 40, even though it did not induce a luteolytic effect upon the corresponding ovary after this time.

Influence of the non-gravid uterine horn on luteal activity during early unilateral pregnancy in the pig was investigated by Anderson, Rathmacher and Melampy (1966). Luteal progesterone content was determined in unilaterally hysterectomized pigs and in animals with a non-gravid horn. Progesterone content was similar in the right and left ovaries at day 13 and day 15 in pigs with an anterior half of the non-gravid right horn (*Figure 5*). A reduction in progesterone content occurred by the sixteenth day in the ovary homolateral to the non-gravid horn as compared with the ovary associated with the gravid horn. Hormone content decreased at a rate of -18 µg/day between day 13 and day 25. In the contralateral ovary it increased at a rate of 42 µg/day during this time. Luteal progesterone remained constant from day 13 to day 25 ($b = -2$ µg/day) in the ovary on the side of unilateral hysterectomy. Simultaneously progesterone increased gradually ($b = 18$ µg/day) in the contralateral ovary in these unilaterally hysterectomized animals. Reduction in luteal progesterone in pregnant pigs with a non-gravid horn corresponded to the time in the oestrous cycle when the uterus may influence the corpus luteum. Unilateral luteal regression observed in these experiments appeared to result from the local

action of the non-gravid horn, whereas the compensatory progesterone production by corpora lutea on the gravid horn side may
have been due to a local stimulatory effect by the gravid uterus.

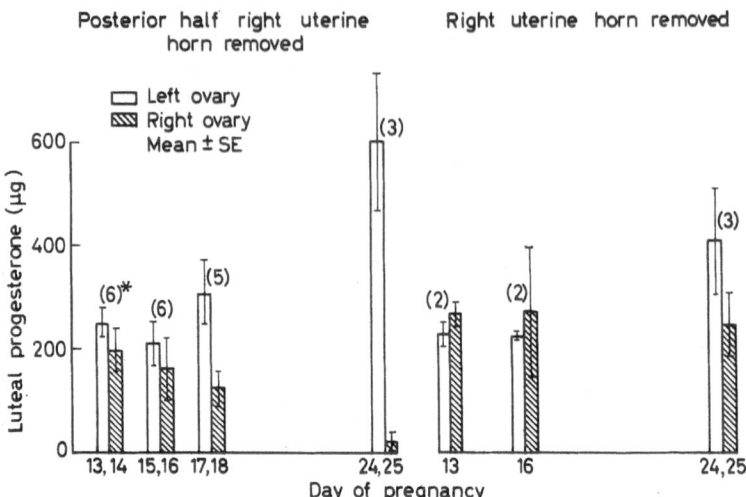

Figure 5. Luteal progesterone content in the left and right ovaries during unilateral pregnancies (left horn) in pigs with an anterior half of the right horn and in unilaterally hysterectomized animals (From L. L. Anderson *et al.*, 1966, by courtesy of the American Physiological Society)

The possibility of compensatory luteal progesterone production in
unilaterally ovariectomized and unilaterally hysterectomized pigs
during early pregnancy has been investigated (Rathmacher, Anderson, Henricks and Melampy, 1967).   Pigs were unilaterally ovariectomized the fifth day after mating.   Cycling animals were unilaterally
hysterectomized at days 8–10 and one-half of these pigs were unilaterally ovariectomized the fifth day after mating.   One group of
pigs served as unoperated controls.   Removal of an ovary or a
uterine horn did not reduce the number of surviving embryos at the
twenty-fifth day.   In the unilaterally ovariectomized pregnant
animals the weight of the remaining ovary increased when compared
with the same (left) ovary in pigs with both ovaries.   Furthermore,
the weight of the one ovary (left) was greater at day 25 in animals
with two intact horns as compared with unilaterally hysterectomized-

pigs. These increases in ovarian weight of unilaterally ovariectomized animals were due partly to follicular growth. Corpora lutea from the one ovary of unilaterally ovariectomized pigs were heavier (21 per cent increase) than those from animals with two intact ovaries. The increased weight of the ovaries and the corpora lutea in unilaterally ovariectomized animals indicated that one ovary compensated during early stages of pregnancy. In these experiments, the presence of only one gravid horn did not affect the compensatory growth of the ovary. There was an increase (13 per cent) in the progesterone content of individual corpora lutea in unilaterally ovariectomized pigs as compared with animals having both ovaries intact. Total luteal progesterone content, however, increased only 8 per cent in unilaterally ovariectomized pigs when the number and weight of corpora lutea were considered. Presence of either one or two gravid horns did not influence total luteal progesterone content.

RNA : DNA ratios for luteal tissue were high in the four experimental groups. These ratios were comparable with those of liver or pancreas. No differences were observed in the ratios of RNA to DNA of unilaterally ovariectomized or unilaterally hysterectomized pigs. RNA content of corpora lutea from unilaterally ovariectomized animals was greater than that found in animals with both ovaries intact. There was an increase in luteal cell number rather than an increase in cell size in unilaterally ovariectomized pigs. Based on histological evidence Corner (1921) concluded that lutein cells in the pig increased in size without division.

There are conflicting reports relative to the effect of unilateral ovariectomy on ovarian function in cycling pigs. For example, Brinkley, Wickersham, First and Casida (1964) reported that in pigs unilaterally ovariectomized at day 1 or 6 of the cycle, a decrease in luteal weight and progesterone content occurred in the remaining ovary by day 14 of the same cycle, whereas Short, Peters, First and Casida (1965) found no change in luteal weight 6 days after unilateral ovariectomy at days 1, 7, or 13 of the cycle.

UTERINE INNERVATION AND OVARIAN FUNCTION

Uterine innervation appears to play a minor role in either direct or indirect action on ovarian function in the pig. Uterine distension by metallic or plastic cylinders did not alter postoperative oestrous intervals (Anderson, 1962). Furthermore, oestrous cycles continued following either uterine denervation (Anderson, Bowerman and Melampy, 1963) or uterine autotransplantation (Anderson *et al.*,

1963; du Mesnil du Buisson and Rombauts, 1963b). Uterine action initiating luteolysis in the pig may be dependent upon neurohumoral mechanisms, but not involve major afferent nerves from the uterus.

THE PITUITARY AND REGULATION OF THE OESTROUS CYCLE AND PREGNANCY

Nalbandov (1961) suggested that pituitary luteotrophin in the cycling pig was released at about the time of oestrus, and further-more, that no further release of a pituitary luteotrophin was neces-sary for the normal life span (about 16 days) of the corpora lutea. On the other hand, in mated animals a secondary, and perhaps continuous release of a pituitary luteotrophin was essential for implantation and the maintenance of the corpora lutea. Exogenous progesterone did not induce luteal regression during the first 12 days of the cycle; however, it was effective in causing luteal regression between days 12 and 16 in previously mated pigs (Sammelwitz, Aldred and Nalbandov, 1961).

It has been reported by du Mesnil du Buisson and Léglise (1963) that corpora lutea develop in pigs hypophysectomized only a few hours after the initial signs of oestrus. These glands were charac-terized by reduced progesterone concentration. Anderson and Melampy (1966) have found similar results in pigs hypophy-sectomized at oestrus; the corpora lutea averaged 257 mg and con-tained 67 µg progesterone/g of luteal tissue (average of 4 pigs) at the twelfth day. Ovarian follicles were absent at this time and by day 20, the corpora lutea had regressed. Brinkley, Norton and Nalbandov (1964) found that a blocking dosage of progesterone beginning either 2 days before ovulation, the day of ovulation or 1 day after did not prevent the luteal development. Anderson, Dyck, Mori, Henricks and Melampy (1967) observed that corpora lutea developed to the twelfth day of the cycle (average weight 305 mg; 64 µg progesterone/g of tissue) when pigs were pituitary stalk-sectioned the day after the first day of oestrus. By day 16 the corpora lutea had declined slightly (average 260 mg), but the progesterone concentration was greatly reduced (3 µg/g). At day 12 and day 16 the follicles were completely regressed in these experimental animals. In mated pigs, the corpora lutea develop during the early phase following hypophysectomy. Du Mesnil du Buisson, Léglise, Ander-son and Rombauts (1964) reported that corpora lutea were normal, as indicated by progesterone content, at the twelfth day in pigs hypophysectomized the fourth day after mating. In pigs pituitary stalk-sectioned the day after mating, pregnancy continued and corpora lutea were maintained (average weight 431 mg; 54 μg progesterone/g tissue) at the twelfth day (Anderson *et al.*, 1967). By day 16 the corpora lutea averaged 292 mg and contained 28 μg progesterone/g of tissue. Again, the follicles were completely regressed at these stages of pregnancy.

It appears that in the pregnant pituitary stalk-sectioned pig the conceptus prevented uterine luteolytic activity. Immediate regression of follicles in pituitary stalk-sectioned pigs indicates that the hypothalamus or other central nervous system neurons control the release of anterior pituitary gonadotrophins in cycling, pregnant and hysterectomized animals. From these experimental results, it may be concluded that if a pituitary luteotrophin is required during the first 12 days of the cycle or pregnancy, only initial pituitary support is essential at oestrus, or even prior to oestrus or ovulation. Follicle-stimulating hormone activity does not appear to be involved in luteal development during this period, as indicated by complete follicular failure in either pituitary stalk-sectioned or hypophysectomized pigs.

The necessity of a pituitary gonadotrophin (luteotrophin) for maintenance of corpora lutea in hysterectomized pigs is clearly established. Du Mesnil du Buisson and Léglise (1963) reported that corpora lutea began to regress within 5 days and regression was complete within 10–11 days in animals hysterectomized at days 4–8 and hypophysectomized 22 days after the beginning of oestrus. Persisting corpora lutea regressed when pigs were hypophysectomized 29, 46, 97 and 99 days after hysterectomy (du Mesnil du Buisson, 1966). Complete luteal regression occurred within 20 days also when the animals were hypophysectomized at oestrus and hysterectomized in the early luteal phase of the cycle (du Mesnil du Buisson, 1966). In hypophysectomized pigs, exogenous gonadotrophins were effective in maintaining corpora lutea when the animals were previously hysterectomized (Anderson *et al.*, 1965). When the uterus remained intact in hypophysectomized pigs, the same gonadotrophin treatments failed to maintain corpora lutea. In pigs hypophysectomized at days 3 and 4, or 10 and hysterectomized at days 10–12, an acid-acetone extract of ovine pituitary (250 mg powder/day), human chorionic gonadotrophin (1,000 I.U./day), equine luteinizing hormone (Armour, 5 mg/day), or bovine luteinizing hormone (NIH, 5 mg/day) given daily from days 12 to 20 maintained the corpora lutea as indicated by weight and progesterone concentration. With these different hormone treatments, the corpora lutea weighed an average of 545 mg (range 359–761 mg) and contained an average concentration of 46 µg progesterone/g tissue (range 20–84 µg). Highest concentration of luteal progesterone was obtained with luteinizing hormone (NIH) in these experimental animals. Luteotrophic action of luteinizing hormone was evident between days 12 and 20 under these experimental conditions; however, initial stages of luteal regression were present by day 27. The progesterone concentration was reduced by one-half during this time in luteinizing hormone- (Armour or NIH) treated hypophysectomized-hysterectomized pigs. Similar reductions in luteal activity were observed in animals given continued injections of human chorionic gonadotrophin or an acid-acetone extract of ovine pituitary from days 20–27, 29 or 33. Corpora lutea were maintained in hypophysectomized-hysterectomized pigs between days 20 and 32 by daily injections of acetone-dried porcine anterior pituitary (50 mg powder/day; Anderson et al., 1967). In 3 animals, the average weight of the corpora lutea was 606 mg and progesterone concentration 84 µg/g. Ovarian follicles were completely regressed in these hypophysectomized-hysterectomized pigs. Bovine or porcine prolactin did not maintain luteal function in hypophysectomized-hysterectomized animals between days 12 and 20 (Anderson et al., 1965; Anderson and Melampy, 1966). These results clearly indicate that exogenous gonadotrophins of pituitary or placental origin effectively maintain luteal function in the pig for a period of at least 10 days when the uterus is removed.

Presence of the non-gravid uterus in the hypophysectomized pig alters the effect of exogenous gonadotrophins on ovarian function. Between days 12 and 20 corpora lutea were partially regressed (range of corpus luteum weight 61–518 mg) and contained very little progesterone (range 2–14 µg progesterone/g of tissue) in hypophysectomized pigs which were given the same dosages of either human chorionic gonadotrophin, ovine pituitary or luteinizing hormone (Armour or NIH; Anderson et al., 1965). Ovarian follicular activity also was absent in these experimental animals by day 20 of the cycle. These results indicate that the uterus inhibits utilization of pituitary gonadotrophin (luteotrophin) by the ovary in the non-gravid pig.

Social and other environmental factors, mediated through the central nervous system, control or modify reproductive activity in the pig (Signoret and du Mesnil du Buisson, 1961; Signoret and Mauléon, 1962; du Mesnil du Buisson and Signoret, 1962). These investigations have been concerned with the behavioural and sexual responses of pigs to visual, auditory and olfactory stimuli.

URINARY OESTROGEN EXCRETION IN CYCLING, PREGNANT AND HYSTERECTOMIZED PIGS

Bredeck and Mayer (1958) reported chemical assays of oestrogens in pregnant sow urine. During the oestrous cycle and pregnancy, the principal urinary oestrogen is oestrone (Raeside, 1961; Velle, 1959). Raeside (1961) reported that oestrone levels reached 20–30 μg/l urine at oestrus and were followed by a decline (< 5 μg/l) during the remainder of the cycle. Rombauts (1962) observed two peaks in urinary oestrone during gestation; a minor peak between day 28 and day 31 (1–2 mg/24 h), and then a rapid increase at day 80 which remained high until after parturition (4–6 mg/24 h).

Bowerman, Anderson and Melampy (1964) determined the concentrations of urinary oestrogens in cycling, pregnant, hysterectomized and ovariectomized pigs. Concentrations of oestrone per 24 h urine collection are shown in *Figure 6*. Although urine samples were analysed routinely for oestradiol, the amounts were usually below 1·4 μg, the minimum sensitivity of the method used. Oestriol was not detected in six urine samples. As shown in *Figure 6*, urinary oestrone in cycling animals increased to a peak just before or with the onset of oestrous behaviour. Immediately following oestrus there was a sharp decline in urinary oestrone. At days 15–18 urinary oestrone levels in pregnant animals were similar to those observed in cycling animals. Between days 18 and 30 of pregnancy, there was a marked increase in urinary oestrone. By day 40 the quantity of oestrone had decreased to a level similar to that of the dioestrous phase. From day 40 to parturition urinary oestrone increased sharply. Pigs were hysterectomized at days 5 and 6 of the oestrous cycle. From days 15–120, the levels of urinary oestrone in these animals remained consistently low. Oestrone was the principal urinary oestrogen of ovariectomized and ovariectomized-hysterectomized pigs injected with oestradiol benzoate alone or in combination with progesterone. Almost one-half of the injected oestrogen was recovered in the urine within 4 or 5 days. Results with ovariectomized and ovariectomized-hysterectomized pigs given oestradiol benzoate alone or in combination with progesterone indicated that the uterus, the progesterone-stimulated uterus, or progesterone alone did not alter oestrogen metabolism. Differences in levels of urinary oestrogen excretion between pregnant and hysterectomized animals may represent altered oestrogen production rather than oestrogen metabolism. Rombauts and du Mesnil du Buisson (1964) found that urinary oestrogens did not increase following hysterectomy at day 70 of gestation in the pig. It was suggested that

307

the large increase in urinary oestrone after day 70 in pregnant animals was produced by the placenta rather than the ovaries. Subsequently, it was reported that the principal oestrogen in pig

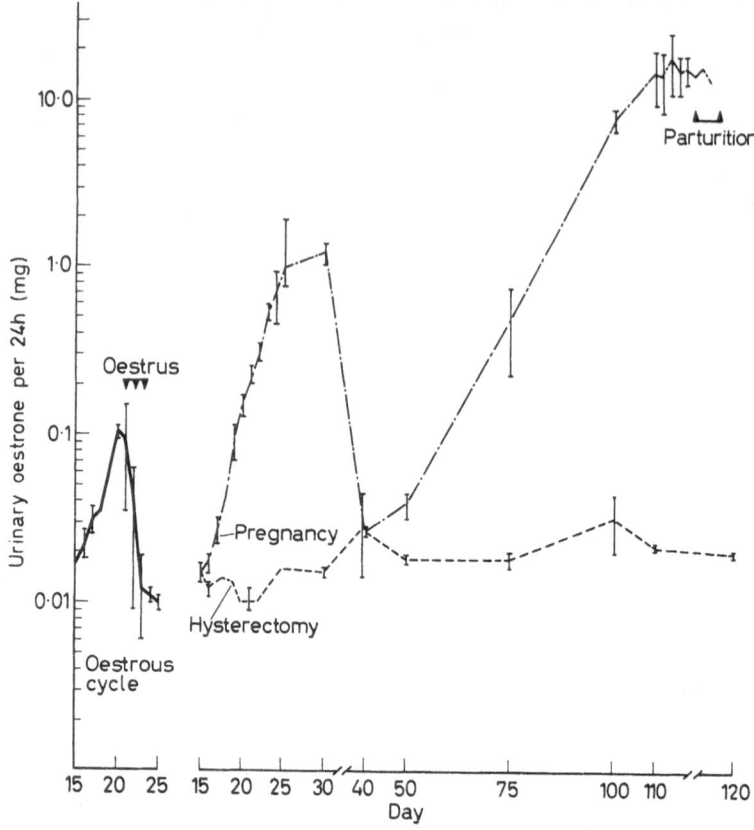

Figure 6. Daily urinary oestrone excretion by cycling, pregnant and hysterectomized pigs. Each value represents the mean and range for 2 pigs (Adapted from data of A. M. Bowerman *et al.*, 1964, by courtesy of the Editor, *Iowa State College Journal of Science*)

placenta is oestrone, and furthermore, that the placenta is the major source of this hormone in late gestation (Rombauts, 1964).

Exogenous oestrogen increased the duration of the oestrous interval (Kidder, Casida and Grummer, 1955) and maintained corpora lutea in the pig (Gardner, First and Casida, 1963). Oestradiol benzoate, given at levels sufficient to maintain corpora lutea in cycling pigs, failed to do so following hypophysectomy (du Mesnil du Buisson, 1966). These results may indicate an indirect action of the hormone on the life span of the corpus luteum.

SUMMARY

Pituitary gonadotrophin content (follicle-stimulating hormone activity and luteinizing hormone activity) and ovarian morphology during the oestrous cycle indicate that gonadotrophins are synthesized and released during the luteal phase, as well as at the onset of behavioural oestrus and ovulation. However, the physiological events culminating in ovulation in the pig appear to be sufficient for the formation and persistence of corpora lutea of the oestrous cycle, whereas prolongation of the life of corpora lutea is associated with a reciprocal uterine–pituitary action. Development and maintenance of corpora lutea during the first 12 days in cycling, pregnant or hysterectomized pigs are dependent upon release of pituitary gonadotrophin at oestrus. Maintenance of luteal function in pregnant or hysterectomized pigs after this period requires continued pituitary gonadotrophin (luteotrophin) support. Corpora lutea are essential for the maintenance of pregnancy; their ablation quickly leads to abortion, or recurrence of oestrus in either cycling or hysterectomized animals.

During advanced gestation and during lactation there is an increase in weight of the anterior pituitary gland as well as an associated increase in follicle-stimulating hormone, luteinizing hormone and somatotrophic hormone activities. Ovarian events do not mimic the gland content of the gonadotrophins during these reproductive stages, whereas the somatotrophic hormone may provide stimuli for growth of the organs associated with reproduction and lactation. Thyrotrophic hormone activity of the pituitary indicates no definitive relationship with these reproductive stages, but it tends to increase in pregnancy and lactation.

The principal urinary metabolite of oestrogens is oestrone and it is excreted in increasing concentration with advancing stages of pregnancy. This oestrone is primarily of placental origin in late pregnancy. In hysterectomized animals urinary oestrone remains consistently low for several months. During the oestrous cycle, urinary oestrone excretion coincides with follicular growth during the pro-oestrual stage, and the onset of oestrous behaviour and ovulation.

The uterus plays an important role in direct and/or indirect regulation of ovarian and pituitary function in the pig. Oestrous cycles occur with such regularity that it may appear that the uterus contributes nothing to the regulation of the cycle, and therefore the major control is by a reciprocal relationship between the ovaries and pituitary. However, the striking changes in ovarian function,

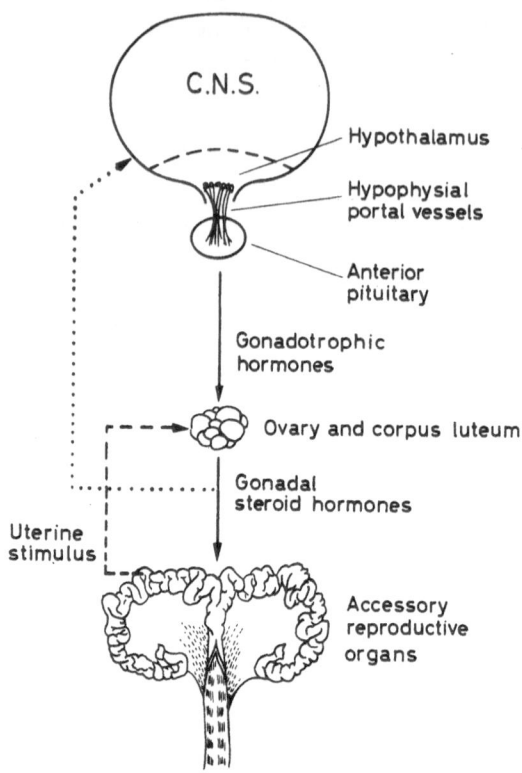

Figure 7. Reproductive relationships between the central nervous system (C.N.S.), pituitary, ovaries and uterus in the pig

particularly luteal maintenance, during early pregnancy (days 12–20) implicate a negative effect of the non-gravid uterus. This luteolytic action of the non-gravid uterus is evident in experimental pigs in which corpora lutea are maintained in the ovary homolateral to the gravid uterine horn, and luteal failure results in the opposite ovary homolateral to the non-gravid horn. If the non-gravid horn is removed in early unilateral pregnancy, the corpora lutea persist in both ovaries. A similar situation exists in hysterectomized and partially hysterectomized pigs. Experimental evidence indicates that the uterus and pituitary appear to regulate the life of the corpora lutea in the pig by two possible mechanisms (*a*) a local effect by the uterus, and (*b*) a systemic effect by the pituitary.

As shown in *Figure 7*, the central nervous system plays a role in regulating the production and release of gonadotrophic hormones. These are responsible for the synthesis and secretion of the ovarian hormones which are involved in uterine as well as central nervous system activity.

The uterus of the unmated pig is the source of a luteolytic stimulus during the progestational phase of the cycle. This luteolytic action is prevented by the conceptus and is absent following hysterectomy; hence, the persistence of corpora lutea and the absence of oestrous cycles.

ACKNOWLEDGEMENT

Supported by USDHEW, NIH Research Grant HD 01168–07 and the American Cyanamid Co., Princeton, New Jersey. Journal Paper No. J-5713 of the Iowa Agricultural and Home Economics Experiment Station, Ames, Iowa. Project No. 1325.

REFERENCES

Anderson, L. L. (1962). *J. Anim. Sci.* **21,** 597
— Bowerman, A. M. and Melampy, R. M. (1963). *Advances in Neuro-endocrinology*, Univ. Ill. Press; Urbana
— Butcher, R. L. and Melampy, R. M. (1961). *Endocrinology* **69,** 571
— — — (1963). *Nature, Lond.* **198,** 311
— Dyck, G. W. and Rathmacher, R. P. (1966). *Endocrinology* **78,** 897
— —Mori, H., Henricks, D. M. and Melampy, R. M. (1967). *Am. J. Physiol.* **212,** 1188
— Léglise, P. C., du Mesnil du Buisson, F. and Rombauts, P. (1965). *C.r. hebd. Séanc. Acad. Sci., Paris* **261,** 3675
— and Melampy, R. M. (1966). Unpublished
— Rathmacher, R. P. and Melampy, R. M. (1966). *Am. J. Physiol.* **210,** 611
Baker, B. (jun), Hollandbeck, R., Norton, H. W. and Nalbandov, A. V. (1956). *J. Anim. Sci.* **15,** 407
Bowerman, A. M., Anderson, L. L. and Melampy, R. M. (1964). *Iowa St. J. Sci.* **38,** 437
Bredeck, H. E. and Mayer, D. T. (1958). *Reproduction and Infertility III Symposium.* New York; Pergamon Press
Brinkley, H. J., Norton, H. W. and Nalbandov, A. V. (1964). *Endocrinology* **74,** 9

Brinkley, H. J., Wickersham, E. W., First, N. L. and Casida, L. E. (1964). *Endocrinology* **74,** 462

Corner, G. W. (1915). *Contr. Embryol.* **2,** 69. Carnegie Institution of Washington

— (1919). *Am. J. Anat.* **26,** 117

— (1921). *Contr. Embryol.* **13,** 117. Carnegie Institution of Washington

Day, B. N., Anderson, L. L., Emmerson, M. A., Hazel, L. N. and Melampy, R. M. (1959). *J. Anim. Sci.* **18,** 607

Duncan, G. W., Bowerman, A. M., Hearn, W. R. and Melampy, R. M. (1960). *Proc. Soc. exp. Biol.* **104,** 17

Erb, R. E., Nofziger, J. C., Stormshak, F. and Johnson, J. B. (1962). *J. Anim. Sci.* **21,** 562

Gardner, M. L., First, N. L. and Casida, L. E. (1963). *J. Anim. Sci.* **22,** 132

Gomes, W. R., Herschler, R. C. and Erb, R. E. (1965). *J. Anim. Sci.* **24,** 722

Hollandbeck, R., Baker, B. (jun.), Norton, H. W. and Nalbandov, A. V. (1956). *J. Anim. Sci.* **15,** 418

Kidder, H. E., Casida, L. E. and Grummer, R. H. (1955). *J. Anim. Sci.* **14,** 470

Masuda, H., Anderson, L. L., Henricks, D. M. and Melampy, R. M. (1967). *Endocrinology* **80,** 240

Mayer, D. T., Glasgow, B. R. and Gawienowski, A. M. (1961). *J. Anim. Sci.* **20,** 66

Melampy, R. M., Henricks, D. M., Anderson, L. L., Chen, C. L. and Schultz, J. R. (1966). *Endocrinology* **78,** 801

du Mesnil du Buisson, F. (1966). *Thesis.*

— (1961a). *Annls Biol. anim. Biochim. Biophys.* **1,** 105

— (1961b). *C.r. hebd. Séanc. Acad. Sci., Paris* **253,** 727

— and Dauzier, L. (1957). *C.r. Séanc. Soc. Biol.* **151,** 311

— — (1959). *Annls Zootech.* Suppl. 147

— and Léglise, P. C. (1963). *C.r. hebd. Séanc. Acad. Sci., Paris* **257,** 261

— — Anderson, L. L. and Rombauts, P. (1964). *Fifth Int. Congr. Anim. Reprod. A.I.* (*Trento*) 571

— and Rombauts, P. (1963a). *Annls Biol. anim. Biochim. Biophys.* **3,** 445

— — (1963b). *C.r. hebd. Séanc. Acad. Sci., Paris* **256,** 4984

— and Signoret, J. P. (1962). *Annls Zootech.* **11,** 53

Nalbandov, A. V. (1961). *Recent Prog. Horm. Res.* **17,** 119

— (1963). *J. Anim. Sci.* **22,** 558

Oxenreider, S. L., McClure, R. C. and Day, B. N. (1965). *J. Reprod. Fert.* **9,** 19

Parlow, A. F. (1961). *Human Pituitary Gonadotropins.* Springfield: Charles C. Thomas

— Anderson, L. L. and Melampy, R. M. (1964). *Endocrinology* **75,** 365

Raeside, J. I. (1961). *Fourth Int. Congr. Anim. Reprod. A.I.* 355. The Hague

Rathmacher, R. P. and Anderson, L. L. (1963). *J. Anim. Sci.* **22,** 1139

Rathmacher, R. P., Anderson, L. L., Henricks, D. M. and Melampy, R. M. (1967). *Endocrinology* (In press)

Robinson, G. E. (jun.) and Nalbandov, A. V. (1951). *J. Anim. Sci.* **10,** 469

Rombauts, P. (1962). *Annls Biol. anim. Biochim. Biophys.* **2,** 151

— (1964). *C.r. hebd. Séanc. Acad. Sci., Paris* **258,** 5257

— and du Mesnil du Buisson, F. (1964). *C.r. hebd. Séanc. Acad. Sci., Paris* **258,** 5076

— Pupin, F. and Terqui, M. (1965). *C.r. hebd. Séanc. Acad. Sci., Paris* **261,** 2753

Sammelwitz, P. H., Aldred, J. P. and Nalbandov, A. V. (1961). *J. Reprod. Fert.* **2,** 387

Short, R. E., Peters, J. B., First, N. L. and Casida, L. E. (1965). *J. Anim. Sci.* **24,** 929

Signoret, J. P. and Mauléon, P. (1962). *Annls Biol. anim. Biochim. Biophys.* **2,** 167

— and du Mesnil du Buisson, F. (1961). *Fourth Int. Congr. Anim. Reprod. A.I.* 171. The Hague

Spies, H. G., Zimmerman, D. R., Self, H. L. and Casida, L. E. (1960). *J. Anim. Sci.* **19,** 101

Steelman, S. L. and Pohley, F. M. (1953). *Endocrinology* **53,** 604

Velle, W. (1959). *Acta vet. scand.* **1,** 19

DISCUSSION

PROFESSOR C. GEMZELL (*Uppsala, Sweden*)

I would like to ask about the doses of pituitary extracts you were using for the maintenance of the corpus luteum since it is suggested that you have to use quite large doses.

MELAMPY

A daily dosage of 50 mg of desiccated porcine anterior pituitary maintained corpora lutea for 14 days in hypophysectomized-hysterectomized pigs. The amount of luteinizing hormone activity in the daily dosage of this preparation was approximately 200 μg as measured by the ovarian ascorbic acid depletion assay.

DR. J. R. GODING (*Babraham, Cambridge*)

Have you any information on secretion rates of progesterone from the pig's ovary? Have you any direct evidence for the luteolytic factor acting on the remote corpus luteum via the pituitary? Could it not be as Dr. Moore and Mr. Rowson have suggested, that the action is via the systemic route without the intervention of the anterior pituitary?

MELAMPY

In answer to your second question I agree with this possibility. With regard to your first question, we collect blood from one ovary by cannulation of the major vein about 2 cm from the ovary. We get about 100 ml in 10 min free flow. Work is in progress in which we are measuring the degree of vascularization or blood flow through the corpus luteum. The degree of vascularization appears to continue beyond the time of maximum progesterone output.

REPRODUCTION IN THE FEMALE MAMMAL

PROFESSOR A. V. NALBANDOV (*Illinois, U.S.A.*)

I was interested in the effect of growth hormone on the uterine endometrium. Do you think this is acting via the ovary or are you implying a direct effect?

MELAMPY

It must be a direct effect because in our search for luteotrophins of pituitary origin, we used NIH bovine growth hormone injected into hypophysectomized animals starting at day 7 or 8 after hypophysectomy. Complete ovarian failure occurred but after 10 days the endometrium grew and appeared to resemble that from an animal at day 25 of pregnancy. Further, in the cow, if the calf is removed immediately at parturition, uterine involution occurs but involution is delayed by giving bovine growth hormone.

DR. D. SMIDT (*Germany*)

Will Professor Melampy give us his views about the possible anatomical pathways involved in conveying the locally-acting luteolytic influence from the uterus to the ovary.

MELAMPY

That's a very good question but I have no answer. An excellent discussion of the vascular arrangements in the pig was published recently (*J. Reprod. Fert.* **9** (1965) 19–27) but it does not really provide a clue in answer to this question.

DR. D. SMIDT

Have you attempted denervation of the uterus to see whether that affects the results?

MELAMPY

Yes, several years ago Dr. Anderson isolated a section of the middle portion of the uterine horn. He denervated that as completely as he could and the only innervation remaining was that in the arterial wall. These animals continued to cycle after denervation as they did before. Also uterine autotransplants in guinea-pigs and pigs allowed cycling.

DR. C. POLGE (*Cambridge*)

As I understood it, you were able in your unilaterally pregnant pigs to maintain pregnancy by giving injections of progesterone daily from day 5–25, yet you did not get regression of the corpora lutea on the pregnant side, they remained highly functional. You also stressed the importance of pituitary luteotrophic support for the maintenance of corpora lutea in the pig and you mentioned that Professor Nalbandov had already demonstrated that giving exogenous progesterone after day 16 of the cycle caused regression of the corpora lutea in pregnant pigs. Can you explain how in your unilaterally pregnant pigs you maintained pregnancy but did not get regression of the corpora lutea.

MELAMPY

I have no explanation of that.

LUTEAL FUNCTION

DR. C. POLGE

What is the dose that you give? Perhaps it is a dose effect.

MELAMPY

I think it is 250 mg, which might not be sufficient to suppress the pituitary. The pituitary might need much more and that could be an explanation.

DR. R. M. MOOR (*Cambridge*)

Do you know when the uterus of the non-pregnant pig becomes lytic? At which stage of the cycle does this take place and what factors lead to the lytic situation?

MELAMPY

The only information we have is that if we hysterectomize after day 16 and sometimes after day 15 then the corpora lutea persist at the next cycle. On days 13 or 14 this does not happen so the luteolytic action has been accomplished by this time.

On a different matter, I would like Dr. Polge to comment with regard to some work he has done on the uterine effects in pregnancy.

DR. C. POLGE

This was some work done in collaboration with Mr. Rowson and Dr. Chang in which we were collecting eggs from pigs soon after fertilization, by flushing the Fallopian tubes. In some animals incomplete collection of eggs was obtained and we thus left varying numbers of fertilized eggs in the Fallopian tube. After recovery, the animals were checked regularly for return to oestrus, and we noticed that in animals in which more than 5 fertilized eggs were left, the animals became pregnant. On the other hand, with between one and four fertilized eggs in the Fallopian tube the animals returned to oestrus, but with an extended cycle of about 27 days. An argument in support of this observation is that it has already been shown that a portion of a non-gravid uterus will result in luteal regression and as Dr. Melampy mentioned in his experiments and also in Dr. du Mensil du Buisson's experiments, only a quarter of the uterus had to remain non-pregnant in order to cause luteal regression. We therefore suggest that four embryos are insufficient to overcome the luteolytic effect of the portion of the uterus which is non-gravid and in a polytocous species like the pig which normally ovulates 15–18 eggs it is necessary during the early stages of pregnancy to have more than four viable foetuses in order to maintain pregnancy over the time when normal luteal regression occurs.

DR. D. SMIDT

What evidence have you that those eggs which remain in the uterus were really fertilized?

DR. C. POLGE

Simply that other eggs recovered were fertilized and usually from our observations it is rare in the pig to get many unfertilized eggs in a group of fertilized eggs. Either about 98 per cent are fertilized or they are mainly unfertilized. The recovered eggs were fertilized and so we assumed that the eggs left behind were also fertilized. Where we obtained unfertilized eggs and left several behind, these pigs had normal cycles following operation, and hence were an effective control.

315

REPRODUCTION IN THE FEMALE MAMMAL

DR. D. SMIDT

You assumed, therefore, that the luteolytic effect of the uterine tissue is only effected in the first 3 or 4 weeks of pregnancy because there are a lot of litters of 1 to 4 piglets.

MELAMPY

Those litters presumably start out with a larger number and the pregnancy wastage is later. I don't think that applies to this situation at all. You may start out with a litter of 12 or 14 and may only farrow 2 but the remainder get the pregnancy established.

DR. J. SPINCEMAILLE (*Belgium*) *translating for* DR. DU MENSIL DU BUISSON (*France*)

I reported earlier about the local influence of the uterus in the maintenance of the corpus luteum in the sow to which Professor Melampy referred. The difficulty of having a pregnancy in the presence of a total sterile horn can perhaps be explained better by the luteolytic action of the uterine horn. Recently we have investigated the influence of oestrogen on the luteolytic mechanism. Firstly in the normal animal, injection of oestradiol benzoate (5 mg/day) for 7 days starting on the twelfth day of the cycle stops the cycle in several cases and maintains the corpus luteum for at least 100 days. Secondly, the same treatment in an animal with a unilateral pregnancy maintains the corpus luteum in both ovaries and maintains pregnancy in nine out of eleven cases. Thirdly, in the animal hypophysectomized at the beginning of the cycle this treatment together with daily injections of 5 mg LH a day until day 20 provoked the maintenance of the corpus luteum. In these hypophysectomized animals neither LH alone nor oestradiol benzoate alone were able to maintain the corpus luteum until day 20. If the uterus was absent, however, LH alone would maintain the corpus luteum until day 20. Our conclusion, at present therefore, is that the luteolytic substance is anti-LH. We can reasonably assume that oestrogen is a luteotrophic substance inhibiting the luteolytic action of the uterus and acting in other cases by blocking the release or secretion of the luteolytic substance of the uterus.

DR. R. DEANESLY (*Cambridge*)

Dr. du Mensil du Buisson told me about his experiments with oestrogens and I tried to get a similar effect in the guinea-pig but was unsuccessful although his theory seems sound.

316

THE EXISTENCE OF A LUTEOLYTIC HORMONE IN THE UTERUS OF THE GUINEA-PIG

B. T. DONOVAN

Department of Neuroendocrinology, Institute of Psychiatry,
De Crespigny Park, London, S.E.5

THE NORMAL GROWTH AND REGRESSION OF THE CORPORA LUTEA

Oestrous cycle

OVULATION and corpus luteum formation recur at approximately 16-day intervals in the guinea-pig, at a time when regression of the previous set of corpora lutea is marked.   Loeb (1911a, b) recorded that the corpora lutea are fully developed from the histological point of view from the sixth day after ovulation, and that regressive changes can be detected on the tenth day.   When the changes in the volume of the corpora lutea during the cycle were traced by Rowlands (1956), Perry and Rowlands (1962a) and Bland and Donovan (1966a), it was found that the maximum size is reached at about the eleventh or twelfth days (*Figure 1*).   Thereafter regression sets in and is rapid.   It is perhaps unfortunate that the morphological changes in the corpora lutea may not be correlated with their functional state for Rowlands and Short (1959) found that the concentration of progesterone in luteal tissue 11–13 days after ovulation (7·7 µg/g) is half that at 6 days (16·2 µg/g) despite the lack of histological modification in the interval.

Pregnancy

Although sterile mating is without effect on the corpora lutea, implantation brings about marked enlargement of these bodies.   In the pregnant guinea-pig the corpora lutea continue to grow until the eighteenth to twentieth day after coitus and remain at that size for the rest of pregnancy (Rowlands, 1956).   The progesterone content of the corpora lutea of pregnant females is much above that of unmated animals at 11–13 days of age (being at the high 6-day level in pregnant females), increases further by 21–23 days and then

317

remains constant for the rest of pregnancy (Rowlands and Short, 1959).

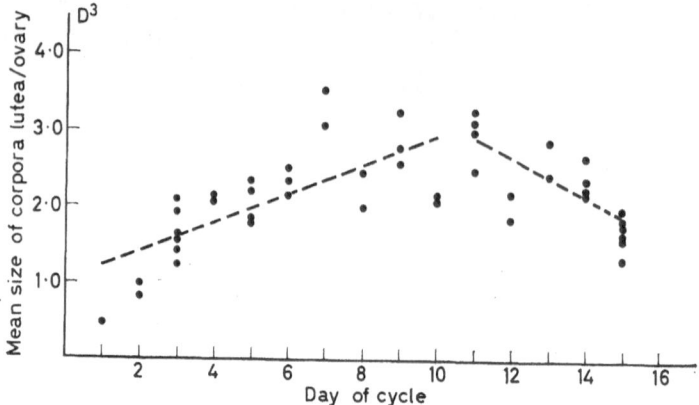

Figure 1. *The growth and regression of the corpora lutea during the oestrous cycle of normal guinea-pigs. The mean size of the corpora lutea in each ovary is plotted and the calculated regression lines included*

FACTORS INFLUENCING THE LIFE SPAN OF THE CORPORA LUTEA

The life span of the corpora lutea can be influenced by removal or distension of the uterus or by the administration of hormones. Before discussing the mechanisms that might be involved it is useful to outline the information that is available.

Hysterectomy

Ever since the comprehensive work of Loeb (1923, 1927) it has been repeatedly confirmed that removal of the uterus prolongs luteal function in the guinea-pig (Desclin, 1932; Herlant, 1933; Klein, 1939; Rowlands, 1961). The luteal bodies not only remain as structural entities in the ovaries, but also secrete progesterone, as is shown by the extension of the oestrous cycle and the finding that extirpation of these organs precipitates a fresh ovulation (Loeb, 1927). When hysterectomy was performed on the fifth day after ovulation, the vagina remained closed for at least 8 weeks and, in 4 of 5 animals, for 8 months (Rowlands, 1961). Removal of the uterus on the tenth day after ovulation caused significant enlargement of the corpora lutea within 10 days and the maintenance of the large size for about 40 days, but then regression set in and the luteal bodies shrank to *post-partum* size about 80 days after hysterectomy. By the fifteenth day after ovulation regression of the corpora lutea is well advanced and the formation of a new set is imminent. Nevertheless, in 24 of 35 guinea-pigs hysterectomized at this time the vagina failed to open as expected and it was apparent that sufficient hormone was secreted by the corpora lutea to prevent oestrus and ovulation for long periods. Even so, enlargement of the corpora lutea only occasionally occurred and was never great enough to restore the volume to that reached earlier in the cycle. Other animals hysterectomized late in the cycle ovulated and the new set of corpora lutea was maintained. Further details have been provided by Butcher, Chu and Melampy (1962a) who found that 10 of 16 guinea-pigs in which the uterus was completely removed, or only the cervix retained, on days 5 or 6 of the cycle came into heat after 63–98 days, while the remainder failed to do so for at least 4 months.

Partial hysterectomy extends luteal function to a lesser degree (Loeb, 1923, 1927; Butcher *et al.*, 1962a; Howe, 1965), for hemihysterectomy lengthens the cycle by a matter of days, although a more marked effect has been reported (Herlant, 1933). The retention of one-half or one-quarter of a uterine horn is enough to allow the resumption of oestrous cycles of variable length and there is a tendency for succeeding cycles to revert to normal. Loeb (1927) reported that the effects of ablation of the upper or lower parts of the uterus were indistinguishable and reasoned that the consequences of hysterectomy did not depend on the extirpation of a specific part of the uterus, or a particular vascular or nervous structure, but on the quantity of uterus removed. When hemihysterectomy was combined with unilateral ovariectomy on the homolateral or contralateral side, Fischer (1965) discovered that those females spayed homolaterally ran oestrous cycles of the normal 16-day length, while the cycles of those contralaterally ovariectomized were about twice as long. Simple hemihysterectomy caused lengthened oestrous cycles as expected but the ovaries of these animals showed a striking histological asymmetry, for the corpora lutea were maintained on the operated side whereas on the control side they had regressed. Fischer (1965) concluded that the uterus exerts a direct inhibitory effect on the corpora lutea, which is mediated locally, and this view is substantiated by measurement of the size of the corpora lutea in the ovaries of animals hemihysterectomized on days 3 or 4 of a cycle and killed on day 14 (Bland and Donovan, 1966b). The size was estimated from the two largest diameters and the number of sections of each corpus luteum (D^3) and the mean volume of the corpora lutea on the operated side ($3\cdot1$ mm³) was greater than normal and much above that of the luteal bodies in the control ovary ($2\cdot0$ mm³) (*Figure 2*).

Similar effects on luteal function to those of hysterectomy can be obtained by chemical destruction of the endometrium but extensive

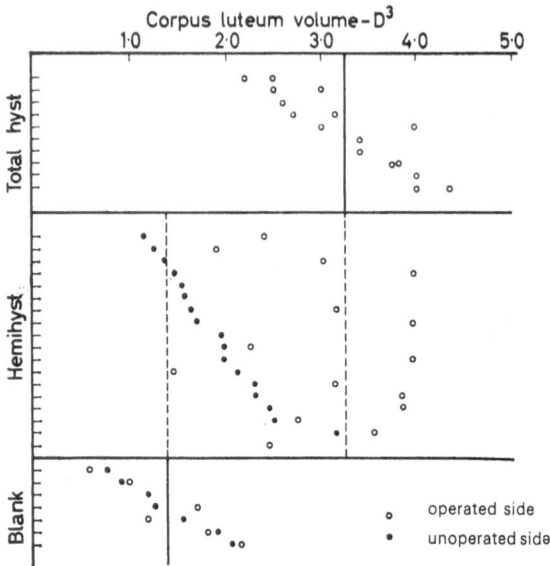

Figure 2. The effect of hysterectomy and hemihysterectomy on the size of the corpora lutea (D^3). The volume of the corpora lutea is increased following total hysterectomy (top) when compared with the normal values (bottom). After hemihysterectomy (middle) the sizes of the corpora lutea on the control side fall within the normal range while those on the operated side show enlargement

traumatization of the uterus in the early part of the cycle and the induction of deciduomata delayed the expected oestrus by only 3–7 days (Loeb, 1927). Conversely, the replacement of uterine tissue after hysterectomy has been found to counter the expected sequel to the operation, although, in some experiments, the autotransplantation of pieces of the uterus to a subcutaneous location at the time of hysterectomy, or subsequent homotransplants of uterine tissue, has not proved to be of value (Loeb, 1927). Such studies are described in more detail later.

Distension of the uterus

In contrast to the extension of corpus luteum function seen with hysterectomy, distension of the uterus of the guinea-pig brings about the opposite result. Donovan and Traczyk (1960, 1962) found that when two cylindrical glass beads (7 × 3 mm) were inserted into each horn of the uterus during the first few days of the cycle, then the length of that cycle was shortened by several days. With progressively delayed introduction of beads the response became variable and bead insertion late in the cycle tended to prolong that cycle. However, animals carrying beads in the uterus for long periods subsequently ran short or normal cycles; cycles were never extended. Moore (1961) observed that when a glass bead 1·5 mm in diameter was sutured into one horn of the uterus on the third day of the cycle, that cycle was shortened to 11·7 days, but when the bead was introduced on the eighth day the cycle lasted 22·4 days. In the work of Donovan and Traczyk (1960, 1962) and of Moore (1961), damage to the endometrium caused by the insertion of a foreign body into the uterus late in the cycle has to be taken into consideration, and, to avoid this complication, distension of the uterus has since been initiated only during the first 4 days of the cycle (Bland and Donovan, 1965a, 1966a). Two beads present in each horn again shortened the oestrous cycle by about 3 days but when a single bead was present in each horn the cycle was curtailed to a lesser extent. Unilateral distension of the uterus did not cause as marked a reduction in cycle length as did bilateral manipulation and the greatest alteration occurred after the placement of 4 beads in one horn. This approximated to the response to one bead present on each side. Two beads unilaterally placed failed to alter the length of the oestrous cycle.

From the similarity in the abridgement of the guinea-pig sexual cycle seen with uterine distension and with ablation of the corpora lutea (in both cases the cycle is cut to about 11–12 days—Loeb, 1911a, b; Dempsey, 1937) it was suggested that dilation of the uterus promoted the physiological inactivation of the corpora lutea (Donovan and Traczyk, 1962). This view has since been substantiated, for despite the slight alteration in the oestrous cycle seen when one bead was present in each horn of the uterus, a marked reduction in the size of the corpora lutea has been detected (Bland and Donovan, 1965b, 1966a). Beads were inserted on days 1–4 of the cycle and the animals killed at various intervals before the next oestrous period. As the results in *Figure 3* indicate, the growth of the corpora lutea proceeded normally up to the ninth day and was

11+   321

then succeeded by abrupt regression over the next 2 days. From the eleventh day regression of the corpora lutea was slow and at a rate very similar to that recorded in normal guinea-pigs. When the effect of unilateral distension of the uterus was examined, an intrigu-

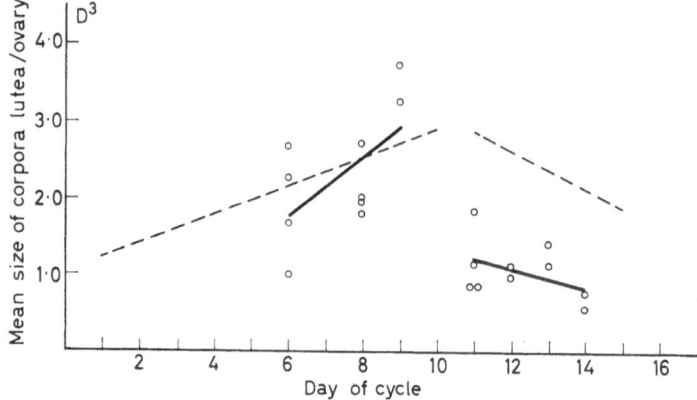

Figure 3. The effect of distension of the uterus on the size of the corpora lutea. One glass bead was inserted into each uterine horn between days 2–4 of the cycle and the animals killed on the days indicated. The growth and regression of the corpora lutea in normal animals is indicated by the broken line (taken from Figure 1) and it will be seen that the growth of the corpora lutea was normal over the first 9 days of the cycle. Between 9 and 11 days an abrupt and premature regression of the corpora lutea occurred

ing response emerged. In this experiment the ovaries were kept separate after autopsy so that the response of the two organs could be compared and an accelerated regression of the corpora lutea was found only in the gonad associated with the distended horn of the uterus (*Figure 4*). The volumes of the corpora lutea in the two ovaries were highly significantly different and the regression curves were parallel.

The influence of the hypophysis

Few studies of the significance of the hypophysis in the control of luteal function in the guinea-pig have been made. Dempsey (1937) killed 3 animals 12 days after hypophysectomy carried out immediately after ovulation and noted that the corpora lutea appeared normal although follicular atresia was apparent. More extended studies were made by Rosenbusch-Weihs and Ponse (1957), who found that the corpora lutea could persist for more than 4 weeks, and occasionally for at least 9 weeks, in hypophysectomized animals.

Although signs of degeneration were detected in some luteal bodies, others appeared quite active. According to Perry and Rowlands (1962b) and Rowlands (1962) the effect of hypophysectomy on the corpora lutea seemed to depend upon the time in the cycle when

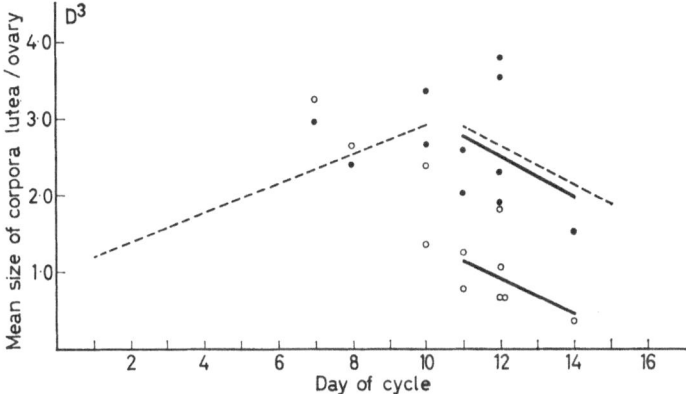

Figure 4. *The effect of unilateral distension of the uterus on luteal size. Two beads were inserted into the right horn of the uterus between days 2–4, and the volumes of the corpora lutea in the right ovary (open circles) compared with those in the left ovary (filled circles) in animals killed on the days indicated. Regression lines for the volumes of the corpora lutea in the right and left ovaries are given and those for normal animals (taken from Figure 1) are indicated by broken lines. Note the difference between the rate of regression of the corpora lutea in the two ovaries*

the operation was performed. After hypophysectomy on the second or fifth day after ovulation the corpora lutea continued to grow but at a slightly slower rate and for a slightly longer time than in the intact female. Maximum size was reached on day 12, when they were indistinguishable histologically from the corpora lutea of intact animals. They did not enlarge further, as after hysterectomy or during pregnancy, but persisted at maximum size for at least 3 weeks. In other guinea-pigs hypophysectomized on day 10 after ovulation the corpora lutea regressed almost as rapidly as in the normal cycle. Decidual reactions could not be elicited from the uterus of animals tested 6 days after ovulation and 4 days after hypophysectomy although implantation occurred in similarly hypophysectomized mated animals, with the corpora lutea growing to pregnancy size in some. In this connection, Deanesly (1961, 1964) has shown that the development of the traumatic deciduoma in the guinea-pig uterus requires more progesterone than does ovo-implantation. Quite recently, Heap, Perry and Rowlands (1965) reported that in non-pregnant guinea-pigs the growth of the corpora lutea was slightly reduced after hypophysectomy during the first 8 days after ovulation and that a maximum size of about $1\cdot7$ mm^3 was attained 30 days later. They confirmed that hypophysectomy carried out on the tenth day of the cycle did not interfere with luteal regression. The progesterone concentration of luteal tissue was higher after hypophysectomy than usually found during the normal cycle or in pregnancy but the plasma progesterone level was lower than at the mid-luteal phase of the oestrous cycle or the equivalent stage of pregnancy. Luteal regression has been observed to follow hypophysectomy of pregnant guinea-pigs when performed between days 34 and 36 of gestation (Pencharz and Lyons, 1934).

Prolactin (known to be luteotrophic in the rat) has failed to extend luteal function in intact guinea-pigs in dosages up to 1 mg (15 I.U.) daily. Oestrus recurred at the expected time (Rowlands, 1962). Amounts of prolactin as high as $2\cdot5$ mg daily were given by Aldred, Sammelwitz and Nalbandov (1961). Treatment began on days 6, 8, 10, or 12 after the first day of vaginal opening and continued until day 20. There was no effect on the life span or size of the corpora lutea. The extent of luteinization in ovarian tissue grafted to the spleen of spayed guinea-pigs has been considered to be increased by treatment with oxytocin and presumed to be due to the release of prolactin (Huntingford, 1963). However, the grounds for this conclusion are not convincing. Donovan (1961) gave large amounts (6 I.U.) of oxytocin daily to adult females and began treatment on each day of the cycle on at least two occasions in the course of a series of 50 treatment cycles. No change in the length of the cycle ensued so that it is unlikely that luteal function was affected.

There is little doubt that the hypophysis of the guinea-pig produces a lactogenic hormone (Nicoll, Bern and Brown, 1966), although the control of the secretion of this hormone may differ from that in the rat and ferret (Everett, 1966; Donovan, 1963, 1965). Aron and Marescaux (1962) reported that involution of the mammary gland occurred after complete hypophysectomy in the guinea-pig and that in cases of partial hypophysectomy contact with the pituitary stalk was essential for continued support of the mammary gland. Pieces of anterior pituitary tissue of equal size but deprived of the influence of the hypothalamus were inadequate for this purpose. By contrast, rat pituitary tissue transplanted to the kidney secretes prolactin and maintains mammary gland function. Grafts of pituitary tissue taken from adult female donors or from newly-born young have been made beneath the kidney capsule of intact female guinea-pigs. No indication of the secretion of a luteotrophic hormone was obtained in that there was no prolongation of the oestrous cycle. However, the grafts took poorly (Donovan, unpublished). Grafts of anterior pituitary tissue made to the eyes of hypophysectomized guinea-pigs have been found to maintain follicular activity in the ovaries and to support constant vaginal oestrus (Schweizer, Charipper and Haterius, 1937).

It has been suggested that the ovaries of the hypophysectomized guinea-pig may be peculiarly refractory to stimulation with gonadotrophins in general, for Perry and Rowlands (1963) found it difficult to obtain good follicular development in hypophysectomized immature animals following treatment with pregnant mare serum, human menopausal gonadotrophin, or human, sheep, horse or guinea-pig pituitary extract. Luteal cysts were produced and there was hypertrophy of the theca interna cells and of the stromal tissues. Previous workers (Jares, 1931; Loeb, 1932; Hayward, Pollock and Loeb, 1939; Guyénot and Ponse, 1939; Hamburger and Pedersen-Bjergaard, 1946) had earlier injected various preparations of gonadotrophin into intact guinea-pigs with confusing results. Nevertheless, some follicular development occurred when up to four whole pituitary glands from adult female guinea-pigs were implanted subcutaneously into immature recipients by Schmidt (1937). Ovulation or corpus luteum formation did not take place unless glands from male donors were used. Follicle stimulation without luteinization was also observed in hypophysectomized guinea-pigs treated with an extract of the urine of an ovariectomized woman (Guyénot, Held and Ponse, 1939); the addition of a bovine pituitary extract or HCG led to corpus luteum formation.

There is some indirect evidence for the participation of the pituitary gland in the control of luteal function. Herlant (1933) found that the cytological changes in the hypophysis which followed hysterectomy in adult guinea-pigs and were associated with persistent corpus luteum activity also ensued in similarly operated immature animals before corpora lutea appeared in the ovaries. This seemed to rule out the possibility that the pituitary gland was responding secondarily to changes in ovarian function but attempts to block the reaction by section of the cervical sympathetic nerves failed. Work of a different kind led Deanesly and Perry (1965) to the view that the hypophysis was important for luteal activity in hysterectomized guinea-pigs. Out of 9 animals receiving, usually over 3 days, 0·375–1·5 mg reserpine, five showed vaginal opening at the end of the injections and up to 6 days later. Large and small corpora lutea were to be found in the ovaries, but the shrinkage of the corpora lutea was not as striking as with progesterone treatment.

The effect of gonadal hormones

Studies with gonadal hormones go far to substantiate the view that the persistence of the luteal bodies seen after hysterectomy is accompanied by the secretion of progesterone. Rowlands (1962) first established that a single injection of 2 μg oestradiol benzoate caused rupture of the vaginal membrane of spayed animals 3 days later, and then observed that 1,000 μg of the hormone caused but a partial response in intact females and was ineffective in hysterectomized animals. In the latter instances the oestrogenic action on the vagina was opposed by the secretions of the corpora lutea. Daily treatment of hysterectomized animals with up to 1 mg stilboestrol from the time of operation failed to inhibit the growth of the corpora lutea despite the occurrence of mucification of the vaginal epithelium. On the other hand, Deanesly and Perry (1965) reported that 2 μg oestradiol benzoate given daily to hysterectomized animals, after a treatment with progesterone had ended, caused vaginal opening after 3 or 4 days. However, the prior effects of progesterone need careful evaluation. Vaginal opening also occurred in pregnant guinea-pigs given amounts of oestrogen sufficient to cause abortion, but the corpora lutea persisted apparently unchanged (Kelly, 1931).

In an attempt to determine whether the release of hypophysial hormones could be inhibited by gonadal hormone, Aldred, Sammelwitz and Nalbandov (1961) and Nalbandov (1961) injected 8 mg/kg per day progesterone into pregnant guinea-pigs. The injections were begun on the day of mating or 4 days later and a loss of ovarian weight and of luteal area took place in those animals treated from the day of mating. In view of the loss of effect of injections begun 4 days after mating it was suggested that an initial outpouring of a luteotrophic hormone was essential for normal luteal function and that release of this hormone could be inhibited only by sufficiently early treatment with progesterone. When progesterone administration was delayed for 4 days, enough luteotrophic hormone was discharged for normal corpus luteum activity. Tablets of progesterone were implanted subcutaneously 15–63 days after hysterectomy by Deanesly and Perry (1965) and the absorption rate was considered to be 1–2 mg/day. From measurement of the size of the corpora lutea it appeared that the corpora showed marked shrinkage and histological regression in some of the group but not in all. The vagina opened after removal of the tablets in 6 of 9 animals with ovulation occurring in 2 and large follicles being found in the ovaries of other animals. It was concluded that the corpora lutea were maintained by pituitary gonadotrophic hormone and that inhibition of the release of this hormone by progesterone caused luteal regression. Much smaller amounts of progesterone were given by Dempsey (1937) for 20 days beginning on the tenth day of the cycle. The release of luteinizing hormone was depressed in that ovulation was inhibited without interference with follicular development and corpora lutea were lacking in the ovaries at the end of the treatment period.

Nerve supply to the genital tract

Since distension of the uterus is influential in causing regression of the corpora lutea in the guinea-pig (page 321) it might be expected that the afferent innervation of the uterus is important in the control of luteal function. This has not been proven although Hill (1962) claimed that oestrous cycles stopped in guinea-pigs suffering denervation of the ovary and suggested that this might follow disturbance of normal uterine glandular activity.

Attempts to denervate the uterus by division of the hypogastric and pelvic nerves were made by Donovan and Traczyk (unpublished) in connection with their studies of the effects of distension of the uterus but interpretation of the results was complicated by the occurrence of severe perineal erosion which made vaginal cycles difficult to follow. Subsequent experiments with mated guinea-pigs indicated that division of the pelvic nerves depressed the incidence of pregnancy, presumably by interfering with the implantation of the conceptuses (Donovan and Traczyk, 1965).

The influence of the placenta

The importance of the placenta in ensuring growth of the corpora lutea during pregnancy is not fully established. In part, this stems from the fact that ovariectomy performed as early as the third to the sixth day after mating is compatible with the maintenance of gestation for about 16 days in some animals (Deanesly, 1963). This implies that the corpora lutea can be dispensed with and that the placenta may produce progestational hormones to a limited but significant degree early in pregnancy. Later on, the secretion of progesterone by the placenta becomes important, for ovariectomy can be performed at midpregnancy without abortion (Herrick, 1928;

Heap and Deanesly, 1964). In animals spayed on day 28 of gestation, or later, the mean concentration of progesterone in arterial
blood plasma was much lower than in corresponding intact females.
A positive effect of the placenta upon the corpora lutea was demonstrated by Klein (1939) who showed that removal of the gravid
uterus was compatible with continued luteal function but that when
the uterus was emptied, oestrus and ovulation quickly followed.
Hypophysectomy on days 34 to 36 of pregnancy caused resorption
of the foetuses within 2 days after operation but animals subjected to
the same operation at 40–41 days delivered viable young at term
(Pencharz and Lyons, 1934).

More direct evidence for the secretion of a luteotrophin by the
placenta has been sought by Bland and Donovan (1965c, 1965d) in
experiments involving the transplantation of conceptuses to extra-
uterine locations (such as the spleen) in non-pregnant female hosts.
Although development of embryos has not been induced outside the
uterus differentiation of trophoblast and placental membranes has
been noted after the insertion of either 6-day blastocysts or older
conceptuses into the spleen. The onset of oestrus is not usually
influenced by these procedures but has been inhibited in 7 animals
in which 11–12 day conceptuses were maintained in the spleen.
The fact that the effective grafts had been in contact with the uterus
for 5 or 6 days may be important in indicating the existence of some
conceptus-uterine interaction. It is also noteworthy that removal of
the conceptuses from guinea-pigs up to the fifteenth day after mating
permitted the return of oestrus at the expected time. In a series of
11 animals from which conceptuses aged between 9 and 15 days were
collected the mean length of the pre-operative control cycle was
$15 \cdot 9 \pm 0 \cdot 1$ days, and the mean length of the cycle after mating was
$17 \cdot 5 \pm 0 \cdot 1$ days (K. P. Bland, unpublished).

THE CONTROL OF LUTEAL FUNCTION

From the information just presented it is clear that the control of
luteal function is vested in the uterus, for consistent changes in the
activities of the corpora lutea have only followed removal or distension of this organ. This is not a novel conclusion for Loeb (1927)
long ago suggested that the uterus produced an internal secretion
governing the life of the corpora lutea. However, it has proved
difficult to provide unequivocal evidence for this idea since the facts
are susceptible to interpretation in several ways (Bland and Donovan,
1966c). Nevertheless, with the realization of the existence of unilateral effects arising from hemihysterectomy or unilateral distension

of the uterus this conclusion now seems inescapable. Indeed, it can now be presumed that the uterus of the guinea-pig produces a luteolytic hormone which acts directly on the ovary, and not through the pituitary gland.

Source of the luteolytic hormone

The endometrium is the uterine structure most likely to produce a luteolytic hormone. In this way secretion of the agent could be timed to occur at the correct phase of the ovarian cycle, for secretion would vary according to the degree of endometrial development. In accord with this view, damage to the endometrium by the injection of fixatives or corrosive agents into the lumen of the uterus caused changes in the life span of the corpora lutea similar to those seen after hysterectomy (Butcher *et al.*, 1962a). The greatest effect was seen in those animals with the most destruction of the endometrium, as evaluated histologically. The induction of deciduomata throughout the uterus, with consequent alteration of endometrial activity, extends the length of the oestrous cycle by 3–7 days (Loeb, 1927). However, endometrial glands need not be concerned with the production of luteolytic hormone. In the work of Butcher *et al.* (1962a) the preservation of an anterior portion of one horn in partially hysterectomized animals favoured the continuation of oestrous cycles, although the myometrium and endometrium became oedematous and the lumen was filled with fluid. In most cases uterine glands were absent and there was no apparent relationship between the presence of uterine glands and the life span of the corpora lutea. Reduction in the thickness of the endometrium following upon ligation of the uterine horns at the cervix with subsequent distension of the uterus failed to modify the oestrous rhythm (Donovan, unpublished). Even so, replacement of endometrial tissue alone could restore oestrous cycles in the hysterectomized guinea-pig (Butcher, Chu and Melampy, 1962b). The best results were obtained with autotransplants of the entire uterus, with oestrous cycle lengths of less than 32 days being observed only in animals with surviving endometrium. It is interesting to note that in this work the anterior portions of the uterine horns were separated from the oviducts and inserted beneath the abdominal oblique muscles. Any luteolytic factor produced by the grafts must therefore have reached the ovaries through the systemic circulation, or have diffused through the abdominal muscles into the peritoneal fluid. Loeb (1927) could not detect any change in luteal function with replacement of uterine tissue but the transplants were inserted beneath the skin.

Extracts of the uterus of the guinea-pig, of an unspecified nature, inhibited the synthesis of progesterone by a homogenate of ovarian tissue (Hess and Cooper, 1965).

Nature of the luteolytic hormone

The nature of the luteolytic factor is quite unknown, but it might be expected to be a protein or polypeptide. Since the uterus is known to produce relaxin, which is a protein, the activity of this material in causing luteolysis has been tested in hysterectomized animals (Bland and Donovan, unpublished). Six International Units were given daily for 5 days and the 3 animals were killed on the day after the last injection. The volumes of the corpora lutea were estimated as in previous work but had not fallen below the expected values for hysterectomized animals, although relaxation of the pubic symphysis had occurred.

Secretion of the luteolytic hormone

When the uterus contains one or more glass beads, release of the luteolytic factor in effective amount probably occurs at about the ninth day of the cycle with prompt regression of the corpora lutea. In normal animals release of the agent takes place later and over a longer period of time. This is indicated by the curves for regression of the corpora lutea in normal animals and by the fact that hysterectomy on the tenth day of the cycle allows subsequent enlargement of the corpora lutea. Even on the fifteenth day of the cycle enlargement of the corpora lutea remained possible after hysterectomy although shrinkage of the luteal bodies had taken place (Rowlands, 1961).

There would seem to be a considerable reserve in the capacity for progesterone secretion by the corpora lutea, for morphological shrinkage of these bodies, as observed with distension of the uterus, is not necessarily accompanied by a fall in the output of progesterone. This is indicated by the relative lack of effect on the oestrous cycle of distension of the uterus with one bead in each horn although an unequivocal regression of the corpora lutea could be measured (Bland and Donovan, 1965a, b, 1966a) and by the observations of Dempsey (1937) and Hermreck and Greenwald (1964) that unilateral ovariectomy in which at least one corpus luteum is left in the remaining ovary has no effect on the oestrous cycle.

Mode of passage of the luteolytic factor from uterus to ovaries

With the realization that each horn of the uterus controls luteal function in the ovary associated with it a question arises concerning the pathway traversed between these two organs. The very existence of a local effect would seem to exclude entry of the material into the blood draining the uterus and entering the systemic circulation, for then considerable mixing and dilution would occur and an action on both ovaries expected. Information concerning the existence of specialized vascular links between uterus and ovaries is lacking and alternative possibilities include diffusion along the Fallopian tube or passage in the lymph vessels connecting the uterus and gonad. Displacement of the ovaries from their normal location seems to break the functional connection between the uterus and corpus luteum. Normal cycles were not resumed following transplantation (on days 5 to 8 of the cycle) of the right ovary to the right kidney and of the left ovary to the left kidney. The operated cycle ended at the expected time but the new set of corpora lutea still persisted in the grafts until autopsy 35 days after the last oestrus (Bland and Donovan, unpublished). In other experiments the Fallopian tube, or the Fallopian tube and the main blood vessels in that region have been interrupted unilaterally in normally cycling animals. Tube section can seemingly be carried out without disturbing the rate of luteal regression, which remains the same in both ovaries, but disruption of the vascular connections favours persistence of the corpora lutea in the ovary on the operated side.

Under certain circumstances both ovaries can be affected by unilateral interference with the uterus. Thus, the placement of four beads in one horn caused a greater shortening of the oestrous cycle than two beads in one horn or one bead in each horn and it is reasonable to suppose that the luteolytic material was released by the distended horn in a sufficient amount to reach the opposite ovary and initiate luteal regression. Conversely, the corpora lutea on the operated side do not persist indefinitely in cases of hemihysterectomy. Since the oestrous cycle is extended only by a matter of days (instead of months as with complete hysterectomy) it would seem that the remaining uterine horn can produce sufficient luteolytic material to affect the ovary on the operated side.

The role of the pituitary gland in luteolysis

The emphasis laid on the existence and importance of a uterine luteolytic hormone in the control of luteal function need not be taken to mean that the pituitary gland is of little value in this mechanism.

The production of the luteolytic hormone by the uterus may be indirectly governed by the hypophysis through the ovaries, so that hypophysectomy arrests the secretion of gonadal hormones and hence the manufacture and release of the luteolytic factor. Butcher *et al.* (1962a) suggested that oestrogen 'may inhibit the production of a substance by the uterus which causes luteal regression in the cycling animal.' Changes in the secretion of gonadal hormones may also help in the explanation of the finding that hypophysectomy carried out on the tenth day of the cycle was compatible with regression of the corpora lutea, while removal of the pituitary gland on the second day led to persistence of these tissues. Perhaps progesterone favoured luteal regression in the work of Aldred *et al.* (1961) by acting on the uterus to enhance luteolysin secretion but this explanation is not applicable to the hysterectomized animals of Deanesly and Perry (1965). In part, the vaginal opening could be accounted for on the basis of an over-secretion of gonadotrophin by the pituitary after release from the inhibiting action of progesterone, with the gonadotrophin causing increased follicular development, high oestrogen secretion and ovulation. Reserpine could act in this way but the shrinkage of the corpora lutea remains inexplicable.

The possible control of luteal function by the pituitary gland is being investigated in another way. It is known that lesions of the anterior hypothalamus cause suspension of ovulation and the onset of constant follicular activity in the ovaries. Oestrogen secretion is manifest by the existence of a constantly open vagina and large follicles are present in the ovaries (Dey, Fisher, Berry and Ranson, 1940; Dey, 1943). In preliminary experiments, Donovan, O'Keeffe and O'Keeffe (unpublished) placed similar lesions in the anterior hypothalamus of hysterectomized guinea-pigs and found that in six of 8 animals vaginal opening occurred post-operatively for periods varying from 1–20 days. However, oestrous smears were not co-existent with vaginal opening: dioestrous smears were collected in all animals throughout the time that the vagina was open except on one occasion in one animal, when the smear was oestrous. The occurrence of vaginal opening was not necessarily associated with regression of the corpora lutea, for hysterectomy-sized corpora lutea were to be found in the ovaries of five of the 8 animals.

Pregnancy

The corpora lutea seem to be exposed to two opposing influences as pregnancy begins—one luteolytic from the uterus and the other luteotrophic from the conceptus. To some extent the conceptus (besides producing a luteotrophic hormone) could favour the extension of luteal activity by arresting the production of the uterine factor. Loeb (1927) commented on the similarity between the corpora lutea of pregnant and of hysterectomized guinea-pigs and suggested that in pregnancy a functional inactivation of the uterine mucosa might be responsible, at least in part, for the change in luteal function. Decidualization of the uterus could be one means of inhibiting the release of the luteolytic factor. Nevertheless, mere prevention of regression of the corpora lutea is apparently not enough to provide the progestational support needed for pregnancy: the positive stimulus of the placental gonadotrophin is required and evidence for the existence of this hormone has been presented earlier.

The dynamic nature of the interaction between luteotrophic and luteolytic factors is well shown in the results of an investigation still in progress (Bland and Donovan, unpublished). Adult females were mated and on the following day two glass beads were introduced into one horn of the uterus, before entry of the fertilized eggs from the Fallopian tubes. Normal implantation took place in the control horn, but conceptuses were never found in the horn containing the glass beads. Damage to the endometrium was not responsible for this failure of implantation on the distended side for control insertion and removal of beads was performed in other animals without disturbing this process. When the volumes of the corpora lutea came to be examined it was particularly interesting to find that the corpora lutea had regressed in the ovary on the distended side, while on the pregnant side the corpora lutea had enlarged. Not only does this finding reconfirm the local nature of the luteolytic action but it also implies that the luteolytic agent can block the luteotrophic action of the placental gonadotrophin.

ACKNOWLEDGEMENT

We wish to express our gratitude to the Population Council for their support of our investigations described in this paper, and to thank Dr. Kathleen Hall for presenting us with some Relaxin from a supply provided by the Warner-Chilcott Company.

REFERENCES

Aldred, J. P., Sammelwitz, P. H. and Nalbandov, A. V. (1961). *J. Reprod. Fert.* **2,** 394–399

Aron, M. and Marescaux, J. (1962). *C.r. Séanc. Soc. Biol.* **156,** 1916–1918

Bland, K. P. and Donovan, B. T. (1965a). *J. Physiol., Lond.* **179,** 34P–35P

— — (1965b). *Nature, Lond.* **207,** 867–869

Bland, K. P. and Donovan, B. T. (1965c). *Acta endocr., Copenh.* Suppl. 100, 78
—— (1965d). *J. Reprod. Fert.* **10**, 189–196
—— (1966a). *J. Physiol., Lond.* **186**, 503–515
—— (1966b). *J. Endocr.* **34**, iii–iv
—— (1966c). In: *Advances in Reproductive Physiology* **1**, 179–214. Ed. by A. MacLaren. London; Logos Press
Butcher, R. L., Chu, K. Y. and Melampy, R. M. (1962a). *Endocrinology* **71**, 810–815
———— (1962b). *Endocrinology* **70**, 442–443
Deanesly, R. (1961). *J. Endocr.* **22**, xxx–xxxi
— (1963). *J. Reprod. Fert.* **6**, 143–152
— (1964). *Proc. 5th Int. Congr. Anim. Reprod.* **2**, 384–387
— and Perry, J. S. (1965). *J. Endocr.* **32**, 153–160
Dempsey, E. W. (1937). *Am. J. Physiol.* **120**, 126–132
Desclin, L. (1932). *C.r. Séanc. Soc. Biol.* **109**, 972–973
Dey, F. L., Fisher, C., Berry, C. M. and Ranson, S. W. (1940). *Am. J. Physiol.* **129**, 39–46
— (1943). *Endocrinology* **33**, 75–82
Donovan, B. T. (1961). *J. Reprod. Fert.* **2**, 508–510
— (1963). *J. Endocr.* **27**, 201–211
— (1965). *Abstr. Pap.* XXIII *Int. physiol. Congr.* 284
— and Traczyk, W. (1960). *J. Physiol., Lond.* **154**, 50P–51P
—— (1962). *J. Physiol., Lond.* **161**, 227–236
—— (1965). *J. Endocr.* **33**, 335–336
Everett, J. W. (1966). In: *The Pituitary Gland* **2**, 166–194. Ed. by G. W. Harris and B. T. Donovan. London; Butterworths
Fischer, T. V. (1965). *Anat. Rec.* **151**, 350
Guyénot, E., Held, E. and Ponse, K. (1939). *Archs Anat. Histol. Embryol.* **26**, 289–345
— and Ponse, K. (1939). *Archs Anat. Histol. Embryol.* **26**, 253–288
Hamburger, C. and Pedersen-Bjergaard, K. (1946). *Acta path. microbiol. scand.* **23**, 84–102
Hayward, S. J., Pollock, J. H. and Loeb, L. (1939). *Am. J. Physiol.* **125**, 113–118
Heap, R. B. and Deanesly, R. (1964). *J. Endocr.* **30**, ii–iii
— Perry, J. S. and Rowlands, I. W. (1965). *Acta endocr. Copenh.* Suppl. 100, 76
Herlant, M. (1933). *C.r. Séanc. Soc. Biol.* **114**, 273–275
Hermreck, A. S. and Greenwald, G. S. (1964). *Anat. Rec.* **148**, 171–176
Herrick, E. H. (1928). *Anat. Rec.* **39**, 193–200
Hess, M. and Cooper, M. E. (1965). *Abstr. 8th Int. anat. Congr.* 51. Wiesbaden
Hill, R. T. (1962). In: *The Ovary* **2**, 231–261. Ed. by S. Zuckerman. London; The Academic Press
Howe, G. R. (1965). *Endocrinology* **77**, 412
Huntingford, P. J. (1963). *J. Obstet. Gynaec. Br. Commonw.* **70**, 929–946

Jares, J. J. (1931). *Anat. Rec.* **49,** 185–189
Kelly, G. L. (1931). *Surgery, Gynec. Obstet.* **52,** 713–722
Klein, M. (1939). *C.r. Séanc. Soc. Biol.* **130,** 1393–1395
Loeb, L. (1911a). *Dt. med. Wschr.* **37,** 17–21
— (1911b). *J. Morph.* **22,** 37–70
— (1923). *Proc. Soc. exp. Biol. Med.* **20,** 441–443
— (1927). *Am. J. Physiol.* **83,** 202–224
— (1932). *Endocrinology* **16,** 129–145
Moore, W. W. (1961). *Physiologist, Lond.* **4,** 76
Nalbandov, A. V. (1961). *Recent Prog. Horm. Res.* **17,** 119–139
Nicoll, C. S., Bern, H. A. and Brown, D. (1966). *J. Endocr.* **34,** 343–354
Pencharz, R. I. and Lyons, W. R. (1934). *Proc. Soc. exp. Biol. Med.* **31,** 1131–1132
Perry, J. S. and Rowlands, I. W. (1962a). In: *The Ovary* **1,** 275–309. Ed. by S. Zuckerman. London; The Academic Press
— — (1962b). *J. Endocr.* **25,** v–vi
— — (1963). *J. Reprod. Fert.* **6,** 393–404
Rosenbusch-Weihs, D. and Ponse, K. (1957). *Revue suisse Zool.* **64,** 271–280
Rowlands, I. W. (1956). *Ciba Fdn Colloq. Ageing* (1956) **2,** 69–85
— (1961). *J. Reprod. Fert.* **2,** 341–350
— (1962). *J. Endocr.* **24,** 105–112
— and Short, R. V. (1959). *J. Endocr.* **19,** 81–86
Schmidt, I. G. (1937). *Endocrinology* **21,** 461–468
Schweizer, M., Charipper, H. A. and Haterius, H. O. (1937). *Endocrinology* **21,** 30–39

DISCUSSION

Dr. G. K. Benson (*Liverpool*)

Could you tell us the source of the pituitary glands which you transplanted to the kidney capsule, and the relationship of the donors to the recipients? Did you make cytological studies on the transplants and how long were they under the capsule before you examined them for viability?

Donovan

These were preliminary experiments. We took pituitary tissue from adult donors or from foetuses and transferred this to the kidney capsule of the intact cycling animals. We observed no effect on the cycle but the pituitary tissues did not persist well and were poorly vascularized. They remained in the capsule for 10–14 days. We were interested as much in storage of hormones as in secretion.

Professor L. E. Reichert (*Atlanta, Georgia*)

What are your reasons for feeling that the luteolytic hormone is a peptide or a protein?

Donovan

I think one can eliminate steroids from knowledge of what the uterus can produce. What alternative substances are there—catacholamines or some other

335

transmitting agent? That doesn't seem very likely—the substance has to persist long enough to travel from the uterus to the ovary. Also from hemihysterectomy experiments substances can pass from the uterine horn to affect the opposite ovary so they must persist for some time within the blood or tissue fluids.

PROFESSOR S. M. McCANN (*Dallas, Texas*)

Is there any evidence from the use of labelled substances placed either in the uterine lumen or in the endometrium to establish a pathway for local action of a luteolytic hormone?

DONOVAN

We have looked for such a pathway but have no evidence so far.

DR. R. M. MOOR (*Cambridge*)

You have shown that removal of the conceptus on day 11 and then transplantation to the spleen will maintain the corpus luteum. Would this not have occurred just by removing the conceptus? We did some work with sheep where we removed the conceptus and this was sufficient to maintain the corpus luteum.

DONOVAN

We do not think this is the same effect. Dr. Bland has removed the conceptuses after day 15 in a mated animal and has found that it came into oestrus at the expected time. You could also conclude from this that a continuing secretion of placental luteotrophin is required for luteal maintenance.

MISS B. J. WEIR (*London*)

Your studies are mainly concerned with the corpus luteum of ovulation but did you notice what happened to the accessory corpora in the guinea-pig ovary if, and when, any were formed.

DONOVAN

We haven't noticed any. Occasionally small corpora lutea are formed but the only way to check that these are accessory is to collect the fertilized eggs and any experiments in which we have collected fertilized eggs, the number has matched the number of corpora lutea or has been less. We have attributed the fewer eggs to faulty technique rather than accessory corpora lutea.

DR. J. SPINCEMAILLE (*Belgium*)

May I ask if the corpus luteum of the hysterectomized guinea-pig remains functional after hypophysectomy?

DR. J. S. PERRY (*Cambridge*)

The original work on hypophysectomy in the guinea-pig which led to the finding that it did not cause the regression of the corpus luteum was undertaken in hysterectomized animals. Following this we established that the corpora lutea continued after hypophysectomy of the cycling animal and we formulated the idea that the corpus luteum, once formed, continued until stopped, and we now refer to this as 'luteolysin'. I am not yet convinced that there is such a substance in the uterus, we may eventually explain the effect of the uterus on the ovary by its effect on oestrogen and I am thus impressed by Professor Nalbandov's data. The timing of operations in this type of work is extremely important.

DONOVAN

I would agree with Dr. Perry except on one point. I find it hard to reconcile evidence of a unilateral effect. The existence of unilateral hysterectomy causing corpus luteum enlargement and unilateral distension of the uterus causing regression indicates some local mechanism. I think it reasonable to consider the existence of a luteolytic factor.

DR. WAYNFORTH (*London*)

Surgical removal of the endometrium in the rat gives some prolongation of the cycle but does not reproduce the hysterectomy effect. Removal of the whole endometrium does not simulate hysterectomy, so a luteolytic factor of the endometrium, if it exists, is not the complete answer.

COMPARATIVE STUDIES ON PROGESTERONE SYNTHESIS *IN VITRO*—THOUGHTS ON CORPUS LUTEUM FORMATION

A. V. NALBANDOV, B. COOK, C. C. KALTENBACH
and P. L. KEYES

Department of Animal Science—Genetics, University of Illinois

DUNCAN, Bowerman, Hearn and Melampy (1960) and Duncan, Bowerman, Anderson, Hearn and Melampy (1961) first showed that porcine corpora lutea when incubated in an *in vitro* system can be caused to synthesize progesterone. Since that time a considerable amount of work has been published dealing with the rate of progesterone synthesis by corpora lutea of rats, cows, and women. The significant difference between Duncan's original work and that dealing with corpora lutea from non-porcine species lay in the fact that both the cow and human corpora lutea responded to the addition to the medium of LH-containing hormones by significant increases in progesterone synthesis while porcine corpora lutea were said to be unable to show such a response. These species differences and the possible causes underlying them were of interest to us.

In our experiments corpora lutea from sheep, cows, and pigs were compared for their ability to synthesize progesterone in simultaneous experiments in which tissues from at least two of the species mentioned were incubated at the same time (Cook, 1966). In each experiment various hormones were added to the incubation medium. In all instances incubation of the luteal tissues was begun within 15–60 min after their removal from the animal, after surgical operation, or, after autopsy. The tissues were sliced and 200–400 mg were incubated for 3 h at 37°C in 5 ml of KRBB at pH 7·4 containing 50 μc of acetate-1-^{14}C. The gas phase was 95 per cent O_2 : 5 per cent CO_2. Hormones, where desired, were added to the incubation medium and the progesterone was extracted by the method of Armstrong, O'Brien and Greep (1964).

The results of one such experiment are shown in Table 1. It is seen that contrary to results reported by Duncan *et al.* (1960, 1961) porcine corpora lutea do respond to LH stimulation by a statistically significant increase in progesterone synthesis. Of considerable

338

interest to us was the finding that while both ovine and bovine LH caused significant increases, the increase in progesterone synthesis produced by porcine LH and by unfractionated porcine pituitary

Table 1

Effect of LH of different sources on progesterone synthesis in porcine corpora lutea
in vitro

Source of LH (2 μg/ml equiv. NIH-LH-B1)	Final progesterone concentration (μg/g)
1. None	100·4
2. Porcine	137·9
3. Ovine	125·0
4. Bovine	121·0
5. Lyophilized whole porcine AP	141·7

$P < 0.01$ to $P < 0.05$: 1 vs. 2, 3, 4, 5
2 vs. 3, 4
5 vs. 3, 4.

glands was greater than that produced by the non-porcine LH. The differences in augmentation of progesterone synthesis between porcine and non-porcine LH were also statistically significant. A summary of a considerable number of such comparisons is presented in Table 2.

Table 2

Summary of several experiments of effect of gonadotrophins on progesterone synthesis in porcine corpora lutea

Hormone	No. of tests	Effect	$P > 0.05$	$P < 0.05 - < 0.01$
Non-porcine LH	8	+ (?)	6	2
Porcine LH	7	+	1	6
Porcine FSH	3	−	1	2
Porcine prolactin	4	−	2	2

These observations show that contrary to the original observations by Duncan, porcine corpora lutea are capable of responding to LH stimulation by increased progesterone synthesis. They also suggest the very interesting possibility that porcine LH has a greater stimulatory effect than non-porcine LH. However, this suggestion of a species specific effect of porcine LH will need to be studied further before it can be accorded the status of a fact. Table 2 also suggests that other hypophysial hormones such as FSH and prolactin may actually depress progesterone synthesis in porcine and in other corpora lutea and it is now certain that they do not have the stimulatory effect obtained by the addition of LH to the medium.

It should be made clear that these *in vitro* studies tell nothing about the ability of LH to prolong the life span of the corpus luteum. They do, however, clearly show that LH is the only hypophysial hormone that can stimulate progesterone synthesis *in vitro*. That it has a similar effect *in vivo* has thus far been demonstrated in sheep in which infusion of 10 mg of LH over a 60 min period causes a significant rise in the rate of progesterone synthesis by the ovary bearing a corpus luteum. Thus it appears that *in vitro* studies of progesterone synthesis may have some relation to the *in vivo* situation although it is clear that the rate of progesterone synthesis *in vitro* is only a tiny fraction of the rate observed from the ovary *in situ*. In our own work with sheep, rabbits, and sows, it is still not clear whether we are normally dealing with a single hypophysial substance which is both luteotrophic and steroidogenic or, whether these activities should be ascribed to two different hormones and that two such hormones should be looked for.

In this connection studies conducted on steroidogenesis on rabbit corpora lutea are of considerable interest (Keyes and Nalbandov, 1967). In these experiments does were mated to fertile bucks. One of each doe's ovaries was x-rayed on day 10 of pregnancy using a dose which had been established to be able to destroy all the follicles including the great majority of the primary follicles. Thus, the only remaining ovarian components were the corpora lutea and the interstitial tissue. On day 20 of pregnancy the normal, non-irradiated ovary was removed and 28 h later the remaining ovary was tapped for a venous blood sample and later removed for the analysis of the corpora lutea for progesterone. In those experiments in which the ability of corpora lutea in the irradiated ovary to maintain pregnancy was tested, it was found that one ovary containing corpora lutea was able to maintain pregnancy. However, when the normal ovary was removed leaving an irradiated ovary containing corpora lutea, nine out of 9 does aborted 30–55 h after removal of the normal ovary. This occurred in spite of the fact that the corpora lutea in the irradiated ovary appeared perfectly normal and were histologically indistinguishable from the corpora lutea in the non-irradiated ovary. In contrast it was found that when identically prepared females with one irradiated ovary were given 2–4 μg of oestradiol per day, pregnancy was maintained for as long as oestrogen was injected in seven of 7 females. This suggested one of three possible mechanisms of action of oestrogen: either it acted directly on the uterus, thus postulating (contrary to previous reports) that both progesterone and oestrogen are needed to maintain pregnancy; oestrogen could conceivably be acting directly on the corpora lutea and in some manner participating in progesterone synthesis or in its secretion into the peripheral circulation; finally, it could be acting on the animal's own hypophysis, eliciting the release of a luteotrophic substance which in turn controlled progesterone synthesis and release.

Conclusions concerning two of these possibilities can be reached from the following experiments. In one case females were treated as above and corpora lutea from irradiated and non-irradiated ovaries were pooled and compared for progesterone concentration. It should be remembered that in this trial one ovary of each female was normal and thus contained oestrogen-secreting follicles. The results show (Table 3) that there was no difference between the progesterone content of the two types of corpora lutea which clearly indicates that irradiation of corpora lutea does not impair their ability to synthesize progesterone.

Table 3

Progesterone concentration of corpora lutea of x-irradiated and non-irradiated ovaries of the same pregnant rabbits

Treatment of ovary	*No. of animals*	*Progesterone concentration* $\mu g/100$ mg *tissue*
Normal	6	8·75
x-Irradiated	7	8·45

In the next experiment the normal ovary was removed and the effects of oestrogen on progesterone synthesis and its release by the irradiated ovary were tested (Table 4). The results clearly show that in the absence of oestrogen treatment no progesterone synthesis by irradiated corpora takes place and that no progesterone can be detected in the venous ovarian blood. The mechanism of action of oestrogen in the synthetic pathway is completely unknown and is being studied. There remains the possibility that oestrogen acts via the hypothalamo-hypophysial system and this pathway is now under investigation.

A few words of caution are in order concerning generalization on the mechanism of corpus luteum formation in domestic animals. Initially, work from this laboratory proposed the hypothesis that in pigs a single pulse of a hormone, probably LH, of very short duration,

Table 4

Progesterone concentration in corpora lutea and in the effluent plasma in x-irradiated ovaries with and without oestrogen injection

Treatment	µg Progesterone in	
	Corpora lutea (/100 mg)	Plasma (/100 ml)
None	2 (5)*	2 (5)
2–4 µg oestradiol	10·85 (4)	35·97 (3)

* Number of animals in pools

is sufficient to cause the corpus luteum to form and to persist and function for its normal life span during the cycle (Sammelwitz, Aldred and Nalbandov, 1961; Brinkley, Norton and Nalbandov, 1964a, b).   In the guinea-pig the duration of action of this hormone was found to be somewhat longer than in the pig, lasting for at least 2 days (Aldred, Sammelwitz and Nalbandov, 1961).   This concept was subsequently confirmed and extended to include the ewe.   In both the ewe (Denamur and Mauleon, 1963), and the pig (du Mesnil du Buisson, 1965), it was found that hypophysectomy on the day of ovulation was compatible with the formation of corpora lutea which were normal in size and in their ability to synthesize progesterone.   We are in complete agreement with the findings on the pig but we are unable to confirm the French work on hypophysectomized sheep (Kaltenbach, 1966).   We find that, if the pituitary is removed from sheep between days 1 to 5 of the cycle, the corpora lutea formed are significantly smaller than normal with a corresponding significant decrease in progesterone content when compared to normal controls.   In those animals in which the hypophysectomy was incomplete, even small bits of hypophysial tissue were able to cause the formation of corpora lutea, normal both in structure and function.   As yet, these experiments have not been extended beyond day 5 of the cycle but it is now perfectly clear that the concept of a hormonal pulse of a short duration will not hold for the sheep.

These experiments strongly suggest that we should be prepared to find that some animals, like the pig, may require a single pulse of short duration of a hypophysial hormone. Others, like the guinea-pig, may require a pulse lasting over 24–48 h for corpus luteum formation and still others, like the sheep, appear to need continuous support lasting at least for the first 5 days of the cycle and perhaps even longer.

These remarks are intended to sound words of caution against premature attempts made by some to propose all-embracing models for mechanisms controlling corpus luteum formation. The evidence is far from complete and we must be fully prepared to contemplate that the problem of corpus luteum formation and function may be solved in one way by the rat, in an entirely different way by some of the other laboratory mammals, and in still other ways by the various domestic animals.

ACKNOWLEDGEMENT

This study was in part supported by NIH Grant AM 6976. We wish to acknowledge the generous gifts of purified hypophysial hormones by the Endocrinology Study Section, NIH.

REFERENCES

Aldred, J. P., Sammelwitz, P. H. and Nalbandov, A. V. (1961). *J. Reprod. Fert.* **2,** 394–399

Armstrong, D. T., O'Brien, J. and Greep, R. O. (1964). *Endocrinology* **75,** 488–500

Brinkley, H. J., Norton, H. W. and Nalbandov, A. V. (1964a). *Endocrinology* **74,** 9–13

— — — (1964b). *Endocrinology* **74,** 14–20

Cook, B. (1966). *Ph.D. Thesis.* University of Illinois

Denamur, R. and Mauleon, P. (1963). *C.r. Séanc. Soc. Biol.* **257,** 527–530

Duncan, G. W., Bowerman, A. M., Hearn, W. R. and Melampy, R. M. (1960). *Proc. Soc. exp. Biol. Med.* **104,** 17–19

— — Anderson, L. L., Hearn, W. R. and Melampy, R. M. (1961). *Endocrinology* **68,** 199–207

Du Mesnil du Buisson, F. (1965). *Proc. 2nd Int. Congr. Endocr., Lond.* **1,** 680–685

Kaltenbach, C. C. (1967). *Ph.D. Thesis.* University of Illinois.

Keyes, P. L. and Nalbandov, A. V. (1967). *Endocrinology* **80,** 938–946

Sammelwitz, P. H., Aldred, J. P. and Nalbandov, A. V. (1961). *J. Reprod. Fert.* **2,** 387–393

REPRODUCTION IN THE FEMALE MAMMAL

DISCUSSION

Dr. R. V. Short (*Cambridge*)

With reference to hypophysectomy of sheep during the cycle, you referred to the original work of Denamur and Mauleon. Denamur, Martinet and I have been continuing this work using normally-cycling rather than prepubertal sheep. I feel we must stand halfway between your view and Denamur's previous ideas. Normally-cycling sheep of the *Isle de France* breed have been hypophysectomized between days 5 and 10 of the cycle. Hypophysectomy on day 5 causes most of the corpora lutea to stop secreting within 5 days, although some carry on for longer. Operation on day 10 results in most of the corpora lutea continuing to secrete until day 14. The pituitary fossa has been serially sectioned in all these animals and in 33, no pituitary tissue has been found; however, I think it is possible that some exists in the pars tuberalis. Have you looked at the pars tuberalis and how important do you think this is? If the length of life of the corpus luteum is prolonged by hysterectomy, and hypophysectomy is carried out when the corpora lutea are 30–40 days old, they always regress completely within 72 h as judged by the lack of progesterone in ovarian vein blood. The first signs are apparent at 48 h. If pituitary stalk section is carried out rather than hypophysectomy, the corpus luteum continues normally for at least 15 days. The conclusion is that something is able to continue to secrete a luteotrophic hormone, after section from hypothalamic connections. What do you think is the importance of the pars tuberalis and have you excluded this from your surgical operation and if so how did you do it?

Nalbandov

I do not know the importance of the tuberalis and we are not able to exclude it from our reasoning. I do not believe that the tuberalis is involved in luteal regression since in cases where the corpus luteum has regressed, invariably there is pars tuberalis tissue left. It is interesting that hypophysectomized hysterectomized animals will tolerate more remnants of pituitary tissue and still the corpus luteum will regress whereas in cycling animals hypophysectomized from days 1–5 the slightest remnant of pituitary tissue will prolong corpus luteum function.

Dr. R. Kilpatrick (*Sheffield*)

We have been measuring progestational steroidogenesis by rabbit ovarian tissue *in vitro*. Following LH addition, there can be a fivefold increase in progestational steroids. We have been measuring both progesterone and 20α-hydroxy-4-pregnane-3-one, which is produced in much larger amounts than progesterone when using whole ovarian tissue containing corpora lutea. When the corpora lutea were separated from the interstitial tissue, progesterone and 20α-hydroxy progesterone production was almost entirely from the interstitial tissue, and no progestational steroids were produced by the corpus luteum.

We have tried preliminary experiments of the effect of oestrogen on separated corpora lutea but have not found steroidogenesis. Have you measured 20α-hydroxy progesterone production and release as well as progesterone?

Nalbandov

We have measured 20α-hydroxy progesterone but I am puzzled by our results. In all cases where we had measured progesterones in effluent blood, without

oestrogen the progestational material is almost exclusively 20α-hydroxy-proges-
terone. If oestrogen is added, the material is progesterone. Has anyone any
explanation for this?

PROFESSOR R. H. MELAMPY (*Iowa, U.S.A.*)

What is the effect of equine LH on pig ovarian slices?

NALBANDOV

We have no information on this point.

DR. J. C. HANCOCK (*Edinburgh*)

Klein demonstrated that oestrogen played a role in the maintenance of the
corpus luteum in the rabbit. The obvious alternative to the existence of a luteal
lysin is that the uterus is actually metabolizing or destroying an agent which
normally would maintain a corpus luteum. Is it possible that the active luteotrophin is oestrogen and the uterus is metabolizing it? Is there any evidence
against this theory?

DR. B. J. DONOVAN (*London*)

The difficulty has been to establish that the uterus metabolizes gonadal hormones
to any extent. This has not been proved.

MR. L. E. A. ROWSON (*Cambridge*)

It is possible to design an experiment to prove this point. If you irradiated both
ovaries containing corpora lutea, and if the uterus was using more oestrogen in the
latter phase of the cycle when the uterus becomes luteolytic you would see whether
an increased utilization of oestrogen occurred at this latter stage. We have grafted
the endometrium into the flank of the hysterectomized sheep and studied the effect
on the corpus luteum. We find they do regress after a longer than normal interval.

DR. R. DEANSLEY (*Cambridge*)

Is there a continued maintenance supply of gonadotrophin rather than a short
burst? In the original hypophysectomized rabbit work the impression arose that
the corpus luteum survived only for 1 or 2 h after the supply of gonadotrophin was
removed. Then came the new idea of gonadotrophin release only at ovulation.
What is the present position?

NALBANDOV

We have to be prepared for a different system in each species, a continuous
supply in the rabbit and possibly the sheep in contrast to a short burst in the pig
and possibly the human.

DR. C. CHANNING (*Cambridge*)

From our evidence in the mare we can maintain corpora lutea *in vitro* without
continuous gonadotrophin. If you take follicles at any stage of the cycle, isolate
the granulosa cells and grow them in amylose, they grow and secrete progesterone
irrespective of gonadotrophin supply, or the stage of the cycle, providing the initial
cells are healthy. It is possible that LH assists breakdown of the theca granulosa
barrier.

345

STUDIES ON THE FORMATION AND
MAINTENANCE OF THE
CORPUS LUTEUM

WILLIAM HANSEL

Cornell University, Ithaca, New York

INTRODUCTION

FACTORS favouring the growth of the corpus luteum and secretion of progestational compounds are said to be 'luteotrophic' while factors causing luteal regression and diminution of progesterone secretion are generally referred to as 'luteolytic'. However, luteal tissue maintenance and progesterone secretion may not always be closely associated. For example, corpora lutea persist 'morphologically' for relatively long periods of time in hypophysectomized rats, but these corpora do not produce progesterone and can be readily differentiated from truly functional corpora on histological grounds. On the other hand, stimulation of bovine luteal tissue growth by methods described below has almost invariably been associated with an increased total progesterone content of the tissue, and often with an increased concentration, as well. Most studies indicate that luteal tissue contents and ovarian vein blood levels of progesterone are correlated.

Although prolactin clearly maintains progesterone production by corpora lutea in the rat and mouse, it has been very difficult to demonstrate a luteotrophic effect of this hormone in other species. Donaldson, Hansel and Van Vleck (1965) concluded that attempts to demonstrate a luteotrophic action of prolactin in guinea-pigs, rabbits, cattle, sheep, goats, swine and monkeys had been unsuccessful or unconvincing.

Consequently, studies on the nature of the mechanisms controlling luteal tissue growth and regression in cattle were undertaken several years ago. The cow proved to be an excellent experimental animal for this work. Relatively large amounts of luteal tissue of known ages can be obtained for analysis with a minimum of effort by removal of corpora lutea through incisions in the anterior vaginal wall. Bovine luteal tissue obtained in this way also proved most adaptable to *in vitro* studies of progesterone synthesis as described

346

below. In addition, we were fortunate enough to find a specific and repeatable method for inhibiting normal development of the corpus luteum in the cow and have been able to use this method to good advantage as an experimental tool to study the effects of various gonadotrophic hormones on corpus luteum growth and progesterone production. This method is described below.

LUTEOTROPHIC EFFECTS OF ANTERIOR PITUITARY HORMONES IN INTACT HEIFERS

Two experiments involving a total of 85 Holstein heifers were conducted to test the luteotrophic effects of various adenohypophyseal hormones. Both experiments were based on the ability of the particular anterior pituitary hormone preparation being tested to overcome the inhibition of growth (Armstrong and Hansel, 1959) and progesterone synthesis (Staples and Hansel, 1961) that occur in the corpora lutea of heifers given daily injections of oxytocin during the first third of the oestrous cycle.

The first experiment (Simmons and Hansel, 1964) included 32 heifers divided into five groups on the basis of the pituitary hormone preparation they received. Beginning on the day of oestrus (day 0) each animal was subjected to a series of three treatment cycles with a normal oestrous cycle intervening between each treatment. The treatments were each 10 days in length and the corpora were collected on the eleventh day. Corpora were removed surgically after the first and second treatments and were collected at slaughter after the third. The treatment cycles to which each animal was subjected were: (1) no treatment, (2) daily injections of oxytocin alone, and (3) daily injections of pituitary hormone plus oxytocin. The pituitary hormone treatments included: (1) bovine growth hormone given at two levels, (2) equine luteinizing hormone (LH) given at two levels, (3) ovine prolactin, (4) human chorionic gonadotrophin (HCG), and (5) a crude aqueous extract of bovine anterior pituitary glands. The levels of each hormone preparation administered are shown in Table 1. The oxytocin was administered subcutaneously at the rate of 0·33 USP units/kg of body weight. A total of 156 corpora were collected from the 32 heifers and analysed for progesterone and Δ^4-pregnene-20-β-ol-3-one (2-β-ol) as described by Staples and Hansel (1961). ^{14}C labelled progesterone was added to each sample in order to determine the percentage recovery and the results were adjusted accordingly.

The second experiment (Donaldson et al., 1965) involved 53 heifers subjected to the same three basic treatments: (1) no treatment,

(2) oxytocin, and (3) oxytocin plus gonadotrophin. However, in this experiment the heifers were treated on days 2–6 inclusive and

Table 1

Mean progesterone values of original corpora lutea from heifers treated with oxytocin and various pituitary hormones

Treatment cycle	Number of corpora lutea	Weight of corpora lutea g	Concen-tration‡ µg/g	Total Progestin§ µg
STH group (25 mg)				
Control	3	6·9	37·9	247
Oxytocin	2	1·6*	11·5†	24
STH + oxytocin	2	3·4	26·7*	152
STH group (75 mg)				
Control	5	5·8	39·2†	222
Oxytocin	5	2·4	11·2†	58*
STH + oxytocin	4	3·4	13·7†	80
LH group (20 mg)				
Control	4	6·6	38·9	253
Oxytocin	4	1·3*	2·8†	5*
LH + oxytocin	4	2·8	7·2†	36*
LH group (50–100 mg)				
Control	4	6·2	36·2	258
Oxytocin	5	0·9*	1·1†	1*
LH + oxytocin	5	0·9*	7·1†	11*
Prolactin group				
Control	6	4·8	50·0	251
Oxytocin	5	3·0	12·9†	72
Prolactin + oxytocin	5	1·9*	5·1†	21*
Pituitary extract group				
Control	5	4·7	74·3*	358
Oxytocin	5	2·4	5·8†	41*
Pituitary extract + Oxytocin	5	7·9	26·9*	369
HCG group				
Control	5	4·8	58·8	290
Oxytocin	5	0·7†	1·1†	1†
HCG + oxytocin	5	11·7†	46·8	608

* Significantly different from the mean of all controls ($P < 0.05$).
† Significantly different from the mean of all controls ($P < 0.01$).
‡ Does not include Δ^4-pregnene-20β-ol-3-one.
§ Includes Δ^4-pregnene-20β-ol-3-one and progesterone in any cyst fluid present.

the corpora removed at laparotomy or slaughter on day 7. Eleven heifers were used once, 17 twice and 25 three times. Twelve corpora were removed after no treatment, 10 after oxytocin, 10 after oxytocin plus bovine LH, 7 after oxytocin plus HCG, 5 after oxytocin plus HCG incubated with 6 M urea at 40°C for 24 h to destroy its LH activity and 5 after oxytocin plus bovine prolactin (Table 2). In

Table 2

Net weight, progesterone concentration, and total progesterone content of corpora lutea removed on day 7 of the oestrous cycle

Treatment No.	Treatment	No. Corpora lutea	Net weight (g)	Progesterone	
				Concentration (μg/g)	Total (μg)
1	Control	12	$4 \cdot 62^1 \pm 0 \cdot 29 \ddagger \cdot \S \cdot \parallel$	$31 \cdot 7^1 \pm 3 \cdot 7 \parallel$	$145^1 \pm 15 \cdot 5 \parallel$
2	Oxytocin (O)	10	$2 \cdot 68^2 \pm 0 \cdot 25$	$19 \cdot 6^2 \pm 5 \cdot 7$	$58^2 \pm 21 \cdot 5$
3	O + Prolactin	5	$3 \cdot 21^2 \pm 0 \cdot 31$	$16 \cdot 9^2 \pm 4 \cdot 7$	$57^2 \pm 18 \cdot 4$
4	O + UHCG*	5	$2 \cdot 73^2 \pm 0 \cdot 36$	$34 \cdot 7^1 \pm 14 \cdot 2$	$100^1 \pm 40 \cdot 7$
5	O + HCG	7	$4 \cdot 72^1 \pm 0 \cdot 78$	$61 \cdot 9^3 \pm 11 \cdot 9$	$334^3 \pm 92 \cdot 6$
6	O + LH†	10	$4 \cdot 85^1 \pm 0 \cdot 62$	$41 \cdot 3^3 \pm 5 \cdot 3$	$212^3 \pm 44 \cdot 9$

* UHCG is human chorionic gonadotrophin incubated with 6 M urea at 40°C for 24 h.
† Luteinizing hormone.
‡ Standard error of the mean.
§ Treatment means with different superscript numbers differ significantly. $(P > 0 \cdot 05.)$
‖ Residual mean squares for net weight, progesterone concentration, and total progesterone are $1 \cdot 89$, $414 \cdot 2$, and $17,559$, respectively.

1, 2 and 3 means with the same superscript numbers do not differ significantly $(P > 0 \cdot 05)$; those with different numbers differ significantly

addition, 16 control corpora were removed at day 4 of the cycle, 16 after oxytocin treatment on days 2 and 3 and 10 after HCG plus oxytocin injections given on days 2 and 3 of the cycle. Thus, a total of 91 corpora were removed and analysed for progesterone as outlined above. The oxytocin was given subcutaneously at a daily rate of 0·33 USP units/kg of body weight, HCG and urea incubated-HCG at a level of 2,000 I.U./day, bovine LH at a rate of 30 or 50 mg/day and bovine prolactin at a rate equivalent to 84 units of NIH standard prolactin per day. The LH and prolactin were suspended in 5 per cent beeswax in sesame oil and given intramuscularly.

The results of these two experiments show that all LH containing preparations tested, with the single exception of equine LH overcame the inhibitory effects of concurrently administered oxytocin on corpus luteum weights, progesterone contents and concentrations. In fact, bovine LH, HCG and crude bovine pituitary extracts significantly increased luteal tissue weights and total progesterone contents over the untreated controls (Tables 1 and 2). In individual instances greatly enlarged corpora weighing as much as 18 g and containing up to 1,200 μg of progesterone resulted from the treatments.

The data in Table 2 also show that inactivation of the LH component of HCG by incubation with 6 M urea (Schmidt-Elmendorff, Loraine and Bell, 1962) abolishes its ability to increase corpus luteum weights and progesterone contents in oxytocin-treated animals. Similar results obtained after urea inactivation of purified bovine LH preparations will be cited in a subsequent section. Incubation with urea in this manner does not inactivate FSH, and any inactivation of prolactin has been shown to be reversible so that it is unlikely that inactivation of either of these hormones contributed to the results under the condition of this experiment.

It is important to point out that none of the other pituitary hormone preparations tested gave evidence of a luteotrophic effect. Neither ovine nor bovine prolactin preparations overcame oxytocin induced luteal inhibition, and a bovine growth hormone preparation was similarly ineffective. Furthermore, observations on the corpora lutea of three additional heifers (not shown in Tables 1 or 2) treated with FSH gave no indication of a stimulatory effect on either luteal tissue weight or progesterone content and concentration.

In a third experiment with intact heifers an attempt was made to prolong the functional life of the corpus luteum and the oestrous cycle in normal heifers by single injections of gonadotrophins (Donaldson and Hansel, 1965). Five heifers were allotted to each of the following treatment groups (1) uninjected controls, (2) bovine pituitary extracts (3 glands), (3) bovine LH (30 units NIH-LH-B2), and (4) bovine LH (30 units) incubated in 6 M urea for 24 h. The test substances were homogenized in Freund's complete adjuvant, made up to a volume of 20 ml and injected subcutaneously at the sixteenth day of the oestrous cycle. Two of the heifers in each of groups 3 and 4 received a bovine LH prepared by the procedure of Ellis (1961) while the remaining three received the NIH-LH-B2* preparation.

* Kindly supplied by the Endocrinology Study Section, National Institute of Health, Bethesda, Maryland.

The results of this experiment (Table 3) again furnish evidence of the ability of a bovine LH preparation and a crude bovine anterior pituitary gland extract to prolong the life span of the corpus luteum and thus lengthen the oestrous cycle. As in the previous experiment, incubation of the purified LH preparation with urea abolished its ability to stimulate the corpus luteum. In order to test the possibility that the results were due to antibodies produced against LH, serum samples were drawn 30–60 days after the injections and tested

Table 3

Prolongation of the bovine oestrous cycle by single injections of LH in Freund's adjuvant at day 16

Treatment	No. of heifers	Length of dioestrous intervals (days)
Untreated	5	$20 \cdot 0 \pm 0 \cdot 7$
AP extract	5	$31 \cdot 0 \pm 2 \cdot 9*$
Bovine LH	5	$36 \cdot 4 \pm 2 \cdot 5*$
Urea-treated LH	5	$21 \cdot 6 \pm 0 \cdot 2$

* Significantly different from the untreated heifers and those receiving urea-treated LH. $(P < 0 \cdot 05.)$

for precipitating antibodies by the micro-immuno diffusion test of Crowle (1958). The test failed to reveal the presence of precipitins when either the bovine pituitary extract or bovine LH was used. The possibility that non-precipitating antibodies were present was not excluded. Wiltbank, Rothlisberger and Zimmerman (1961) reported prolongation of the functional life of the bovine corpus luteum as a result of HCG injections.

LUTEOTROPHIC EFFECTS IN HYSTERECTOMIZED HEIFERS

Several studies concerning the luteotrophic effects of bovine anterior pituitary hormones were carried out with hysterectomized Holstein heifers. In the first experiment (Malven and Hansel, 1964) the uteri and corpora lutea were simultaneously removed from 7 heifers between the ninth and eleventh days of the cycle. The heifers were then observed closely for oestrus to determine if ovaries of hysterectomized heifers are capable of functioning, and slaughtered 14–31 days after hysterectomy.

In the second experiment 13 heifers were hysterectomized on the eleventh day of the cycle and the corpora present in their ovaries

Table 4

Ovulation and corpus luteum formation after simultaneous hysterectomy and corpus luteum removal

Heifer No.	Days after hysterectomy		Number of post operative ovulations	Luteal tissue present at slaughter			
	Oestrus	Slaughter		Description	Net wt. (g)	Total progesterone (µg)	Total 20-β-ol (µg)
Total hysterectomy							
280*	3	14	2	2 Cystic corpus luteum	5·4	102	0
					4·9	204	0
290		14	0	No luteal tissue	4·0		
A288	4	30 (Died)	1	Cystic corpus luteum	5·5	0	0
A294		30	0	Luteinized follicle wall		138	0
A310		30	1	Cystic corpus luteum	3·3	106	67
Subtotal hysterectomy							
A309	6	31	1	Solid corpus luteum	5·4	166	92
A292	3, 25	28	2	Regressed corpus luteum	0·5	0	0
				Recent corpus luteum	1·0†		

* Daily oxytocin injections (days 4 to 10 after hysterectomy).
† Sample lost during progestin analysis.

marked for future identification by injecting them with sterile charcoal suspended in sterile physiological saline. Seven heifers received 10 daily injections of crude aqueous extracts of bovine anterior pituitary tissue beginning 10 days after hysterectomy. The daily dose varied between 1·0 and 1·8 gland equivalents per day. Three hysterectomized heifers received 10 daily injections of the crude aqueous extracts of bovine anterior pituitary tissue incubated with 6 M urea as described above to inactivate the LH component. The remaining three heifers served as uninjected controls. The progesterone contents of the corpora removed in both experiments were determined as above.

Corpora lutea normally persist in a functional state for long periods of time in hysterectomized heifers, as they do in many other species. However, the ovaries of such animals are capable of functioning, as indicated by the nearly normal performance of most of the animals subjected to simultaneous hysterectomy and corpus luteum removal (Table 4). However, it should be pointed out that 2 of the 7 heifers failed to ovulate during the period between hysterectomy and slaughter. One of these animals formed functional luteal tissue in the wall of a large luteinized follicle that was equivalent in weight and progesterone content to some of the corpora formed by other heifers after ovulation.

Administration of crude bovine anterior pituitary extracts significantly increased the weights and progesterone contents of the corpora remaining after hysterectomy (Table 5). Incubation of the crude extracts with 6 M urea to destroy the LH component abolished the stimulatory effect, as it did in the intact heifers.

It should be noted in passing that the results of this experiment provide no evidence for a luteolytic hormone of pituitary origin. It was hoped that a luteolytic effect might be 'unmasked' by injecting crude bovine anterior pituitary extract devoid of the luteotrophic effects of the contained LH. Such was not the case; the progesterone contents and weights of the corpora that received the urea-incubated extracts were almost identical to the values for the untreated hysterectomized animals (Table 5). Obviously, such evidence does not necessarily disprove the existence of a pituitary luteolytic principle, but it must be considered as strong negative evidence.

IN VITRO LUTEOTROPHIC EFFECTS OF ANTERIOR PITUITARY HORMONES

A series of experiments have been carried out testing the ability of purified bovine LH* to stimulate progesterone synthesis in bovine luteal tissue slices under a variety of conditions with a view to developing an *in vitro* assay technique suitable for measuring blood levels of this hormone (Seifart and Hansel, 1965 and 1966). The

Table 5

Effects of hysterectomy and hypophysial extracts on bovine corpora lutea

Treatment	No. corpus luteum	Av. wt. (g)	Av. progesterone concentration (μg/g)	Av. progesterone content (μg)	Av. 20-β-ol content (μg)	No. of induced corpora
Corpora lutea removed at hysterectomy	7	5·9	38·9*	213	0	—
Corpora lutea remaining 20 days after hysterectomy						
Non-injected	3	4·8	52·4	249	41	0
Crude hypophysial extract (1·0–1·8 gland equivalents/ day for 10 days)	7	7·4	53·7	419†	38	3
Urea-incubated hypophysial extract (2·0 gland equivalents/day	3	4·4	55·1	239	19	0

* One sample lost during analysis.
† Includes corpus from one heifer which contained no progesterone.

luteal tissue was taken at day 11 of the cycle, sliced and incubated with known amounts of LH in Krebs-Ringer bicarbonate buffer or bovine serum.† The progesterone determinations in these experiments were made by a modification of the thin layer chromatography techniques described by Armstrong, O'Brien and Greep (1964).

Net *in vitro* synthesis of progesterone in the incubated slices of luteal tissue was increased by the addition of purified bovine LH to the incubation medium. The regression between stimulatory levels of LH added to the incubation medium and the resulting progesterone

* NIH-LH-B2 kindly supplied by the Endocrinology Study Section, USPHS.
† Details of these procedures available upon request.

synthesis was highly significant and linear between 0·005 and 1·0 µg per incubation flask of 5 ml. (*Figures 1* and *2*).

In order to determine the specificity of the increased progesterone synthesis to LH, two experiments were conducted in which all six anterior pituitary hormones, each at two concentrations were assayed alone and in combination. It was found that FSH, TSH,

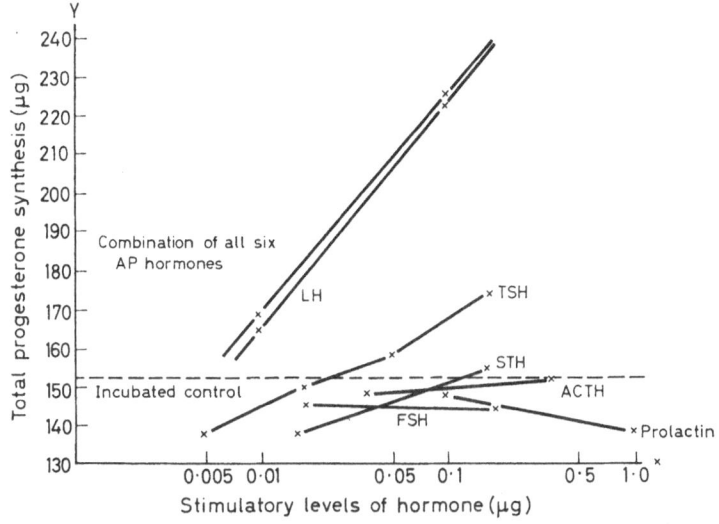

Figure 1. Influence of anterior pituitary hormones individually and in combination on progesterone synthesis in vitro

ACTH, STH and prolactin all failed to evoke a significant response. The regression lines for LH alone and the combination of all six anterior pituitary hormones were significant, but they were not significantly different from one another (*Figures 1* and *2*). This result indicates that there are no interactions and that the observed response is specific for LH. The slight response to the TSH preparation in the first experiment (*see Figure 1*) was probably due to LH present as a contaminant, since TSH and LH are notoriously difficult to separate. In the second experiment (*Figure 2*) neither the level of TSH used nor the urea-incubated TSH gave a response above the incubated control level. In incubation studies with bovine serum it was found that it is absolutely necessary to add 30 m mole nicotinamide in order to inhibit a potent nuclease present in the serum.

Omission of this inhibitor will result in a greatly reduced synthesis, presumably by destruction of the obligatory co-factor, NADPH.

It was found that the response to LH observed in serum was distinctly higher than that observed in Krebs-Ringer buffer. Dilution of the serum 50 per cent with buffer reduced this response pro-

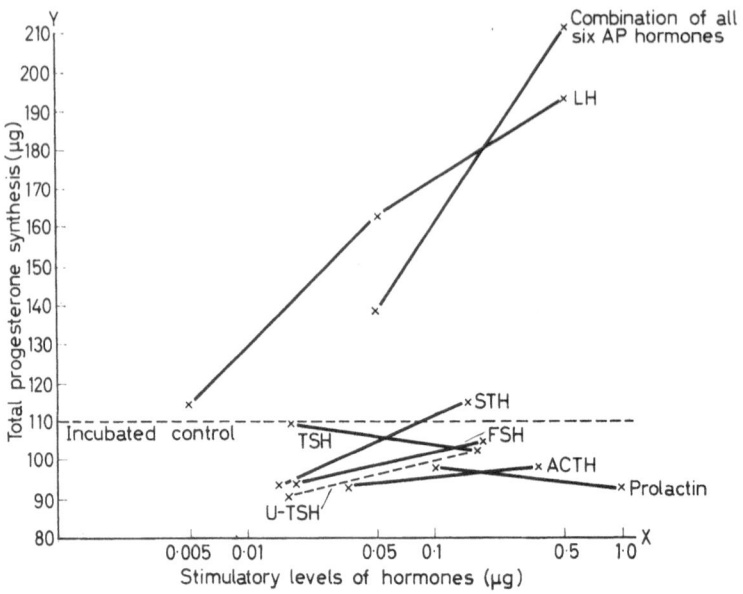

Figure 2. Influence of anterior pituitary hormones individually and in combination on progesterone synthesis in vitro

portionately. Since it was important to know whether this effect was due to endogenous LH, or to the fact that serum constitutes a more complete incubation medium, an experiment was conducted in which 0·25 ml horse antibovine LH was added to the incubation medium. The anti-LH was produced by injecting the same purified bovine LH preparation into mares at a level of 5 mg three times weekly for 6 weeks. One millilitre of the mare's serum used was sufficient to neutralize the effects of 80 µg of NIH-LH-B2 in the rat ventral prostate assay. It was found that this anti-LH completely abolished the effect of LH in the incubation medium, and that it reduced all values approximately to that obtained with Krebs-Ringer control buffer (*Figure 3*). These data suggest that the observed difference between serum and Krebs-Ringer buffer is due primarily to endogenous LH.

The remarkable sensitivity of this response, its apparent specificity and the demonstrated ability of anti-bovine LH to abolish the stimulatory effect of LH all provide evidence for the ability of bovine LH to promote progesterone synthesis in bovine luteal tissue.

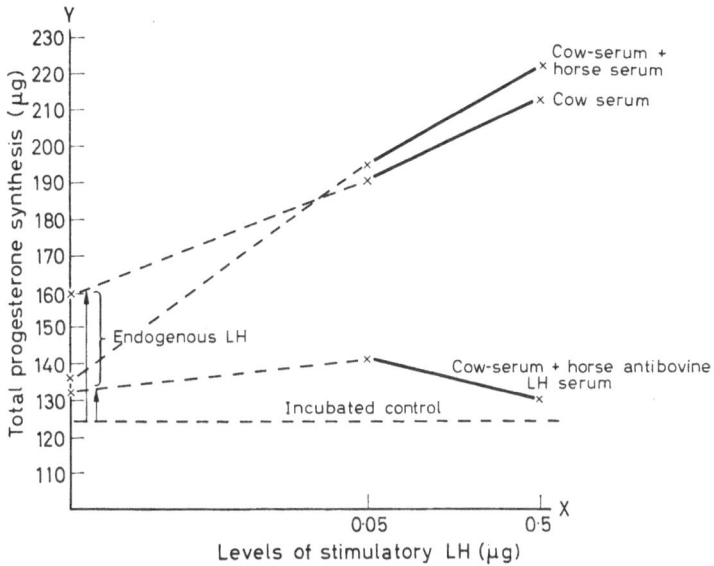

Figure 3. Influence of horse anti-LH serum on the effect of stimulatory LH on progesterone synthesis in vitro

Collectively, these *in vitro* findings and the *in vivo* results previously cited indicate that LH is the major luteotrophic principle in cattle. They do not completely rule out the possibility of synergism of several hormones in the luteotrophic response.   None of the pituitary hormone preparations (including the purified bovine LH) used were entirely pure for the hormone they represented.   However, the LH preparations did not contain measurable quantities of FSH, and there are no indications of a luteotrophic effect of prolactin.   The fact that bovine pituitary LH and HCG are both luteotrophic, despite the fact that they have very different origins also seems to reduce the possibility of important synergistic effects.

LUTEOLYTIC EFFECTS IN HYSTERECTOMIZED HEIFERS

If one accepts these arguments that LH is the major luteotrophic hormone, the question of whether luteal regression is due to a decrease in plasma LH level or to the intervention of an active luteolytic mechanism capable of causing regression in the presence of continued high levels of LH becomes of great interest. The last of these concepts has received considerable attention and the uterus appears to play a luteolytic role in many species. However, the former idea appears to deserve more consideration than it has received. Indeed, some of the data obtained in the present studies suggest that a lowered LH secretion rate, which may result from either reduced hormone synthesis or sequestration of the hormone by the pituitary basophils can result in a degree of luteal regression under certain experimental conditions.

Because of the many indications that the uterus exerts a localized luteolytic effect (*see* Anderson and Melampy, 1965) and because no one has yet succeeded in extracting a luteolytic principle from uterine tissues (*see* Malven and Hansel, 1965, for example) our attention has been directed to studies on luteal regression in hysterectomized heifers. In one of these experiments, summarized in Table 6, the effects of exogeneous oestradiol, oxytocin, LH and antibovine LH on luteal tissue weights and progesterone contents have been studied in 42 hysterectomized heifers. The heifers were hysterectomized at the tenth day of their oestrous cycles at which time the corpora lutea were marked with sterile charcoal. Treatments were begun 21 days later and continued for either 6, 15 or 30 days. Thus, corpora in the 6-day treatment groups were 36 days old; those in 15- and 35-day treatment groups were 45 and 60 days of age, respectively. All animals were slaughtered at the end of treatment, the corpora lutea were removed and their progesterone contents determined as previously described. Comparisons were subsequently made with corpora from appropriate control animals (Table 6).

The results indicate that the corpora of the hysterectomized heifers were slowly regressing during this period; the 60-day old corpora (30-day controls) contained significantly less ($P < 0.05$) total progestins than the 36-day old corpora (6-day controls). Oestradiol injections and antibovine LH significantly reduced the weights and progestin contents of the corpora lutea. The results obtained with oestradiol agree with those obtained by Kaltenbach, Niswender, Zimmerman and Wiltbank (1964), and are probably best explained on the basis of the demonstrated ability of oestrogens to depress LH secretion in several species (*see* Ramirez, Abrams and McCann, 1963 for example). The reduction in both luteal weight and progestin content caused by anti-bovine LH again furnishes proof of the luteotrophic nature of LH in cattle. Equine LH significantly

Table 6

Effects of hysterectomy and various treatments on bovine corpora lutea

Group	Treatment*	Days on treatment	Luteal tissue weight (g)	Total luteal progesterone plus 20-β-ol† (μg)	Total luteal progesterone (μg)	Progesterone concentration (μg/g)
1	Control (5)§	6	5·71	348 ± 47‡	228 ± 32	41·0 ± 4·0
2	Control (5)	15	5·57	288 ± 29	200 ± 18	36·7 ± 2·9
3	Control (5)	30	4·13	162 ± 42	124 ± 30	30·1 ± 2·5
4	Oestradiol (5)‖	6	4·45	175 ± 22‖	139 ± 12	31·7 ± 3·0
5	Oestradiol (5)‖	15	3·44‖	131 ± 15‖	110 ± 14	32·3 ± 4·4
6	Antibovine LH (2)	6	3·02‖	124	98	32·1
7	Equine LH (5)	15	5·63	433 ± 66‖	330 ± 49‖	58·2 ± 6·6¶
8	Oxytocin (5)	15	5·23	296 ± 45	243 ± 34	46·2 ± 3·3
9	Oxytocin (5)	30	4·81	241 ± 51	186 ± 39	36·6 ± 5·6

* Hysterectomies performed on the tenth day of the cycle; treatments begun 21 days later.
† Δ⁴ pregnane-20-β-ol-3-one.
‡ Standard error of the mean.
§ Numbers in parentheses represent the number of animals in each group.
‖ Significantly different from respective control (*P* < 0·05).
¶ Significantly different from respective control (*P* < 0·01).

increased both the progestin content and concentration in the corpora lutea of hysterectomized heifers. This is a most interesting result, since equine LH given at the same level but for a shorter period of time was the only LH containing preparation that failed to exhibit a luteotrophic effect in the experiments with intact heifers. Oxytocin injections, even for 30 days had no luteolytic effect in hysterectomized heifers, as they do in intact animals. The presence of the uterus is necessary for the oxytocin response.

It should be pointed out that neither the anti-bovine LH nor the oestradiol injections caused complete regression of the corpora lutea in the hysterectomized heifers. In contrast, injections of oestradiol in amounts ranging from 700 μg to 5·0 mg/day into 5 intact heifers for similar periods of time caused somewhat greater reduction in both luteal tissue weights and progesterone contents, and a return to oestrus by the last day of treatment in most cases.

The studies of Donaldson *et al.* (1965) cited earlier also provided evidence that luteal regression occurs when pituitary LH levels are reduced. In these studies LH assays were carried out by the method described by Florsheim, Velcoff and Bodfish (1959), which is based on the augmentation of radiophosphorus uptake by chick testes.

The results (Tables 7 and 8) show that oxytocin administered 3–4 h before the animals were slaughtered at 6 h after the beginning

Table 7

The effect of oxytocin on the LH content of anterior pituitary tissue of heifers on days 0, 4 or 7 of the oestrous cycle

Day	Number of glands measured	μg NIH-LH-S1/mg *dried tissue*	
		Control Mean ± S.E.	Oxytocin Mean ± S.E.
0	10	3·0 ± 0·33	1·7 ± 0·47
4	10	9·1 ± 0·62	10·5 ± 2·54
7	10	19·6 ± 8·34	8·3 ± 1·41

of oestrus caused a further reduction in pituitary gonadotrophin levels, which are already at minimal levels for the entire cycle. No difference was found in the gonadotrophin levels of control and oxytocin-treated heifers at day 4. At day 7 the gonadotrophin level was reduced by about one half in the oxytocin-treated heifers. This difference, however, was not significant at the 5 per cent level of probability.

When the correlation coefficients between pituitary gonado-trophin levels and total progesterone in the corpora lutea were calculated (Table 8) a large negative correlation ($-0·75$) was found for the control animals and a large positive correlation ($0·96$) in the oxytocin-treated heifers. These results suggest that the initial stimulus for corpus luteum development results from the ovulatory release of LH. The corpora removed at day 4 did not respond to exogenous HCG given on days 2 and 3, possibly because they are already responding at a maximal rate to a level of gonadotrophins already in excess of requirements at this time.

Oxytocin given at the beginning of oestrus appeared to cause gonadotrophin release, an observation in agreement with the previous demonstration that oxytocin given at this time hastens ovulation (Hansel, Armstrong and McEntee, 1958). The data further suggest that by day 7 the oxytocin-treated heifers were incapable of synthesizing or releasing sufficient gonadotrophin to maintain normal corpus luteum size and progesterone production. The large positive correlation coefficient between pituitary gonadotrophin content and corpus luteum progesterone in the oxytocin-treated

Table 8

Correlation between pituitary gonadotrophin* levels and total progesterone content† of the corpus luteum in glands collected on days 4 or 7 of the cycle

Day 4				Day 7			
Control		Oxytocin		Control		Oxytocin	
Pit. Gonad.	Total Prog.	Pit. Gonad.	Total Prog.	Pit. Gonad.	Total Prog.	Pit. Gonad.	Total Prog.
6·9	63	5·3	24	3·6‡	186	5·0	5
8·7	41	7·7	37	9·1	164	6·7	18
9·5	33	8·4	18	15·1	223	7·7	24
9·7	26	10·8	71	19·2	171	8·9	39
10·6	42	20·0	50	50·9	114	13·4	151
Correlation coefficients							
−0·76		+0·49		−0·75		+0·96	

* µg/mg dried anterior pituitary.
† µg.
‡ The hypothalamus of this heifer was accidentally stimulated with a needle during collection of cavernous sinus blood about 30 min before slaughter.

animals suggests that pituitary gonadotrophin levels may reflect plasma levels under these conditions, and that these levels are directly controlling and limiting progesterone synthesis in the corpora lutea.

The large negative correlation between pituitary gonadotrophins and corpus luteum progesterone content in the untreated heifers probably represents storage of gonadotrophin in the pituitary. Thus, pituitary levels may not reflect plasma levels at day 7 in untreated animals. It is worthy of note that exogenous gonadotrophin (HCG or pituitary extracts) will increase corpus luteum size and progesterone content at this time, suggesting that plasma LH levels may be limiting luteal development at this time in the normal animal.

These results indicate that reduction in LH secretion rate by either oestradiol or oxytocin injections or inactivation of circulating LH by anti-bovine LH can cause regression of bovine corpora lutea, but it remains to be demonstrated that such a reduction in LH secretion rate occurs in normal animals at the time the corpus luteum is regressing. All available data suggest that the pituitary LH content increases during the time of luteal regression in the cow. Therefore, in order to defend the LH withdrawal theory, one must postulate that the adenohypophysis sequesters large amounts of the hormone during the period of luteal regression and the ability of the pituitary to do this has yet to be demonstrated. Ultimately, the question must be answered by accurate measurements of blood levels of LH during this critical period in order to determine if a decline actually occurs. The fact that neither oestrogen nor antibovine LH serum caused luteal regression to the degree that it occurs in the normal cyclic animal again suggests that the uterus may play a luteolytic role, but the nature of this mechanism remains obscure.

SUMMARY OF CONCLUSIONS

A series of experiments conducted over a period of several years have provided the following evidence that LH is the major luteotrophic principle in cattle:

(1) Purified bovine LH and other LH-containing preparations overcame the inhibitory effects of concurrently injected oxytocin on luteal tissue weight, progesterone content and concentration.

(2) Incubation of these preparations with urea, a procedure known to selectively inactivate LH abolished their ability to overcome the inhibitory effects of oxytocin.

(3) Prolactin of ovine or bovine origin, FSH and Growth Hormone were all incapable of overcoming oxytocin-induced luteal inhibition.

(4) Single injections of purified bovine LH in Freund's adjuvant given at mid-cycle significantly prolonged the functional life span of corpora lutea and lengthened the oestrous cycle. Incubation of the LH with urea abolished the effect.

(5) Crude anterior pituitary extracts and an equine LH preparation significantly increased corpus luteum weights and progesterone contents in hysterectomized heifers. Incubation of the crude anterior pituitary extracts with urea abolished the stimulatory effect

due to the contained LH. However, these urea incubated extracts did not possess any luteolytic properties.

(6) Purified bovine LH in amounts as low as 0·005 μg stimulated progesterone synthesis *in vitro* by bovine luteal tissue slices. None of the other anterior pituitary hormones tested gave this effect. Addition of small amounts of anti-bovine LH serum from mares injected with purified bovine LH abolished the LH stimulation.

(7) Corpora lutea of hysterectomized heifers were caused to regress by the administration of the same anti-bovine LH serum and by oestradiol injections.

(8) The question of whether corpus luteum regression in normal animals is due to a decline in LH secretion rate, the intervention of an active luteolytic mechanism involving the uterus, or both, remains unanswered.

REFERENCES

Anderson, L. L. and Melampy, R. M. (1965). In: *Proc. Conf. on Estrous Cycle Control in Domestic Animals.* USDA Misc. Publ. 1005, pp. 64–77
Armstrong, D. T. and Hansel, W. (1959). *J. Dairy Sci.* **42,** 533
— O'Brien, Judith and Greep, R. O. (1964). *Endocrinology* **75,** 488
Crowle, A. J. (1958). *J. Lab. clin. Med.* **52,** 784
Donaldson, L. E. and Hansel, W. (1965). *J. Dairy Sci.* **48,** 903–904
— — and Van Vleck, L. D. (1965). *J. Dairy Sci.* **48,** 331
Ellis, S. (1961). *Endocrinology* **69,** 554
Florsheim, W. H., Velcoff, S. M. and Bodfish, R. E. (1959). *Acta Endocr., Copenh.* **30,** 175
Hansel, W., Armstrong, D. T. and McEntee, K. (1958). Recent Studies on the Mechanism of Ovulation in the Cow. *Proc. IIIrd Symposium on Reprod. and Infertility.* Ed. by F. X. Gassner. 63 p. Pergamon Press
Kaltenbach, C. C., Niswender, G. D., Zimmerman, D. R. and Wiltbank, J. N. (1964). *J. Anim. Sci.* **23,** 995
Malven, P. V. and Hansel, W. (1964). *J. Dairy Sci.* **47,** 1388–1393
— — (1965). *J. Reprod. Fert.* **9,** 207
Ramirez, V. D., Abrams, R. M. and McCann, S. M. (1963). *Fedn Proc. Fedn Am. Socs exp. Biol.* **22,** 506
Schmidt-Elmendorff, H., Loraine, J. A. and Bell, E. T. (1962). *J. Endocr.* **24,** 153
Seifart, K. and Hansel, W. (1965). *Fedn Proc. Fedn Am. Socs exp. Biol.* **24,** 320
— — (1966). (Unpublished data)
Simmons, K. R. and Hansel, W. (1964). *J. Anim. Sci.* **23,** 136
Staples, R. E. and Hansel, W. (1961). *J. Dairy Sci.* **44,** 2040
Wiltbank, J. N., Rothlisberger, J. A. and Zimmerman, D. R. (1961). *J. Anim. Sci.* **20,** 827

DISCUSSION

Professor R. J. Fitzpatrick (*Liverpool*)

What dosage of oxytocin did you use in this work? From your publications it appeared to be a large dose compared to that required to elicit responses in the uterus or mammary gland. If so, it raises the possibility that the high dose stimulated the release of prolactin in a similar manner to that postulated for endogenous oxytocin. Might not a similar mechanism result in the suckling stimulus being luteotrophic in the cow?

Hansel

The dose of oxytocin used was high, namely 15 I.U./100 lb body weight compared to 5 I.U. or less required to initiate milk ejection in the cow. We were not interested, however, in the circulating levels of oxytocin but were using it as a tool to terminate the corpus luteum. The correlation between progesterone content of the corpora lutea and pituitary gonadotrophin content in oxytocin-treated animals is good evidence that part of the effect is a decrease in the pituitary content of LH.

Dr. D. W. Schomberg (*Cambridge*)

Have you found any corpora lutea which did not respond *in vitro*?

Hansel

We have not found any which did not respond at all but we have had some that did not respond as well as expected.

Mr. L. E. A. Rowson (*Cambridge*)

When you gave high levels of oxytocin in the early part of the cycle you got a tremendous output of gonadotrophin. Do you ever get an extra ovulation?

Hansel

We do get new ovulations and corpora lutea forming after day 10 but not as early as day 7.

Dr. R. V. Short (*Cambridge*)

Until one can ascertain that the concentration of hormone in a gland *in vivo* is an accurate reflection of secretion rate by that gland, one cannot infer that changes in glandular content reflect changes in production or secretion rate. In the guinea pig the concentration of progesterone in the corpus luteum may bear no relation to the rate of secretion. There seems to be some contradictory evidence regarding the luteotrophic effect of LH. There is evidence in several species that the corpus luteum is stimulated *in vitro* by LH and HCG. In six successive experiments in 3 cows we have failed to find any increase in peripheral blood progesterone levels following a single intravenous injection of HCG in doses ranging from 5,000 to 15,000 I.U. HCG. I think you have produced some very compelling evidence that LH may be luteotrophic but at the present time I would be prepared to acknowledge it only as a possibility.

Hansel

I appreciate your comments and acknowledge that some of the points you mention are valid. However, there is considerable evidence showing good correlation between ovarian venous progesterone and that in the corpus luteum in the cow, ewe and gilt. With regard to the intravenous infusions of a single large dose of HCG I think you are not infusing it slowly enough and it is infused into the jugular vein, rather than the ovarian artery.

PROFESSOR A. W. NALBANDOV (*Illinois, U.S.A.*)

A single injection of LH into sheep may have no effect on progesterone levels in peripheral blood but the infusion of the hormone over a 30–60 min period results in an increased progesterone level in the ovarian venous blood. I am not saying, however, that this is evidence for LH being luteotrophic in the sheep.

SHORT

Do you believe that the stalk-sectioned pituitary gland of a sheep is capable of secreting LH, for such a gland is certainly capable of prolonging the functional life of a corpus luteum?

HANSEL

This raises quite a problem. The portal vascular system of the pituitary gland can regenerate within a few days and it is difficult to be certain that this has not occurred in any given case. There is no clear evidence that the pituitaries of stalk-sectioned animals cannot produce some LH.

EVENING DISCUSSION

PROFESSOR A. V. NALBANDOV (*Illinois, U.S.A.*)

It has become quite clear in the course of our discussions that the problem of corpus luteum maintenance is far from solved, and a tremendous amount of work has yet to be done. The important factors that have emerged so far indicate that there are two basic concepts:

1. Those workers who believe in a uterine luteolytic factor.
2. Those that are sceptical about a humoral factor and are more willing to believe in a neural connection between the uterus and the ovaries, perhaps through the hypothalamo-hypophyseal complex.

Professor Hansel has demonstrated that LH is a luteotrophic hormone, but he has not established that it is a stereogenic hormone. It is clear that prolactin ought to be looked at with a great deal of scepticism as a luteotrophic hormone and definitely does not qualify as a stereogenic hormone. I had occasion to inspect the work of Eli Romanoff (Worcester Foundation) who had some disturbing results concerned with sterogenesis in the perfused whole ovary of the cow. He found that when he added prolactin to the perfusate, the corpus luteum produced copious progesterone. When the corpus alone is incubated prolactin does not have this effect. This is the difficulty in comparison between the whole ovary and a corpus luteum preparation. Perhaps in the studies of the whole ovary a complex system is acting. I hope that later Dr. du Mesnil du Buisson will comment concerning the effect of oestrogen on the pig corpus luteum.

DR. B. T. DONOVAN (*London*)

I think your division of people into definite groups is wrong, it depends on the species in question. Our work on the guinea-pig indicates the presence of a luteotrophic factor in the uterus, whereas parallel work in the ferret causes us to come down very heavily in favour of pituitary luteotrophic factors which, so far as we can tell, is identifiable with prolactin. We can replace the pituitary gland by prolactin administration and maintain the corpus luteum: whereas other hormones will not, so I think your division is arbitrary and unreal. One aspect which has been neglected is the changing response of the corpora lutea, that of the early cycle may be different from the corpus luteum later in the cycle.

DR. I. ROTHCHILD (*Cleveland, Ohio, U.S.A.*)

·I think it is possible to explain the duration of the corpus luteum in the gilt in relation to pregnancy by assuming that the corpus luteum is dependent on a mixture of prolactin and LH throughout its lifetime and the establishment of the embryo in the uterus reduces the luteolytic effect of the uterus. The uterus would interfere with the ability of prolactin to maintain the corpus luteum. After 25 days this factor no longer counts and LH and prolactin might act. Du Buisson has shown that if he removes all the foetuses from one horn late in pregnancy, and all but one from the other horn but leaves the horns intact, then there is an abortion, but if he removes the empty horn, the one remaining foetus is maintained indicating that there is a luteolytic effect in the empty horn.

DR. F. DU MESNIL DU BUISSON (*France*)

I have reported previously the different reasons why I think the uterus exerts a local control on the maintenance of corpora lutea in the sow and I should like to thank Professor Rothchild for mentioning this work.

The reason why it is difficult for pregnancy to be established in the presence of a sterile uterine horn can be explained by the luteolytic effect of the latter.

My last experiments with P. Léglise and P. Rombauts were concerned with the effects of oestrogen treatment on this luteolytic mechanism.

1. In the normal, cycling animal injection of large doses of oestradiol benzoate (5 mg/day for 7 days starting on the twelfth day of the cycle) results in cessation of cycles and maintenance of the corpus luteum for at least 100 days.

2. In the animal with one pregnant and one sterile uterine horn the same treatment causes maintenance of corpora lutea of both ovaries and continuation of gestation (9 cases out of 11).

3. In the hypophysectomized, cycling animal with an entire uterus, injection of 5 mg oestradiol benzoate given with 5 mg LH daily from day 12 until day 20 causes maintenance of corpora lutea of normal size.

In the hypophysectomized animal at the beginning of the cycle with a normal uterus, treatment with LH alone or with oestradiol benzoate alone is incapable of causing maintenance of the corpus luteum until day 20.   On the contrary, however, with L.L. Anderson, P. Léglise and P. Rombauts at Jouy-en-Josas we showed that in the hysterectomized animal injection of LH alone caused maintenance of the corpus luteum until day 20; we concluded that the luteolytic factor acts as an anti-LH agent at the level of the ovary.

In the experiment cited above (Point 3) oestrogen is not acting by modifying a gonadotrophic factor from the hypophysis because the hypophysis is absent.   It is therefore, reasonable to suggest that oestrogen has a luteotrophic effect in that it inhibits the luteolytic action of the uterus and that it acts by blocking the secretion or excretion of the luteolytic substance.

These experiments led us to propose the following scheme for the maintenance of the corpus luteum in the sow:

It is not logical to distinguish between the corpora lutea of the oestrous cycle until day 14 and the corpora lutea of pregnancy from this time until parturition.   We separate the two types not at day 15 but at day 25.

During the first 25 days of gestation the corpus luteum is at first independent of the hypophysis, then from day 15 it is dependent upon the secretion of LH and responsive to the luteolytic action of the uterus.

Later in pregnancy or pseudopregnancy, LH has no effect, and it is possible that a factor of the same nature as prolactin comes into play at this time.

As we said at the *2nd Endocrinology Congress* in London (1964), inhibition of the luteolytic action of the uterus by the embryo, which occupies the whole uterus by day 15, is perhaps the signal for pregnancy.

Dr. B. T. Donovan

This means that if the pig is mated and becomes pregnant the luteolytic factor is not produced or does not act.   But the blastocysts are thin and narrow and don't embed in the wall of the uterus until later.   How is the production of the luteolytic factor prevented?

Dr. F. du Mesnil du Buisson

It is possible that the embryo sac is capable of preventing the production of the luteolytic factor.   Melampy and Anderson have shown as I have, by hysterectomy, that the embryo is not an essential factor in the maintenance of the corpus luteum.

PROFESSOR R. M. MELAMPY (*Iowa, U.S.A.*)

We used the marbles which we obtained from your department in Paris and obtained the same effect as you did.

DR. I. ROTHCHILD

The interesting thing that arises from the sheep work of Rowson and Moor is that if you remove the embryos after the thirteenth day the luteolytic effect has almost disappeared.

DR. B. T. DONOVAN

This means that the embryo is producing some substance which is acting on the endometrium perhaps to affect the release of the luteolytic factor.

DR. R. M. MOOR (*Cambridge*)

Yes, we think that that is what happens.

DR. B. T. DONOVAN

This leads us to another question, the fact seems to have been neglected that the embryo might produce luteotrophic hormone. Is there any evidence in the pig for the production of placental gonado-trophin?

DR. I. ROTHCHILD

You can define a luteotrophic hormone as a substance which has a possible stimulatory action on the corpus luteum or a substance which prevents an inhibitory effect of another substance. If something maintains the corpus luteum it does so by either of the afore-mentioned mechanisms. The same argument is valid for possible luteolytic substances or processes. I think this is where we have all been misled by looking at the corpus luteum of different animals—for a substance which has a particular effect in one animal and which may not have that effect in another, and this would not mean that it is luteotrophic in one and not in the other, and the same is true of the luteolytic effects of substances.

DR. B. T. DONOVAN

Equally you can say that a substance is not luteotrophic, but anti-luteolytic.

DR. R. M. MOOR

It seems that if you have corpora lutea in the pig of different ages you don't get regression of the two sets of corpora lutea at the same time. This suggests that the age of the corpora lutea might be an important factor in the pig. In the sheep very clearly the uterus is capable of destroying corpora lutea whatever their age at day 16. If a corpus luteum is induced late in the cycle it will regress at day 16 in spite of being only a few days old.

In the sheep if one has a normal 16-day oestrous cycle and then induce ovulation at days 5, 10 or 13, the original corpus luteum will grow and regress and the new corpora lutea will also grow and regress at the same time. We think that whatever the luteolytic effect of the uterus is, it is so powerful that it can cause regression of a corpus luteum whatever the age. This doesn't happen in the pig, in fact the second set seem to continue.

DR. I. ROTHCHILD

There seems to be a real difference between the pig and the sheep.

DR. F. DU MESNIL DU BUISSON

What is the controlling mechanism between the ovary and the uterus? If the uterus is removed to a wall graft, the cyclical behaviour of the gilt is maintained. In this condition it is difficult to explain how the ovary is given information from the transplanted uterus. It would appear that the uterus is acting via the pituitary.

PROFESSOR A. V. NALBANDOV

I always prefer to think that it is a neural signal but it is difficult to see how a neural connection can exist in the above experiments.

VI ARTIFICIAL CONTROL OF REPRODUCTION

CONTROL OF THE OVARIAN CYCLE IN THE SHEEP

T. J. ROBINSON

Department of Animal Husbandry, University of Sydney, Sydney, N.S.W.

INTRODUCTION

PUBLISHED work in this field has been reviewed recently by Hammond Jun. (1961) and Lamond (1964). This paper summarizes work conducted by the author and associates in the University of Sydney during the past 6 years, preliminary reports of which have been published (Robinson, 1964, 1965; Shelton, 1964a, 1965). The remainder is in the press as a research monograph to be published by the Sydney University Press. Reference will be made to the individual contributions within that monograph.

The programme has involved the following steps:

1. A critical evaluation of an oral progestagen ('Provera', MAP) for the synchronization of oestrus in the entire cyclic ewe.

2. The evaluation of a number of progestagens of potential value, using the progestagen-oestrogen interaction on oestrous behaviour in the spayed ewe. All tests were made using intramuscular injection.

3. The evaluation in the entire ewe of progestagens tested in the spayed ewe in order to determine the validity of this test.

4. A comparison of the physiological effects of several progestagens in spayed and entire ewes when given by intramuscular injection or by intravaginal deposition, as a first step in the development of an intravaginal tampon impregnated with a progestagen and which could be inserted and removed at will.

5. The development and preliminary testing of such a tampon or 'artificial corpus luteum', which took the form of a progestagen impregnated sponge which was inserted intramuscularly or intravaginally.

6. The evaluation of a number of progestagens at various dose levels when administered in sponges inserted intravaginally.

7. The determination of the effect on the oestrous cycle of non-impregnated intravaginal sponges.

8. An examination of the vaginal flora following the use of intravaginal sponges with or without several bactericides.

373

9. The determination of the time of ovulation following withdrawal of progestagen impregnated intravaginal sponges.

10. The estimation of the proportion of ova fertilized following such treatment.

11. The estimation of the rate of sperm transport following such treatment.

12. The estimation of the rate of absorption of progestagen from intravaginal sponges.

13. The evaluation of the practical application of progestagen impregnated sponges, with or without PMS for the advancement of the breeding season in crossbred and purebred ewes.

14. The evaluation of such sponges for the synchronization of oestrus for large-scale artificial insemination of Merino ewes.

All told, the studies, commenced in 1960, have involved observations on 20,000 ewes in 50 tests, most with a factorial design. Standard methods of analysis involving appropriate transformations have been used throughout. The tests were conducted on the McCaughey Memorial Institute, Jerilderie, N.S.W., the University of Sydney Animal Husbandry Farms, Camden, N.S.W., and on many private properties in New South Wales, South Australia and Western Australia. Even on the private properties, most experimental designs were complex and in some instances involved laparotomies on large numbers of animals for ovarian examination or ovum recovery. The interest and co-operation of all concerned is most gratefully acknowledged.

Because of the great bulk of material, the following account is a summary of results without details of methods, but with references to the detailed papers.

EVALUATION OF AN ORAL PROGESTAGEN ('PROVERA', MAP) IN THE ENTIRE CYCLIC EWE

This test, conducted in the autumn of 1962, used 900 ewes, of which 600 were incorporated into a factorial experiment and 300 were control ewes (Lindsay, Moore, Robinson, Salamon and Shelton, 1967). One factor introduced was the dose of progestagen but, as this had no demonstrable effect on the most important characters studied, all 900 ewes could be regarded as incorporated in a simple 3×2 factorial experiment with $n = 150$.

Both methods of progestagen treatment—daily injections of progesterone or feeding of MAP—were effective in blocking oestrus but the release of oestrus was more effective with progesterone. The timing was more precise than with oral treatment (*Figure 1*) and more ewes came into oestrus following cessation of treatment

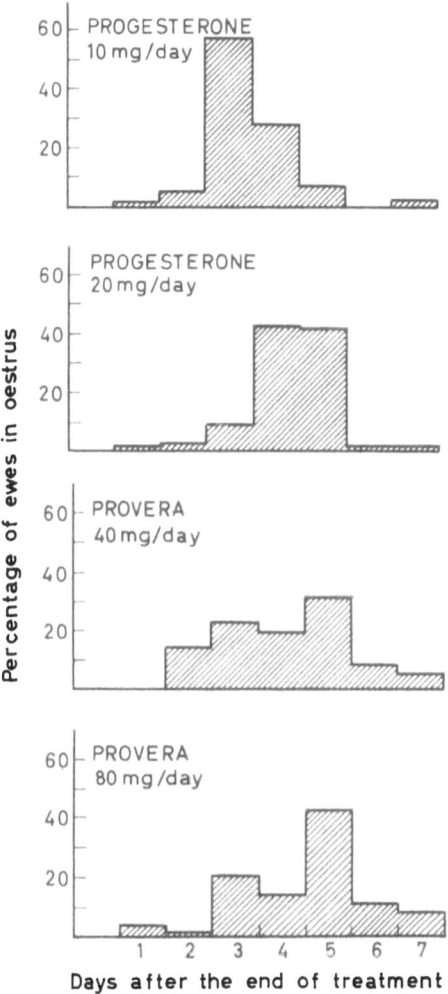

Figure 1. The effect of type and dose of progestagen on the interval between the last progestagen treatment and the occurrence of oestrus (From D. R. Lindsay *et al.*, 1967, by courtesy of Sydney University Press)

(Table 1). Fertility to natural service showed a linear decline for control, progesterone, and MAP treatments, but the decline following artificial insemination (0·1 ml undiluted semen) was much greater. The overall differences between treatments and method of insemination and the treatment × method interaction were highly significant ($P < 0·01 - < 0·001$).

Table 1

Summary of oestrous and lambing results following the use of injected progesterone or fed 'Provera' (MAP)

(From D. R. Lindsay *et al.*, 1967, by courtesy of Sydney University Press)

Progestagen treatment	Method of mating	Number of ewes			Percentage lambed
		Treated	In oestrus	Lambed	
Nil	A.I.	150	144	96	66·7
	Natural	150	137	116	84·7
Progesterone	A.I.	150	116	61	52·6
	Natural	150	125	89	71·2
'Provera' (MAP)	A.I.	150	82	16	19·5
	Natural	150	93	62	66·7
Total		900	697	440	63·1

Progesterone—10 or 20 mg i/m in oil daily for 16 days.
Provera (MAP)—40 or 80 mg orally in sheep nuts daily for 16 days.

The lack of precision of time of oestrus, the low proportion of ewes in oestrus, and the exceedingly low conception rate to artificial insemination following the oral use of MAP, coupled with problems of management and cost, seem to rule out this method of approach.

THE EVALUATION OF A NUMBER OF PROGESTAGENS IN THE SPAYED EWE

Shelton (1964a, 1965) has published brief reports of tests of progestagens using the spayed ewe. Full reports have been published (Shelton, 1964b; Shelton, Robinson and Holst, 1967) on 7 progestagens which were thoroughly tested against progesterone. These and three others, tested in entire ewes, are shown in Table 2 and *Figure 2*.

The basis of the test is the progesterone-oestrogen interaction on oestrous behaviour (Robinson, 1954a, b, 1955; Robinson and Moore, 1956; Robinson, Moore and Binet, 1956). The object of the tests, which commenced in January 1960, was to find a highly potent progestagen which, in its reaction pattern, exhibited the same

Table 2

Progestagens tested

S.E	Progesterone:	Pregn-4-ene-3,20-dione
S.E.	SC-10363:*	17α-acetoxy-6-methylpregna-4,6-diene-3,20-dione
S.E.	SC-9880:*	17α-acetoxy-9α-fluoro-11β-hydroxypregn-4-ene-3,20-dione
S.E.	SC-9022:*	17β-hydroxy-21-methyl-21-methylene-19-nor-17α-pregn-4-en-3-one
S.E.	SC-9392:*	3β-acetoxy-21-methyl-21-methylene-19-nor-17α-pregn-4-en-17β-ol
S.E.	SC-11800:*	3β,17β-diacetoxy-17α-ethynyloestr-4-ene
S.E.	'Provera'†: (MAP)	17α-acetoxy-6α-methylpregn-4-ene-3,20-dione
S.	'Chlormadinone'‡ (CAP):	17α-acetoxy-6-chloropregna-4,6-diene-3,20-dione
E.	SC-10230:*	17α-acetoxy-21-fluoro-6-methylpregna-4,6-diene-3,20-dione
E.	SC-10017:*	17α-acetoxy-21-fluoro-6α-methylpregn-4-ene-3,20-dione
E.	'Enovid':	17α-ethynyl-17β-hydroxyoestr-5(10)-en-3-one (with 1·5 per cent of the 3-methyl-ether of ethynyloestradiol)

S.E. Tested in spayed and entire ewes.
S.   Tested in spayed ewes only.
E.   Tested in entire ewes only.
* Searle.   † Upjohn.   ‡ Eli Lilly.

inter-relationship with oestrogen as progesterone itself, this relationship to be manifested by an exactly comparable behavioural and vaginal oestrous response to injected oestradiol benzoate (ODB). The three main variates—the dose of progestagen, interval from final progestagen treatment to the injection of oestrogen, and dose level of oestrogen—were introduced to provide a multi-dimensinal biological model of known dimensions as far as progesterone was concerned. The unknown progestagen must fit this model exactly in order to be of sufficient value for further investigation in the intact animal for the control of ovulation.

1. PROGESTERONE

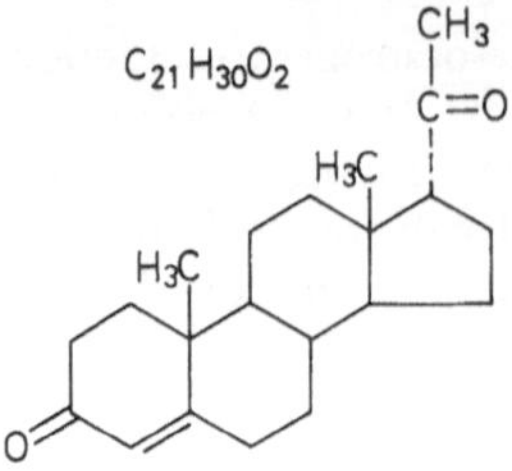

$C_{21}H_{30}O_2$

2. SC-10363 (SEARLE)

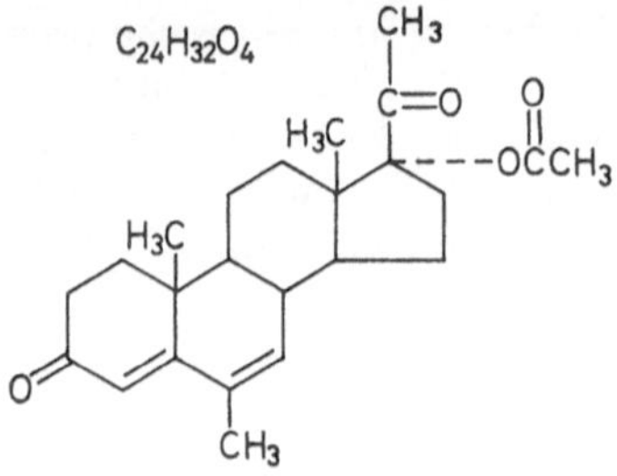

$C_{24}H_{32}O_4$

3. SC-9880 (SEARLE)

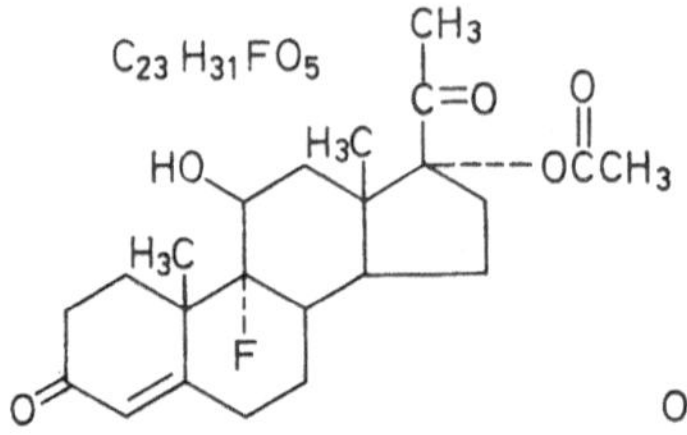

$C_{23}H_{31}FO_5$

4. SC-9022 (SEARLE)

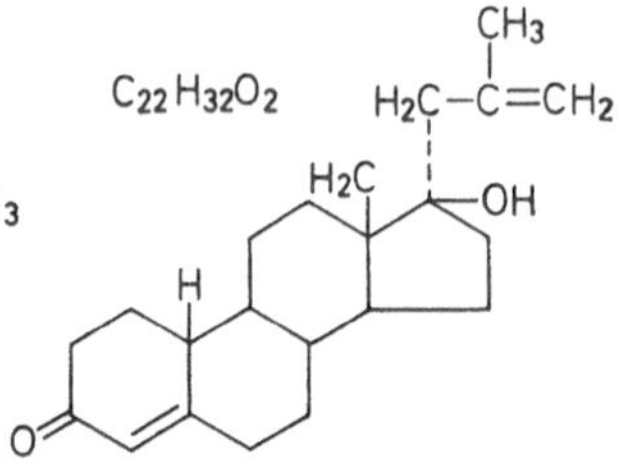

$C_{22}H_{32}O_2$

Figure 2a

5. "PROVERA"; MAP (UPJOHN)

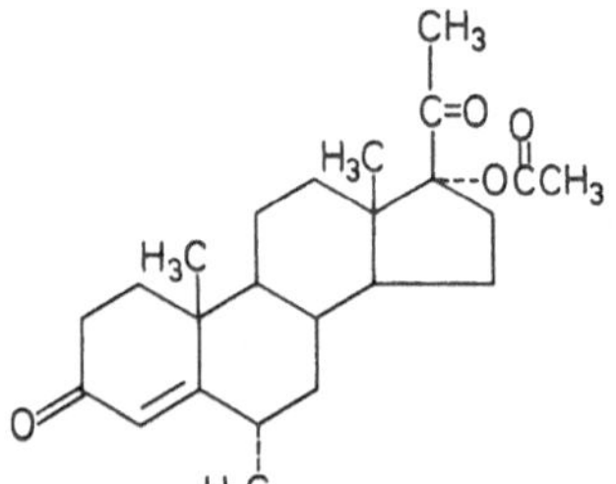

6. SC-9392 (SEARLE)

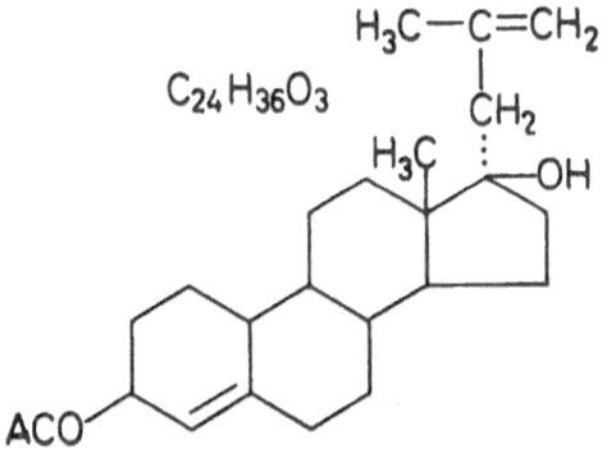

$C_{24}H_{36}O_3$

7. SC-11800 (SEARLE)

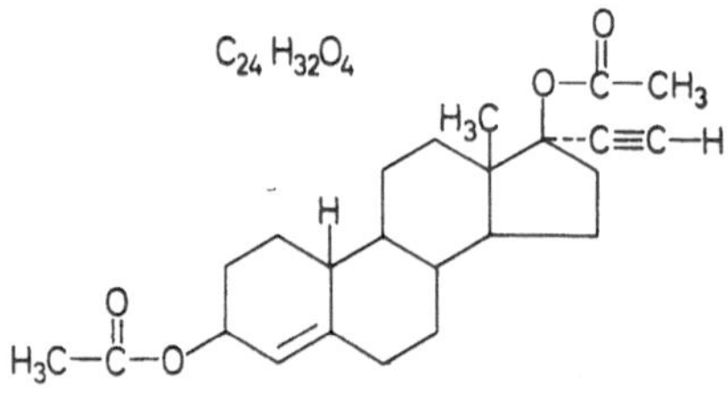

$C_{24}H_{32}O_4$

8. SC-10230 (SEARLE)

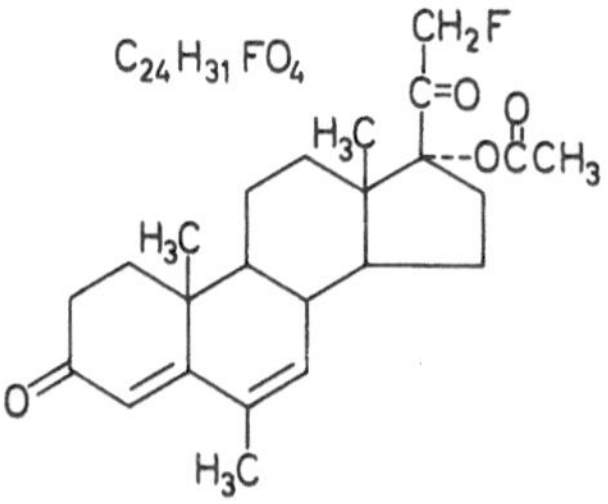

$C_{24}H_{31}FO_4$

Figure 2b

378

CONTROL OF THE OVARIAN CYCLE

Fifteen tests were conducted in a flock of 160 spayed ewes using progesterone, SC-10363, SC-9880, SC-9022, SC-9392, SC-11800, 'Provera' (MAP), and 'Chlormadinone' (CAP). Five compounds —SC-10363, SC-9880, SC-9022, MAP and CAP—at doses as low as 0·1 or 0·4 mg/day were highly active as oestrus-conditioning agents. Of these, SC-10363, MAP and CAP appeared to owe some of their

9. SC-10017 (SEARLE)

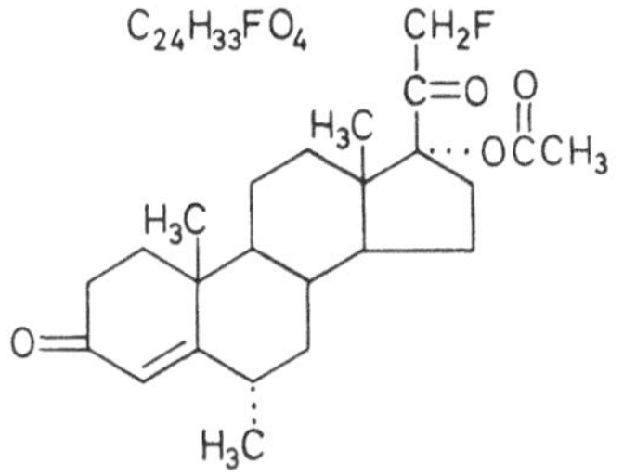

$C_{24}H_{33}FO_4$

10. "ENOVID" (SEARLE)*

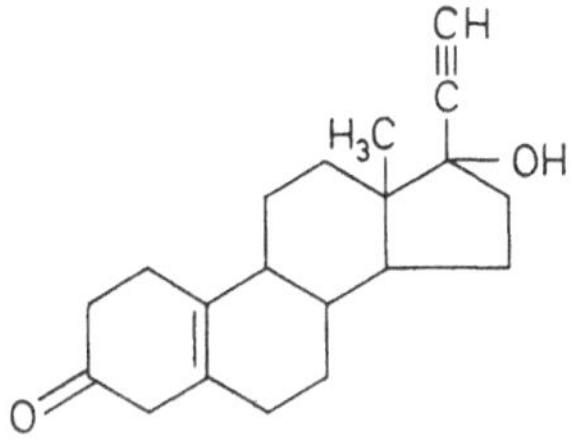

*Norethynodrel + 1·5%
3-Methyl-Ether of
ethynyloestradiol

11. "CHLORMADINONE", CAP

(ELI LILLY)

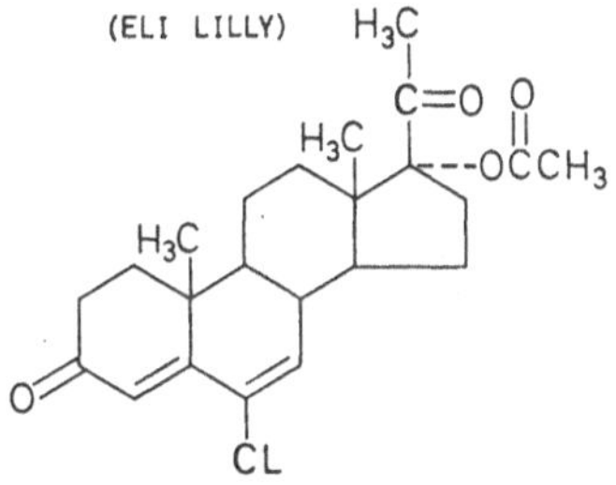

Figure 2c

high potency to prolonged duration of activity. By contrast, SC-9880, which was effective at 0·4 mg/day, was 20 to 25 times as active as progesterone and had an identical activity profile. SC-9022 was less intensively studied but appeared comparable. The characteristics of SC-9392 and SC-11800 differed widely from those of progesterone.

The characteristics of SC-10363, which was intensively studied in

379

Table 3

The behavioural response of the spayed ewe to ODB following no progestagen, or
progesterone or SC-10363 for 12 days
Factorial 3×3; $n=28$; $N=252$

(From J. N. Shelton *et al.*, 1967, by courtesy of Sydney University Press)

Main effect	Total ewes	Ewes in oestrus
Progestagen		
Progesterone	84	48
SC-10363	84	37
Nil	84	32
P linear		<0.01
quadratic		N.S.
Dose of ODB		
-1	84	11
0	84	41
$+1$	84	65
P linear		<0.001
quadratic		N.S.
Total	252	117

Notes: 1. ODB injected *2 days* after final progesterone; *4 days* after final SC-10363.
2. Doses of ODB. Progesterone and SC-10363–10·0, 15·6, 24·3 µg; Nil–15·6, 24·3, 38·0 µg.

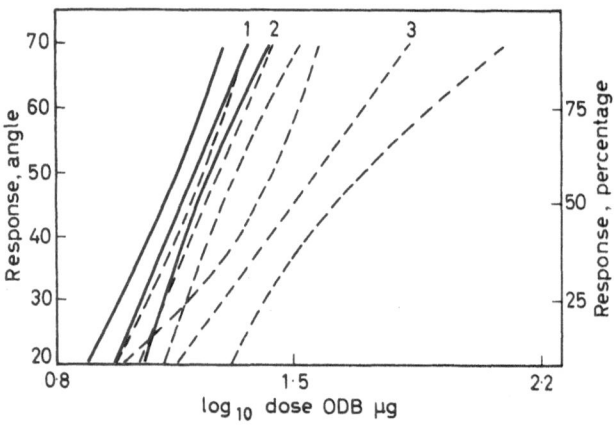

*Figure 3. The behavioural responses to ODB given 2 days after progesterone pretreatment
(1), 4 days after SC-10363 pretreatment (2), and without pretreatment (3)* (From J. N.
Shelton *et al.*, 1967, by courtesy of Sydney University Press)

5 tests, are shown in Table 3.   Significantly fewer were served of the unprimed than of the primed ewes despite the higher dose of oestrogen.   The response to ODB given *4 days* after daily priming (0·1 mg) with SC-10363 was slightly less than that when given *2 days* after progesterone $(0·05 < P < 0·10)$.   An earlier test had shown the response to be much less at 2 days than at 4 days.   The slopes of the dose-response lines following progesterone and SC-10363 were indistinguishable from one another and were steeper than that of the unprimed ewe (*Figure 3*).

Table 4

The behavioural response of the spayed ewe to ODB following SC-9880, MAP or CAP for 12 days

Two tests—Factorial $2^3 \times 3$; $n = 5$; $N = 120$
(From J. N. Shelton *et al.*, 1967, by courtesy of Sydney University Press)

Main effect	Total ewes	Ewes in oestrus	
		(1) SC-9880 MAP	(2) SC-9880 CAP
Type of progestagen			
SC-9880	60	35	42
Other	60	35	27
P		N.S.	< 0·01
Dose of progestagen (mg/day)			
0·1	60	39	32
0·4	60	31	37
P		N.S.	N.S.
Time to ODB (days)			
2	60	37	32
4	60	33	37
P		N.S.	N.S.
Dose of ODB (µg)			
15·6	40	10	9
24·3	40	25	25
38·0	40	35	35
P linear		< 0·001	< 0·001
quadratic		N.S.	N.S.
Total	120	70	69

Note: There were several significant interactions.

These data indicate:

1. That SC-10363 is highly potent as a conditioner of oestrus.

2. That it is longer acting than progesterone in that optimum response is obtained at 4 rather than at 2 days.

3. Despite manipulation of time/dose relationships in the 5 tests, no combination of time and dose gave a behavioural response *quite* as good as that obtainable following progesterone.

The characteristics of MAP and CAP, as compared with the 9α fluoro steroid, SC-9880, are shown in Tables 4, 5 and 6, which

Table 5

Significant interactions between type and dose of progestagen and time of injection of ODB

Number of ewes in oestrus $(n = 15)$

(From J. N. Shelton *et al.*, 1967, by courtesy of Sydney University Press)

Dose of progestagen (mg)	Type of progestagen and time of injection of ODB			
	SC-9880		MAP	
	2 days	4 days	2 days	4 days
0·1	10	9	10	10
0·4	13	3	4	10
Total	23	12	14	20

$p < 0·001$

illustrate the factors introduced into the test, as finally evolved. From Table 4, the overall performances of MAP and SC-9880 appear indistinguishable. However, the significant interactions in Table 5 show them to be different. MAP exhibits its greatest activity when ODB is given 4 days after cessation of treatment, whereas SC-9880 results in greatest activity at 2 days. CAP, was not as effective as SC-9880 at any combination of dose or time interval. The overall response following its use was less (Table 4), and from Table 6 it is seen to have its maximum activity at 4 rather than at 2 days, and to have a less steep dose-response line.

The characteristics of SC-9880, as compared with progesterone, are shown in Table 7. In all 6 tests in which it was studied, it behaved exactly like progesterone except that the optimum dose was

Table 6

Significant interactions between type of progestagen and time of injection and dose of ODB

Number of ewes in oestrus

(From J. N. Shelton *et al.*, 1967, by courtesy of Sydney University Press)

ODB	Type of progestagen	
	SC-9880	CAP
Time of injection $(n=30)$		
2 days	22	10
4 days	20	17
P	<0.05	
Dose of ODB (μg) $(n=20)$		
15·6	5	4
24·3	17	8
38·0	20	15
P	$\rightleftharpoons 0.05$	
Total	42	27

Table 7

The behavioural response of the spayed ewe to ODB following SC-9880 or progesterone for 12 days

Factorial 2×3; $n=24$; $N=144$

(From J. N. Shelton *et al.*, 1967, by courtesy of Sydney University Press)

Main effect	Total ewes	Ewes in oestrus
Type of progestagen		
Progesterone	72	35
SC-9880	72	36
P		N.S.
Dose of ODB (μg)		
10·0	48	5
15·6	48	27
24·3	48	39
P linear		<0.001
quadratic		N.S.
Total	144	71

Notes: 1. There were no significant interactions.
2. ODB injected *2 days* after final progesterone or SC-9880.

0·4 mg as compared with 10 mg. It combined a potency some 25 times that of progesterone with a comparable duration of activity. In this it appeared unique, with the possible exception of SC-9022 which also appeared short acting but not as potent.

THE EVALUATION OF PROGESTAGENS IN THE ENTIRE EWE

Nine progestagens (Table 2) were compared in 3 tests using 228 entire cyclic ewes for their capacity to inhibit ovulation during treatment and to release it, with oestrus, after cessation (Shelton and Robinson, 1967a). All treatments were by injection. The performances of SC-10363 and of SC-9880 were predictable from the spayed ewe data. SC-10363 was highly effective as an inhibitor of ovulation but appeared too long acting to permit effective release. SC-9880 was effective as an inhibitor of ovulation at the same dose as that required for conditioning in the spayed ewe—0·4 mg/day. The onset of oestrus and ovulation following cessation of injections was reliable and precise, with time relationships comparable to progesterone. The other progestagens tested were much less effective.

Table 8 includes the approximate potencies, relative to that of progesterone, of the 10 progestagens used. Note the consistency of the relative potency of SC-9880 as determined by the Clauberg, spayed ewe, and entire ewe tests.

Table 8

The relative potencies of various progestagens used

Progestagen	Relative potency estimated by		
	Clauberg	Spayed ewe	Entire ewe
Progesterone	1	1	1
SC-10363	25	100	25
SC-9880	20	20–25	20–25
SC-9022	25	20–25	< 20
SC-9392	10	$\ll \ll 5$	$\ll \ll 5$
SC-11800	2·5	< 5	≤ 5
SC-10230	50	No data	20–25
SC-10017	60	No data	20–25
'Enovid'	0·5	No data	< 1
'Provera' (MAP)	50	20–50	No data
'Chlormadinone' (CAP)	100	> 10	No data

COMPARISON OF INTRAMUSCULAR INJECTION AND INTRAVAGINAL APPLICATION OF PROGESTAGENS

Four tests were conducted in spayed and two in entire ewes, in which the activities of progesterone, SC-10363 and SC-9880 were compared when given by intramuscular injection or intravaginal application in oil or propylene glycol solution (Shelton and Robinson, 1967b; Shelton and Moore, 1967b).

Table 9 shows the importance of the solvent. More spayed ewes

Table 9

Ewes in oestrus following priming by intravaginal deposition of progestagen (From J. N. Shelton and T. J. Robinson, 1967b, by courtesy of Sydney University Press)

Main effect	Ewes treated	Ewes in oestrus
Type of progestagen		
Progesterone	48	17
SC-10363	48	18
P		N.S
Solvent		
Peanut oil	48	13
Propylene glycol	48	22
P		<0·05
Dose of ODB (μg)		
10·0	32	1
15·6	32	11
24·3	32	23
P linear		<0·001
quadratic		N.S.

were served following the injection of ODB when either progesterone or SC-10363 was administered intravaginally in propylene glycol than in oil $(P < 0·05)$. The steepness of the dose-response line, and the MED of 16·4 μg applicable to ODB following progesterone in propylene glycol, indicate the absorption of physiological quantities of steroid. The data for SC-9880 were less conclusive (Table 10). Fewer ewes were served following intravaginal application than intramuscular injection $(P < 0·01)$. However, the slope of the dose-response line was sufficiently steep as to indicate some absorption.

The conclusions concerning the absorption of SC-10363 from the

13+

vagina, and the importance of the solvent used for deposition, were confirmed in entire cyclic ewes (Table 11). Progesterone (10 or

Table 10

Ewes in oestrus following priming by intramuscular injection or intravaginal deposition of SC-9880

(From J. N. Shelton and T. J. Robinson, 1967b, by courtesy of Sydney University Press)

Main effect	Ewes treated	Ewes in oestrus
Method of administration		
SC-9880—i/m (oil)	60	26
SC-9880—i/vag (P.G.)	60	13
P		<0·01
Dose of ODB (µg)		
10·0	40	1
15·6	40	11
24·3	40	27
P linear		<0·001
quadratic		N.S.

Table 11

Control of oestrus in entire ewes by daily intramuscular injection or intravaginal deposition of progestagen

(From J. N. Shelton and N. W. Moore, 1967b, by courtesy of Sydney University Press)

Progestagen	Route and solvent	Ewes treated	Ovulation	
			Blocked	Released
Progesterone	i/m —Oil	48	47	44
	—P.G.	48	46	40
	i/vag—Oil	48	6	5
	—P.G.	48	20	12
SC-10363	i/m —Oil	48	47	27
	—P.G.	48	47	11
	i/vag—Oil	48	32	7
	—P.G.	48	42	20
Total		384	287	176

20 mg) injected intramuscularly in oil or in propylene glycol effectively blocked oestrus, and release within a week of cessation of treatment was good. SC-10363 (0·4 or 0·8 mg) similarly administered was equally good at blocking oestrus but release was poor. Progesterone (20 or 40 mg) intravaginally was much less effective than SC-10363 (0·8 or 1·6 mg) intravaginally and, for both progestagens, administration in propylene glycol was much more effective than in oil.

These experiments demonstrated the possibility of absorption from the vagina. Fertilization and lambing data were obtained and these confirmed the low fertility to be expected following intramuscular treatment of a long acting progestagen such as SC-10363. Fertility of ewes in oestrus following intravaginal treatment was much higher.

THE DEVELOPMENT AND PRELIMINARY TESTING OF REMOVABLE PROGESTAGEN IMPREGNATED PLASTIC INSERTS

The evidence of effective absorption from propylene glycol, but not from oil, suggested that the former solvent, being miscible with water, was absorbed leaving the insoluble steroid in a microcrystalline suspension on the vaginal wall. While never tested, this concept led to the development by the author (Robinson, 1964, 1965) of a sponge impregnated with progesterone or SC-9880. A plastic sponge was used because it has a large surface area upon

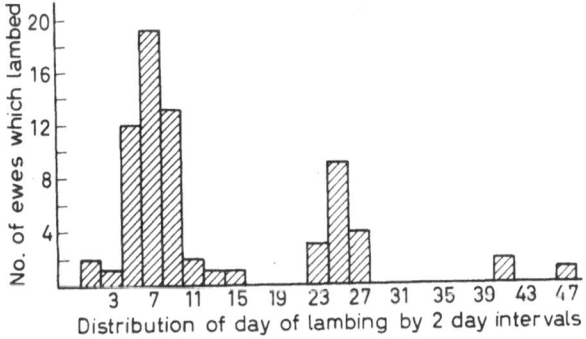

Figure 4. Distribution of lambing of all 70 ewes which lambed of the 72 treated with impregnated sponges. Test 1; Jerilderie; August–September 1964 (From T. J. Robinson, 1967, by courtesy of the Editor, Nature)

which to impregnate progestagen in a microcrystalline form, and it is inert and soft. In the first test, sponges were inserted subcutaneously or intravaginally. In the second, only intravaginal insertion was used. Tables 12 and 13 summarize the results and *Figure 4* shows the distribution of lambing.

Table 12

Results of preliminary experiments using progestagen impregnated sponges
(Data for Progesterone and SC-9880 pooled)
Natural mating
(From T. J. Robinson, 1965, by courtesy of the Editor, *Nature*)

Method of insertion	Ewes treated	Sponges intact	Oestrus		Ewes lambed
			Blocked	Released	
Subcutaneous	36	36	31	26	18
Intravaginal	36	30	29	29	22

Table 13

Results of preliminary experiments using progestagen impregnated sponges
(All sponges used intravaginally)
Artificial insemination
(From T. J. Robinson, 1965, by courtesy of the Editor, *Nature*)

Progestagen	Ewes treated	Sponges intact	Oestrus		Ewes lambed
			Blocked	Released	
Progesterone	54	46	46	30	17
SC-9880	52	51	51	51	37

THE EVALUATION OF A NUMBER OF PROGESTAGENS USED INTRAVAGINALLY IN SPONGES

The next step was the testing in intravaginal sponges of five progestagens—SC-10363, SC-9880, SC-9022, SC-11800 and 'Provera' (MAP)—each at 3 dose levels: 450 ewes were involved. This test indicated SC-9880 to give the most precise results, so this compound was then used at 4 dose levels in a second test involving 480 ewes. Ewes were artificially inseminated at the first oestrus following withdrawal of the sponges, using 0·1 ml milk diluted semen (Robinson, Moore and Holst, 1967a).

Results for these tests are summarized in Table 14. SC-9880

Table 14

Comparison of 5 progestagens used in intravaginal sponges for the blocking and release of oestrus and ovulation in cyclic Merino ewes

(Data pooled for dose of progestagen)

(From T. J. Robinson *et al.*, 1967a, by courtesy of Sydney University Press)

(90 ewes per treatment = 450)

Progestagen	Number of ewes			Percentage lambed†
	Available*	Oestrous	Lambed	
SC-10363	90	29	17	58·6
'Provera' (MAP)	89	67	29	43·3
SC-11800	85	19	8	42·1
SC-9022	85	47	20	42·6
SC-9880	85	77	38	49·4
P	—	<0·001	<0·001	N.S.
Total	434	239	112	46·9

* 16 ewes lost sponges (3·6 per cent).
† Of ewes inseminated.

Table 15

Comparison of 4 doses of SC-9880 in intravaginal sponges used in cyclic Merino ewes

(From T. J. Robinson *et al.*, 1967a, by courtesy of Sydney University Press)

(120 ewes per treatment = 480)

Dose SC-9880 (mg)	Number of ewes			Percentage lambed†
	Available*	Oestrous	Lambed	
5	117	67	21	31·3
10	116	94	36	38·3
20	118	96	45	46·9
40	119	101	43	42·6
P		<0·001	<0·001	N.S.
Total	470	358	145	40·5

* 10 ewes lost sponges (2·1 per cent).
† Of ewes inseminated.

impregnated sponges were significantly more effective than any others in the proportion of ewes in which the time of oestrus was effectively controlled (77 of 85, 91 per cent). The next most

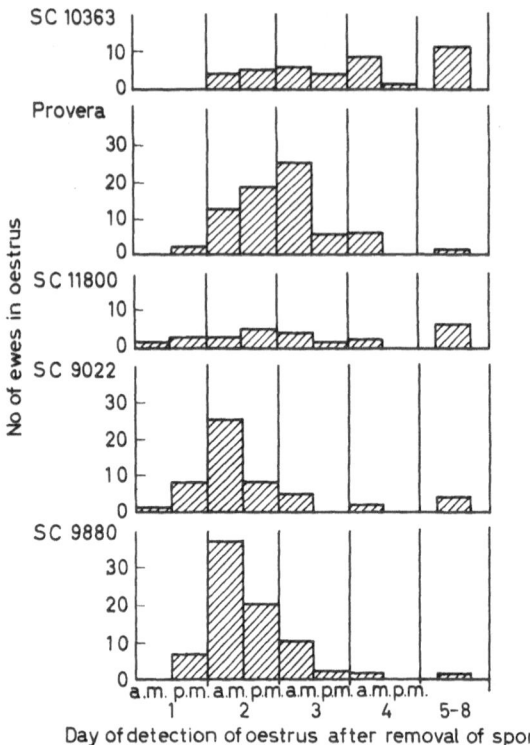

Figure 5. *The distribution of time of onset of oestrus of ewes from which progestagen-impregnated sponges were withdrawn a.m. on day 0. Three doses pooled for each progestagen* (From T. J. Robinson *et al.*, 1967, by courtesy of Sydney University Press)

effective steroid was MAP (67 of 89, 75 per cent). The time of oestrus following the use of SC-9880 was more precise than that following any other progestagen, and both the time of onset and the overall precision were affected by dose (*Figure 5*). Further, the behaviour pattern of each progestagen used in sponges followed exactly that predictable from the spayed ewe test and the various tests in entire ewes. For these reasons attention thereafter was concentrated solely upon SC-9880. Table 15 presents the main results for the next test, which involved 4 dose levels of SC-9880 in sponges.

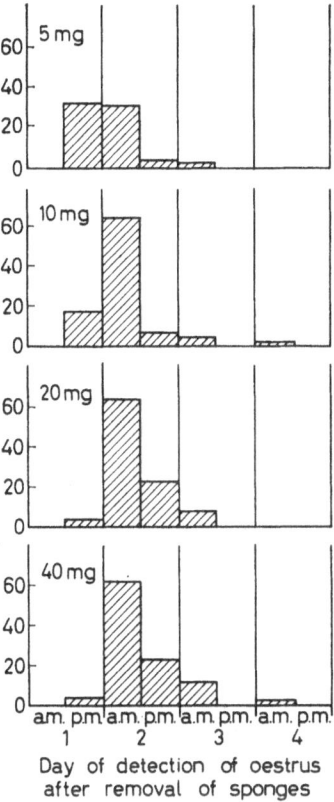

Figure 6. *The distribution of time of onset of oestrus of ewes from which were withdrawn intravaginal sponges impregnated with 4 dose levels of SC-9880. Sponges were withdrawn a.m. on day 0* (From T. J. Robinson *et al.*, 1967, by courtesy of Sydney University Press)

Doses of 10, 20 and 40 mg were effective for the blocking and release of oestrus, while 5 mg was too low. There was a highly significant effect of dose on time of onset of oestrus ($P < 0.001$). Forty-eight per cent of oestrous ewes came into heat on the first day following withdrawal of 5 mg sponges, compared with 18 per cent for 10 mg and 4 per cent for 20 or 40 mg sponges. At these higher doses,

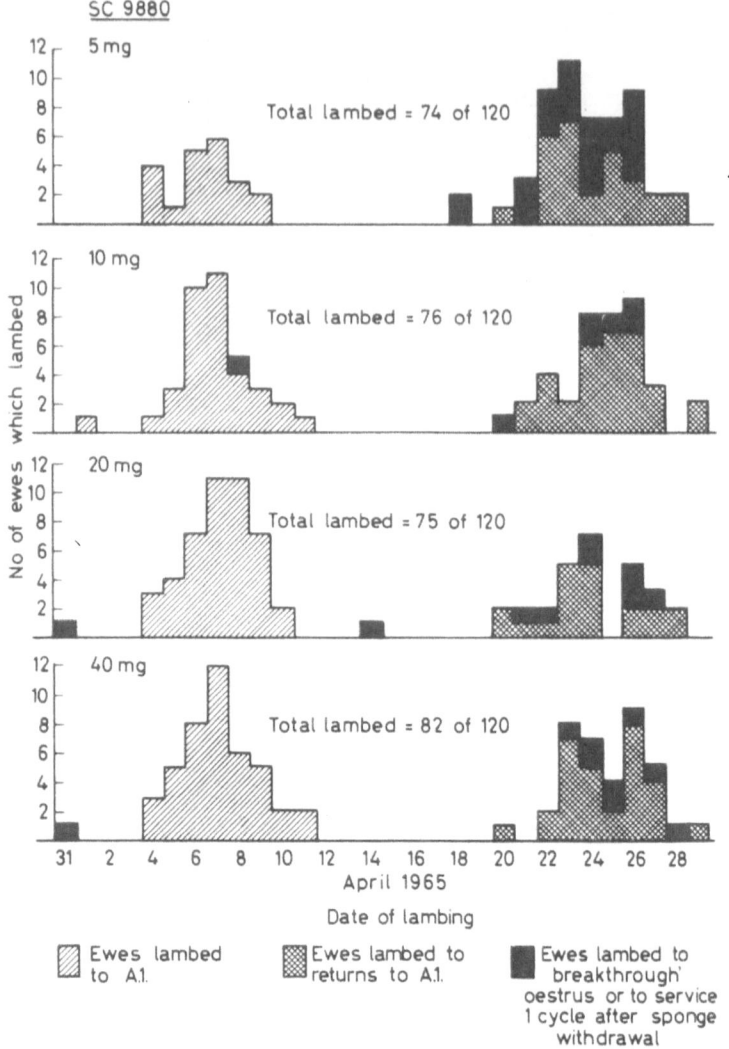

Figure 7. The distribution of dates of lambing of ewes from which were withdrawn intravaginal sponges impregnated with 4 dose levels of SC-9880. Sponges were withdrawn on 5 November, 1964 (From T. J. Robinson *et al.*, 1967, by courtesy of Sydney University Press)

83 per cent of treated ewes were served within 4 days after sponges were withdrawn, and of these, 86 per cent were detected on the second day (*Figure 6*).

The mean lambing rate following first oestrus (ewes lambed of ewes inseminated) was 46·9 per cent for the first test and 40·5 per cent for the second (Tables 14 and 15). When ewes which lambed to the second oestrus were added, the percentages increased to 66·9 and 64·0 per cent. Of the 5 progestagens compared in the first test, SC-9880 resulted in the highest (75·6 per cent) and SC-11800 the lowest (54·4 per cent) lambing ($P < 0·05$). In the second test, there was a slight but insignificant difference in the lambing rate between the highest (68·3 per cent) and the lowest (61·7 per cent) doses.

The patterns of lambing are shown in *Figure 7*. Good control was effected by most dose levels of SC-9880 and MAP and by the highest dose of SC-9022. Poor control was effected by SC-10363 and the lower doses of SC-9022, while SC-11800 effected little, if any, control. Conception rates of 40 to 50 per cent to a single insemination using diluted semen are to be expected in Merino sheep inseminated in the Spring months, and an overall fertility of 65–75 per cent of ewes lambed of all ewes in a flock following two oestrous periods is normal, under Australian conditions, for matings at this time of the year.

THE EFFECT ON THE OESTROUS CYCLE OF NON-IMPREGNATED INTRAVAGINAL SPONGES

The presence of the non-impregnated sponges had no effect on the random distribution or the percentage incidence of oestrus (45 per cent) compared with a control flock of 50 untreated ewes (50 per cent detected in oestrus). Of the 108 ewes detected during an 18 day observation period, 26 (24 per cent) were in oestrus during the 4 days following sponge withdrawal. This represents 22 per cent of the total period of observations (Moore and Robinson, 1967).

The use of sponges impregnated with 500 mg progesterone resulted in a significant increase in the total number of ewes detected in oestrus—139 v 108 ($P < 0·01$) and in a high concentration of oestrus within 4 days after treatment ($P < 0·001$).

It was concluded that the presence in the vagina of a sponge does not affect ovarian activity and so differs from that of a foreign body in the uterus (Moore and Nalbandov, 1953; Nalbandov, Moore and Norton, 1955).

THE VAGINAL FLORA FOLLOWING THE USE
OF INTRAVAGINAL SPONGES

The author is indebted to Associate Professor R. V. S. Bain of the Department of Veterinary Pathology and Bacteriology, University of Sydney, for carrying out bacteriological examinations of sponges after insertion in the vagina for 16 days. Vaginal infections have not been a problem except in maiden ewes in which sponges have been inserted forcibly. When care is used in insertion, infections are exceedingly rare; of the order of 0·1 per cent.

Aerobic cultures were made from 40 sponges on blood agar plates. Anaerobic cultures were not made as these had given no extra information in earlier tests (*see* Robinson, 1965). In addition, vaginal smears were taken from 64 ewes after sponge withdrawal, 16 following each of 4 different bactericide treatments incorporated into the sponges (Table 16).

Table 16

Vaginal smears classed positive for three types of bacteria and for polymorphs (From N. W. Moore and T. J. Robinson, 1967, by courtesy of Sydney University Press)

| Bactericide | Number of smears | | | | |
	Total	Gram + ve bacteria	Gram − ve bacilli	Strepto- cocci	Polymorpho- nuclear cells
Nil	16	16	12	4	10
Duocillin	16	13	12	2	11
Zephiran	16	5	14	1	10
Savlon	16	7	11	4	14
P	—	<0·01	N.S.	N.S.	N.S.
Total	64	41	49	11	45

Growth in culture from all sponges was heavy and there was no marked effect of any bactericide nor of the presence or absence of progesterone. The prevailing bacteria were *Corynebacteria* and Gram negative bacilli of which those tested were *Enterobacteriaceae*.

The vaginal smears showed a higher incidence of polymorpho-nuclear cells in impregnated than in non-impregnated sponges ($P < 0·001$). This was probably due to the irritating effect of the large quantity of progesterone present. It is pertinent that more of these sponges were lost than of the non-impregnated $(P < 0.05)$ and that loss of 'Savlon' coated sponges was high. Since the presence of pathogens is not a problem. it appears unnecessary to use such an ointment when inserting sponges.

THE TIME OF OVULATION FOLLOWING WITH-DRAWAL OF INTRAVAGINAL SPONGES

The time of ovulation relative to the time of onset of oestrus was determined in 180 ewes, 90 of which were subjected to laparotomy, 9 at each of 10 intervals of time. They were cyclic Merino ewes treated with intravaginal sponges at 3 dose levels. The experiment was conducted in late Summer, 1965. Laparotomies were conducted at time intervals of 0, 24, 36, 48, 54, 60, 66, 78 and 84 h after withdrawal of sponges (Robinson and Smith, 1967a).

At the time of sponge withdrawal, only one animal showed any evidence of ovarian activity. At 24 h, activity was evident in 4 animals, at 48 h in 8, and thereafter in all but one of the 54 ewes examined. Ovulation commenced between 36 and 48 h for the 10 and 20 mg doses and was general by 60 h. It commenced between 60 and 66 h for the 40 mg dose and was general by 78 h. These times correspond to ovulation occurring about 12–24 h after the onset of oestrus. There was a clear effect of dose on time of oestrus and of ovulation, and the response to 20 mg was particularly precise.

Detailed examination of the observed times of onset of oestrus and of ovulation indicated no deviation from normally accepted time relationships.

THE FERTILIZATION OF OVA RELEASED FOLLOWING WITHDRAWAL OF INTRAVAGINAL SPONGES

The proportion of ova fertilizable following synchronization of oestrus with SC-9880 impregnated intravaginal sponges was studied in two identical tests conducted simultaneously in two locations (Moore, Quinlivan, Robinson and Smith, 1967). The pattern in each test was identical and the pooled main results are shown in Table 17.

Overall, the percentage fertilized after insemination at the first oestrus—58·1—was significantly lower than at the second oestrus—79·5 $(P < 0.01)$. Further, the interaction between method of insemination and cycle was significant $(P < 0.05)$. The percentage of eggs fertilized after A.I. at the first oestrus using undiluted semen was indistinguishable from that at the second cycle with all methods of insemination (80·8 v 79·5 per cent). However, it was much lower for natural service—where the rams were used so intensively as to be producing highly diluted semen (Salamon, 1962)—and for A.I. using diluted semen.

Table 17

Fertilization of ova at first and second oestrus following withdrawal of SC-9880 impregnated intravaginal sponges

(From N. W. Moore *et al.*, 1967, by courtesy of Sydney University Press)

(Ova recovered from 23 to 28 ewes in each group)

| Cycle | Percentage ova fertilized | | |
	Natural service	A.I.— undiluted	A.I.— diluted
First	52·0	80·8	39·1
Second	78·6	77·8	82·1

Clearly, although normal fertilization rates can be achieved at the first oestrus, this is a highly sensitive situation and factors such as dilution can drastically reduce the proportion of ova fertilized.

SPERM TRANSPORT AFTER INSEMINATION FOLLOWING WITHDRAWAL OF INTRAVAGINAL SPONGES

The sensitivity of fertilization rates, particularly at the first oestrus following sponge withdrawal, raises the question of the normality of sperm transport following such treatment. This has been studied by Quinlivan and Robinson (1967). The numbers of spermatozoa recoverable *in vivo* from the Fallopian tubes of untreated control ewes were compared with those recoverable at the first oestrus after withdrawal of SC-9880 impregnated sponges.

A characteristic was enormous variability in sperm numbers, and a highly skew pattern of distribution. The mean numbers recovered from the control ewes were higher than from the treated ($P < 0.001$) due mainly to a sixfold difference at 24 h. The control ewes showed the classical pattern of increasing tubal spermatozoa with passage of time to 24 h (Mattner, 1963) followed by a marked fall at 36 h.

This investigation provided clear evidence that failure of sperm transport contributes to the sub-fertility associated with artificial insemination following the use of progestagen impregnated sponges.

RATE OF ABSORPTION OF SC-9880
FROM INTRAVAGINAL SPONGES

This phase of the investigation was conducted in collaboration with the Department of Organic Chemistry, University of Sydney. Preliminary assays of residual SC-9880 or progesterone involved drying of the sponges, ethanol extraction, application to a silica gel column,

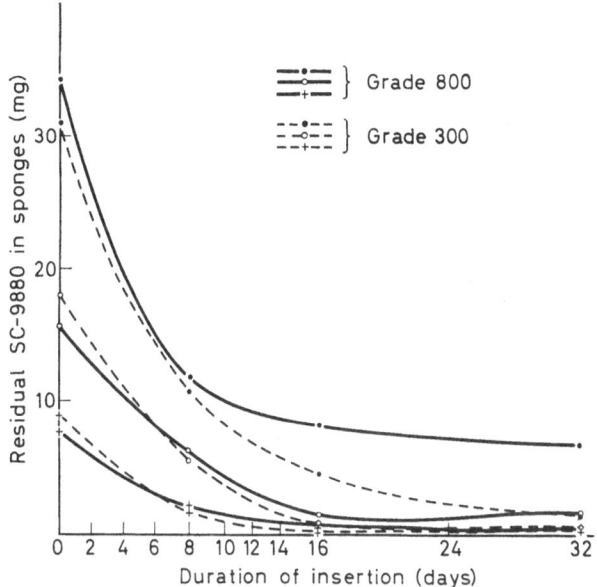

Figure 8. The quantity of residual SC-9880 estimated in sponge after 0, 8, 16 and 32 days insertion (From J. Morgan et al., 1967, by courtesy of Sydney University Press)

and subsequent elution, drying and weighing. This method was fairly satisfactory for progesterone but less so for SC-9880.

Gas-liquid chromatography (G.L.C.) proved a more accurate method of analysis and the following results were obtained using this technique. For quantities in excess of 0·5 mg, the estimated accuracy was ± 3 per cent. The accuracy was poor below 0·5 mg (Morgan, Lack and Robinson, 1967).

Figure 8 shows the curves of residual SC-9880 in sponges of two densities and impregnated with approximately 10, 20 and 30 mg steroid. These were prototype commercial sponges and the errors in the initial quantity impregnated were much greater than in sponges prepared in our laboratory. One hundred and twenty were extracted following insertion for 0, 8, 16 and 32 days. SC-9880 was absorbed less rapidly from the denser sponges and there appeared to be a slight effect of initial dose. These effects were so small however as to permit pooling of all the data for the calculation of the following regression equation:

$$y = 2 \cdot 0000 - 0 \cdot 0756x$$

where $y = \log_{10}$ percentage SC-9880 in the sponge
$\quad\quad x =$ number of days inserted.

This regression shows 16 per cent of the SC-9880 present at any time to be absorbed over a 24 h period, and the half-life of sponges to be 4 days. After a 16 day insertion period, 4 to 9 per cent of the initial dose can be expected to remain, that is, $0 \cdot 4$–$0 \cdot 9$ mg for a 10 mg sponge or $1 \cdot 2$–$2 \cdot 7$ mg for a 30 mg sponge. Hence from the sixteenth to the seventeenth day, between $0 \cdot 06$ and $0 \cdot 14$ mg is absorbed from a 10 mg sponge, and $0 \cdot 19$ and $0 \cdot 43$ mg from a 30 mg sponge.

The residual steroid present in 10 mg sponges after such a duration of insertion is marginal. A daily dose of the order of $0 \cdot 4$ mg by injection is required for optimum suppression of ovulation and conditioning for an oestrous response. By contrast, a 30 mg dose should still be providing adequate progestagen after 16 days.

EVALUATION OF THE PRACTICAL APPLICATION OF PROGESTAGEN IMPREGNATED SPONGES FOR THE ADVANCEMENT OF THE BREEDING SEASON OF THE ANOESTROUS EWE

It is clear that a sponge, impregnated with SC-9880 and inserted intravaginally, duplicates the effect of daily injections of progesterone. Hence such treatment could be expected to react with PMS injection in the anoestrous sheep in a manner similar to that observed following a course of treatment with progesterone. The following predictions therefore can be made:

1. Use of intravaginal sponges with PMS in British breeds in deep anoestrus would result in a single oestrus with ovulation in a proportion of ewes, but would not induce a series of cycles.

2. Use with PMS in British breed × Merino ewes in mid-anoestrus might result in a series of cycles in a proportion of ewes.

3. Use with or without PMS in British breed ewes in late anoestrus might initiate a series of cycles and so advance the onset of full breeding activity.

The two latter predictions have been examined in a series of 4 tests, and found to be valid.

The first test involved 421 anoestrous Merino × Border Leicester ewes, of which 366 were treated and 55 were controls. The original design involved 408 ewes in a $2 \times 2 \times 3$ factorial with $n = 34$, but 42 were found to be in advanced pregnancy and were discarded. The test involved a comparison between progesterone and SC-9880 administered by intramuscular injection (12 mg and 0·4 mg/day) or by intravaginal sponge (800 mg and 30 mg), and three times of injection of PMS (750 I.U.) relative to the cessation of progestagen treatment. It was conducted in the mid-anoestrous period of September–October, 1964.

The main results are presented as percentages in Tables 18 and 19.

Table 18

Effect of progestagen and method of administration with PMS in anoestrous Merino × Border Leicester ewes
(Results expressed as percentages—Total ewes = 366)

(From N. W. Moore and P. J. Holst, 1967, by courtesy of Sydney University Press)

Observation*	P	Progesterone		SC-9880	
		i/m	i/vag	i/m	i/vag
1. Ewes served—1st cycle	< 0·001	84·8	71·0	64·4	81·9
2. Ewes served—2 cycles	< 0·01	95·7	87·1	82·8	90·4
3. Eggs fertilized	< 0·01	64·0	41·2	25·0	70·6
4. Ewes lambed—1st cycle	< 0·01	38·0	30·5	28·7	37·0
5. Ewes lambed—2 cycles	N.S.	62·0	52·7	48·3	54·3

* Based on all ewes for 1, 2, 4, 5 ($n = 87$–94).
  Based on eggs recovered for 3 ($n = 12$–25).

Of the 366 ewes treated, 326 or 89 per cent came into oestrus, 277 immediately after treatment and an additional 49 one cycle later. Also, 100 ewes returned to service, so that a total of 149 was served at the second cycle, indicative of an initiation of cycles. Of the 277 first mated, 113 (41 per cent) lambed, and an additional 86 (58 per cent) lambed as a result of returns and first matings at the second cycle. The lambing results for two cycles thus were 199 ewes lambed of 366 treated—54·4 per cent.

The most effective treatment combinations were progesterone intramuscularly or SC-9880 in intravaginal sponges, followed by PMS, 0 or 24 h after the final injection or withdrawal of sponges.

399

Such treatments resulted in the initiation of several cycles and some 60 per cent of ewes lambed following two cycles (Moore and Holst, 1966).

Table 19

Effect of time interval to PMS in anoestrous Merino × Border Leicester ewes
(Results expressed as percentages—Total ewes = 366)

(From N. W. Moore and P. J. Holst, 1967, by courtesy of Sydney University Press)

Observation*	P	Interval to PMS (h)		
		0	24	48
1. Ewes served—1st cycle	N.S.	75·3	81·8	70·2
2. Ewes served—2 cycles	N.S.	88·4	90·1	88·7
3. Eggs fertilized	N.S.	54·5	64·3	38·1
4. Ewes lambed—1st cycle	< 0·001	39·7	41·3	12·1
5. Ewes lambed—2 cycles	< 0·05	57·0	62·0	44·4

* Based on all ewes for 1, 2, 4, 5 ($n = 121$–124).
Based on eggs recovered for 3 ($n = 21$–28).

The second, third and fourth tests used 501 British breed ewes in the late summer period of January–February, 1965.   In the second test, oestrous ewes were artificially inseminated, in the third they

Table 20

Summary of two tests using artificial insemination or hand mating of anoestrous Dorset Horn ewes treated with 10 mg SC-9880 impregnated sponges with or without 750 I.U. PMS

(Sponges removed January 4–6, 1965)
(From T. J. Robinson and J. F. Smith, 1967b, by courtesy of Sydney University Press)

Gonado-trophin	Number of ewes				Number of lambs	Percentage	
	Treated	Oestrous		Lambed two cycles		Ewes lambed	Lambs born
		First cycle	Second cycle*				
PMS	128	122	61	70	103	54·7	80·5
No PMS	126	92	77	67	85	53·2	67·5

* Returns to service + first services.

400

were hand mated to individual rams in pens, and in the fourth they were paddock mated (Robinson and Smith, 1967b).

SC-9880 impregnated sponges containing 10 mg were inserted for from 17 to 19 days. In those ewes to which PMS was given, 750 I.U. was injected subcutaneously on the day of removal of the sponges. Tables 20 and 21 summarize results from the three tests. Of 128

Table 21

Paddock mating of anoestrous Border Leicester and Southdown ewes treated with 10 mg SC-9880 impregnated sponges without PMS

(Sponges removed February 14, 1965)

(From T. J. Robinson and J. F. Smith, 1967, by courtesy of Sydney University Press)

Breed	Number of ewes				Number of lambs	Percentage	
	Treated	Oestrous		Lambed two cycles		Ewes lambed	Lambs born
		First cycle	Second cycle				
Border Leicester	14	10	?	10	13	71·4	92·9
Southdown	97	78	?	70	83	72·2	85·6

Dorset Horn ewes treated with sponges plus PMS in early January, 122 came into oestrus within 3 days and were artificially inseminated or hand mated, and 50 lambed. One cycle later, 61 were served— 5 first services and 56 returns. Hence cycles were initiated, and 70 ewes (54·7 per cent) lambed following these two cycles and produced 103 (80·5 per cent) lambs.

Of 126 Dorset Horn ewes treated at the same time with sponges but with no PMS, 92 came into oestrus within 3 days and 23 lambed. One cycle later 77 were served—13 first services and 64 returns. Again, cycles were initiated, and 67 ewes (53·2 per cent) lambed following these two cycles and produced 85 (67·5 per cent) lambs.

Of 111 Border Leicester and Southdown ewes treated with sponges in January–February, and with no PMS, 88 came into oestrus within 3 days and 62 lambed. Cycles were initiated but accurate data for services one cycle later are not available. Eighty ewes lambed (72·1 per cent) following these two cycles and produced 96 (86·5 per cent) lambs.

401

It seems that intravaginal sponges impregnated with SC-9880 can be used for advancing the breeding season. Whether or not PMS should be used will depend on factors such as the breed, nutritional status and stage of the breeding season, factors which will determine the relative depth of anoestrus. None of the flocks used in these 4 tests was in deep anoestrus. The crossbred ewes treated in September–October showed a 14·5 per cent incidence of oestrus in the 55 control ewes observed for one month. Treatment with progestagen—given either intramuscularly or intravaginally—together with PMS, resulted in cyclic activity in at least a proportion of animals. The British breed ewes were anoestrus. In each test, rams had been run with the ewes for 1–3 months prior to treatment and no oestrous activity was apparent. However, sponges were withdrawn 2 weeks after the longest day, in early January, for the Dorset Horns. This breed could be expected to be approaching breeding activity in January and although the onset of full activity was more dramatic with PMS, the use of sponges alone initiated cyclic activity so that the net result after two cycles was not very different from that when PMS was used. Similarly, the Border Leicester and Southdown ewes treated 6 weeks later with sponges, without PMS, commenced to exhibit synchronized cyclic activity.

EVALUATION OF THE PRACTICAL APPLICATION OF PROGESTAGEN IMPREGNATED SPONGES IN THE CYCLIC MERINO EWE

Twelve tests involving 12,940 Merino ewes have been conducted in three Australian States. Three were conducted in the Spring of 1964 and the remaining 9 in the Summer and Autumn of 1965. Progesterone and SC-9880 have been compared in sponges, the latter at several dose levels. All but one test have involved artificial insemination (Robinson and Moore, 1967; Robinson, Salamon, Moore and Smith, 1967b). Each test has been conducted by a member of the research team and most have had balanced factorial designs. This has enabled an analysis to be made of the effect on fertility of factors such as individual rams, day of insemination, inseminator, dose of SC-9880, and method of insemination (artificial or natural).

Tables 22 and 23 summarize the most important results from the Spring inseminations, which were conducted using pooled semen from groups of 5 rams. The SC-9880 impregnated sponges (30 mg) were much more satisfactory than the progesterone sponges (500 mg). Fewer were lost, the incidence of oestrus was higher following with-

drawal (78·9 v 62·6 per cent), and the overall lambing result following two cycles was better (75·0 v 58·3 per cent of available ewes
lambed).

Insemination of all ewes 48–52 h after withdrawal of sponges,

Table 22

Gross summary of results of Spring inseminations of Merino ewes

(From T. J. Robinson *et al.*, 1967a, by courtesy of Sydney University Press)

| Progestagen | Number of ewes | | | | Ewes lambed | | Percentage —Two cycles |
	Treated	Lost sponges etc.	Avail-able	In oestrus	A.I.	Two cycles	
1. Progesterone	1424	495	929	553	276	542	58·3
2. SC-9880	536	33	503	397	229	377	75·0
3. Progesterone	1428	212	1216	791	314	No data	No data

Table 23

Results of spring insemination of Merino ewes on the basis of method of A.I.

(From T. J. Robinson *et al.*, 1967a, by courtesy of Sydney University Press)

| Basis of insemination | Number of ewes | | | | Percentage —Two cycles |
| | Available | Lambed | | | |
		A.I.	Second cycle	Total	
Progesterone					
Undiluted—Oestrous ewes	218	71	60	131	60·1
Undiluted—Time basis*	230	84	64	148	64·3
Diluted—Oestrous ewes	238	48	83	131	55·0
SC-9880					
Diluted—Oestrous ewes	270	124	78	202	74·8
Diluted—Time basis*	233	105	70	175	75·1

* All ewes inseminated 48–52 h after withdrawal of sponges.

regardless of whether detected in oestrus or not, resulted in an overall result at least as good as insemination only of ewes when teased. There appeared to be a slight reduction in conception rate of the order of 13 per cent when the pooled semen was diluted, but this was confounded with day of insemination.

The 50 per cent conception rate, based on ewes lambed of ewes inseminated with diluted semen, obtained following the use of SC-9880 impregnated sponges is quite normal for a Spring insemination (Salamon and Robinson, 1962) as is the 75 per cent lambing rate for two cycles. The distributions of time of onset of oestrus and of lambing are shown in *Figures 9* and *10*.

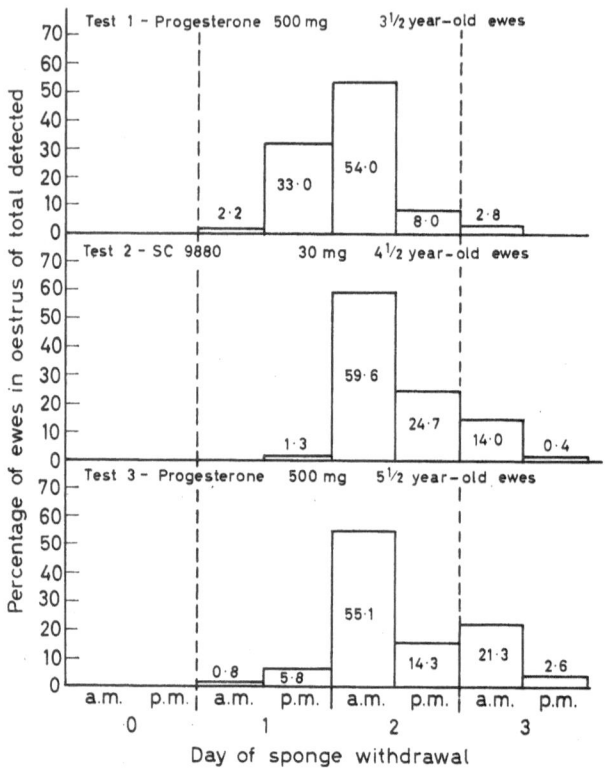

Figure 9.   The percentage distribution of time of onset of oestrus of ewes from which progestagen impregnated intravaginal sponges were withdrawn a.m. on day 0 (From T. J. Robinson et al., 1967a, by courtesy of Sydney University Press)

These were the only tests in which pooled semen was used. In subsequent tests, ewes were classed to individual rams, and as will be seen, there were enormous individual differences between rams, and conception rates to the first insemination have been lower than in these tests.

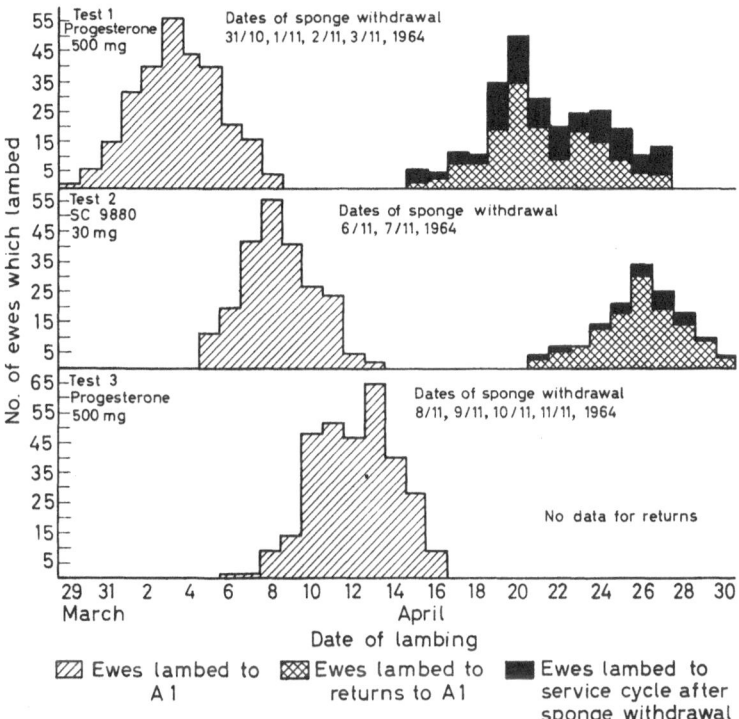

Figure 10. *The distribution of dates of lambing of ewes from which progestagen-impregnated intravaginal sponges were withdrawn over a 3 or 4 day period* (From T. J. Robinson *et al.*, 1967a, by courtesy of Sydney University Press)

Table 24 shows the distribution of time of onset of oestrus following treatment with 3 dose levels of SC-9880 in Summer and Autumn. The number of ewes detected in oestrus may be affected by the method of detection, libido of rams, and by other factors. When, for example, oestrus was detected by individual rams, the percentage detected fell from 85 per cent at the first day to 67–68 per cent at the third and fourth days $(P < 0.01)$, and there were marked differences between rams (58–93 per cent; $P < 0.001$).

405

The day of oestrus at the second oestrus following sponge withdrawal was still highly synchronized, whether in ewes which returned to service or which were unmated at the first oestrus. There was a normal distribution about day 18–19 for 10 mg, and day 19–20 for 20 mg sponges. When sponges were withdrawn over 3 or 4 days, 85 to 90 per cent of ewes came into oestrus over a 4 or 5 day period

Table 24

Percentage distribution of oestrus
Sponges withdrawn a.m. on day 0
(From T. J. Robinson *et al.*, 1967, by courtesy of Sydney University Press)

Dose SC-9880 (mg)	Number of observations	Day						Total
		1		2		3		
		a.m.	p.m.	a.m.	p.m.	a.m.	p.m.	
10	2,847	4·6	28·3	53·6	11·2	2·2	0·1	100·0
20	1,481	0·0	12·9	64·1	20·8	2·0	0·2	100·0
40	88	0·0	6·8	53·4	39·8	—	—	100·0

Table 25

Summary of lambing data—Summer and Spring inseminations of Merino ewes
(From T. J. Robinson *et al.*, 1967b, by courtesy of Sydney University Press)

Test	Ewes inseminated	Ewes lambed		Percentage lambed— Two cycles
		First cycle	Two cycles	
1. 'Wanganella' Stud A A.I.	396	64	196	49·5
Stud B A.I.	294	59	105	35·7
Stud C A.I.	479	19	—	—
Stud D A.I.	264	69	—	—
2. 'Mulureen' A.I.	1,937	328	541	27·9
3. 'Coonong' A.I.	1,452	408	871	60·0
4. 'Salisbury' A.I.	1,065	439	868	81·5
5. 'Old Cobran' A.I.	956	258	601	62·9
6. 'Ledgworth' A.I.	793	270	588	74·1
7. 'Blanket Flat'				
Natural mating	183	106	153	83·6
8. 'Hazeldean' A.I.	354	136	274	77·4
Natural mating	354	101	263	74·3
9. 'Ashrose' First cycle A.I.	247	90	152	61·5
Second cycle A.I.	210	—	168	80·0

commencing 18 days after withdrawal for 10 mg and 19 days after for 20 mg sponges.

Table 25 presents a gross summary of lambing data for two oestrous periods. The percentage of ewes lambed of ewes inseminated at the first controlled oestrus has ranged from 4·0 to 44·7 per cent. For two oestrous periods the range was 27·9 to 81·5 per cent. For natural mating, the percentages for the one test conducted were 57·9 and 83·6.

Factors introduced into the tests which affected lambing rates were individual rams, dose of SC-9880, class of ewe, and, where all ewes were inseminated on a time basis, whether or not oestrus was exhibited. In none of these tests, in which individual rams were used, were the results as satisfactory as those obtained in the Spring inseminations using pooled diluted semen.

CONCLUSIONS

The main hypothesis upon which this work has been based has been that maximum potential fertility of the ewe at a mating period depends upon a finely balanced endocrine relationship, and in particular upon a balance between progesterone and oestrogen. Any manipulative measure aimed at bringing under control the time of ovulation and oestrus must maintain this balance if full fertility is to be achieved.

The available data on ovarian hormone output and on endocrine interactions involved in the suppression and release of ovulation and in the induction of oestrous behaviour in the sheep point to alternating progesterone-oestrogen phases of the cycle. Despite the conclusion of Short, McDonald and Rowson (1963) that oestrogen is produced with progesterone during the luteal phase of the cycle, recent work using more refined techniques—made possible by gas–liquid chromatography (G.L.C.)—suggests that only progesterone is produced while the *corpus luteum* is functional. The blood level falls rapidly 2 days before oestrus, at which time oestrogen appears. The ovarian blood level of oestrogen rises rapidly, reaches a peak several hours before the onset of oestrus, and then falls rapidly (Brown, J. B. and Moore, N. W., personal communication).

A very fine balance between progesterone and oestrogen is involved in the manifestation of oestrous behaviour. Further, there is an annual rhythmic change in sensitivity to progesterone, at least in regard to inhibition of ovulation (Lamond, 1964), and to oestrogen (Raeside and McDonald, 1959; Reardon, 1959; Reardon and Robinson, 1961). A progesterone–oestrogen regime adequate to produce oestrous behaviour in the Autumn (the normal breeding season) is inadequate in the Spring. Further, the pattern of female mating behaviour has a seasonal rhythm. In the Autumn, oestrous females will actively seek out males in order to mate, whereas in the Spring they will not do so. They will only mate if sought out by the ram (Lindsay, D. R., and Scaramuzzi, R. J., personal communication).

Apart from this annual rhythmic change in sensitivity to ovarian hormones, there is evidence that environmental factors such as level of nutrition, environmental temperatures, and size of mating areas may greatly affect the manifestation of oestrus. There is ample field evidence in Australia that drought or semi-drought conditions with consequent poor nutrition may greatly reduce the incidence of oestrus in Merino flocks which are normally bred in the Spring months. While failure of ovulation could be implicated, 'silent oestrus' (ovulation without oestrus) unquestionably plays a major role. The excessively high temperatures in Western Australia probably affected the manifestation of oestrus while the work of Inkster (1957) and Lindsay and Robinson (1961) has shown the importance of mating area.

Social factors may also be important. Older ewes tend to have longer and more intense oestrous periods (McKenzie and Terrill, 1937; Lambourne, 1956) and may compete more successfully than young inexperienced ewes for available rams (Lindsay and Robinson, 1961).

These are important considerations for many reasons, among them the apparent relationship between the occurrence of oestrus and the chances of conception following forced mating or artificial insemination. The conception rate of sows in oestrus, and which are calm when inseminated—characteristic of a full oestrous response—is much higher than that of sows which are agitated (Du Mesnil du Buisson, 1961). Restall (1961) increased the lambing percentage of artificially inseminated ewes by permitting them to run with vasectomized rams for about 6 h after insemination. Despite the abundant evidence in the present investigations that ovulation occurred in the vast majority of animals on cessation of progestagen treatment, conception rates of non-oestrous animals inseminated on a time basis were much lower than those of oestrous ewes.

The concept that a finely poised endocrine balance is involved in the manifestation of a full oestrous response, associated with ovulation and efficient fertilization and subsequent development of pregnancy, lay behind the use of the spayed ewe assay test as a method

408

for screening progestagens. The results presented show the great value of this test for predicting the capacity of compounds of unknown characteristics for the blocking and release of ovulation, with oestrus, in the entire ewe. Progesterone, SC-9880, SC-10363, and 'Provera' (MAP), each of which was effective to some degree in conditioning the spayed ewe to respond to oestrogen, but which differed in optimum time relationships between cessation of progesterone treatment and injection of oestrogen, behaved in quite a predictable manner in the entire sheep whether given by daily intramuscular injection, by intravaginal deposition, or by intravaginal sponge. SC-9022 was not so intensively studied but, apart from a dose discrepancy, its performance in the entire sheep was predictable from that in the spayed ewe test. Compounds which failed to condition the spayed ewe failed also to block and release ovulation in an effective manner. The test can be used on a year round basis, with a simple adjustment of the dose of oestrogen in the Spring to allow for lowered sensitivity. The doses of progestagen can be based on data from the Clauberg test, and with those compounds of short activity (e.g., SC-9880) the relationship between the Clauberg, spayed ewe, and entire ewe tests is remarkably close. The important point about the test is its predictive capacity concerning both potency and duration of activity relative to that of progesterone.

Of the progestagens tested, the 9α fluoro compound, SC-9880, appeared unique. It coupled high potency with short duration of activity and in the whole series of experiments its physiological performance did not deviate from that of progesterone itself, except that it was 20 to 25 times more active. Other compounds, notably 'Provera' (MAP) and SC-10363, had comparable potency but they differed in their durations of activity. 'Chlormadinone' (CAP), which differs from SC-10363 only in the presence of a Cl atom instead of a CH_3 group on C_6, behaved remarkably like SC-10363 in the spayed ewe test and hence was not studied further.

These studies have not been designed to relate chemical structure with function. The relationship between structural and physiological characteristics of SC-10363 and CAP may be fortuitous. Of all the compounds tested, only two, SC-9880 and SC-9022, combined high or moderately high potency with short duration of activity, the former being some 4 to 8 times the more active. They appear quite unrelated chemically, SC-9880 being 17α-acetoxy-9α-fluoro-11β-hydroxypregn-4-ene-3,20-dione, while SC-9022 is 17β-hydroxy-21-methyl-21-methylene-19-nor-17α-pregn-4-en-3-one. The mechanism by which 9α fluoridation may increase the potency of steroids has been discussed by several authors. Mechanisms suggested have included an inhibition of metabolic inactivation (Bergstrom and Dodson, 1960) and a resistance to inactivation in the liver (Lipschutz, Figueroa, Jadrijevic and Girardi, 1957; Bush, Meigs and Hunter, 1962; Bush, 1962). Neither of these suggestions is satisfactory for SC-9880 as it does not appear to rely for its high activity in the sheep upon an inhibition of inactivation. It appears inactivated as rapidly as is progesterone. Fried and Borman (1958) have suggested that a negative inductive effect of the halogen in the 9α position increases the acid dissociation of the 11β hydroxyl group. Devine and Lack (1966) believe that this may increase the hydrogen bonding power of the hydroxyl group to an enzyme donor site. Such a mechanism could account for high *specific* activity unrelated to slow inactivation. This theory has been tested by the preparation of a series of 9α-halo-11β-hydroxyprogestagens with increasing electronegativity of the halogen atom. Three of these have been tested in the spayed ewe (Holst, P. J. and Moore, N. W., personal communication) and have been found to couple high potency with short duration of activity.

The circumstances under which absorption of steroids can take place from the vagina are important. In view of the difficulties experienced in obtaining such absorption when SC-9880 is deposited intravaginally in solution, the high rate of absorption (16 per cent/day) from sponges is a little surprising. For the first few days after insertion quite large quantities—4–5 mg/day—are absorbed from 30 mg sponges. Failure of absorption of SC-9880 from oily solutions deposited intravaginally is understandable, but not from propylene glycol solutions. The impregnated sponge provides a most effective means of administration, but the rate of absorption indicated in this study is rather too high. Modification of the method of preparation to provide larger crystal sizes should slow down absorption and provide a bigger margin for error after sponges have been inserted for a full cycle. Certainly the data on rate of absorption—available only after completion of all the tests—throw considerable light on causes of relative failure in some tests. It is clear that in many of those conducted in 1965 the doses of SC-9880 used were below optimal and that this almost certainly contributed to the high incidence of ovulation without oestrus and to the very low fertility obtained in some tests.

Table 15 shows the importance of dose of progestagen relative to the blocking of ovulation, the conditioning of the ewe to exhibit an oestrous response, and subsequent fertility of ewes following artificial

410

insemination. All doses of SC-9880 used in intravaginal sponges (5, 10, 20, 40 mg) were effective in blocking both ovulation and oestrus. Ovulation was controlled in 101, 106, 111, 116 ewes of 120 at each dose. However, the incidence of oestrus associated with the first ovulation after cessation of treatment was 67, 94, 96, 101 ewes. Hence the incidence of 'silent heat' at the time of the controlled ovulation was high at the 5 mg dose (34 ewes) and low at the 10, 20, 40 mg doses (12, 15, 15 ewes). The percentage of oestrous ewes which lambed to a single insemination with diluted semen clearly was related to the general overall effectiveness of treatment. Lambing rates were 31·3, 38·3, 46·9, 46·2 per cent for the 5, 10, 20, 40 mg doses ($0.05 < P < 0.10$). These latter data failed to attain significance but they are typical of the general pattern exhibited throughout the whole series of investigations. The minimum dose of progestagen needed to block ovulation does not result in maximum fertility at the time of the first controlled ovulation, as postulated by Lamond (1964). Ovulation can be blocked by doses of progestagen too low to condition for oestrus while an excessively high dose appears to spread the time of onset of oestrus and of ovulation without increasing the occurrence of oestrus or conception. There appears to be an optimum dose which in the case of SC-9880 used in sponges prepared as in these investigations, and inserted for about 16 days, lies between 20 and 40 mg. A 30 mg sponge is highly effective. Such a sponge can be expected to contain 2–7 mg SC-9880 after 16 days insertion, and to release 0·3–0·7 mg from steroid during day 16–17. This is a physiological dose, and it results in the majority of ewes coming into oestrus on the second day after sponge withdrawal (86 per cent of oestrous ewes) and, by simple extrapolation of the 20 and 40 mg data, ovulation can be expected between 60 and 72 h after withdrawal. From the data available there seems no advantage in increasing the dose above 30 mg.

The time of oestrus and of insemination relative to that of cessation of progestagen treatment appears to be important, regardless of whether the steroid is fed or injected or given in an intravaginal sponge. When fed or injected, maximum conception rates were obtained in ewes in oestrus and mated or inseminated 2–4 days after cessation of treatment. When given by intravaginal sponge, maximum conception rates were obtained 2 days after sponge withdrawal.

The inference is that, outside these optimum times, the progesterone–oestrogen relationship characteristic of oestrus and ovulation is sufficiently abnormal as to depress fertility. Many factors could be involved, and data on fertilization are pertinent. Normal fertility, based on fertilization of eggs and lambs born, was achieved after artificial insemination or natural service at oestrus controlled by progesterone given intramuscularly in oil. By contrast, injection in propylene glycol was followed by poor fertilization. Deposition of SC-10363 intravaginally, either in peanut oil or in propylene glycol, resulted in good fertility but intramuscular injection was followed by poor fertilization and lambing rates. Similar results involving different injection media and consequent rates of absorption of progestagen had been reported earlier (Robinson, 1960).

Further evidence of the importance of the progesterone–oestrogen relationship comes from the work of Moore and Holst (1967) with the anoestrous crossbred ewe. Two-thirds of eggs recovered from mated ewes were fertile when PMS was given 0 or 24 h after cessation of progestagen treatment and half the ewes lambed. By contrast, only 38 per cent of eggs were fertilized and 17 per cent of ewes lambed of those mated when PMS was given 48 h after cessation of progestagen.

The proportion of failure attributable to faults in the environment of the tract affecting sperm and ovum transport, fertilization, and ovum development and that attributable to faulty time relationships between oestrus and ovulation, is not clear. Generally speaking, fertilized ova have been recovered at the expected stages of development, and shifts in cell stage attributable to treatment have been slight. It is clear that, when failure has occurred, it has been due to failure of fertilization, as shown by the close parallels between fertilization rates and lambing performances. This applies equally to the cyclic ewe in which the time of ovulation and oestrus is controlled by progestagen treatment and to the anoestrous animal in which ovulation is induced with PMS following a similar period of treatment.

The data of Moore and Holst show overall fertilization and lambing rates of 53·5 and 40·8 per cent in normally anoestrous crossbred ewes, with parallel treatment effects, thus:

PMS given   0 h after progestagen: 66·7 and 52·7 per cent
  ,,   ,,   24 h after progestagen: 64·3 and 50·5 per cent
  ,,   ,,   48 h after progestagen: 38·1 and 17·2 per cent

The overall discrepancy of 12·7 per cent represents about 24 per cent of fertilized eggs, and such a loss must be accepted as normal. Brambell (1948) concludes, from an analysis of the data of Henning (1939), that approximately 40 per cent of fertilized sheep ova are lost, while Dutt (1951) estimates a 24 per cent loss. Further, in the ewe with controlled oestrus, the loss of fertilized eggs must occur early, as the distribution of oestrus in ewes which return to service is identical with that of those unmated at the first cycle. Further evidence that the reproductive tract is normal for the survival of fertilized eggs comes from the egg transfer work of Shelton and Moore (1966a). The survival rate of eggs transferred to ewes with controlled cycles was indistinguishable from that obtained in untreated ewes.

The faults contributing to lowered rates of fertilization—whether internal environment or faulty timing—appear to be largely overcome by using sufficiently large numbers of spermatozoa. The ova themselves appear quite normally fertilizable and none of our evidence supports the views of Foote and Waite (1965) concerning the incidence of abnormal ova following the control of ovulation. In the first report of the use of progestagen impregnated sponges (Robinson, 1965), excellent conception rates were obtained. A large excess of semen was used for artificial insemination (c. 0·2 ml = 800×10^6 spermatozoa) and, for natural mating, an excess of rams. It seems from the present series of tests that the norms for sperm numbers cited by Emmens and Robinson (1962) may not apply to the controlled oestrous sheep. It appears further that dilution *per se* may reduce fertility. Some of this work suggests a 12 per cent drop in fertility associated with dilution rates as low as 1 : 1–1 : 1·5. This is confirmed by the fertilization data of Moore *et al.* (1967). Ovum transport was normal and, when from $470–700 \times 10^6$ spermatozoa were used, 81 per cent of eggs recovered were classed as fertilized. When from $140–220 \times 10^6$ spermatozoa were used at a 1 : 2 dilution in heated milk the percentage was 39. For naturally mated ewes, with hard worked rams, the figure was 52 per cent. Control ewes mated or inseminated in identical fashions yielded a uniformly distributed 80 per cent of fertilized eggs. The definitive experiment designed to sort out the effects of dilution and of sperm numbers has yet to be done, but it does seem that the use of diluted semen at the first controlled ovulation results in a marked reduction in the rate of fertilization. Dilution may be done artificially as in an artificial insemination programme, or it may be the result of excessive use of rams following on synchronization of oestrus. Salamon (1962) has shown that the ejaculate of a ram used 11 times a day is comparable in volume and sperm numbers to a sample used for artificial insemination after about a 1 : 3 dilution.

An altered pattern of sperm transport is involved in the failure of fertilization of the controlled oestrous ewe. Not only are fewer tubal sperm present at 24 h after insemination—the approximate time of fertilization—than in the normal ewe, but the distribution of numbers present differs markedly. Fifty-five per cent of the control ewes had > 6,400 sperm present as compared with only 5 per cent of the treated animals. By contrast, 28 per cent of the treated ewes had < 200 tubal sperm as compared with only 10 per cent of normal animals. These differences in distribution are highly significant.

The enormous variation between rams in the fertilizing capacity of their semen following dilution cannot be explained. No semen was used unless it appeared satisfactory for density and motility. Variation was characteristic of all tests in which individual rams were used and was greater at the first controlled oestrus than at the second. Pooling of the semen of several rams prior to dilution evened out the variation and gave levels of fertility nearer the upper than the lower limit of individual rams. Hence the inability of some semen to withstand the handling and dilution techniques is a negative characteristic and it seems that the inclusion of such semen in a pooled sample does no positive harm to the sample. Inability of the semen of some bulls to withstand dilution and freezing is a well known phenomenon (*see* Melrose, 1962) and such bulls are culled from cattle artificial insemination programmes. The commonly accepted norms for fertility of bulls used in such programmes are based therefore on highly selected and tested individuals. It would be interesting to know the conception rates obtainable in a cattle artificial insemination programme using bulls selected at random and used without prior testing for fertility. An intensive study of factors which affect fertilizing capacity of ram semen is urgently required.

Several of the tests permitted valid comparisons between two experienced inseminators. Overall, the performance of each was identical but differences between inseminators on different days did occur. Day to day variations occurred also in the percentage of ewes detected in oestrus by the teaser rams and in the mean conception rate of ewes inseminated. These are important factors in the planning and execution of experiments such as those reported here.

Remaining conclusions deal primarily with the technology of the use of impregnated sponges. They owe their effectiveness to the continual release of progestagen at a rate which is uniform relative to the quantity present in the sponge at any one time. A non-impregnated sponge in the vagina does not affect the oestrous cycle. Absorption through the vaginal wall does not depend upon the use of a carrier such as the base of an ointment, and is quite effective when the sponges are inserted dry. Vaginal infection is not a problem, and only some 1 in 500 to 1,000 ewes have had serious vaginal infections on sponge withdrawal. The use of an ointment is not necessary and may, in fact, be a disadvantage in that excessive loss of sponges may be associated with its use. Impregnation of sponges with a broad spectrum bacteriostat would probably be an advantage. Another possible technical improvement, currently being studied in this laboratory, would be to irrigate the vagina with a suitable mild disinfectant. This could have the dual effect of controlling the vaginal flora and of washing out residual progestagen, so providing a more uniform end-point to treatment.

The rate of absorption—16 per cent per day—from the present sponges is excessively high and wasteful. A slower rate would result in more uniform treatment and a bigger margin for error. Thus, a 30 mg sponge releasing 8 per cent steroid per day would release about 2·4 mg during the first day, 1·2 mg during the eighth, and 0·7 mg during the sixteenth day, by which time about three-quarters of the total would have been absorbed. The use of different sponge materials and methods of impregnation should be investigated, together with further work on sponge sizes.

There are considerable breed differences in the anatomy of the vagina, which in maiden ewes of some British breeds is exceedingly difficult to penetrate. An absolute prerequisite for the successful use of the sponge technique is that the vagina must not be damaged during insertion and removal. This poses a problem in maiden ewes of some breeds. If the sponges are not fully inserted into the anterior vagina they fall out. If they are fully inserted, vaginal damage may occur. The use of small sponges and applicators and the testing of alternative sponge materials is necessary. The texture of the sponge appears to be important in relation to retention. Losses of progesterone impregnated sponges are much higher than those of sponges with SC-9880, due presumably to their much harder texture caused by the large quantity of progesterone which must be incorporated.

Although further work is still required, it can be concluded that the use of intravaginal sponges impregnated with an appropriate dose of a suitable progestagen, for example, 30 mg of SC-9880, is of practical value. It is a most valuable research tool and has a place in practical sheep breeding. The action of such a sponge left in place for one cycle, 16 or 17 days, is very similar to that of daily injections of progesterone. The release of ovulating hormone from the pituitary is inhibited while the sponge remains in place and the normal action of progesterone in sensitizing the animal to respond to oestrogen is duplicated. The effect on the reproductive tract needs to be studied. Following withdrawal of the sponge the pituitary inhibition is removed. In mature sheep in the breeding season, ovulation with oestrus occurs with remarkable precision. This is particularly valuable for an artificial insemination or hand-mating programme. In immature sheep in which 'silent oestrus' is common, and in mature Merino ewes in the spring months, the incidence of oestrus with ovulation is greatly increased. Mature crossbred ewes in anoestrus may be induced to ovulate with 750 I.U. PMS given at, or shortly after, the time of sponge withdrawal and about 50 per cent have conceived to natural service. Regular cycles may be induced in a high proportion. Data for British breed ewes in deep anoestrus are not available but it could safely be assumed that they would react to progestagen sponge-PMS in the same way as to progesterone injection-PMS treatment. That is, there would be a single ovulation-heat period and no initiation of cycles. The breeding season of such ewes may be advanced by treating in late anoestrus, and PMS may not be necessary. Progestagen treatment and withdrawal when the annual curve of reproductive activity is on the up-grade seems to result in a gonadotrophin 'surge' and the initiation of a series of synchronized cycles. This has a great potential value in practice in that the majority of ewes in a flock can be brought suddenly into full breeding activity at a time when normally only sporadic activity would be evident. The consequent uniform and predictable lambing dates should assist in rearing and selection programmes.

Fertility at the first oestrus still poses a problem. 'Normal' fertility *can be* but not always *is* achieved. This raises the question of what is 'normal' fertility, particularly that associated with a large scale artificial insemination programme. Pending further work on factors involved in the sub-fertility associated with the first ovulation after sponge withdrawal, it is suggested that inseminations be delayed until the second cycle, if a high initial conception rate is required and it is necessary to use diluted semen. For small scale operations using undiluted semen or natural mating, the first oestrus should be used. The first oestrus may also be used for large scale operations using diluted semen, if a conception rate of 30–40 per cent is acceptable. In this case it can be done on a time basis, 48 h after sponge withdrawal. All ewes should be inseminated regardless of whether or not they have been teased. This technique is simple and involves a minimum of drafting. It may be possible to eliminate the teasers altogether but this must await further study on the possible relationship between fertility following artificial insemination and the act of mating with the sterile teaser rams.

ACKNOWLEDGEMENTS

The author acknowledges the financial assistance received for these investigations from the University of Sydney, the Trustees of the McCaughey Memorial Institute, the Sheep and Wool Research Committee of the Australian Wool Board, and G. D. Searle and Co. Ltd. The majority of the experimental steroids used were graciously donated by Searle.

DISCUSSION

The discussion follows the paper by W. Hansel on page 439.

REFERENCES

Bergstrom, C. G. and Dodson, R. M. (1960). *J. Am. chem. Soc.* **82,** 3479–3480

Brambell, F. W. R. (1948). *Biol. Rev.* **23,** 370–407

Bush, I. E. (1962). *Pharmac. Rev.* **14,** 317–445

— Meigs, R. A. and Hunter, S. (1962). *J. Endocr.* **24,** ii–iii

Devine, A. B. and Lack, R. E. (1966). *J. chem. Soc.* 1902

Du Mesnil du Buisson, F. (1961). *Annls Zootech.* **10,** 57–67

Dutt, R. H. (1951). *J. Anim. Sci.* **10,** 1075

Emmens, C. W. and Robinson, T. J. (1962). In: *The Semen of Animals and Artificial Insemination,* pp. 205–251. Ed. by J. P. Maule. C.A.B. Tech. Comm. No. 15.

Foote, W. C. and Waite, A. B. (1965). *J. Anim. Sci.* **24,** 151–155

Fried, J. and Borman, A. (1958). *Vitams. Horms.* **16,** 303–374

Hammond, J. (jun.) (1961). In: *Control of Ovulation,* pp. 162–176. Ed. by C. A. Villee. Oxford, London, N.Y., Paris; Pergamon Press

Henning, W. L. (1939). *J. agric. Res.* **58,** 565–580

Inkster, I. J. (1957). *Proc. N.Z. Soc. Anim. Prod.* **17,** 72–76

Lambourne, L. J. (1956). *Proc. Ruakura Fmrs'. Conf. Week* 1956, 16–20

Lamond, D. R. (1964). *Anim. Breed. Abstr.* **32,** 269–285

Lindsay, D. R. and Robinson, T. J. (1961). *J. agric. Sci.* **57,** 137–140, 141–145

— Moore, N. W., Robinson, T. J., Salamon, S. and Shelton, J. N. (1967). *In the press**

Lipschutz, A., Figueroa, S., Jadrijevic, D. and Girardi, S. (1957). *Nature, Lond.* **180,** 508–509

McKenzie, F. F. and Terrill, C. E. (1937). *Res. Bull. Mo. agric. Exp. Stn.,* No. 264

Mattner, P. E. (1963). *Aust. J. biol. Sci.* **16,** 688–694

Melrose, D. R. (1962). In: *The Semen of Animals and Artificial Insemination,* pp. 205–251. Ed. by J. P. Maule. C.A.B. Tech. Comm. No. 15

Moore, N. W. and Holst, P. J. (1967). *In the press**

— Quinlivan, T. D., Robinson, T. J. and Smith, J. F. (1967). *In the press**

— and Robinson, T. J. (1967). *In the press**

Moore, W. W. and Nalbandov, A. V. (1953). *Endocrinology* **53,** 1–11

Morgan, J., Lack, R. E. and Robinson, T. J. (1967). *In the press**

Nalbandov, A. V., Moore, W. W. and Norton, H. W. (1955). *Endocrinology* **56,** 225–231

Quinlivan, T. D. and Robinson, T. J. (1967). *In the press**

Raeside, J. I. and McDonald, M. F. (1959). *Nature, Lond.* **184,** 458–459

Reardon, T. F. (1959). 'Quantitative Studies of Oestrogens in the Ewe.' *M.Sc.Agr. Thesis,* Univ. Sydney.

— and Robinson, T. J. (1961). *Aust. J. agric. Res.* **12,** 320–326

Restall, B. J. (1961). *Aust. vet. J.* **37,** 70–72

Robinson, T. J. (1954a). *Endocrinology* **55,** 403–408

— (1954b). *Nature, Lond.* **173,** 878

— (1960). *Proc. N.Z. Soc. Anim. Prod.* **20,** 42–53

— (1964). *Proc. Aust. Soc. Anim. Prodn.* **5,** 47–52

— (1965). *Nature, Lond.* **206,** 39–41

— and Moore, N. W. (1956). *J. Endocr.* **14,** 97–109

— — (1967). *In the press**

— — and Binet, F. E. (1956). *J. Endocr.* **14,** 1–7

— — and Holst, P. J. (1967a). *In the press**

— Salamon, S., Moore, N. W. and Smith, J. F. (1967b). *In the press**

— and Smith, J. F. (1967a). *In the press**

— — (1967b). *In the press**

Salamon, S. (1962). *Aust. J. agric. Res.* **13,** 1137–1150

— and Robinson, T. J. (1962). *Aust. J. agric. Res.* **13,** 52–68

Shelton, J. N. (1964a). *Proc. Aust. Soc. Anim. Prodn.* **5,** 43–46

— (1964b). 'An Investigation into Factors affecting Fertility of the Sheep with particular Reference to Artificial Insemination'. *Ph.D. Thesis,* Univ. Sydney

— (1965). *Nature, Lond.* **206,** 156–158

— and Moore, N. W. (1966a). *J. Reprod. Fert.* **11,** 149–151

— — (1967b). *In the press**

— and Robinson, T. J. (1967a). *In the press**

— — (1967b). *In the press**

— — and Holst, P. J. (1966). *In the press**

Short, R. V., McDonald, M. F. and Rowson, L. E. A. (1963). *J. Endocr.* **26,** 155–169

* In: *The Control of the Ovarian Cycle in the Sheep. A Research Monograph.* Ed. by T. J. Robinson. Sydney University Press, 1967.

CONTROL OF THE OVARIAN CYCLE IN CATTLE

WILLIAM HANSEL

Cornell University, Ithaca, New York

INTRODUCTION

A_RAPIDLY 'exploding' world population and an increasing con-
sumer preference for leaner cuts of meat demand rapid improvement
in the efficiency of rates of gain in our beef cattle.   The development
of larger animals with leaner carcasses containing more protein and
less fat seems imperative.   Accomplishment of these goals within a
minimum time requires the use of artificial insemination techniques
and maximal utilization of semen from highly selected bulls transmitting the desired characteristics.   However, progress in applying
artificial insemination to beef cattle has been extremely slow because
of the practical problems of carrying out insemination under range
conditions and the low conception rates frequently encountered in
artificially bred beef animals.   Clearly, the development of oestrous
cycle synchronization which can be used under practical management conditions could play a major role in facilitating the use of
artificial insemination and improved selection techniques in the beef
cattle industry.   It seems clear that the greatest potential application of artificial cycle regulation is in beef cattle.

The major application of cycle regulating techniques in areas
where dairying is a highly developed industry is likely to be limited,
for the most part, to heifers.   Many dairy heifers are currently bred
naturally while at pasture to bulls of questionable genetic merit.
The resulting calves are usually culled, and thus contribute nothing
to genetic progress.   Since the average cow in the large commercial
dairy areas produces only 3·5–4·0 calves in her lifetime this is
obviously a wasteful process.   Less rigid culling can be carried out
because of the reduced number of calves from which to select.   If
heifers can be grouped according to age and size and their cycles
synchronized they can be conveniently bred artificially to bulls of
known genetic merit.

Oestrous cycle synchronization should prove particularly useful in
underdeveloped areas of the world where transportation and re-
frigeration facilities are limited and husbandry practices are poorly

developed. Artificial insemination, as usually practiced, is difficult under these conditions. Synchronization of the cycles of all of the breeding animals within a locality for insemination at a predetermined time would appear to offer many advantages under these conditions. In fact, even under conditions of intensive husbandry there are real advantages from the standpoint of efficient labour utilization in being able to predict when oestrus will occur in groups of cattle.

Inability to obtain sufficient numbers of recipient animals at the proper stage of the oestrous cycle has been a major deterrent to experiments involving ovum collection, culture, and transfer in cattle. The availability of successful methods for synchronizing the cycles of relatively large numbers of animals may provide the impetus for more experimentation in this area. It should also be noted that the precision of many other types of experiments might be improved by using animals whose oestrous cycles had previously been synchronized.

EARLY ATTEMPTS AT SYNCHRONIZATION

It has long been known that enucleation of the bovine corpus luteum by manipulation of the ovary through the rectal wall results in oestrus and ovulation in a high percentage of the treated animals within 3–5 days. Although this treatment is subject to the criticisms that most of the 'persistent' corpora lutea removed in this way in practice are not really persistent and that adhesions are likely to develop subsequently in and around the oviducts of many treated animals, the fact remains that corpus luteum removal is an effective method for recycling cattle. The procedure may be said to represent the first attempt at artificial oestrous cycle regulation. The fertility of animals bred at the first oestrus after corpus luteum removal is quite variable, but many reports indicate that 50 per cent, or more of the treated animals may conceive.

Many attempts were made between 1948 and 1960 to regulate the oestrous cycle by subcutaneous injections or implants of progesterone, either alone or in combination with gonadotrophic hormone preparations. These attempts have been adequately reviewed on previous occasions (Anderson, Shultz and Melampy, 1964 and Hansel, 1959) and only a few of these papers need be cited. Following the original demonstration by Dutt and Casida (1948) of synchronization of oestrus in ewes by daily injections of progesterone during the breeding season, several investigators carried out similar studies in cattle. Ulberg, Christian and Casida (1951) showed that daily injections of 25–50 mg of progesterone effectively suppressed oestrus and ovulation until 3–6 days following the last injection. Trimberger and Hansel (1955) obtained similar results with daily subcutaneous injections of 50, 75, or 100 mg of progesterone. However, these workers also reported a marked reduction in the conception rate of cattle bred at the first post-treatment oestrus (12·5 per cent) as compared to control animals (64 per cent), and an increase in certain ovarian abnormalities associated with first oestrus.

Many additional experiments (*see* Anderson *et al.*, 1964), carried out prior to 1960, verified and extended these observations. For example, Nellor and Cole (1956) gave beef heifers single injections of 560 mg of progesterone in a starch suspension and followed this with single injections of equine gonadotrophin 15 days later. Oestrus and ovulation occurred 1–4 days after the gonadotrophin injection, but the fertility of the heifers treated in this way was low (17 per cent). Similarly, Ulberg and Lindley (1960) injected 0·5–10·0 mg of oestradiol benzoate into beef heifers soon after the end of a series of progesterone injections. Again, most of the animals returned to oestrus within a few days after the oestrogen administration, but no improvement in conception rate was noted in the oestrogen-treated animals.

ORAL PROGESTATIONAL COMPOUNDS AND SYNCHRONIZATION

The first reports of synchronization of oestrous cycles in ruminants by orally effective progestational compounds appeared in 1960. Hansel and Malven (1960) reported effective synchronization of the cycles of 32 Hereford cows after feeding 500 to 968 mg of 6-methyl-17-acetoxyprogesterone (MAP)* per animal per day for 20 days. Only 25 per cent of the treated animals conceived when bred at the synchronized oestrus, a result probably related to the excessive dose levels used. Nellor, Ahrenhold and Nelson (1960) also reported inhibition of oestrus and ovulation when 0·4–1·8 mg MAP/kg of body weight was fed to beef heifers for periods of 15–20 days. Oestrus occurred 4–5 days after the end of treatment with 0·4 mg and the time from the end of treatment to oestrus increased as the dose increased. Conception rates of the treated animals were not reported. This work, as well as the early attempts to synchronize

* Kindly supplied by the Upjohn Co., Kalamazoo, Michigan.

the oestrous cycles of swine (Nellor, 1960) was undoubtedly influenced by the earlier reports of Pincus and his associates and others (*see* Hartman, 1963) on the use of orally active steroids as ovulation inhibitors in rabbits, mice and humans.

A surprising number of papers concerning the use of orally active steroids of this type for oestrous cycle synchronization of cattle, sheep and swine have appeared during the intervening 5 years, and work on the problem seems to be continuing at an accelerated pace. Much of this work has been summarized in the *Proceedings of a Conference on Oestrous Cycle Control in Domestic Animals* held at the University of Nebraska in 1964.† Numerous steroid and non-steroid compounds have now been tested for their ability to synchronize oestrous cycles and, more recently, unique methods for administering the necessary drugs by single applications have been devised. The first compound studied (MAP) has been placed on the market and others may be expected to reach the market in the future.

CYCLE REGULATION IN DAIRY HEIFERS

The first reports of successful synchronization in dairy heifers fed MAP appeared in 1961 (*see* Hansel, 1961). Later Van Blake, Brunner and Hansel (1963) reported successful synchronization after feeding a more potent progestational agent, 6-chloro-Δ^6-dehydro-17-acetoxyprogesterone (CAP)* to Holstein heifers. In summarizing the results of these and other studies with dairy heifers Zimbelman (1965) and Hansel (1965) concluded that the conception rates in heifers bred artificially at the synchronized oestrus are rather variable (26–70 per cent), but generally below the rates obtained in comparable untreated heifers bred artificially under optimal conditions. Heifers failing to conceive when bred at the synchronized oestrus returned to oestrus after normal cycle intervals and normal conception rates were obtained.

The minimal effective dose for MAP has been established as 180–200 mg per animal per day. CAP is approximately 20 times more potent; only 10 mg per animal per day is required. The optimal length of the hormone feeding period appears to be 18 days.

Heifers may be fed either MAP or CAP individually by adding the drug to, or mixing it with, the animals daily grain ration. Group feeding of heifers is also possible, provided the drug is mixed with a

† U.S. Department of Agriculture Miscellaneous Publication 1005, 1965.
* Kindly supplied by the Syntex Laboratories, Palo Alto, California and Eli Lilly Laboratories, Greenfield, Indiana.

minimum of 1·4 kg of grain per animal per day and adequate feeder space is provided. Consistent daily intakes of either progestational compound is an important factor in both the degree of synchronization obtained and the fertility of the treated animals.

CAP effectively synchronized the oestrous cycles when fed to 3-year old lactating cows (Van Blake *et al.*, 1963); similar results might be expected with MAP feeding.

The results that might be expected as a result of feeding an oral progestogen of this type in rather carefully controlled field trials are illustrated in Table 1. The 41 Holstein heifers involved were

Table 1

Conception rates in Holstein Heifers synchronized by feeding MAP and bred artificially on two consecutive days

Location of heifers	No. of heifers	No. pregnant after insemination on days 3 and 4 post-treatment	No. pregnant after insemination at the following oestrus
Farm A	21	16 (76·2%) *	20 (95·2%)
Farm B	20	9 (45·0%)	15 (75·0%)
Total	41	25 (61·0%)	35 (85·4%)

* Includes two heifers that were noted in oestrus and rebred on the fifth and seventh days after the end of the feeding period.

located at two farms in the same area. Rectal palpations made on all heifers prior to the trial indicated that all had been in oestrus at least once prior to the trial and that all had normal reproductive tracts. The 21 heifers at farm A were group-fed MAP at the rate of 200 mg per heifer per day for 18 days in an average of 1·4 kg of a grain mixture consisting of ground corn (83 per cent), soybean oil meal (15 per cent), and minerals (2 per cent). At the same time the heifers at farm B were fed the same amount of MAP individually for the same length of time. The MAP, in this case was simply placed on top of the grain ration each day. All heifers were artificially inseminated on the third and again on the fourth day after cessation of the drug feeding, regardless of whether or not they were in oestrus. In addition, any animals seen in oestrus on days 5, 6 or 7 were bred again. Heifers returning to oestrus approximately one cycle later were rebred artificially, and all animals were checked for pregnancy by rectal palpations 74 days after the initial inseminations. The feeding of MAP was carried out by the dairymen, and local breeding technicians made all inseminations.

The conception rate obtained in the heifers on farm A was quite good and the results suggest that there may be some advantage to inseminating on two successive days. However, the results in the heifers on farm B are no better than those usually obtained after single inseminations of the synchronized animals. There were no obvious causes for the different conception rates at the two farms, although the owner of the heifers at farm A did find one heifer in oestrus on the fifth and one on the seventh day after the end of the feeding period. These two heifers were rebred and conceived. It may be practical to inseminate dairy heifers in this way with only a minimum of time spent in checking for oestrus.

CYCLE REGULATION IN BEEF CATTLE

Zimbelman (1965), among others, has pointed out the wide variations that have been reported in beef cattle bred at synchronized oestrus and emphasized the need for comparisons of compounds or procedures to be made within the same experiment and with adequate numbers of animals. An experiment in which an attempt was made to meet these criteria was recently reported by Hansel, Donaldson, Wagner and Brunner, 1966.

The objectives of the experiment were (1) to compare the effects of feeding MAP and CAP in pellets or in liquid feed on oestrous synchronization and subsequent fertility after artificial insemination with frozen semen, and (2) to study the effect on fertility of oxytocin injected at the time of insemination. Four-hundred and thirty-two crossbred beef cows were randomly assigned to 36 groups in a $3 \times 2 \times 2 \times 3$ factorial experiment consisting of three treatments (MAP, CAP, control), oxytocin or no oxytocin, pellets or molasses-urea liquid feed and three bulls.

MAP was fed at an average level of 240 mg per animal per day for 18 days. CAP was fed at an average level of 10 mg per animal per day for 18 days. The liquid ration contained water, molasses, urea, salt, fish oil, corn screenings, kelp, dicalcium phosphate, a trace mineral mixture, terramycin, phenothiazine and a vitamin supplement. Ground corn cobs fed free choice served as source of roughage. The CAP and MAP were dissolved in minimal amounts of ethanol and added directly to the liquid mixing vat each day. The amount of the liquid ration fed was adjusted so that complete consumption was obtained in each lot each day. The pellets, which were fed in addition to the liquid ration, consisted of:

Pellets	Per cent
Molasses	5·0
Soybean meal	7·0
A trace mineral concentrate	0·05
Salt	0·5
Limestone	0·5
Dicalcium phosphate	0·5
Wheat middlings	17·5
Bran	10·0
Hominy	19·95
Oat feed	20·0
Alfalfa meal	19·0

The MAP and CAP were thoroughly mixed into the feed prior to pelleting so that 1·63 and 1·77 kg respectively, of pelleted feed contained the required daily amounts of hormone for each animal. The cows were group-fed and the control animals received 1·86 kg of pelleted feed containing no drug per animal per day. Care was taken to provide adequate feeder space for both the pellets and the liquid feed.

All cows were ear tagged prior to the experiment and accepted if they were not pregnant and the reproductive organs appeared normal. Animals with gross abnormalities of the uterus were excluded from the experiment, but animals with apparently cystic follicles or corpora lutea were not discarded. The incidence of cystic follicles was less than 1 per cent. Many cystic corpora lutea appear to contain nearly normal amounts of progesterone, although the average content is significantly reduced (Hansel, 1964).

All cows had calved 50 days or more prior to the beginning of the feeding period, and the calves were removed from the cows prior to the beginning of the experiment.

Oxytocin* was injected subcutaneously at a rate of 40 U.S.P. units per cow immediately after insemination. Frozen sement from three Charolais bulls (bulls A, B and C) was used. All inseminations were made by one technician using standard procedures. The control animals were bred during a 20-day period beginning 13 July 1964; the MAP and CAP fed groups came into oestrus and were bred during a 7-day period beginning 23 July 1964. Cows returning to oestrus were rebred during a 10-day period beginning 11 August,

* Armour's POP kindly supplied by M. E. Davenport, Armour Pharmaceutical Co., Kankakee, Ill.
† Obtained from the American Breeders Service, Madison, Wisconsin.

1964, and all cows were checked for pregnancy 9–10 October, 1964. The results of this experiment are summarized in Table 2 and *Figure 1.* A total of 354 of the 432 cows came into oestrus and were

Table 2

Oestrous cycle synchronization results and conception rates in beef cattle fed orally active progestogens (1964)

Treatment	No. of cows	No. synchronized or bred	Percentage of synchronized cows pregnant		Percentage of all cows pregnant	
			To first breeding	After second breeding	To first breeding	After second breeding
MAP	136	117	48·7	73·5	41·9	68·9†
CAP	138	115	34·8	72·2	28·9	67·4†
Controls	134	122*	59·8	67·2	54·4	61·9‡
Total	408	354	48·0	70·9	41·6	66·0

* Bred during a 20-day period.
† Includes cows not seen in oestrus during the synchronization period and that came in oestrus and were bred for the first time one cycle later.
‡ Excludes cows that failed to conceive to first breeding but did not return to oestrus during the 10-day period during which the second breedings were carried out.

bred during the experimental period. Twenty-four of the remainder lost their identification tags during the experiment, 50 were not seen in oestrus for the first service and 4 proved to be pregnant from a previous breeding. Thus, excluding the 24 cows on which identification was lost, 117 of 136 MAP-fed cows and 115 of 138 CAP-fed cows came into oestrus during the synchronization period. Some of the MAP and CAP-fed cows not seen in oestrus during the synchronization period came in oestrus approximately one cycle length later, indicating that they had 'silent' ovulations during the synchronization period.

The distribution of animals in oestrus on each day for each treatment is shown in *Figure 1.* Twelve control cows failed to come into oestrus during the 30-day period allotted for their breeding.

MAP gave slightly better synchronization than CAP; 75·2 per cent of the synchronized cows were in oestrus by the end of the third day after withdrawal of MAP as compared with 54·8 per cent of the CAP-fed cows. The occurrence of oestrus in CAP-fed cows tended to have a bimodal distribution. More of the cows fed pellets were in oestrus on the second day after hormone withdrawal than those fed the hormone in the liquid ration.

Pregnancy rates for MAP, CAP and control groups are given in Table 2, expressed as a percentage of cows synchronized and as a percentage of total cows in each treatment group. Conception rate at first service in the MAP-fed cows (48·7 per cent) was not signifi-

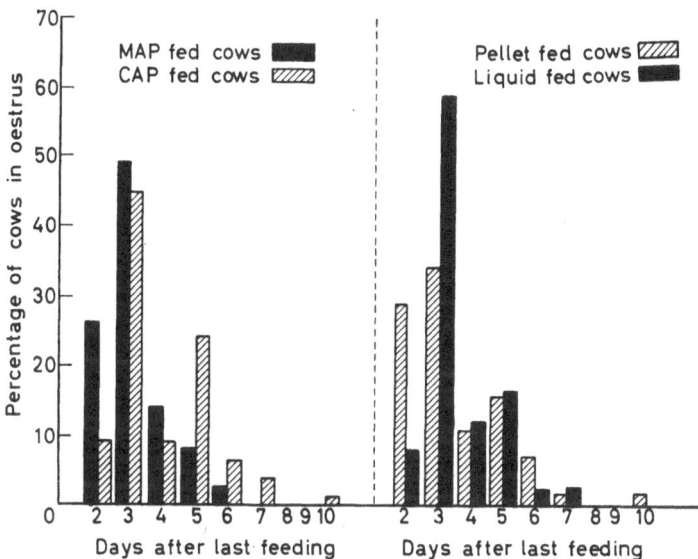

Figure 1. The occurrence of oestrus in MAP and CAP fed cows after cessation of treatment. The data represent the percentage of synchronized animals in oestrus each day

cantly different from controls (59·8 per cent), but the conception rate of the CAP-fed cows (34·8 per cent) was significantly lower than that of either the MAP or control cows $(P < 0·05)$.

The conception rate in cows fed pellets (52·1 per cent) was not significantly higher than that found in cows fed liquid feed (44·4 per cent). The pregnancy rate in oxytocin treated cows was 43·2 per cent, which was not significantly different (but approached significance at the 5 per cent level) from the 53·3 per cent obtained in controls.

Bull A had higher $(P < 0·05)$ fertility (58·1 per cent) than bulls B (44·6 per cent) or C (41·9 per cent); their conception rates on second (non-treatment) services were 76·2, 73·2 and 59·1 per cent, respectively.

Synchronization of oestrus was satisfactory with both compounds and both methods of feeding. However, the cows not detected in

427

oestrus, but which apparently ovulated during the synchronization period as judged by rectal palpations, occurrence of metoestrous bleeding, or return to oestrus one cycle later represent a major problem. In part, this may be a simple problem of oestrus detection resulting from the large number of animals in oestrus at one time. However, it was evident in earlier studies that ovulation unaccompanied by oestrus was characteristic of cattle fed relatively high levels of progestogens, particularly CAP (Van Blake *et al.*, 1963). Cows ovulating without showing oestrus in the present experiments may well have consumed more than the calculated doses of CAP and MAP during the feeding period.

The slightly poorer synchronization observed with CAP than with MAP may result from its greater physiological activity. This may be reflected in a longer half-life and a longer (and more variable) period of inhibition upon the hypothalamic centres controlling oestrous behaviour and luteinizing hormone release (Sawyer, 1964). Pelleted feeds pass through the digestive tract more rapidly than non-pelleted rations, and this fact probably accounts for the greater number of pellet-fed than liquid-fed cows in oestrus on the second day after hormone withdrawal.

Since fertility at the synchronized oestrus was significantly higher in MAP-fed than in CAP-fed cows ($P > 0.05$), it is concluded that a difference exists between the two compounds in this respect. The lowered fertility in the CAP-fed animals emphasizes the need for further basic research on the effects of these compounds on normal reproductive processes. Fertility of the MAP-fed cows was not significantly different from that obtained in control cows ($P > 0.05$). However, these results and those obtained in previously reported experiments indicate that conception rates at first service in MAP-fed cattle are often approximately 10–15 per cent lower than in untreated cattle. Conception rates at second service were uniformly high in all groups indicating that there were no prolonged effects of the progestogens on fertility.

Conception rates at first service in both treated and untreated animals in the present experiment appear low when compared to non-return rates reported for dairy cattle bred artificially. However, non-return rates are normally somewhat higher than pregnancy rates based on rectal palpations, and some reports imply that fertility in beef cattle is likely to be quite variable and often lower than that of dairy cattle. The effect, if any, of the liquid molasses-urea ration fed to all groups on fertility, is unknown.

The pelleted feed containing the progestogens was fed in addition to the liquid molasses-urea ration. The added nutrients supplied by the pelleted feed did not result in improved fertility, since the difference in conception rate (7·7 per cent) in favour of the pellet-fed cows was not statistically significant ($P > 0.05$).

Oxytocin administered to Holstein heifers at the beginning of oestrus hastens ovulation (Hansel, Armstrong and McEntee, 1958) and may also play a role in sperm transport (Van Demark, 1958). It has been reported to improve fertility in swine bred artificially under certain conditions (Stratman, Self and Smith, 1959). These observations suggested that oxytocin administered to beef cattle at the time of insemination might improve fertility. However, the conception rate at first service was reduced (10·1 per cent) in the oxytocin treated cattle; this reduction approached significance at the 5 per cent level. Other ovulation hastening agents, for example human chorionic gonadotrophin, given early in oestrus have also been shown to reduce fertility (Hansel, McEntee and Armstrong, 1960).

CAUSES OF REDUCED FERTILITY IN PROGESTERONE-TREATED CATTLE

The exact causes of the lowered conception rates in cattle synchronized by oral or injected progesterone remain obscure. Results suggest that fertility is highest in cattle bred after synchronization by orally active compounds having the least physiological activity. As an example, fertility is higher in beef cattle bred after synchronization with MAP than CAP; the latter compound suppresses oestrus when fed at a level of 0·026 mg/kg of body weight per day as contrasted to a level of approximately 0·500 mg/kg for MAP. Conception rate is lower still when cattle are bred following synchronization by subcutaneous or intramuscular injections of progesterone itself. The rapidity and completeness with which the residual progestogen is removed from the blood and tissue fluids after the end of the feeding or injection period may be important factors influencing fertility of the treated animals.

It seems unlikely that the lowered fertility is caused by inadequate progesterone secretion by the corpora lutea formed following the ovulations associated with the synchronized oestrus. Some very preliminary figures shown in Table 3 suggest that the corpora lutea of MAP- and CAP-fed animals contained approximately as much progesterone as the corpora of untreated animals at 3 and 10 days after oestrus. Histological examination of the endometrium from these same animals indicated that the CAP- and MAP-fed heifers had somewhat more glandular development than the controls at 3 days post-oestrus; no differences among the three groups were observed at 10 days. To some extent, the differences noted in endo-

Table 3

Weights and progesterone contents of corpora lutea removed from heifers after oestrous cycle synchronization by oral progestogens*

Stage of cycle at which corpora lutea removed	Treatments prior to oestrus					
	Untreated		Fed MAP 18 days		Fed CAP 18 days	
	Wt. of corpus (g)	Total progesterone (μg)	Wt. of corpus (g)	Total progesterone (μg)	Wt. of corpus (g)	Total progesterone (μg)
3 days post-oestrus	1·1	24·3	0·79	11·4	0·94	13·7
10 days post-oestrus	4·7	159	5·5	118	6·8	197

* Each treatment group consisted of 4 animals, two of which were slaughtered at 3 and two at 10 days post-oestrus.

metrial development at 3 days may have been due to increased oestrogen secretion by the treated heifers, as Zimbelman (1965) has suggested.

Furthermore, a large margin of safety appears to exist in regard to the amount of progesterone produced by a normal animal early in pregnancy. For example, Staples and Hansel (1961) found that normal 15-day bovine corpora lutea from pregnant heifers contained nearly three times the amount of progesterone found necessary to support normal embryo development to that time. The recent report of Shelton and Moore (1966) indicating that fertilized ova survive quite well when transferred into the uteri of ewes previously synchronized by progesterone injections indicates that a similar case can be made for sheep.

These somewhat limited observations serve to focus attention on events occurring at, or shortly after ovulation, as possible causes of the reduced fertility in progestogen-treated cattle. Some preliminary observations in our laboratory have suggested that ovulation time might be more variable, and, on the average somewhat earlier in MAP-fed heifers than in untreated heifers. The subject needs more careful investigation, since Wiltbank (1965) reported that ovulation occurred at about the normal time in heifers synchronized by feeding a progestogen, but that the length of oestrus was shortened by approximately 4 h.

Too few ova have been recovered from cattle bred after synchronization to know whether there is an interference with either fertilization or ovum transport. Wiltbank (1965) presented some preliminary data suggesting that a lower percentage of ova were fertilized in progestogen-treated heifers than in normally cycling animals. Ovum transport does not appear to be hastened since the few ova recovered have been found in the oviducts at 3 days post-oestrus. The alternative possibility that entry of fertilized ova into the uterus may be delayed as a result of the treatments appears not to have been investigated.

OTHER MODES OF ADMINISTRATION OF CYCLE SYNCHRONIZING COMPOUNDS

Since beef animals in many parts of the world are kept under range conditions where feeding compounds such as MAP or CAP would pose some practical problems, an increasing amount of attention has been given to the development of synchronization techniques requiring only a single application. Nevertheless, it must be pointed out that the nutritive condition of a large portion of the beef animals in these same areas is such that a 3–4 week period of supplemental grain feeding is necessary for good reproductive performance. None of the compounds used for synchronization is likely to cure post-partum anoestrus due to an inadequate energy intake. The data of Wiltbank, Rowden, Ingalls, Gregory and Koch (1962) are most impressive on this point. One can make a good case for feeding the progestogens under many practical conditions.

Single injections of progestogens such as CAP and MAP are not practical since the injected compounds retain their activity for long and variable periods of time. Several attempts have been made to inject or implant these materials in such a way that the residues can be removed with a minimum of surgical manipulation after about 18 days. None of the attempts reported have been very successful. Similarly, it has not been possible to administer the progestogen intravaginally impregnated on sponges as Robinson (1964) has done for ewes, because sponges will not remain in the bovine vagina for predictable periods of time. It may be possible to design such a sponge for cattle, but none has, as yet, been reported.

One of the most intriguing ideas along this line is to administer the progestogen in a heavy bolus that will remain in the rumen and release the drug over an approximate 18-day period. Preliminary studies in our laboratory indicate the feasibility of this idea, but considerable additional experimentation will be necessary to develop a bolus that will give the proper rate and duration of drug release. Progestogens, such as CAP and MAP are soluble in dimethyl sulphoxide (DMSO) and this remarkable chemical has the ability to carry these compounds through the intact skin. Results to date indicate that oestrus is inhibited during daily treatments of 10–20 mg of CAP applied to the animal's back in DMSO. However, results with smaller daily doses of CAP in DMSO have been inconsistent, and a rather large proportion of the treated heifers have ovulated following treatment without showing signs of oestrus.

NON-STEROIDAL DRUGS FOR OESTROUS CYCLE REGULATION IN CATTLE

An increasing awareness of the possibility of regulating the oestrous cycles of farm animals by non-steroidal compounds has led to the testing of many of the antifertility compounds, now being produced in such great profusion by many drug companies, particularly as the high cost of the steroid compounds, such as those discussed above may well be a deterrent to their widespread use.

Although one non-steroid, 1-alpha methylally-thiocarbamoyl-2-methyl-thiocarbamoylyhydrazine (ICI-33,828) has been shown to control the oestrous cycles of swine when fed for 9–21 days (Gerrits and Johnson, 1964 and Polge, 1964), no effective non-steroid for ruminants has yet been reported.

ICI-33,828* fed to 15 ewes at a level of 100 mg/animal per day for 18 days, did not produce good cycle synchronization in a preliminary trial in our laboratory. However, in the same experiment, 15 similar ewes were fed another non-steroid, 1-[p-(β-diethylaminoethoxy)phenyl]-1,2-diphenyl-2-chloroethylene citrate (MRL-4 or Clomiphene)† in two equal feedings at a level of 150 mg/ewe per day for 18 days. The results of this treatment were somewhat unexpected and most interesting. Feeding of the compound at this level did not inhibit oestrus; 13 of the 15 ewes came in oestrus during the 18-day period and 2 ewes came in oestrus twice during the period at 15- and 16-day intervals (Table 4). Surprisingly, 11 of the 15 ewes returned to oestrus within 24 h after the last feeding, even though some of them had been in oestrus as recently as 3 days earlier

* Kindly supplied by Ayerst Laboratories, New York.

† Kindly supplied by Hess and Clark Research Department, Division of Richardson-Merrell, Inc., Ashland, Ohio.

and while still receiving the drug. One of the remaining 4 ewes came in oestrus 15 days after feeding, suggesting that a 'silent ovulation' may have occurred during the synchronization period.

Table 4

Oestrous response of clomiphene treated ewes (1965)

Ewe No.	Day oestrus shown during treatment period	Day oestrus shown after end of treatment†	Day returned to oestrus post-breeding
2	14	1	—
3	14	1	—
4	7	1	—
5	1	1	—
9	2 and 17	—	(?)
11	11	1	—
16	3*	—	—
17	—	—	—
19	5	1	—
21	1 and 17	1	—
28	6	1	—
32	—	15	—
33	3	1	6‡
34	11	2	6‡
37	3	1	—

* Oestrus suspected, but not definite.
† All ewes in oestrus were bred naturally.
‡ There is some doubt as to whether these ewes remained in oestrus for 6 days or returned to oestrus after 6 days.

Two ewes did not come in oestrus, either during the feeding period, or during a 10-day period thereafter, and the remaining ewe was in oestrus on the seventeenth day of the feeding period and did not return to oestrus until 16 days later.

Unfortunately, the exact hour at which oestrus was first shown is not known, but at least some of the ewes came in oestrus before the time when they would have received the drug had the feeding been continued. The inference is that the synchronization did not result from withdrawal of the drug, but rather from its continued feeding for a 16-day period. Eleven ewes showed slight mammary gland enlargement, and 4 ewes showed marked mammary gland development at the end of the feeding period.

Subsequently, 10 Holstein heifers were fed Clomiphene at a level of 800 mg/heifer per day for 18 days. Although this dose is roughly comparable on a body weight basis to that fed to the ewes, the results obtained were quite different. No heifers came in oestrus during the feeding period, and none for at least 9 days after cessation of the treatment. The cycles were not synchronized after the end of the treatment. The ovaries of all treated heifers became remarkably small and inactive during the feeding period. Two of the 10 heifers showed a noticeable increase in mammary gland size at the end of the treatment period. Experiments in which heifers are fed lower levels of this compound are under way.

These very different results obtained with Clomiphene in these two species are characteristic of the rather confusing reports in the literature concerning use of the drug in other species. At relatively low doses, the drug exhibits anti-oestrogenic effects, thus inhibiting the normal feed-back of oestrogen on pituitary gonadotrophin secretion. Therefore, the superovulatory effect frequently reported following low doses of the drug results from augmented gonado-trophin (probably luteinizing hormone) secretion brought about by antagonism of endogenous oestrogen (Barr and Paulsen, 1965 and Coppola, Leonardi and Ringler, 1966). The results obtained with the ewes may be explainable on this basis. If so, synchronization by potent anti-oestrogens given for periods of approximately one cycle length may be possible. The timing relationships cited above suggest that synchronization of this type is quite different from that produced by progestogens, such as MAP and CAP. Clearly, additional work is needed to explore this interesting possibility.

At the same time, it must be recognized that relatively large doses of Clomiphene suppress ovulation and ovarian function by interfering with luteinizing hormone release, as Coppola et al. (1966) have clearly shown. The results obtained after feeding the drug to heifers may represent such an effect. Cattle require considerably less MAP or CAP per unit of body weight than sheep for the inhibition of oestrus, and it seems likely that the 800 mg/day dose of Clomiphene was sufficiently large to inhibit luteinizing hormone release in the treated heifers to the point where both ovarian follicles and corpora lutea regressed. Conceivably, administration of lower doses of this drug for 18-day periods or longer may result in cycle synchronization in cattle, as well as sheep.

CAN CYCLE SYNCHRONIZATION BE ACHIEVED WITHOUT THE USE OF HORMONES OR DRUGS?

Armstrong and Hansel (1959) prevented the normal formation and function of the corpus luteum and caused precocious oestrus in cycling heifers by daily injections of oxytocin during a critical period from days 2–6 inclusive of the cycle. Later, Hansel and Wagner (1960) produced similar results by inserting into the uterus catheters held in place by inflated balloons during the same critical period. Kendrick (1964) has produced similar results by adding a virus to the semen introduced into the uterus at the time of artificial insemination.

This mechanism was used, in combination with progestogen feeding, to produce synchronization after a 10-day, rather than an 18-day treatment period in Holstein heifers (Hansel, 1961). Animals at the eighth day of the cycle, or later at the beginning of the treatment period will, of course, be synchronized by a 10-day progestogen feeding period. Those animals in the first few days of their cycles at the beginning of the treatment period can also be synchronized after a 10-day MAP-feeding period, provided the development of their corpora lutea is arrested by either oxytocin injections or uterine dilatation.

Yamauchi, Nakahora, Kaneda and Inui (1966) have extended these results and suggest that synchronization can be produced by uterine distension alone. These workers injected a viscous gel-like substance* through the cervix into the uterus of cows at various stages of the oestrous cycle. The length of the cycle was reduced, as expected, when the injections were made during the early luteal phase. On the contrary, treatments during the late luteal phase lengthened the cycle to 24–28 days. In 23 of 25 cycles in which the treatments were performed between the early luteal phase and the late luteal phase, ovulations occurred 7–12 days after the treatment (Table 5). The authors suggest that the treatment is effective in synchronizing oestrus in animals at all stages of the cycle, except during the limited periods of pro-oestrus, oestrus and very early post-oestrus. The authors further suggest that animals bred after the shortened cycles caused by uterine distension during the early luteal phase are fertile, a result that agrees with our own observations. Unfortunately, these observations were based on only 34 treatment cycles in 9 cows, and further work is needed before any definite conclusions can be drawn.

CONCLUSIONS

1. Satisfactory oestrous cycle synchronization was achieved by feeding either MAP at a level of 180–200 mg/animal per day or CAP at a level of 10 mg/day for 18 days to both dairy heifers and beef

---

* Gelceptor, F., The Eisai Co., Ltd. Bunkyo-ku, Tokyo.

Table 5

Effect of distending the uterus with a viscous gel-like substance* on oestrous cycle length of the cow

| Time of injection | | No. of cases | Modification of cycle length‡ | | | | | |
Phase	Day		Shortened Cycle length	Shortened Mean interval, injection to ovulation	Unchanged Cycle length	Unchanged Mean interval, injection to ovulation (days)	Lengthened Cycle length	Lengthened Mean interval, injection to ovulation
Post-oestrus	1	4	8	7·0	20, 21, 25	21·0		
Early luteal phase†	2	5	9, 9, 10, 10	7·5	22	20·0		
	2 & 5 or 6	5	9, 10, 12, 13, 14	9·6				
	6	4	14, 14, 17, 17	9·5				
Functional	12	4			20, 21, 22, 22	9·3		
	12 & 16	2			20	8·0	28	16·0
Late luteal phase	14 & 16	3					24, 25, 25	10·7
	17 or 18	2					27, 28	10·5
Pro-oestrus		2			19, 20	1 5		

*Gel-ceptor F, the Eisai Co. Ltd., Bunkyo-ku, Tokyo.
† Mean length of 14 cycles was reduced to 12·5 days ($P < 0·01$).
‡ Mean length of 5 cycles was lengthened to 25·8 days ($P < 0·01$).
(Mean length from injection to ovulation in cases outlined by bold lines is 9·3 ± 1·6 days.)

cattle. Incorporation of the drugs into a liquid ration, and a pelleted or a ground grain ration were equally satisfactory.

2. Under ideal conditions, a conception rate of 50–60 per cent was attained in both dairy heifers and mature beef cattle artificially inseminated at the synchronized oestrus following MAP feeding.

3. Conception rates were lower (approximately 35 per cent) in beef cattle inseminated at the synchronized oestrus following CAP feeding.

4. Normal conception rates were obtained in both MAP- and CAP-fed animals artificially inseminated at the second oestrus after treatment.

5. A conception rate of 60 per cent was obtained in 41 Holstein heifers artificially inseminated on two consecutive days (days 3 and 4) after MAP feeding.

6. Administration of oxytocin at the time of insemination did not improve conception rates in artificially bred beef cattle.

7. Several methods of administering these progestational steroids as a single dose were discussed; none of these methods have yet been perfected.

8. The possibility of producing synchronization in sheep and cattle by feeding a potent anti-oestrogen was investigated and some encouraging results obtained.

9. Data were cited to indicate that some degree of cycle synchronization can be attained in cattle by uterine distension alone.

ACKNOWLEDGEMENT

The author wishes to acknowledge the help of M. A. Brunner, L. E. Donaldson, M. O. Kane, P. V. Malven and H. Van Blake in carrying out the experiments cited. The financial assistance of the New York Artificial Breeders Cooperative, The Syntex Laboratories, Eli Lilly and Co. Laboratories, and the Upjohn Co. is gratefully acknowledged. Part of the funds were provided by the Regional Research Project NE-41, Endocrine Factors Affecting Reproduction in Dairy Cattle.

REFERENCES

Anderson, L. L., Schultz, J. R. and Melampy, R. M. (1964). 'Pharmacological Control of Ovarian Function and Estrus in Domestic Animals'. In: *Gonadotropins: Their Chemical and Biological Properties and Secretory Control.* Ed. by H. H. Cole. San Francisco and London; W. H. Freeman and Co. (*Proc. VIth Biennial Symposium on Animal Reprod.*)
Armstrong, D. T. and Hansel, W. (1959). *J. Dairy Sci.* **42,** 533

Barr, A. and Paulsen, C. A. (1965). *Clin. Res.* **13,** 129 (Abstr.)

Coppola, J. A., Leonardi, R. G. and Ringler, I. (1966). *J. Reprod. Fert.* **11,** 65–71

Dutt, R. H. and Casida, L. E. (1948). *Endocrinology* **43,** 208

Gerrits, R. J. and Johnson, L. A. (1964). *Proc. Vth Congr. on Anim. Reprod. and A.I. Trento,* Italy. 2nd Sect., pp. 455–459

Hansel, W. (1959). 'The estrous cycle of the cow.' In: *Reproduction in Domestic Animals.* Ed. by H. H. Cole and P. T. Cupps. New York and London; Academic Press

— (1961). *J. Dairy Sci.* **44,** 2307–2314

— (1964). *Proc. Vth Int. Congr. Anim. Reprod. and A.I. Trento,* Italy. Vol. 7, 318 pp.

— (1965). *United States Dept. of Agric. Misc. Publ.* 1005. (*Proc. Conf. on Estrous Cycle Control in Domestic Animals*)

— and Malven, P. V. (1960). *J. Anim. Sci.* **19,** 1324

— and Wagner, W. C. (1960). *J. Dairy Sci.* **43,** 796–805

— Armstrong, D. T. and McEntee, K. (1958). *Proc. IIIrd Symp. Reprod. and Infertility,* pp. 63–74. Ed. by F. X. Gassner. New York and London; Pergamon Press

— Donaldson, L. E., Wagner, W. C. and Brunner, M. A. (1966). *J. Anim. Sci.* (In press)

— McEntee, K. and Wagner, W. C. (1960). *Cornell Vet.* **50,** 497–502

Hartman, C. G. and Leatham, J. H. 'Oogenesis and Ovulation', Chap. V. In: *Mechanisms Concerned with Conception.* Ed. by C. G. Hartman (*Proc. Conf. on Physiol. Mechanisms Concerned with Conception,* West Point, N.Y. 1959). New York; The MacMillan Co.

Kendrick, J. W. (1964). *Proc. Vth Int. Congr. Anim. Reprod. and A.I. Trento,* Italy. Vol. 5, pp. 161–165

Nellor, J. E. (1960). *J. Anim. Sci.* **19,** 412–420

— and Cole, H. H. (1956). *J. Anim. Sci.* **15,** 650–661

— Ahrenhold, J. E. and Nelson, R. H. (1960). *J. Anim. Sci.* **19,** 1331

Polge, C. (1964). *Proc. Vth Int. Congr. on Anim. Reprod. and A.I. Trento,* Italy. 2nd Sect., pp. 388–393

Robinson, T. J. (1964). *Proc. Aust. Soc. Anim. Prodn.* **5,** 47–52

Sawyer, C. H. (1964). 'Control of secretion of gonadotropins'. In: *Gonadotropins: Their Chemical and Biological Properties and Secretory Control.* Ed. by H. H. Cole. San Francisco and London; W. H. Freeman and Co.

Shelton, J. N. and Moore, N. W. (1966). *J. Reprod. Fert.* **11,** 149–151

Staples, R. E. and Hansel, W. (1961). *J. Dairy Sci.* **44,** 2040–2048

Stratman, F. W., Self, H. L. and Smith, V. R. (1959). *J. Anim. Sci.* **18,** 634

Trimberger, G. W. and Hansel, W. (1955). *J. Anim. Sci.* **14,** 224–232

Ulberg, L. C., Christian, R. E. and Casida, L. E. (1951). *J. Anim. Sci.* **10,** 752

— and Lindley, C. E. (1960). *J. Anim. Sci.* **19,** 1132–1142

Van Blake, H., Brunner, M. A. and Hansel, W. (1963). *J. Dairy Sci.* **46,** 459–462

Van Demark, N. L. (1958). *Int. J. Fert.* **3,** 220

Wiltbank, J. N. (1965). *Conference on Estrous Cycle Control in Domestic Animals.* U.S. Dept. Agric. Misc. Publ. 1005, pp. 58–59. Lincoln; Univ. of Nebraska

Wiltbank, J. N., Rowden, W. W., Ingalls, J. E., Gregory, K. E. and Koch, R. M. (1962). *J. Anim. Sci.* **21,** 219–225

Yamauchi, M., Nakahora, T., Kaneda, Y. and Inui, S. (1966). Effects of uterine distention on the estrous cycle of the cow. (In press)

Zimbelman, R. G. (1965). Evaluation of some methods for controlling the bovine estrous cycle. *Proc. Conf. on Estrous cycle Control in Domestic Animals,* pp. 17–27. United States Dept. of Agric.' Misc. Publ. 1005

DISCUSSION OF PAPERS BY T. J. ROBINSON AND W. HANSEL

DR. M. C. CHANG (*Worcester, Mass., U.S.A.*)

Could you speculate on why the effect of MAP is quite different from that of progesterone?

ROBINSON

Our general experience has been that the progestogens which are long acting result in lowered fertility. Progesterone itself can be made to give different results if used in different media which affect the length of its action.

DR. M. C. CHANG

Did you inject Provera into your sheep?

ROBINSON

No, but we have administered it by intra-vaginal sponge. It is not as good as SC-9880 for synchronization but fertility in animals coming into oestrus was comparable.

HANSEL

Most workers using MAP in sheep have given either 50 or 60 mg/day. You have used 40 mg and 80 mg. The 50 or 60 mg levels seem to give rather better synchronization than you indicate. Have you tried these levels?

ROBINSON

No.

DR. M. J. K. HARPER (*Macclesfield*)

Dr. Walpole and I have been working with compounds closely related to cronolone (SC-9880) in rats and have shown that these compounds are a mixture of isomers, one oestrogenic and one an anti-oestrogen. If cronolone is similar then I wonder if the results you get in sheep could be due to these particular properties.

DR. P. J. HEALD (*London*)

Could Professor Hansel comment on the fate of these compounds in the rumen when they are fed to the animal. Since the rumen possesses such a high and varied metabolic activity, is it possible that the compounds fed are not, in fact, those causing cessation of ovulation.

HANSEL

This is difficult to answer. As a general statement however, it seems that the compounds are absorbed from the rumen and it is likely that they are effective as such.

DR. J. P. BENNETT (*Godalming*)

I would like to comment regarding Dr. Heald's question. We have some work on similar compounds which indicates that the metabolism is rapid and I feel that speed of metabolism in the rumen may be an important feature.

DR. P. E. LAKE (*Edinburgh*)

Would Dr. Robinson comment on the technique of putting the sponges in the vagina. We have had considerable difficulty in the hen.

ROBINSON

The technique is very simple. We merely insert a plastic tube into the vagina, push the sponge into the vagina through the tube and remove the tube. It is almost 100 per cent successful in the sheep but more difficult in cattle. Here it is 85 per cent successful and the problem is the highly contractile tract of the cow.

PROFESSOR H. KARG (*Munich, Germany*)

We have used tampons in goats with good results. In the pig there are some problems due to necrosis of the epithelium. In cattle it seems that if the tampon is inserted just below the cervix it is not ejected.

DR. J. P. BENNETT

Have you ever tried putting a thread impregnated with material through the skin instead of a vaginal plug?

HANSEL

We have not. I believe that other workers have not been too successful due to residue remaining under the skin after removal of the thread.

MISS B. J. WEIR (*London*)

I was interested to hear about DMSO which is useful for administering to animals not easily injected. Could you elaborate further on this?

HANSEL

No. Perhaps someone else may wish to comment.

PROFESSOR H. KARG

We have applied this daily to the backs of swine and have obtained reasonable synchronization in a small number.

440

CONTROL OF THE OVARIAN CYCLE

DR. G. K. BENSON (*Liverpool*)

I understand DMSO has been withdrawn from general use.

HANSEL

Yes, because it may cause cataract in some animals. This may have been exaggerated out of proportion.

PROFESSOR G. E. LAMMING (*Nottingham*)

Have either of the speakers any idea how the compounds work? We have examined sheep at 2, 3 and 8 days post mating after synchronization and the animals ovulating do not do so in relation to the oestrous period. Could these compounds affect the oestrous-producing centre and the ovulating centre of the hypothalamus in different ways? This may explain the infertility which results from the use of these compounds.

ROBINSON

I agree on one point. We have considerable evidence that the time relation between oestrus and ovulation is abnormal in some cases. However, using sponges, we have no evidence for a marked discrepancy between the time of onset of oestrus and ovulation.

HANSEL

I agree with Professor Robinson and Professor Lamming that two different centres are involved and different compounds may inhibit one to a greater extent than the other.

PROFESSOR W. VELLE (*Oslo, Norway*)

We have obtained good synchronization in ewes and heifers after only a 10-day period of treatment with MAP. Regarding Professor Hansel's comments on the possibility of an increased oestrogen secretion under gestagen treatment we have found that in the boar there is a suppression of urinary oestrogen secretion following MAP and CAP treatment.

Have you any records of the strength of the heat symptoms in treated heifers? We have noticed that the first heat after treatment appears much stronger than previous heats.

HANSEL

I believe you can get synchronization with a 10-day treatment period but we have done little within 10 days. With regard to the strength of heat symptoms, there appears to be an excessive amount of mucus found in the vagina of these animals. Whether this is an indication of stronger symptoms I hesitate to say.

DR. W. JÖCHLE (*Germany*)

We have used a 10-day treatment in gilts, sheep and goats with similar results to those obtained by 16–18 day treatment.

DR. P. G. HIGNETT (*Glasgow*)

Would the speakers comment on the effect of these progestational compounds

441

on a pre-existing functional corpus luteum. Would Professor Hansel comment on reports that the use of oestrogens with progestogens has improved fertility.

HANSEL

The corpora lutea formed after progesterone synchronization appear to be normal. There is less information available on what happens to the existing corpora lutea. I have the impression that they are normal in MAP-fed heifers. However, Dr. Velle's remark that he can get synchronization after a 10-day period indicates that this may not be the case. Here one would expect that the corpora lutea of these animals treated early in the cycle would still be persisting. I also have an unsubstantiated impression that CAP-fed heifers do undergo some premature luteal regression.

In reply to the last question, in my own experience there is no improvement by giving oestrogen to sheep or cattle after a progesterone feeding period. However, the Squibb system of injection of oestrogen before a 10-day oral progesterone period shows considerable promise.

ROBINSON

We were unable to demonstrate an effect on fertility by following progesterone treatment with either oestrogen or testosterone. The question of the functional life of the corpus luteum in the treated animal is a neglected one and we are currently investigating this with attempts to detect peripheral levels of progestagens by gas-liquid chromatography.

DR. C. POLGE (*Cambridge*)

Dr. Hunter has observed that in the pig, insemination of animals induced to ovulate during the luteal phase when progesterone levels are high gives rise to abnormalities of fertilization, particularly polyspermy. Similar effects were observed in the ageing egg from the following oestrus which could be a result of an increase in progesterone due to the corpus luteum following ovulation. Further, injection of progesterone shortly before ovulation in animals ovulating during oestrus gives rise to similar abnormal fertilizations. It is possible, therefore, that this could happen in your animals if you get a prolongation of progestational effects. This type of abnormality will only be detected if the eggs are examined in detail.

HANSEL

What doses of progesterone were you using?

DR. C. POLGE

100 mg progesterone in oil subcutaneously about 24–36 h before ovulation.

ROBINSON

Data on the relationship between fertilization and ultimate lambing, and egg transfer experiments do indicate that in the sheep, the basic problem is one of failure of fertilization. We have no evidence that the eggs are abnormal or that polyspermy occurs although we have not looked at this directly.

DR. M. C. CHANG

Have you considered examination of the endometrium after application of the tampons on different days of the cycle. I feel the endometrium is important.

CONTROL OF THE OVARIAN CYCLE

ROBINSON

We are at present looking at the effect on the endometrium following treatment for different periods of time in relation to spermatozoa survival and transport.

HANSEL

We have analysed data from cows in terms of conception rate from animals on treatments at different stages of the cycle and have found no significant differences.

DR. J. SPINCEMAILLE (*Belgium*)

It has been our experience that intra-uterine infusions of one litre of physiological saline can induce ovulation and heat in the mare. Has Professor Hansel evidence that uterine distension in the cow can have a luteolytic effect? Is it possible that it may also have an indirect gonadotrophic effect?

HANSEL

I think this is related to what happens in the cow, where palpations cause a number to come into oestrus 6–7 days later.

EFFECTS OF PROGESTAGEN-TREATMENT
ON THE PREPUBERTAL OVARY
OF THE PIG

E. SCHILLING

*Max-Planck-Institut für Tierzucht und Tierernährung (3051)
Mariensee, West Germany*

THERE IS considerable interest at the present time in the use of hormones to produce temporary or permanent suppression of oestrus and ovulation. This paper deals with the effects of progestagens on the prepubertal ovary of the pig, a field which has received little attention until recently, and there are many unanswered questions concerning the response of the prepubertal ovary to hormone treatment.

In young female domestic animals, the effects of progestagens have been tested by Röstel, Jöchle and Schilling (1964) and by Jöchle and Schilling (1965). It was shown, that injections of norethisterone in premature gilts caused an 8–12 weeks' delay in the onset of puberty. No female pig came into heat during the fattening period up to a body weight of 110 kg. Female pigs, fed in larger groups, showed better daily gain than untreated controls.

Additionally, we tested the effects of norethisterone-treatment in neonatal piglets. In these animals also we expected a retardation of puberty. Ten milligrammes of norethisteron-enanthate (NE) was administered parenterally in female piglets on the tenth day after birth and a second injection between the sixteenth and eighteenth week of life. None of these 6 animals reached puberty up to the time of slaughter at a body weight of 110 kg. Their uteri were considerably reduced in size and the smaller ovaries contained Graafian follicles, but no corpora lutea. In the ovaries of two treated animals remarkable histological changes could be demonstrated. All the larger follicles showed atresia resulting from an undulating hyperplasia of the granulosa-layer which ultimately filled the whole of the follicular antrum (*Plate Ic*). This kind of atresia seems to be unusual. The idea of permanent sterility following NE-treatment of young piglets does not support the general opinion held of the relative insensibility of the postnatal ovary. This had been demonstrated in rodents, in which the ovaries are relatively undeveloped at birth. The early stages of ovogenesis and folliculogenesis are said to be nearly autonomous (Hisaw, 1947; Brambell, 1956; Franchi,

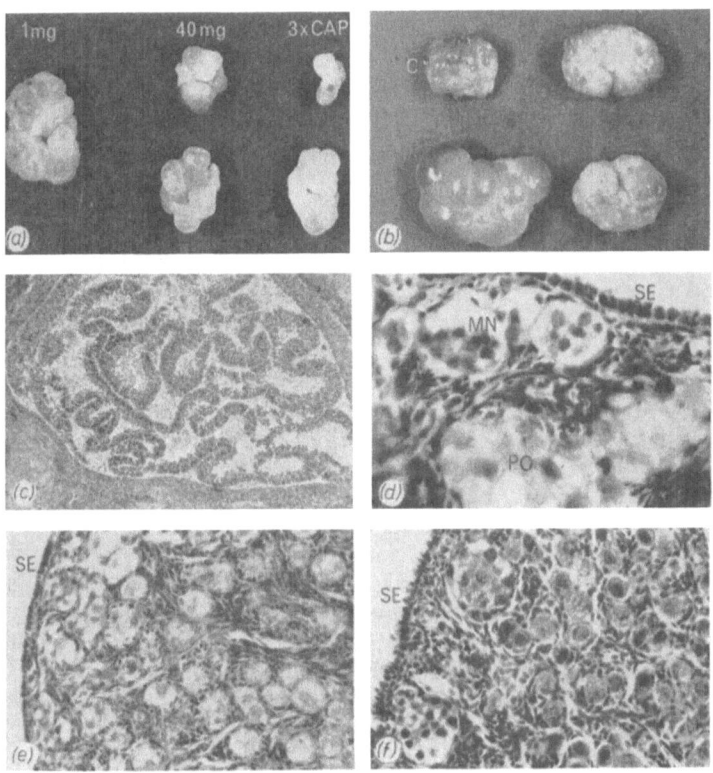

PLATE I

(Reduced ⅔ on reproduction)

a. *Ovaries of CAP-treated pigs* (15 weeks old). 1 mg *(normal size)* 40 mg—3 × *CAP (Magn. 1·5 ×)*

b. *Ovaries of NE-treated pigs* (22 weeks old), 10 mg norethisterone—C = Control (Magn. normal size)

c. *Graafian follicle after* 2 × *NE-treatment in the ovary of a* 7 *months old pig.* *Note the undulating granulosa layer (Magn. 80 ×)*

d. *Pig ovary*—10 days *post natum: Neoformation of oöcytes and primary follicles.* *SE = surface epithelium,* *MN = Multi-nuclear follicles,* *PO = polyovular follicles (Magn. 400 ×)*

e. *Pig ovary*—10 days *post natum: Proliferative stages of oögenesis in the peripheral zone many well-developed primary follicles in the inner cortex (Magn. 160 ×)*

f. *Pig ovary*—10 days *post natum. More intensive oögenesis and primary follicles with incomplete surrounding by stromacells (Magn. 240 ×)*

445

Mandl, Zuckermann, 1962) and gonadotrophins are unnecessary. When Graafian follicles and the theca-layer are formed, then the ovary will become responsive to hypophysial hormones.

The undifferentiated ovary of a new-born animal, however, is not physiologically unresponsive to exogenous hormones. The embryonic ovary can be influenced by hormones of the mother or by direct hormone-treatment of the embryos (Turner, 1939; Amoroso, 1955; Stange and Drescher, 1956). In several recent investigations it was shown that androgens have a definite effect on the ovaries of new-born rats and mice. Pfeiffer (1936) transplanted testes of new-born rats into female litter-mates, and a persistent anovulatory oestrus occurred after puberty. Barraclough (1961) injected a single dose of testosterone-propionate in female rats at the second, fifth, tenth and twentieth day after birth. Hormone-treatment administered before the tenth day induced permanent sterility, and he concluded that there must be a period of steroid-sensitivity between birth and the tenth day of age. Pfeiffer believed a masculinization of the pituitary had occurred, which caused an imbalance in gonadotrophin secretion. Other investigators believe that the developing hypothalamic–hypophysial interrelationship may be influenced (Barraclough, 1961; Sawyer, 1964; Hansel, 1964). According to Takasugi (1953), permanent sterility can also be induced by treating new-born female rats with androgens, oestrogens, progesterone, desoxycorticosterone or cholesterone. The sterility of new-born rats after oestrogen-treatment is believed to be caused by suppression of hypophysial gonadotrophins possibly involving complete suppression of LH and partial suppression of FSH secretion.

Progestogens, even when given in high doses, proved ineffective to induce permanent sterility in mice and rats (Pincus, 1965). An interesting experiment was carried out by Logothetopoulos, Sharma and Kraicer (1961). They injected MAP (6α-methyl-17α-hydroxy-progesterone acetate, 'Provera') daily from birth up to 120 days of life. They observed atrophy in weight of most of the endocrine glands, especially the adrenals and ovaries. The extremely small ovaries did not contain corpora lutea, and follicles were found with only one or two layers of granulosa cells. During the 50 days after cessation a slow regeneration took place and more multi-layered follicles appeared. Following a 15-day treatment with PMS, the ovaries enlarged distinctly and ovulations with formation of corpora lutea occurred. This indicates that the hypothalamic or hypophysial cells did not lose permanently the sensitivity to their trophic hormones. The histological findings in the ovaries were similar to

446

those observed after hypophysectomy in rats (Dean and Greep, 1946). Thus MAP must have inhibited the production of gonadotrophins by the pituitary.

The process of folliculogenesis in the pig is not completed at birth. All authors agree that large numbers of oögonia and oöcytes in nests or clusters can be seen in neonatal ovaries. It was observed that these groups of oöcytes are isolated by invading stroma-cells surrounding each egg by a single layer of cells, and thus forming the typical primary follicles. The peripherally located egg-nests are said to have disappeared completely after 4–5 weeks of life. (Casida, 1935; Wetli, 1942; Hadek and Getty, 1959.) At the same time an intensive development of multi-layered follicles commences. These follicles grow to a diameter of more than 400 μ and the first Call-Exner-bodies can be seen about 8–10 weeks after birth. Vesicular follicles appear on the surface of the ovary between the twelfth and sixteenth week of life. At this stage the ovary of the pig has finished the developmental processes and will remain unchanged until puberty.

Casida (1935) treated young female pigs with gonadotrophic hormone preparations, and the ovaries responded only when vesicular follicles were present, confirming the results obtained in rodents. After hormone-treatment of 16-week-old gilts, follicular growth, ovulations and corpora lutea-formation occurred. On the other hand, no macroscopic reactions could be observed in pigs treated at 6 weeks of age. However, in the peripheral cortex many egg-nests, normally absent in untreated pigs of the same age, could be seen. The conclusion may be thus drawn that in the prepubertal ovary of the pig the processes of folliculogenesis are autonomous and the ovaries are unresponsive to gonadotrophic hormones as shown in laboratory animals.

On the basis of these facts we were doubtful about the general validity of the histological findings in the two NE-treated ovaries mentioned above. In a larger experiment, therefore, we again investigated the effects of progestogens on the undeveloped ovaries of new-born pigs. The main problem was to study the developmental processes until puberty by histological methods. Furthermore, we wanted to clarify whether or not sterility could result from an early postnatal progestogen treatment.

METHODS

Chlormadinone (CAP), a very effective human contraceptive, was used chiefly in these experiments. Additional trials have been

447

carried out with NE and both substances were prepared as long-acting compounds. Ten days after birth single injections were administered in doses of 1 mg, 10 mg and 40 mg CAP. Several animals were also treated with a second injection (40 mg) at the fifth week and a third (40 mg) at the eleventh week of life. The dose of NE was 10 mg/animal at the tenth day after birth.

The effect of these steroids has been measured by histological examination at the tenth, fifteenth and twenty-second week after birth and in ovaries of adult animals more than 6·5 months old.

RESULTS

More treated animals reached sexual maturity as indicated by ovulations and formation of corpora lutea than the controls at the time of slaughter (Table 1). It may be concluded that a single injection of Chlormadinone early after birth did not inhibit follicular

Table 1

Percentage of animals ovulating at slaughter (6·5–7·5 months) following treatment with progestogen 10 days after birth

Treatment	Number of animals	With corpora lutea	Percentage of mature animals
1 mg CAP	11	9	82
10 mg CAP	6	5	83
40 mg CAP	6	5	83
10 mg NE	11	7	64
Controls	105	48	46

Table 2

Weight of ovaries in treated and untreated animals at the fifteenth week of life

Treatment	Number examined	Average weight of ovary
Controls	2	2·00 g
1 mg CAP	2	2·05 g
10 mg CAP	2	0·60 g
40 mg CAP	3	0·60 g
3 × CAP*	2	0·40 g
NE	2	0·25 g

* Injection of CAP at 10 days, 5 weeks and 11 weeks after birth.

448

development, and possibly there was even a stimulation which caused an earlier onset of puberty. At the fifteenth week of life, however, a distinct inhibition in growth of the ovaries of treated animals could be observed (Table 2). While controls showed large ovaries caused by an intensive development of Graafian follicles, the ovaries of animals treated with CAP or NE in doses higher than 1 mg, were small. The surface of these ovaries was smooth, with only a few Graafian follicles (*Plate Ia*). There was no effect of dosage, except that ovaries treated with 1 mg showed normal weight and numbers of vesicular follicles. The inhibition in growth of the ovaries lasted only for a limited period. At 22 weeks of age the ovaries of all hormone-treated animals, including those receiving repeated injections, had increased in size and had reached or exceeded normal weight. The ovaries of NE-treated animals were larger than those of controls, and in one case an intensive enlargement of Graafian follicles could be seen (*Plate Ib*). This may have been caused by a higher secretion of gonadotrophic hormones, probably as a result of a rebound-effect. Contrary to our working hypothesis, an acceleration rather than a retardation of puberty could be observed. In this experiment some hormone-treated pigs ovulated earlier than controls, and in several cases became pregnant at a very early age. This result, however, is a preliminary one and more investigations with other progestogens and different methods of application are planned.

We next investigated whether the postnatal treatment with long acting progestogens had influenced oögenesis or follicle-development. First we had to determine whether oögenesis was still taking place at the time of hormone-treatment on the tenth day after birth. The histological examinations demonstrated that the proliferative processes of oögenesis were not completed by this time. Evidence of intensive neoformation of oöcytes and follicles could be observed in the peripheral cortex (*Plate Id*). Groups of dark-stained nuclei, believed to be derived from oögonia, were lying just beneath the surface-epithelium. These clusters or nests, which have been named 'multi-nuclear follicles' by Hadek and Getty (1959), will be transformed into so-called 'polyovular follicles', groups of oöcytes with large and translucent nuclei, and separated only by thin cell-membranes. Invading stroma-cells subdivide these groups of oöcytes, forming primary follicles. The inner part of the cortex is always filled with primary follicles surrounded by flat follicle cells but there exist differences in number and distribution of the primary follicles. In the one type many primary follicles could be seen, the

15+   449

peripheral zone with multi-nuclear and polyovular follicles was narrow (*Plate Ie*). The other type showed a more undeveloped character with fewer primary follicles, and often with incomplete or missing follicle-cells, but a broad zone of proliferative elements of oögenesis (*Plate If*). The same two types of follicle-development were also found in the ovaries of new-born pigs. This statement means that hormone-treatment at the tenth day after birth must act upon different stages of ovarian development which probably may later influence the hormone-dependent reactions.

At the tenth week of life the ovarian cortex of each untreated pig was enlarged by the formation of multi-layered follicles. These reached a large diameter, but had not yet formed an antrum. In the subepithelial zone there were primary follicles, isolated from each other and surrounded by flat follicle-cells (*Plate IIa*). Oögonia, multi-nuclear or polyovular follicles or other cell-elements of oögenesis were absent.

The ovaries of all treated animals differed by their thinner cortex. Only a reduced number of small multi-layered follicles had developed, and many primary follicles, mostly with missing epithelial cells, filled the cortex. These follicles were often distributed in nests or cords, and the invasion of stroma cells and the isolation of oöcytes were incomplete. The most important observation was the appearance of the germinal elements, which resembled those seen in the ovaries of new-born animals (*Plate IIb* and *c*). Multi-nuclear and polyovular follicles, nests of oöcytes were located in the peripheral zone of the cortex, and we observed different numbers of single or grouped cuboidal cells with a dark stained nucleus and an acidophilic cytoplasm, lying just beneath the surface epithelium. Surrounding follicle cells were often absent. In several sections these cells were distributed in a small continuous layer, representing a 'germinal-epithelium'. Probably these cells were undeveloped, and were more primitive primary follicles. Thus the effects of postnatal progestogen treatment at the tenth week can be summarized as follows: The processes of oögenesis had been maintained until the tenth week of life, but the formation of multi-layered follicles was inhibited. The primary follicles often showed incomplete or missing layers of surrounding stroma cells. An influence of different dosages could not be demonstrated.

At the fifteenth week of life the ovaries of normal pigs are generally enlarged by an intensive development of many Graafian follicles. The surface of the organ is interrupted by vesicles. In the peripheral spaces between them, groups of primary follicles are present.

No elements of oögenesis can be observed. In treated animals, except those of the 1 mg group, the number of Graafian follicles was strongly reduced (*Plate IId*), and many primary and small multi-

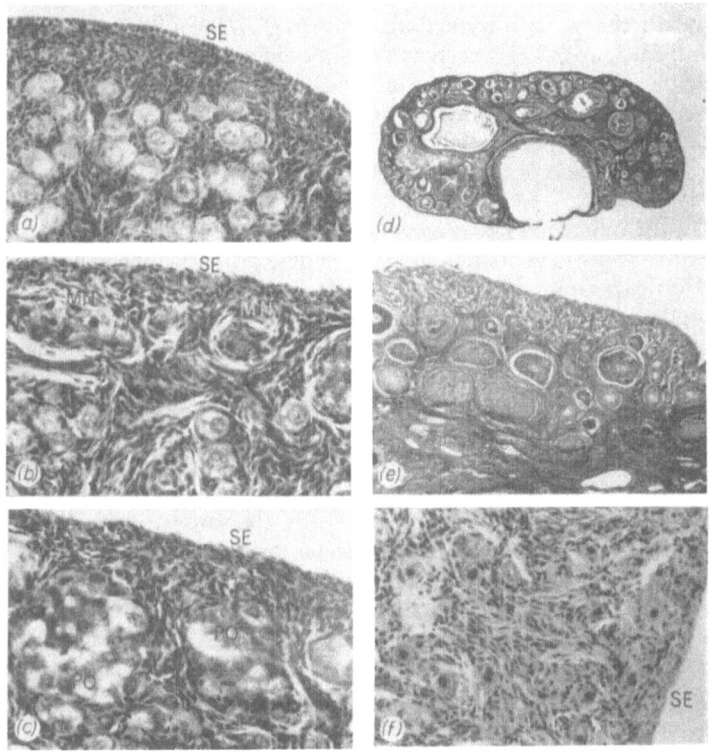

PLATE II
(Reduced ⅔ on reproduction)

Left row: Ovarian cortex of CAP-treated pigs, 10 weeks of age (SE = surface epithelium)

a. *Control animal: normal developed primary follicles only with no multi-nuclear and poly-ovular follicles (160×)*

b. *Treated with 1 mg CAP: multi-nuclear follicles (MN) in the peripheral zone and primary follicles (200×)*

c. *Treated with 10 mg CAP: multi-nuclear (MN) and polyovular follicles (PO) in the peripheral zone*

Right row: Ovarian sections of CAP-treated pigs, 15 weeks of age

d. 10 mg CAP: 2 large vesicular follicles only (Magn. 10×)

e. 10 mg CAP: continuous layer of primary follicles and small multi-layered follicles without antrum formation (Magn. 40×)

f. 10 mg CAP: primitive, primary follicles with dark nuclei and missing epithelium (Magn. 160×)

451

layered follicles without antrum-formation filled the cortex (*Plate IIe*). These types of follicles were more frequent than in controls. In all ovaries, however, the cuboidal undeveloped primary follicles, with their dark nuclei and their acidophil cytoplasm were again observed beneath the surface-epithelium (*Plates IIf, IIIa*). There were few multi-nuclear follicles or polyovular follicles. The conclusion can be drawn that the inhibition of ovarian development lasted longer than 15 weeks post-natum in progestagen-treated animals when given dosages higher than 1 mg CAP. In all treated ovaries, however, the same primitive type of subepithelial germinal cell was present as seen at 10 weeks.

In the untreated mature ovary of swine primary follicles are only the precursors of vesicular follicles. These primary follicles normally lie in the peripheral cortex, mostly located in loose groups and surrounded by a flat layer of stroma cells. The nuclei of the egg-cells are enlarged and translucent. Other elements of folliculogenesis cannot be found.

The ovaries of progestogen-treated animals, with or without corpora lutea, did not only show normal primary follicles in small numbers, but also greater numbers of the primitive type of primary follicles, mentioned before (*Plate IIIa*). These germ cells have been found in varying proportions beneath the surface-epithelium in single or double loculi and sometimes in discrete groups. These cells can often be observed migrating deeper into the cortex to transform themselves into multi-layered follicles. At this time a granulosa layer is formed firstly by stroma cells (*Plate IIIb*): The transformation to Graafian follicles must be normal, as it can be convincingly demonstrated by pregnancies of gestagen-treated pigs. We must assume therefore that not only the typical primary follicle is the dormant kind of germ-cell from which follicular development starts. Additionally a more primitive type of primary follicle, occasionally also seen in normal ovaries, may be preserved during ovarian development and early oögenetic processes. Also Watzka (1957) believed that germ-cells can be changed into a disguised form. The origin and fate of these germinal cells should be studied more intensively. In several sections of adult ovaries, however, structures similar to the multi-nuclear follicles normally seen in undeveloped ovaries of new-born pigs have been observed (*Plate IIIc*).

At this time we are unable to decide whether the postnatal gestagen treatment produced some pathological development of follicles. One observation seems worthy of mention; that is, the proportion of multi-layered follicles with two or more eggs was higher in treated animals than in controls. We observed these among the primary follicles in ovaries of 10-week-old pigs as well as in the multi-layered follicles of pigs 15 weeks of age (*Plate IIId*) and in mature animals, too (*Plate IIIf*). In one animal, 15 weeks old and treated with 10 mg Chlormadinone, we discovered a Graafian follicle containing five eggs each surrounded by corona radiata cells (*Plate IIIe*). There are several theories about the origin of such

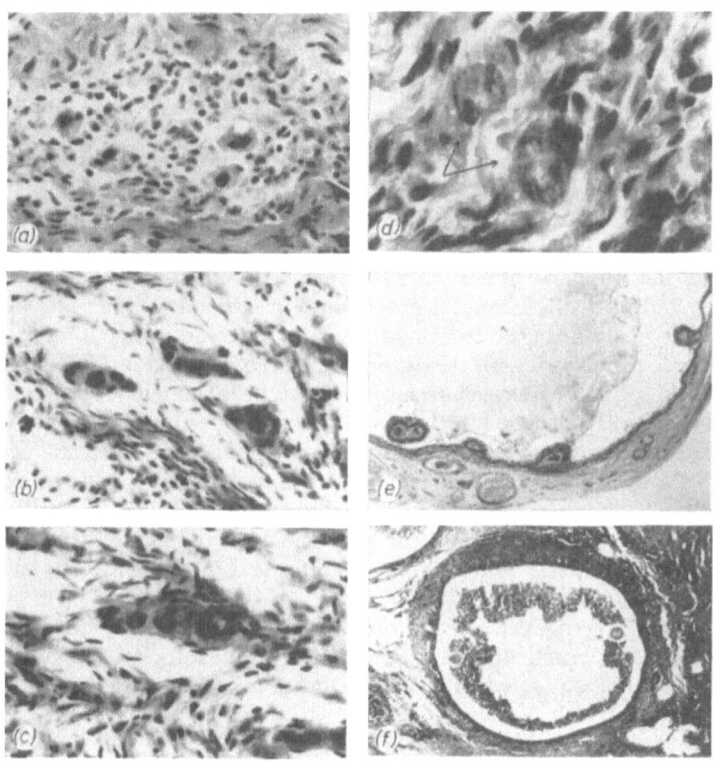

PLATE III
(Reduced ⅔ on reduction)
Left row: Sections of ovarian cortex of CAP-treated pigs (7 months old)
a. *Primitive, incomplete primary follicles in the peripheral cortex* (*Magn. 250×*)
b. *Early formation of follicular epithelium (inner cortex)* (*Magn. 300×*)
c. *Group of germinal cells (multi-nuclear follicle) no cytoplasmatic membranes* (*350×*)
Right row: Follicles with biovular eggs in CAP-treated ovaries
d. *Biovular eggs in multi-layered follicles, pig, 15 weeks of age* (*Magn. 60×*)
e. *Several eggs in a Graafian follicle of a 15 weeks-old pig* (*fat stained section—magn. 60×*)
f. *2 egg-cells in a Graafian follicle, pig, 7 months old* (*Magn. 20×*)

follicles which contain more than one egg (Brambell, 1956). They are said to arise by a second division of the oöcytes, by fusion of adjacent neighbouring follicles or by an incomplete isolation of oöcytes by the stroma cells. This incomplete separation of oöcytes by the failing invasion of stroma cells was characteristic for ovaries of all progestogen-treated animals.

SUMMARY

Postnatal treatment with Chlormadinone and probably NE inhibited ovarian development for more than 15 weeks of life. During this period no Graafian follicles or larger multi-layered follicles developed, and the well-known intensive atresia of oöcytes and primary follicles was noticeably reduced. Neoformation of follicles could be seen for a period longer than 10 weeks. When the inhibiting action of the progestogens had ceased, a gradual regeneration took place connected with the formation of normal vesicular follicles and the onset of puberty was earlier than normal. No sterility resulted from the neonatal progestogen treatment. Generally, the histological findings were similar to those observed in hypophysectomized animals (Paesi, 1949; Ingram, 1953; Burkl and Kellner, 1954). A more detailed discussion of this problem will be published elsewhere (Radermacher, 1966; Schilling and Radermacher, 1967).

In the ovaries of progestogen-treated animals of all ages germinal cells and histological structures representing relative undeveloped stages were preserved. These included the nestlike or cordlike distribution of follicles and the presence of incompletely developed primitive follicles, which can be transformed into normal Graafian follicles.

These investigations of the prepubertal ovary, therefore, in addition to their practical application, point to some interesting problems of folliculogenesis, which will be studied intensively in our future work.

REFERENCES

Amoroso, E. C. (1955). *Br. med. Bull.* **11**, 117–125
Barraclough, Ch. A. (1961). *Endocrinology* **68**, 62–67
Brambell, F. W. R. (1956). 'Ovarian Changes'. In: *Marshall's Physiology of Reproduction.* Ed. by A. S. Parkes, 3rd edn. London: Longmans Green
Burkl, W. and Kellner, G. (1954). *Acta anat.* **23**, 49–57
Casida, L. E. (1935). *Anat. Rec.* **61**, 389–396
Dean, H. W. and Greep, R. O. (1946). *Am. J. Anat.* **79**, 117–120

Franchi, L. L., Mandl, A. M. and Zuckerman, S. (1962). *The Ovary,* Vol. 1, 1–88. London; The Academic Press

Hadek, R. and Getty, R. (1959). *Am. J. vet. Res.* **20,** 578–584

Hansel, W. M. (1964). In H. H. Cole: *Gonadotropins* 160–170. London; Freeman & Co.

Hisaw, F. L. (1947). *Physiol. Rev.* **27,** 95–119

Ingram, D. L. (1953). *J. Endocrin.* **9,** 307–311

Jöchle, W. and Schilling, E. (1965). *J. Reprod. Fert.* **10,** 287–288

Logothetopoulos, B., Sharma, B. and Kraicer, J. (1961). *Endocrinology* **68,** 417–430

Paesi, F. J. A. (1949). *Acta endocrin., Copenh.* **3,** 89–104

Pfeiffer, C. A. (1936). *Am. J. Anat.* **58,** 195

Pincus, G. (1965). *The Control of Fertility.* The Academic Press

Radermacher, R. (1966). *Diss. vet. med.* Berlin

Röstel, W., Jöchle, W. and Schilling, E. (1964). *Proc. V. Int. Congr., Reprod. A.I. Trento,* Vol. III, 474–478

Sawyer, Ch. H. (1964). In H. M. Cole: *Gonadotropins* 113–159. London: Longmans Green

Schilling, E. and Radermacher, R. (1967). *Zbl. Vet. Med.* In press

Stange, H. H. and Drescher, J. (1956). *Arch. Gynäk.* **187,** 693–699

Takasugi, N. (1953). *Annotnes zool. jap.* **26,** 52

Turner, C. D. (1939). *J. Morph.* **65,** 353–381

Watzka, M. (1957). *Das Ovarium.* Hdb. mikr. Anat. VII. Heidelberg; Springer Verlag

Wetli, W. (1942). *Diss. vet. med.* Zürich

VII. HORMONAL CONTROL OF
UTERINE REACTIONS

THE REACTION OF UTERINE ENDOMETRIUM ON SPERMATOZOA AND EGGS

M. C. CHANG

Worcester Foundation for Experimental Biology,
Shrewsbury, Massachusetts, U.S.A.

INTRODUCTION

ONE OF our contemporary authorities on the physiology of the uterus, Reynolds (1949), quoted an English physician, Tyler Smith, who wrote in the *Lancet* of 1848 that the uterus 'supports the race in the same way that the stomach and heart support the individual'. As the uterus is the obvious organ to reproduce young or to develop embryos, the historical and cultural aspects of the uterus can be traced back to very ancient times and to the very primitive tribes (Plaut, 1959). Its biological importance and its physiological function have been dealt with by Marshall in the second edition of *The Physiology of Reproduction* (1922), and in the third edition by Parkes (1952, 1956 and 1960), also by Alan, Danforth and Doisy in the second edition of *Sex and Internal Secretions* (1939) and the third edition by Young (1961). One important function of the uterus has been rather neglected, however, that it also supports spermatozoa for a certain length of time. This function is perhaps just as important as that it 'supports the race' for generations or supports the embryos for several months. I shall deal with the reaction of the uterus on spermatozoa and on early eggs by reporting some of my own observations and the contributions of other learned colleagues in the field of animal reproduction.

It is known that only one sperm is needed to fertilize an egg although hundreds of millions of sperm are deposited in the female tract at mating, and that the sperm will pass through the uterus to perform their ultimate function. In this respect the function of the uterus and other parts of the female tract is to eliminate the vast number of unnecessary sperm and to maintain, to nurse, and to develop the fertilizing capacity of a few selected ones during this short, but important, interval. What happens to those redundant sperm and what happens to those sperm destined to encounter and to fertilize the eggs in the female tract? This is of importance to the workers interested in the physiology of sperm, and to those interested in the physiology of the uterus.

FATE OF REDUNDANT SPERM

A rabbit ejaculate contains from 50 to 700 million sperm with 2–15 per cent sperm heads (Chang, 1956). The number of sperm which can be recovered from both uterine horns 2–48 h after mating is not

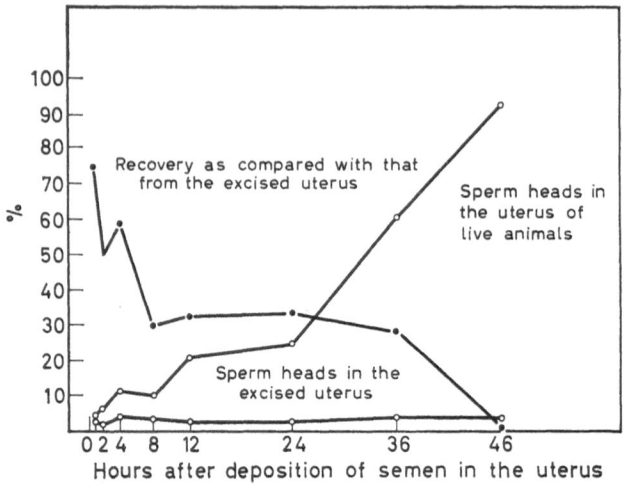

Figure 1. Percentage of spermatozoa recovered from the uterus of live animals (From M. C. Chang, 1956, by courtesy of the Editor, *Annali Ostetricia e Ginecologia*)

more than 2 million. That is, only about 1 per cent of the sperm in an ejaculate gets into the uterine horns of a rabbit. The number of sperm is higher 4–12 h after mating, but many sperm heads are found in the uterus 48 h after mating. When fresh semen is injected into the uterine lumen, whether or not both ends of the uterine horns are ligated, only 4–20 per cent of injected sperm can be found 6 h later. If the same volume of fresh semen is injected into each uterine horn of the same animal and one horn is removed and kept *in vitro* at body temperature, then the disappearance and disintegration of sperm (separation of head and tail) only occurs in the uteri of live animals. These reactions take place rapidly from 1–12 h, slow down from 12–36 h and then increase from 36–48 h (*Figure 1*). The disintegration and disappearance of sperm in the live animal are attributed to the contraction of the uterus and to certain bio-chemical reactions of the uterine endometrium on sperm but not to the motility of sperm or the reaction between seminal plasma and uterine endometrium.

It has been reported that in the rabbit no more than 2 million sperm are found in the whole uterine horns at a given time, from 1–28 h after mating (Braden, 1953). In some laboratory animals, such as the rat and hamster, the uteri are distended and filled with sperm soon after mating. The relative number of sperm getting into the uteri in different species is obviously dependent on the anatomy and physiology of the cervix and the pattern of mating.

The phagocytosis of sperm in the peritoneal cavity or in the female tract has been reported (Hoehne, 1914; Hoehne and Behne, 1914; Yochem, 1929; Merton, 1939; Pasutin, 1952) and a detailed study of this phenomenon in the mouse and rat was given recently by Austin (1957). He reported that in mice killed 2–6 h after mating about 0·1 ml of uterine fluid, containing 20 million sperm but very few leucocytes, can be obtained. Fourteen hours after mating leuco-cytes increase to 25 million and the sperm decrease to 11 million, most of which undergo phagocytosis. About 20 h after mating the uterine content is evacuated in spite of the continued presence of a vaginal plug. Phagocytosis of sperm in the rat, however, is less extensive.

Based on this observation it appears that in about a 10 h period 9 million sperm disappear and 25 million leucocytes appear in the uterine lumen. Assuming that the infiltration of leucocytes to the uterine lumen is a gradual process and assuming that one leucocyte is capable of ingesting at least one sperm, it seems that 9 million sperm have been ingested in a period of 10 h, probably by more than 9 million leucocytes. Since 20 million sperm disappear from the uterus 20 h after mating it is possible that all the 20 million sperm can be ingested in 20 h.

In this connection, I should like to turn to some work done in our laboratory. Yanagimachi and Chang (1963) found that in one uterine horn of an unmated hamster the highest number of leuco-cytes, an average of 30,000, is found soon after oestrus or at the early stage of metoestrus. However, in mated hamsters the highest number of leucocytes, an average of 1·5 million, is found about 12–14 h after mating (*Figure 2*). The appearance of leucocytes in the uterine horn is about 6 h earlier than, and about five times more than in an unmated hamster.

We know that, in the hamster, the number of sperm in one uterine horn soon after mating is 40 million and the highest number of leucocytes reached is 1·5 million about 12 h after mating. Since a large number of sperm are discharged into the vagina and no sperm are found in the uterus 18 h after mating, it implies that not all the

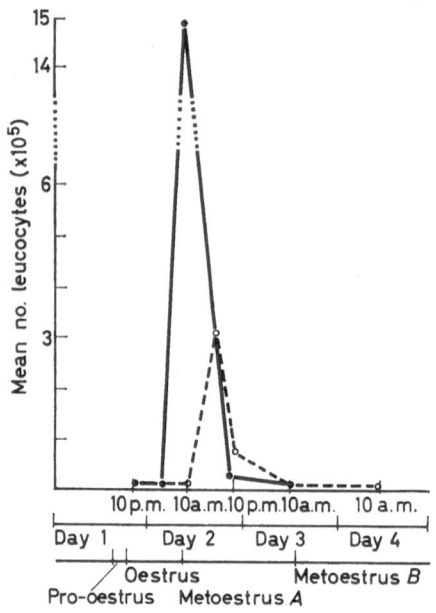

Figure 2. Number of leucocytes appearing in the uterine lumen of unmated (○) and mated (●) female hamsters (From R. Yanagimachi and M. C. Chang, 1963, by courtesy of the Editor, *Journal of Reproduction and Fertility*)

sperm are ingested by leucocytes, because ingestion of 40 sperm by each leucocyte in 18 h is hard to comprehend, so I am not inclined to say that all the unnecessary sperm that get into the female tract will be ingested by leucocytes. Furthermore, Austin (1957) concluded that the leucocytic invasion of the uterine lumen was due to the distension of the uterus with fluid, not due to the presence of sperm. We have found, however, that the number of leucocytes increases tremendously with the injection of sperm or seminal vesicle fluid into the uterus, but not by injection of the same amount of saline. The phagocytosis of rabbit sperm by polymorphonuclear leucocytes is also reported in the rabbit uterus (Menge, Tyler and

Casida, 1962) and the penetration of sperm into the uterine glands in various species is also claimed (Austin, 1960a). Here I should like to stress that, during their passage through the uterus, a large number of sperm may be discharged, disintegrated by some chemical reaction, or ingested by leucocytes, probably in 1–2 days. The spermatozoa certainly have a hard struggle in a short time to reach the site of fertilization. There is, however, a species difference, probably dependent on other physiological activities during evolution. For instance, the mare has a long heat period of 7 days, and the ferret ovulates about 30–40 h after mating, so the horse sperm can survive for 7 days (Day, 1942; Burkhardt, 1949) and the ferret sperm can survive for 5 days in the female tract (Chang, 1965).

Concerning the phagocytosis of sperm in the uterus, I should like to refer to an electron microscopic study of phagocytosis by Bedford (1965). Besides the ultrastructural details of the phagocytosis, this study revealed a very interesting phenomenon related to the reaction of uterine endometrium on spermatozoa. At oestrus the uterine leucocytes are capable of ingesting an intact sperm even when the sperm is still motile. Leucocytes of the pseudopregnant uterus *in vivo*, leucocytes of the pleural cavity *in vivo*, and leucocytes from the uterus of oestrus rabbits *in vitro*, are unable to ingest an intact sperm. The sperm only adheres to the leucocytes. So it appears that some changes occur in the surface membrane of sperm in the oestrous uterus which renders the intact sperm head acceptable to uterine leucocytes. The infiltration of granulocytes in the uterus of oestrous and oestradiol-treated rabbits (Lamming, 1961) and the presence of an 'acid soluble fraction' in the uterine flushing during pseudopregnancy (Haynes, 1963), have been also reported.

The significance of these observations is that the reaction of uterine leucocytes is not a simple reaction between leucocytes and sperm but this reaction is dependent on other uterine ccmponents which are probably controlled by female hormones. Considering that resistance to uterine infection is higher in the uteri of oestrous animals than in the uteri under the influence of progesterone (Rowson, Lamming and Fry, 1953; Lamming, Seaman and Woodbine, 1955) and that capacitation of sperm cannot be achieved in the uteri of pseudopregnant rabbits (Chang, 1958), the reaction of leucocytes of an oestrous uterus on spermatozoa, the antibiotic reaction of a uterus at oestrus, and the capacitation of sperm in the uterus may have some close correlations. In this connection the presence of proteolytic enzymes in the uterine endometrium, as reported by my associates (Albers, Bedford and Chang, 1961), may also play a role for the elimination of unnecessary sperm or for the capacitation of sperm. As for the elimination of redundant sperm, its biological significance is still obscure. It may have some immunological significance to increase the tolerance of maternal organisms and for the survival of embryos. It may be one of the mechanisms to prevent polyspermy as suggested by Braden and Austin (1954a).

In order to determine whether the vagina, the uterus, or the Fallopian tube, is a better environment for the survival of sperm, we (Chang and Pincus, 1964) injected rabbit semen into different parts of the female tract of rabbits 28–30 h before ovulation. It is found that the proportion of eggs fertilized is higher when 20–46 million sperm are injected into the uterus (35 per cent) rather than into the tubes (7 per cent) or the vagina (14 per cent). Since 27–36 per cent of the eggs are fertilized when 114–240 million sperm are inseminated into the vagina while only 0–14 per cent of the eggs are fertilized when 2–46 million sperm are inseminated into the vagina, it appears that the fertilizing capacity of spermatozoa, or the fertilizing life of sperm, is correlated to the number of sperm inseminated. This is probably because a larger number may have the chance to contain better spermatozoa. Based upon the assumption that a larger number of sperm may contain better sperm, the higher proportion of eggs fertilized, when a given number of sperm is injected into the uterine horns rather than into the vagina or the tubes, may be due to the possibility that a larger number of sperm were reserved when injected into the uterine horns and that many sperm leaked out when injected into the tubes or into the vagina. It is not necessarily valid, however, that the uterus is a better place for the survival of sperm or for the maintenance of their fertilizing life.

CAPACITATION OF SPERM IN THE FEMALE TRACT

The elimination of redundant sperm in the female tract is of academic interest and, as far as we know, is of no definite functional importance. Of functional importance, however, is the biochemical changes of those sperm that are destined to perform their ultimate function of fertilizing the eggs.

It was demonstrated 15 years ago (Austin, 1951; Chang, 1951) that rabbit spermatozoa undergo certain physiological changes in the female tract before they are capable of penetrating the zona pellucida for fertilization. This phenomenon was also observed in the rat and denoted by Austin (1952) as capacitation of sperm. Capacitation appears to be a general phenomena in all mammals so far studied (rat, Noyes, 1953; mouse, Braden and Austin, 1954b; hamster, Chang and Scheaffer, 1957; sheep, Mattner, 1963; ferret, Chang and Yanagimachi, 1963; Yanagimachi, 1966), and no evidence has been presented for any species that spermatozoa fresh

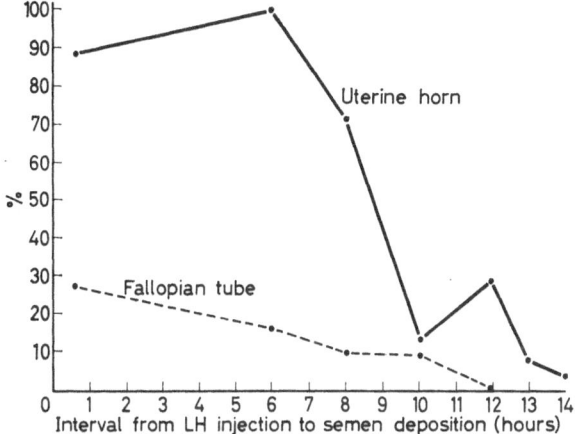

Figure 3. Proportion of eggs fertilized following deposition of semen into the Fallopian tube and the uterine horn (From *Figure 3*, C. E. Adams and M. C. Chang, 1962, by courtesy of the Wistar Institute of Anatomy and Biology)

from the male tract can rapidly penetrate the eggs without undergoing a period of capacitation (Noyes, 1959). Furthermore, authentic fertilization of mammalian eggs *in vitro* can only be achieved by using capacitated sperm (Dauzier, Thibault and Wintenberger, 1954; Chang, 1959; Thibault and Dauzier, 1960; Bedford and Chang, 1962a; Suzuki and Mastroianni, 1965). Even in the experiments performed by Yanagimachi (1966) in our laboratory on the fertilization of hamster eggs *in vitro*, it has been shown that epididymal sperm takes a longer time to penetrate the eggs than sperm recovered from the uterus, which shows that a certain length of time is required for the capacitation of hamster sperm *in vitro*.

It has been shown that capacitation of rabbit sperm can be achieved in the Fallopian tubes of pseudopregnant rabbits but not in the uteri of pseudopregnant rabbits (Chang, 1958) and that capacitation is more efficient in the uteri than in the Fallopian tubes of the rabbits (*Figure 3*, Adams and Chang, 1962). In double

465

mating of rabbits, Dziuk (1965) reported that the spermatozoa that have resided in the female tract for a longer time have a better chance of fertilizing the eggs than those residing in the female tract for a short time. Although capacitation of rabbit sperm is claimed to have been achieved in the interior chamber of eyes, in the colon or in the bladder, and by incubation of ejaculated sperm in the excised uterine horn and in the presence of red cells *in vitro* (Noyes, Walton and Adams, 1958) we (Bedford and Chang, unpublished) were unable to repeat these experiments to achieve capacitation in other organs *in vivo* or in the female tract *in vitro*. Only recently, Kirton and Hafs (1965) have reported the capacitation of rabbit sperm *in vitro* in the presence of β-amylase.

Due to the complexities in the experimental procedures for the assay of capacitation, such as the lack of accurate methods to determine the time of ovulation, the variation of sperm samples, and the possibility of partial capacitation *in vivo* or *in vitro* as well as the completion of capacitation *in situ*, the inconsistency of results in different laboratories is to be expected. The important object of such a study however, is to find out what the mechanism of capacitation is and also the physiological process of capacitation.

Biochemically it has been reported by Hamner and Williams (1963) that a fourfold increase of oxygen uptake by rabbit spermatozoa occurs following their incubation in the rabbit uterus. Mounib and Chang (1964) confirmed these results and found that both endogenous and exogenous respiration was higher in the *in utero* incubated than in the fresh sperm. The possibility of an enhancement of a hexose monophosphate shunt in the *in utero* incubated sperm was also revealed. Whether the enhancement of the hexose monophosphate shunt in the *in utero* incubated sperm is to provide more energy, to supply pentose that may be required for nucleic acid synthesis, or to accelerate other biochemical reactions for sperm to penetrate the egg, remains to be determined.

Morphologically, Austin and Bishop (1958a; 1958b) believe that the shedding or elevation of acrosome in some rodent sperm during their sojourn in the female tract, causes the exposure of perforatorium and may elevate certain mucolytic enzymes, such as hyaluronidase (Austin, 1960b), thus enabling the sperm to penetrate the zona pellucida. Adams and Chang (1962), however, were unable to distinguish any changes in the acrosome of rabbit spermatozoa obtained from the Fallopian tubes, examined under a phase-contrast microscope. Bedford (1963) considered that the morphological changes of acrosome of tubal and uterine sperm were attributable to the effect of ageing. In a later paper, Austin (1963) reported certain ultrastructural differences between rabbit epididymal sperm and the sperm in the zona pellucida, examined under an electron microscope. He concluded that the rabbit spermatozoa differs from the sperm of other rodents in showing no obvious acrosome change, although the possibility of ultrastructural changes in association with hyaluronidase is not excluded.

At the present time the physiological processes of capacitation which enable the sperm to penetrate the egg are still obscure. The capacitation of rabbit sperm in the presence of β-amylase *in vitro* has been reported (Kirton and Hafs, 1965). We (Thorsteinsson and Chang, unpublished) found, however, a very low concentration of amylase in the rabbit endometrium as compared with rabbit blood serum (1 : 40), a higher concentration of amylase in the endometrium of the oestrous uterus than in the pseudopregnant uterus (2 : 1), and yet a relatively higher amylase activity in the lumen of the pseudopregnant uterus than the oestrous uterus. Due to a very low proportion of eggs fertilized by amylase-incubated sperm we (Chang and Thorsteinsson, unpublished) are inclined to believe that the role of amylase during capacitation is still not certain, especially when consideration is given to the involvement of other enzymic activities in the female tract.

In view of the following facts: (*1*) an acrosome reaction of the sperm occurs in the lower organism when sperm are close to the eggs (Dan, 1956; Colwin and Colwin, 1957); (*2*) the activation and fertilization of eggs cannot take place by injection of live spermatozoa into sea urchin eggs (Hiromota, 1962) and (*3*) no change of sperm head can be detected when sperm were mechanically pressed into the cytoplasm of the rabbit eggs (Chang and Hunt, 1962), it appears that the reaction of sperm before fertilization in lower organisms and the capacitation of sperm in the female tract in the higher organisms, may be analogously, and basically, a similar phenomenon which must take place before the sperm are capable of penetrating and activating the egg.

In passing, I should like to mention the so-called decapacitation factor. I have reported (Chang, 1957) that when capacitated rabbit sperm, recovered from the uterus 12 h after mating, is treated with rabbit, human or bull seminal plasma, the sperm is incapable of penetrating the eggs when deposited into the tubes of recently-ovulated rabbits. If, however, capacitated sperm treated with seminal plasma is deposited into the rabbit tubes 6 h before ovulation, these decapacitated sperm can be recapacitated in the female

467

tract. Bedford and Chang (1962b) found that the decapacitation factor in the seminal plasma can be removed by high-speed centrifugation. It is probably conjugated with protein, and is relatively heat stable. Weinman and Williams (1964) reported the presence of a decapacitation factor in the epididymal secretion. I should like to point out here that the capacitation of sperm is a positive process to develop the fertilizing power of sperm while decapacitation is a negative process to inhibit, or to destroy whatever happens to the sperm during capacitation. These two processes may be quite different. For instance, capacitation cannot be achieved in the uterus of pseudopregnant rabbits but capacitated sperm cannot be decapacitated by the treatment with the uterine washing of pseudopregnant rabbits or by incubation in the pseudopregnant uterus for a certain length of time (Chang and Adams, unpublished). To consider uncapacitated (epididymal or ejaculated sperm) as decapacitated sperm is to confuse the issue because the process of capacitation is not necessarily a process to eliminate the decapacitation factor.

REACTION OF UTERINE ENDOMETRIUM ON THE EARLY EGGS

The reaction of the uterine endometrium on the eggs at their late stages of development or vice versa is in the scope of anatomy and physiology of implantation, which has been dealt with by authorities in this field (cf. Eckstein, 1959; Enders, 1963). Here I should like to review some of my published and unpublished works concerning an adverse effect of uterine endometrium on early eggs.

In a study of the importance of the synchronization between the age of eggs and that of the corpus luteum and thus of the uterine endometrium I transplanted 254 fertilized rabbit eggs at the 2-cell stage (Chang, 1950). They were recovered from the tubes, and placed into the uterus of 1-, 3-, or 4-day pseudopregnant rabbits. Of the transferred eggs, only one, which was in a 1-day pseudopregnant rabbit, developed and came to term. In order to determine what happened to the transferred eggs I examined another group of recipient rabbits 2 or 4 days after transfer. I found that of the 96 transferred eggs, all the 45 recovered were degenerated or degenerating, at 4- to 16-cell stages. It is obvious that newly fertilized eggs, whose normal environment is the lumen of Fallopian tubes, are unable to survive and develop in the uterus. This is not necessarily due to the endometrium not being ready for the survival of eggs because some of the recipients used were before ovulation and some were at different days of pseudopregnancy.

This not only happens in the rabbit, but also in the mouse, because two-celled tubal mouse eggs rarely or never come to term when transplanted into the mouse uterus (Fekete and Little, 1942; Beatty, 1951). In the rat, Noyes and Dickman (1960) report that no development is observed when 2-day eggs are transferred into the uteri of 2-day pseudopregnant rats, and only 3 per cent develop when transferred into the 3-day uteri, whereas 65 per cent of 4- and 5-day eggs develop into young when transferred into the uteri of 4-day pseudopregnant rats.

Later work by ourselves (Chang, 1964, 1966b; Chang and Yana-gimachi, 1965; Chang and Harper, 1966) and by others (rabbit, Greenwald, 1961, 1963; rat and mouse, Banik and Pincus, 1962; guinea-pig, Deansley, 1963; Pincus, 1965) all show that application of oestrogen or oestrogen-related compounds soon after mating has the effect of speeding up the transport of eggs from the tube to the uterus and to expel the eggs from the uterus to the vagina. I have evidence to show that those eggs transported into the uterus at early stages will degenerate in due time if they are not expelled from the uterus (*Figure 4*). I was also able to demonstrate that not only does oestrogen have the effect of speeding up the egg transport but pro-gesterone, if applied before ovulation (*Plate I*, Chang, 1966b) or before the transplantation of eggs into the Fallopian tube (Chang, 1966c), also has the same effect. This observation may have certain practical values for the control of fertility.

In this connection I should like to stress that the uterine lumen, due to its chemical composition and physiological activities, is not a suitable place for the survival of early eggs as is the Fallopian tube. I have reported (Chang, 1955), that when recently-ovulated un-fertilized rabbit eggs are transferred from the tubes of unmated rabbits into the uteri of mated rabbits the recovery rate of eggs from the uteri is very low 1 day after transfer (11 out of 82). Most of them probably disintegrate, and when recovered 6 h after transfer most are fertilized at the pronuclear stages but are in poor shape with their zona pellucida swollen and their cytoplasm not in a healthy condition. The swelling and disintegration of the zona pellucida is also observed when newly-fertilized rabbit eggs are cul-tured in rabbit uterine fluid (Chang, unpublished).

The metabolism of the uterine endometrium differs from that of the Fallopian tube (Mounib and Chang, 1965). It has also been observed that the environment of the tubal lumen is quite different from that of the uterine lumen. For instance, two-celled ferret eggs survive and develop when transferred into the tubes of rabbits but

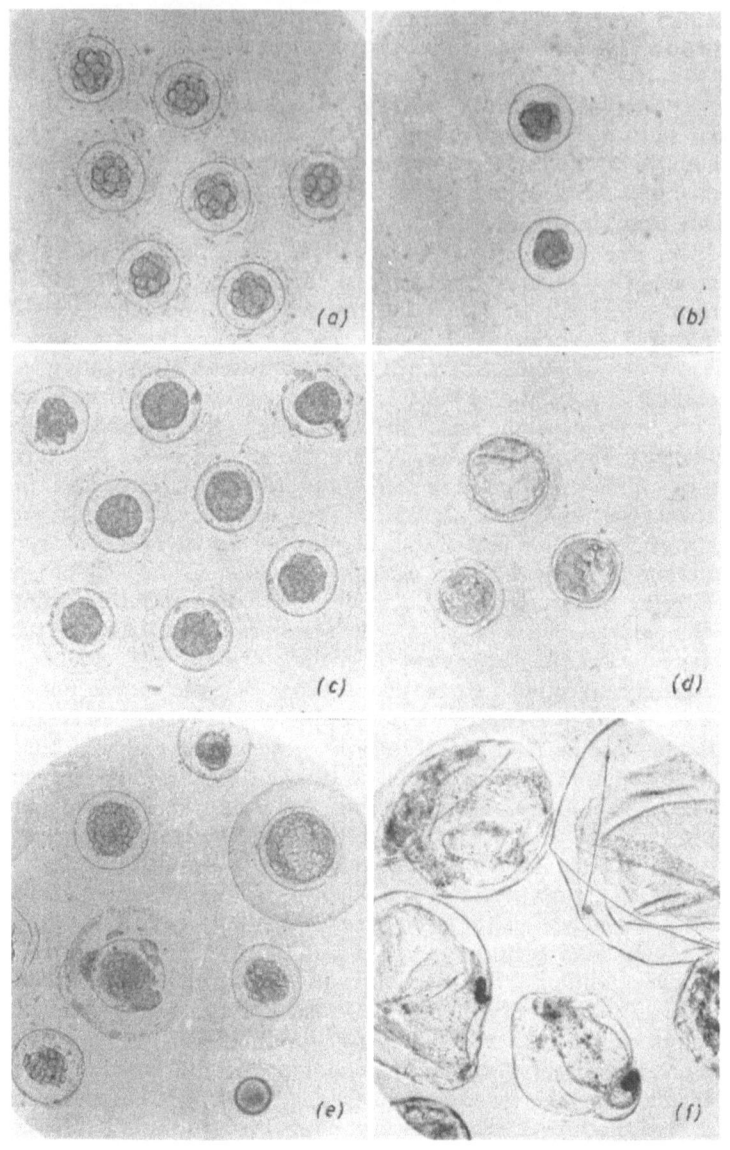

PLATE I

470

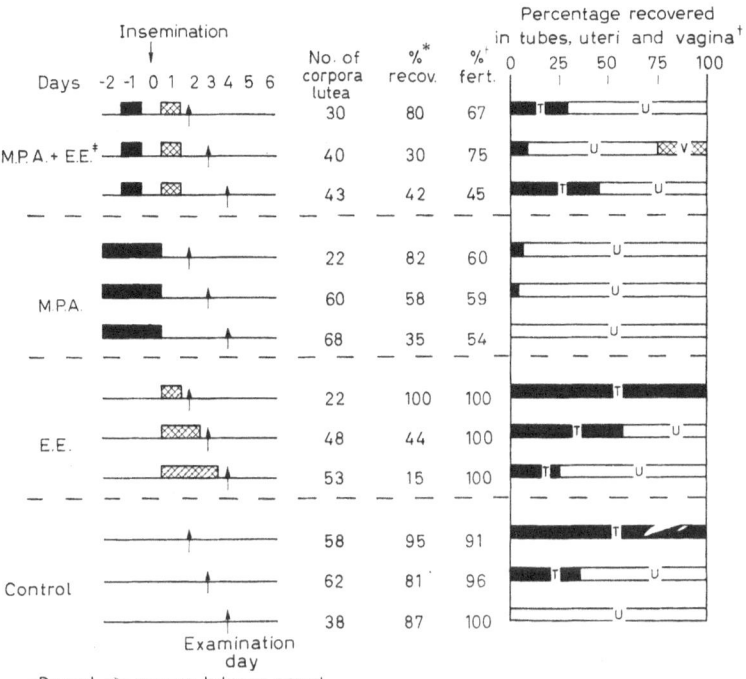

Figure 4. Effects of oral administration of 2 mg medroxyprogesterone acetate and 0·1 mg ethinyloestradiol on transportation and degeneration of rabbit eggs

PLATE I
(Reduced ¼ on reproduction)

Morphology of rabbit eggs recovered from the tube or uterus following the oral administration of Medroxyprogesterone acetate (MPA) and/or Ethinyloestradiol (EE). (a) A group of normal eggs recovered 2 days after insemination from the tubes of a rabbit fed with 0·1 mg EE on day 1; approx. ×63. (b) Two eggs recovered 2 days after insemination from the uterus of a rabbit fed with 2 mg MPA on day −1 and 0·1 mg EE on day 1; approx. ×63. (c) A group of eggs recovered 3 days after insemination from the uterus of a rabbit fed with 2 mg MPA on day −1 and 0·1 mg EE on day 1. Showing retarded cleavage and poor shape; approx. ×63. (d) Three eggs recovered 4 days after insemination from the uterus of a rabbit fed with 0·1 mg EE on days 1, 2, and 3. Showing retarded cleavage and poor shape; approx. ×63. (e) A group of eggs recovered 4 days after insemination from the uterus of a rabbit fed with MPA on days −2, −1, and 0 showing retarded cleavage and poor shape; approx. ×63. (f) A group of eggs recovered 6 days after insemination from the uterus of a rabbit fed with 2 mg MPA 3 days before and after insemination; approx. ×26
(From M. C. Chang, 1966b, by courtesy of the Editor, *Endocrinology*)

471

they degenerate after reaching the rabbits' uteri.   Ferret blastocysts survive and develop when transferred into the rabbits' tubes but they degenerate soon after their transfer into the uteri of pseudopregnant

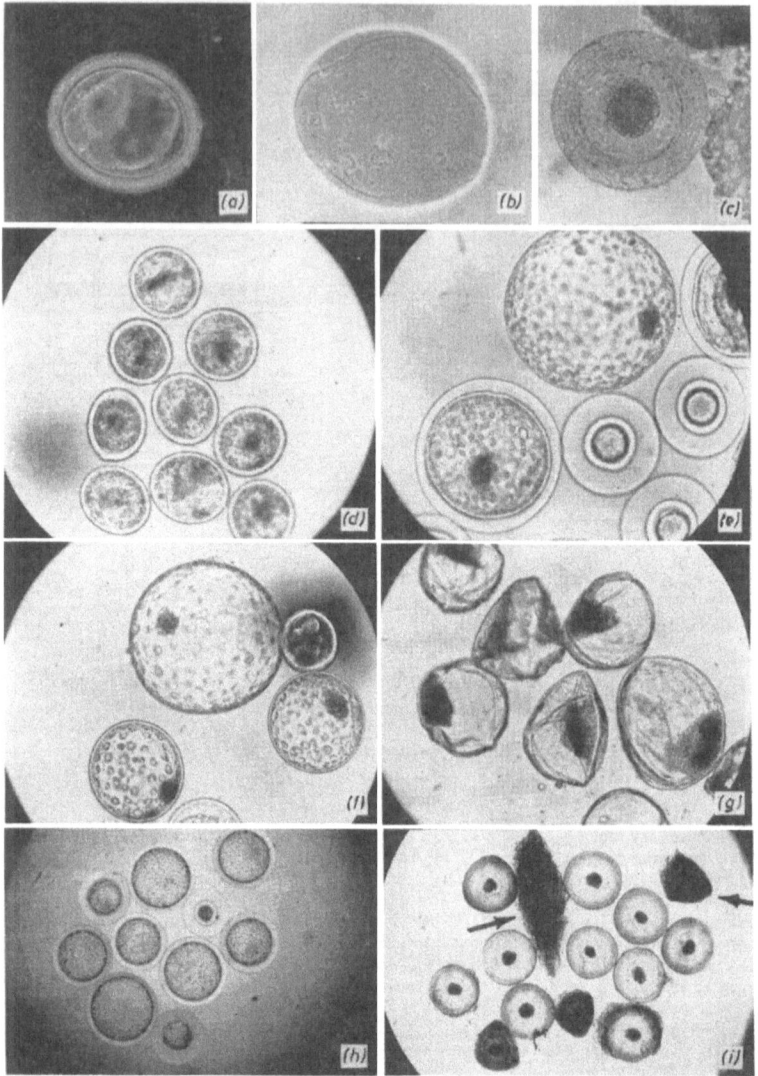

PLATE II

472

rabbits. On the other hand, fertilized rabbit eggs, although they are remarkable for their resistance to manipulation *in vitro* and *in vivo* cannot survive either in the ferret tube or in the ferret uterus (*Plate II*, Chang, 1966a). Recent work by Kirby (1965) has shown that mouse blastocysts recovered from the uterus will develop while the cleaving eggs obtained from the tubes give rise only to tropho-blast and extra-embryonic membranes when transplanted to an extra-uterine site. In this respect, it appears that not only the environment of eggs but the age of eggs also play a role in determin-ing their future development in a new environment. On the other hand, it may indicate that once the eggs have adapted themselves in the uterus they can survive better in other locations.

SUMMARY

Published and unpublished works concerning the reaction of the uterus on spermatozoa and early eggs observed by the author are reported and discussed in the light of the contributions of other workers. Besides the maintenance of embryos or foetuses, the uterus is not only a passage way for sperm, but can, on the one hand, eliminate the redundant spermatozoa, while on the other hand it sustains and develops the fertilizing power of selected ones. Dis-charge of sperm by uterine and cervical activities, phagocytosis of sperm by uterine leucocytes, and other biochemical reactions in the uterine lumen to disintegrate the unnecessary sperm, are major features to eliminate the redundant sperm but the biological signifi-cance in this respect is still obscure.

Certain aspects of the capacitation of sperm in the uteri of mammals with its analogous reaction between sperm and eggs in the lower organisms and the morphological and physiological observa-

PLATE II
(Reduced ¾ on reproduction)
*Reciprocal transplantation of eggs between ferret and rabbit. (a) A ferret egg cleaved into 8–12 cells, recovered from the rabbit Fallopian tube 2 days after transfer; approx. × 130. (b) The same egg as shown in (a) expanded after being stained; approx. × 130. (c) A degenerated rabbit egg recovered from the ferret uterus 4 days after transfer into the ferret ovarian capsule; approx. × 130. (d) A group of ferret blastocysts recovered 6 days after mating; approx. × 63. (e) Three ferret blastocysts recovered from the rabbit uterus, 2 days after the transfer of 6-day ferret eggs in the rabbit Fallopian tube. The other 3 eggs are rabbit eggs; approx. × 63. (f) Three ferret blastocysts recovered 7 days after mating. The small egg is an unfertilized but parthenogenetically cleaved egg; approx. × 63. (g) A group of degenerated ferret blastocysts recovered from the rabbit uterus, 4 days after the transfer of 7-day ferret blastocysts; approx. × 63. (h) A group of early rabbit blastocysts recovered 4 days after mating; approx. × 26. (i) A group of degenerated early rabbit blastocysts recovered from the ferret uterus, 4 days after the transfer of 4-day rabbit blastocysts. The dark arrows indicate unfertilized ferret eggs; approx. × 26 (From M. C. Chang, by courtesy of the Editor, J. Experimental Zoology **44** (1966a))*

tions of capacitated sperm are reported and discussed but the basic mechanism of capacitation and decapacitation is still to be revealed.

The adverse effect of the uterine lumen on the unfertilized or newly fertilized eggs, due to the physiological difference between the tube and the uterine endometrium, rather than the age of endometrium, and the degeneration of eggs by speeding up their transport from the tube to the uterus, is also reported and discussed.

ACKNOWLEDGEMENT

The work reported here was supported by a grant from the Population Council, grants (GM-10529 and GM-08167) and a Career Award (K6-HD-18,334) from the Public Health Service, U.S.A.

Thanks are due to Mrs. Pauline Taparausky for assistance during the preparation of this manuscript.

REFERENCES

Adams, C. E. and Chang, M. C. (1962). *J. exp. Zool.* **151,** 159–166
Albers, H. J., Bedford, J. M. and Chang, M. C. (1961). *Am. J. Physiol.* **201,** 554–556
Allen, E., Danforth, C. H. and Doisy, E. A. (Editors) (1939). *Sex and Internal Secretions,* 2nd edn. Baltimore; Williams & Wilkins
Austin, C. R. (1951). *Aust. J. Sci. Res.* B **4,** 581–596
— (1952). *Nature, Lond.* **170,** 326–327
— (1957). *J. Endocr.* **14,** 335–342
— (1960a). *J. Reprod. Fert.* **1,** 151–156
— (1960b). *J. Reprod. Fert.* **1,** 310–311
— (1963). *J. Reprod. Fert.* **6,** 313–314
— and Bishop, M. W. H. (1958a). *Proc. R. Soc.* B **149,** 234–240
— — (1958b). *Proc. R. Soc.* B **149,** 241–248
Banik, U. K. and Pincus, G. (1962). *Proc. Soc. exp. Biol. Med.* **111,** 595–602
Beatty, R. A. (1951). *Nature, Lond.* **168,** 995–996
Bedford, J. M. (1963). *J. Reprod. Fert.* **6,** 245–255
— (1965). *J. Reprod. Fert.* **9,** 249–256
— and Chang, M. C. (1962a). *Nature, Lond.* **193,** 898–899
— — (1962b). *Am. J. Physiol.* **202,** 179–181
Braden, A. W. H. (1953). *Aust. J. biol. Sci.* **6,** 693–705
— and Austin, C. R. (1954a). *Aust. J. biol. Sci.* **7,** 543–551
— — (1954b). *Aust. J. biol. Sci.* **7,** 552–565
Burkhardt, J. (1949). *J. agric. Sci.* **39,** 201–203
Chang, M. C. (1950). *J. exp. Zool.* **114,** 197–225
— (1951). *Nature, Lond.* **168,** 697–699
— (1955). *La Fonction Tubaire et Ses Troubles,* pp. 40–52. Paris; Masson & Cie
— (1956). *Annali Ostet. Ginec.* **4,** 74–86

Chang, M. C. (1957). *Nature, Lond.* **179**, 258–259
— (1958). *Endocrinology* **63**, 619–628
— (1959). *Nature, Lond.* **184**, 466–467
— (1964). *Fert. Steril.* **15**, 97–106
— (1965). *J. exp. Zool.* **158**, 87–100
— (1966a). *J. exp. Zool.* **161**, 297–304
— (1966b). Endocrinology **79**, 939–948
— (1966c). *Nature, Lond.* **212**, 1048–1049
— and Harper, M. J. K. (1966). *Endocrinology* **78**, 860–872
— and Hunt, Dorothy M. (1962). *Anat. Rec.* **142**, 417–426
— and Pincus, G. (1964). *Proc. 5th Int. Congr. Anim. Reprod., Trento* **3**, 377–380
— and Scheaffer, Diane (1957). *J. Hered.* **68**, 107–109
— and Yanagimachi, R. (1963). *J. exp. Zool.* **154**, 175–183
—— (1965). *Fert. Steril.* **16**, 281–291
Colwin, A. L. and Colwin, Laura H. (1957). 'The Beginnings of Embryonic Development', *A.A.A.S. Publication No. 48*, 135–168
Dan, J. C. (1956). *Int. Rev. Cytol.* **5**, 365–393
Dauzier, L., Thibault, C. and Wintenberger, S. (1954). *C.r. hebd. Séanc. Acad. Sci., Paris* **238**, 844–845
Day, F. T. (1942). *J. agric. Sci.* **32**, 108–111
Deansley, R. (1963). *J. Reprod. Fert.* **5**, 49–57
Dziuk, P. J. (1965). *J. Reprod. Fert.* **10**, 389–395
Eckstein, P. (Editor) (1959). *Implantation of Ova.* Cambridge University Press
Enders, A. C. (Editor) (1963). *Delayed Implantation.* University of Chicago Press
Fekete, E. and Little, C. C. (1942). *Cancer Res.* **2**, 525–530
Greenwald, G. S. (1961). *Fert. Steril.* **12**, 80–95
— (1963). *J. Endocr.* **26**, 133–138
Hamner, C. E. and Williams, W. L. (1963). *J. Reprod. Fert.* **5**, 143–150
Haynes, N. B. (1963). *J. Reprod. Fert.* **6**, 331–332
Hiromata, Y. (1962). *Expl. Cell Res.* **27**, 416–426
Hoehne, O. (1914). *Verh. dt. Ges. Gynäk.* **15**, 514
— and Behne, K. (1914). *Zentbl. Gynäk.* **38**, 5
Kirby, D. R. S. (1965). 'Pre-implantation Stages of Pregnancy', pp. 325–329. *Ciba Fdn Symp.* London; Churchill
Kirton, K. T. and Hafs, H. D. (1965). *Science N.Y.* **150**, 618–619
Lamming, G. E. (1961). *J. Reprod. Fert.* **2**, 517–518
— Seaman, A. and Woodbine, M. (1955). *Nature, Lond.* **175**, 126
Mattner, P. E. (1963). *Nature, Lond.* **199**, 772
Marshall, F. H. A. (1922). *Physiology of Reproduction*, 2nd edition. London; Longmans, Green
Menge, A. C., Tyler, W. J. and Casida, L. E. (1962). *J. Reprod. Fert.* **3**, 396–404
Merton, H. (1939). *Proc. R. Soc. Edinb.* **59**, 207–218

Mounib, M. S. and Chang, M. C. (1964). *Nature, Lond.* **201,** 943–944
— (1965). *Endocrinology* **76,** 542–546
Noyes, R. W. (1953). *West. J. Surg. Obstet. Gynec.* **61,** 342–349
— (1959). *J. Dairy Sci. Suppl.* **43,** 68–83
— Walton, A. and Adams, C. E. (1958). *J. Endocrin.* **17,** 374–380
— and Dickman, Z. (1960). *J. Reprod. Fert.* **1,** 186–196
Parkes, A. S. (Editor) (1952, 1956, 1960). *Marshall's Physiology of Reproduction,* 3rd edn. London; Longmans, Green
Pasutin, G. V. (1952). *Konevodsto* **22,** 12, also *Anim. Breed. Abstr.* **21** (1953) 54
Pincus, G. (1965). 'Pre-implantation Stages of Pregnancy', pp. 378–396. *Ciba Fdn Symp.* London; Churchill
Plaut, A. (1959). *Ann. N.T. Acad. Sci.* **75,** 389–411
Reynolds, S. R. M. (1949). *Physiology of the Uterus.* N.Y.; Paul B. Hoeber Inc.
Rowson, L. E. A., Lamming, G. E. and Fry, R. M. (1953). *Nature, Lond.* **171,** 749
Suzuki, S. and Mastroianni, L. (1965). *Am. J. Obstet. Gynec.* **93,** 465–471
Thibault, C. and Dauzier, L. (1960). *C.r. hebd. Séanc. Acad. Sci., Paris* **250,** 1358–1359
Weinman, D. E. and Williams, W. L. (1964). *Nature, Lond.* **203,** 423–424
Yanagimachi, R. (1966). *J. Reprod. Fert.* **11,** 359–370
— and Chang, M. C. (1963). *J. Reprod. Fert.* **5,** 389–396
Yochem, D. E. (1929). *Biol. Bull. mar. biol. Lab. Woods Hole* **56,** 274–297
Young, W. C. (Editor) (1961). *Sex and Internal Secretions,* 3rd edn. Baltimore; Williams & Wilkins

DISCUSSION

PROFESSOR T. J. ROBINSON (*Sydney, Australia*)

I was most interested to hear Dr. Chang comment that in some circumstances the uterine environment was not adequate for capacitation, and was disappointed by his remark that failure of fertility following exogenous progesterone or oestrogen treatment before or after oestrus was due to an alteration in the transportation pattern of the ovum rather than due to failure of fertilization. In the sheep with controlled oestrus our failure or relative failure at the first ovulation is definitely due to a failure of fertilization. We have attempted to look at the sperm transport pattern and we have found a very confusing picture. Dr. Moore has been studying the fertilization of ova at the controlled oestrus compared with normal oestrus using different doses of spermatozoa. In the treated ewes, there were only 2 or 3 spermatozoa surrounding the ova whether it was fertilized or unfertilized. In the control animals no matter how many spermatozoa he inseminated there were masses of spermatozoa surrounding the ova. It is important to know whether this is due to a defect in the transport of spermatozoa, or in capacitation or in their capacity to link with the egg and I am very grateful to Dr. Bedford for raising this matter. We have inseminated controlled-ovulation animals on a time basis, some exhibited overt oestrus while a proportion did not, but they all ovulated at approximately the same time. The conception rate in sheep showing oestrus has always been three times that of the non-oestrous sheep despite the fact that insemination were done at the same time with the same number of spermatozoa and that ovulation has occurred at the same time as far as we can tell by laparotomy. This again suggests that a slight endocrine imbalance involved in the failure to exert overt oestrus has in some way affected capacitation or the transport of spermatozoa, and having listened to this paper I am beginning to favour failure of capacitation rather than failure of transport.

D. B. MORTON (*Liverpool*)

Dr. Chang has shown that sperm may be recovered from the rabbit uterus up to 48 h after insemination. He has also shown that sperm disintegrate fairly quickly during the oestrous phase; and Dr. Haynes has shown that few sperm remain in the ligated uterus of an oestrous rabbit after 5 h and none after 12 h. Some preliminary experiments have suggested that a cervical sperm reservoir may exist in the rabbit as others have shown in the sheep. Can sperm become capacitated in the cervix and could this reservoir be the source of sperm found in the uterus up to 24 h after fertilization?

DR. M. C. CHANG

Dr. Thibault found that the vagina of the rabbit was able to capacitate sperm, and so why not the cervix. Intact sperm are ingested easily during oestrus which suggests that the uterine environment during oestrus is quite different from that in pseudopregnancy. It also suggests that hormones affect the uterine environment and therefore reflect on phagocytosis. As for the transportation and capacitation of sperm in the female tract of animals treated with progesterone referred to by Dr. Robinson, experiments are in progress in our laboratory to determine these points.

THE INFLUENCE OF THE UTERINE ENVIRONMENT UPON RABBIT SPERMATOZOA

J. M. BEDFORD

Royal Veterinary College, Royal College Street, London, N.W.1*

INTRODUCTION

IN THE invertebrate and non-mammalian vertebrate species in which fertilization is external, ejaculated spermatozoa are subjected only to a relatively simple aquatic environment, prior to their union with the ovum. In the class Mammalia and certain additional groups, on the other hand, fertilization takes place within the female, and thus the fertilizing spermatozoon is necessarily exposed for varying periods of time to the complex environment afforded by the secretions of the female reproductive tract.

Investigation of the physiology and biochemistry of fertilization in the latter group presents obvious difficulties which, unfortunately, are reflected in our present poor understanding of the physiology of spermatozoa within the uterus and oviducts, and of the fertilization process itself. In spite of such difficulties it has, nevertheless, become apparent that important relationships have developed independently in several groups of animals between the ascending spermatozoa, on the one hand, and the 'host' environment of the female on the other. In the case of elasmobranchs, certain reptiles, and the domestic fowl, for instance, the female may store and preserve the physiological integrity of spermatozoa for prolonged periods after mating. Although bat spermatozoa have been found to remain viable for up to 5 months in the female tract (Wimsatt, 1942, 1944), such lengthy sperm storage is rare in mammals and in most of the species which have been investigated, spermatozoa retain their physiological integrity for only a relatively short period within the female. This contrasts with the period of several weeks for which mature spermatozoa remain fertile (though quiescent) in the male. There is, however, considerable variation between species in the duration of the viable life of spermatozoa in

* *Present address:* Dept. of Anatomy, Columbia College of Physicians and Surgeons, New York, N.Y.

the female mammal. Mouse spermatozoa remain fertile for about 6 h only after insemination (Merton, 1939), whereas some ferret spermatozoa remain fertile for up to 126 h (Chang, 1965), and horse spermatozoa for perhaps even longer (Day, 1942; Burkhardt, 1949). Recently, in drawing attention to the striking degree of synchrony which exists in different species between mating, ovulation, and the fertile life of the spermatozoon, Chang (1965) has raised the question of the relative importance of the intrinsic physiological character-istics of the spermatozoon, on the one hand, and those of the female environment on the other, in determining the viability of sperma-tozoa in the uterus and oviducts.

Mammalian spermatozoa possess both glycolytic and oxidative mechanisms for the production of energy, and it is possible that spermatozoa are able to utilize potential nutrients present in the cervical, uterine and oviducal secretions. Recently, for instance, it has been reported that human spermatozoa can metabolize hexo-samines which are found in human cervical secretions (Terner, 1965). It must be clearly stated, however, that an essential role for such potential nutrients in the economy of the fertilizing sperma-tozoon, has not been demonstrated as yet.

Functionally, perhaps the most striking example of the relation-ship between mammalian spermatozoa and the female environment, is the 'capacitation' phenomenon which occurs in the female tract. Until 1951, the tacit assumption had been made that spermatozoa ejaculated by fertile male mammals are fully prepared for entry into the ovum. In that year, however, it was shown experimentally, both by Austin and by Chang, that rat and rabbit sperm require a period of several hours incubation in the female tract before they may penetrate and achieve syngamy with the ovum. It is now clear that some essential functional change, termed 'capacitation', occurs in the spermatozoon under the influence of the female tract, but the biological significance of capacitation, and the changes which take place in the spermatozoon as a concomitant of this pro-cess remain still somewhat of an enigma. It seems likely, however, from reports, in the sheep (Mattner, 1963), and ferret (Chang and Yanagimachi, 1963), that some period of capacitation in the female tract will prove to be mandatory for the spermatozoa of all or most eutherian species. Our present knowledge of the capacitation pro-cess in its several aspects, has been reviewed by Chang (1959a) and more lately by Austin (1964).

It is the intention in the present discussion to consider the implica-tions of some recent experiments which were performed in the hope of clarifying further, the functional relationship between mammalian spermatozoa and the female tract through which they must pass. These experiments have been carried out using rabbit spermatozoa and broadly involve three different aspects of capacitation, namely, the significance of capacitation in establishment of the sperm/ovum contact, the specificity of the female environment for successful capacitation, and finally, the influence of capacitation on the duration of the fertile life of sperm in the female tract.

CAPACITATION AND SPERM/OVUM CONTACT

The first barrier presented to rabbit spermatozoa at the site of fertilization is the cell mass which comprises the cumulus oöphorus and immediately surrounds the ovum as the corona radiata. The cells of the cumulus oöphorus normally begin to disperse after 2–3 h but the cells of the corona persist for a further 3–4 h, and thus for virtually the whole of the fertile life of the ovum (Chang, 1952; Adams and Chang, 1962a). The matrix which binds the follicular cells around the freshly ovulated ovum consists of a hyaluronic acid complex; it has, therefore, been logically assumed that sperm passage as far as the zona pellucida is facilitated by hyaluronidase released from the sperm head. Austin (1960), in studying sperm penetration of the cumulus in the rat, rabbit, and hamster, noted, however, that under *in vitro* conditions mature epididymal or ejaculated spermatozoa were apparently unable to penetrate the matrix of the cumulus. In cumulus masses recovered from mated animals, on the other hand, spermatozoa were seen to be moving quite freely between the cumulus cells. These observations led Austin to suggest that capacitation allows the release of hyaluronidase from living spermatozoa, and this acts to facilitate the passage of spermatozoa through the cumulus to the surface of the zona pellucida. The above hypothesis has been tested *in vivo*, as described below.

Experimental

Spermatozoa flushed with Hanks solution from rabbit uteri 12–14 h after natural mating, were concentrated by mild centrifugation, and aliquots of 0·03 ml containing approximately 110×10^3 motile spermatozoa were inseminated via the fimbria into one tubal ampulla of female rabbits about $1\frac{1}{2}$ h after ovulation. Likewise, (c. 360×10^3) spermatozoa from the vas deferens were inseminated into the contralateral tube. Ova were collected from the Fallopian tubes of these females about 5 h later, when all or most of the corona cells had fallen away from the zona surface. At this time the epididymal spermatozoa could have completed only about one half of the period of 10 h required for the completion of capacitation in the Fallopian tube alone (Adams and Chang, 1962b). The ova were fixed for a short period in acetic alcohol, stained lightly with 1 per cent lacmoid and, using positive phase optics, were examined for evidence of fertilization, and more particularly for the number of spermatozoa in association with each ovum. The spermatozoa were counted by progressively focusing through the depth of the ovum. Spermatozoa become passively enmeshed in the muco-protein coat which forms around rabbit ova 7–8 h after ovulation; for this reason any ova in either group which showed evidence of the formation of this mucoprotein coat were discarded. A proportion of the spermatozoa in association with the fertilized ova were situated in the perivitelline space, but in nearly all ova the majority counted were on or within the zona pellucida (*Plate Ia*).

From the results in Table 1 it is clear that the number of sperma-

Table 1

The influence of capacitation on sperm/ovum contact

Sperm sample	Mean No. of sperm × 10³/ tube	Time of insem. before (−) or after (+) ovulation (hours)	No. expts.	No. ♀s	Ova recovered	% Ova fertilized	Mean No. sperm/ovum (range)
Uterine (capaci-tated)	100 ± 15·7*	+ 1–1·5	11	17	66/76	78	31·6 ± 5·7* (1–140)
Epidi-dymal	360 ± 16·0	+ 1–1·5	11	17	83/98	0	1·9 ± 0·22 (0–6)
Epidi-dymal	460 ± 19·8	− 9–11	36	36	254/315	96	103·4 ± 24·2 (14– > 250)

All ova exposed to spermatozoa for 5–7 h.
* Standard error.

tozoa, in association with the ova exposed to capacitated sperma-tozoa recovered from a mated female ($31·6 \pm 5·7$), was significantly greater ($P < 0·01$) than the consistently small number ($1·9 \pm 0·22$) sticking to the ova which had been exposed in the contralateral tube, to spermatozoa taken directly from the vas deferens. This difference is further emphasized by the fact that, consistently, relatively more vas deferens spermatozoa were inseminated into each animal. It is unlikely that the complete absence or very low number of spermatozoa on ova exposed to samples from the vas deferens could have been due to loss from the tube, since the instillation of similar numbers of these spermatozoa 9–11 h before ovulation resulted in the presence of high numbers ($103 \cdot 4 \pm 24 \cdot 2$) on ova collected 7–8 h after ovulation.

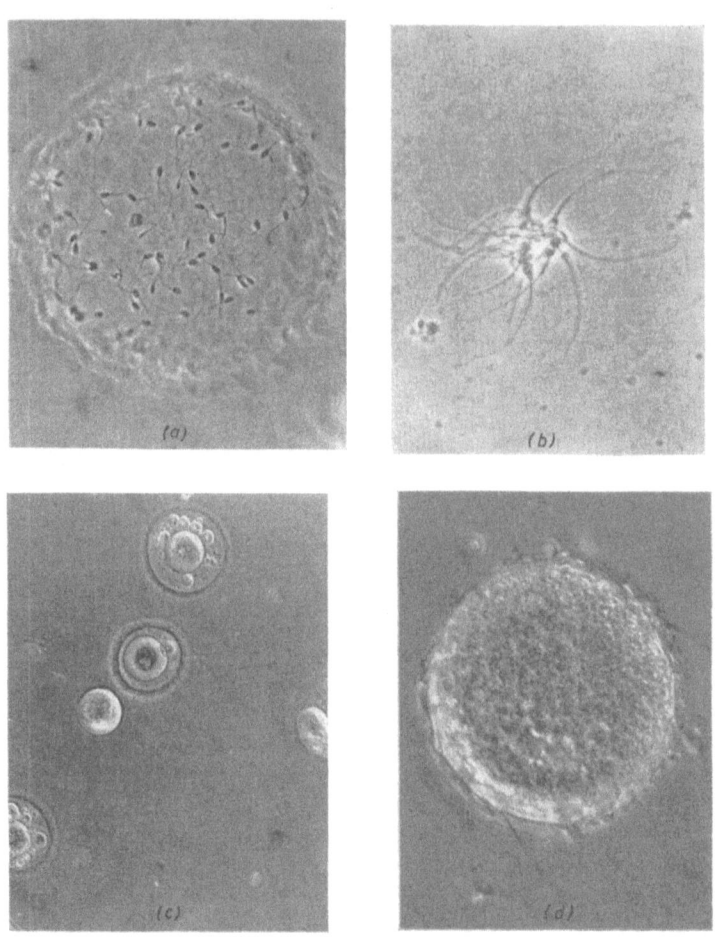

Plate I
(Reduced ⅔ on reproduction)

Discussion

The above results support the interpretation of the *in vitro* observations made by Austin (1960), and make it clear that, *in vivo*, the process of capacitation must confer upon rabbit spermatozoa the ability to reach the zona pellucida during the short fertile life of the ovum. It is equally clear that before capacitation virtually no spermatozoa can establish contact with the surface of the ovum during this crucial period, even when present in much greater numbers than after natural mating. Capacitation in the rabbit may therefore be thought of as facilitating the establishment of the sperm/egg contact, as well as endowing the spermatozoon with the ability to penetrate the zona pellucida.

It is well known that in the absence of the antiagglutinins of seminal plasma, rabbit spermatozoa *in vitro* will stick avidly by the head to each other, to leucocytes in the medium, and to the naked ovum (*see Plate Ib*). Unfortunately, the present experiments do not show whether the large difference in the respective abilities of epididymal and uterine spermatozoa to make contact with the ovum *in vivo*, is due solely to differences in their reaction to the barrier presented by the follicular cells around the ovum, or whether capacitation possibly also changes the adhesive properties and thus the ability of the spermatozoon to adhere to the zona surface *in vivo*. It is hoped that this point will be clarified in further experiments.

The changes which occur in the sperm head as a concomitant of capacitation, must for the moment remain uncertain. Austin (1960) has stated that in rodent spermatozoa penetrating the cumulus the acrosome cap is often disrupted, and he considers that hyaluronidase is released as a consequence of such acrosomal changes. In the rabbit, however, spermatozoa recovered from the uterus 12 h after mating, which *in vitro* fertilization experiments have shown to be

PLATE I

a. *Rabbit ovum fixed in acetic alcohol and stained with lacmoid. When visualized with positive phase optics, the spermatozoa become clearly outlined and allow accurate counts of sperm numbers to be made. Phase contrast × 350* (From Figure 4, J. M. Bedford, by courtesy of the Editor, *Journal of Experimental Zoology*)

b. *Motile rabbit spermatozoa agglutinated by the head region in physiological media— spermatozoa will also stick to other cells in this way when suspended in high electrolyte media, and particularly so in the presence of normal serum. Phase contrast × 750* (From Figure 1, *Expl Cell Research* **38** (1965) 655, by courtesy of Academic Press Inc.)

c. *Group of rabbit ova incubated in the rat uterus with rabbit spermatozoa, for 4 h, followed by culture in Ringer serum. Note that cytolysis has occurred, and that no spermatozoa are present within the zona pellucida. The zona pellucida has been lost completely from one ovum. Phase contrast × 180*

d. *Ovum incubated in the rat uterus with rabbit spermatozoa. The zona pellucida has been lost, and rabbit spermatozoa are adherent to the vitelline surface. Phase contrast × 350* completely capacitated, show no significant changes in fine structure of the acrosome cap (Bedford, 1964a). Evidence that capacitation may well involve some change in the state of the plasma membrane at the sperm surface, has been obtained by demonstration of the reversible inhibition of the capacitated state by some constituent of seminal plasma (Chang, 1957; Bedford and Chang, 1962b). Further evidence is provided by the differential phagocytic response of uterine leucocytes to intact (capacitated) uterine sperm on the one hand, and intact epididymal or ejaculate sperm on the other (Bedford, 1965), and by a difference in the response of the plasma membrane covering the acrosome, to hyperosmotic fixatives (Bedford, 1964b). It must, therefore, be emphasized that although the rabbit sperm head which is penetrating or has penetrated the zona pellucida, is divested of the head plasma membrane and most or all of the acrosomal substance (Hadek, 1963; Austin, 1963) it is by no means certain that the plasma membrane is ruptured, or that any fine structural alterations occur in the acrosome cap of the rabbit sperm, before contact with the zona pellucida has been established. It seems probable that this question, which at present obscures our understanding of the nature of capacitation, will be settled satisfactorily only when significant numbers of spermatozoa within the cumulus have been visualized in the electron microscope.

Note added in proof:

Subsequent investigation with the electron microscope has shown that in most rabbit sperm approaching the zona pellucida, between the cells of the corona radiata, the acrosome cap disintegrates by a process of fusion and vesiculation between the plasma membrane and the outer membrane of the acrosome cap (Barros, Bedford, Franklin, and Austin, 1967), leaving only the 'equatorial' region intact. Sperm heads beginning to penetrate or lying within the zona pellucida, have already been divested of most of the outer acrosome membrane and contents of the acrosome cap (Bedford, 1967); thus in penetrating rabbit spermatozoa it is the inner acrosomal membrane which first 'interacts' with the substance of the zona pellucida.

SPECIES SPECIFICITY OF THE CAPACITATION ENVIRONMENT

A thorough understanding of the environmental factors necessary for complete capacitation of rabbit spermatozoa would be of great help in identifying the important changes which occur in spermatozoa as a concomitant of capacitation. At present, however, our understanding of the environmental requirement for capacitation is far from complete. It is clear that the rate of capacitation may vary in different environmental situations; rabbit spermatozoa, for instance, become capacitated within 6 h in the oestrous uterus, but require at least 10 h to reach this state in the Fallopian tubes (Adams and Chang, 1962b). Capacitation may be accomplished in the uterus of the immature or ovariectomized animal (Chang, 1958; Noyes, Walton and Adams, 1958) but evidently will not occur within 12 h in the progesterone-dominated uterus (Chang, 1958). The fact that rabbit spermatozoa may be capacitated successfully in the oviduct of the progesterone-treated or pseudopregnant female serves, on the other hand, to demonstrate the differential response of the uterus and oviduct to circulating hormones. It is probable that failure of capacitation in the pseudopregnant uterus results from a deficiency of some essential factor, rather than from the presence of some inhibitory substance, since suspension of capacitated sperm in concentrated flushings from the pseudopregnant uterus does not appear to affect their fertilizing ability (unpublished observations).

The specificity of the factors which bring about capacitation is an open question. Experimental results reported by Noyes *et al.* (1958) appear to show that rabbit spermatozoa may become capacitated, not only within the female tract, but also in such abnormal situations as the isolated bladder, the colon, and the anterior chamber of the eye. In view of the more recent information that capacitation takes at least 10 h in the Fallopian tube (Adams and Chang, 1962b) it must be accepted from the results of Noyes *et al.* (1958) that some phase of the capacitation process can occur in the absence of secretions of the reproductive tract. It should be remembered, however, that in these experiments the fertilizing ability of the spermatozoa was assessed by the common method of transferring the spermatozoa from the experimental environment into the oviducts of rabbits, at or soon after the time of ovulation; the experimental spermatozoa would thus have spent some 4–6 h in the tube before the termination of the fertile life of the ovum, which loses its integrity between 4–8 h after ovulation (Chang, 1952; Adams and Chang, 1962a). This allows the possibility that only partial capacitation may have been achieved in the bladder, etc., the whole process being completed subsequently within the recipient Fallopian tube. This point requires further investigation.

Most of the attempts to achieve capacitation *in vitro* have not been successful. Recently, however, Kirton and Hafs (1965) have obtained at least partial capacitation by incubation of rabbit spermatozoa *in vitro* in rabbit uterine fluid, and even more interestingly, in simple physiological solutions which contained β-amylase. In considering these important results, and in particular the fertilization percentage achieved, it must remain for the present uncertain whether the spermatozoa were completely capacitated *in vitro*, since the experimental spermatozoa were assessed by transference into the oviduct of a recipient female some 2–4 h before the end of the fertile life of the ovum.

Although nothing is known of the species specificity of the capacitation process, the results of Kirton and Hafs (1965) do indicate that such specificity is not important for all phases of this functional change in the spermatozoon. In the hope of clarifying this point, attempts have been made to capacitate rabbit spermatozoa in the uteri of clearly unrelated species such as the cat, ferret, guinea-pig and rat.

ATTEMPTS TO CAPACITATE RABBIT SPERMATOZOA

Experimental

Epididymal spermatozoa were collected from adult males and suspended in Hanks solution. Approximately 4–8 million of these spermatozoa were then inseminated directly into each uterine horn, which in some instances was ligated close to the cervix. The results in the case of the cat, ferret, and guinea-pig proved negative. Rabbit spermatozoa were found to survive for not more than 12, 9 and 6 h in the cat, ferret, and guinea-pig respectively, and samples of living spermatozoa removed before these times did not fertilize ova when transferred to rabbit oviducts in the usual way soon after ovulation. The situation in the rat, however, proved to be quite different, in that rabbit spermatozoa survived well for prolonged periods in the rat uterus. The results of several experiments performed using the uterus of the oestrous rat will comprise the remainder of this discussion.

In the first series of experiments, oestrous rats were inseminated directly into the uterine cornu with 0·02–0·03 ml of Hanks solution containing between 4–8 million epididymal rabbit spermatozoa, and the uteri were then ligated close to the cervix. These spermatozoa were recovered $5\frac{1}{2}$–18 h later, and aliquots were placed in the oviducts of rabbits some $1\frac{1}{2}$–$2\frac{1}{2}$ h after ovulation. The tubal ova were assessed for evidence of fertilization after 10–24 h. The results of these experiments (Table 2) demonstrate that, with the exception of one ovum fertilized at 9 h, no ova were fertilized by spermatozoa which had resided less than 14 h in the rat uterus. Rabbit spermatozoa which had been incubated in the oestrous rat uterus for more than 14 h, on the other hand, were able to fertilize a significant proportion of ova, 94 per cent of 24 ova being fertilized by sperm which

Table 2

Studies on the capacitation of rabbit spermatozoa in the rat uterus

Treatment of rabbit spermatozoa in rat uterus			Experimental sperm inseminated 12–13 h post ov. inj.		Control sperm inseminated 0–5 h post ov. inj.	
Time rabbit sperm inc. in rat uterus (h)	No. of expts.	No. of sperm/tube ($\times 1{,}000$)	Ova recovered	% fertilized	Ova recovered	% fertilized
5·25	1	200–350	6/24 (2)	0	13/15 (1)	100
6·25	1	50–100	5/8 (1)	0	5/7 (1)	20
9·25	1	100–220	4/8 (1)	25	7/7 (1)	100
10·50	1	100–220	6/9 (1)	0	5/6 (1)	60
12·50	4	31–350	41/43 (4)	0	7/9 (2)	70
14·0	3	45– 60	26/39 (4)	3	—	—
15·0	3	50	23/26 (3)	48	8/8 (1)	50
17·0–17·50	3	60–465	16/24 (3)	94	18/22 (3)	90
18·0–18·50	3	32–550	31/34 (4)	32	9/11 (2)	88

had been in the rat for 17 h. These findings seemed to indicate that rabbit spermatozoa may be capacitated in the rat uterus, though the process in the rat clearly takes 2·5–3 times longer than in the rabbit. The completeness of capacitation in the rat was brought into doubt by the finding that in some animals many ova were not fertilized, though large numbers of spermatozoa were present on the zona surface. Furthermore, of those eggs fertilized, very rarely were supplementary spermatozoa seen in the perivitelline space; this seemed surprising in view of the large numbers of spermatozoa inseminated into the oviducts.

Although it was clearly apparent that some phase of capacitation had been accomplished in the rat, it seemed feasible that the ova fertilized may have been penetrated by rat-incubated spermatozoa which had been able to complete their capacitation in the rabbit oviduct before termination of the fertile life of the ovum. Further experiments were therefore performed, which, by avoiding the environment of the rabbit oviduct, might show whether or not complete capacitation had been achieved in the rat uterus. In the first experiments rabbit spermatozoa were incubated in fluid-filled rat uteri for 20–25 h. Thereafter, freshly ovulated rabbit ova were introduced into the rat uterus; these ova were removed 4–5 h later and were incubated for several hours in Hanks solution containing 200 per cent heated rabbit serum. As is shown in Table 3, none of the 38 ova were fertilized and none appeared to have been penetrated,

Table 3

The incubation of rabbit ova and spermatozoa in the uterus of the rat

No. expts.	Conc. of motile sperm in utero	Incubation in rat uterus (h)		Ova fertilized
		Rabbit sperm	Rabbit ova	
4	1·8–4·3	24–29	4–5	0/38

though many spermatozoa were adherent to the zona pellucida. In some instances severe cytolysis had occurred (*Plate Ic*) and in several ova the zona pellucida appeared to have been almost completely dissolved (*Plate Id*). It is obvious from these findings that the environment of the rat uterus is inimical to rabbit ova. Chang (1955) found this to be the case also when rabbit ova were incubated in the rabbit uterus, but in Chang's experiments ova were penetrated and fertilized normally, although ovum degeneration occurred also in the homologous uterus.

In a second series of experiments, attempts were made to fertilize rabbit ova *in vitro* with spermatozoa previously incubated in the female tract of the rat. The rabbit spermatozoa were removed from the rats after 20–26 h intra-uterine incubation, centrifuged at 1,000 rev/min for 4 min, and resuspended in either Hanks solution containing 5 per cent heated serum, or acidic saline+5 per cent heated serum, as used previously in *in vitro* fertilization experiments

(Chang, 1959b; Bedford and Chang, 1962a). The spermatozoa were then incubated with freshly ovulated ova in rocking Carrel flasks, in an identical manner to that used in the successful *in vitro* fertilization experiments quoted above. In Table 4 it can be seen

Table 4

Fertilization *in vitro* with rabbit spermatozoa incubated in the uterus of the rat

No. *expts.*	*Rabbit sperm inc. in rat uterus* (h)	*Conc. of sperm* in vitro	*Ova fertilized*
5	20–26	140,000–450,000/ml	0/72

that no ova were fertilized *in vitro* by rat-incubated rabbit spermatozoa. Although bearing the stigma of purely negative results, the complete failure of fertilization in both of these experimental series, in the absence of the environment of the rabbit Fallopian tube, indicated that rabbit spermatozoa cannot be fully capacitated in the heterologous environment of the rat uterus, even though this environment would support the motility of rabbit spermatozoa for much longer periods (see below). This point was confirmed in further experiments in which spermatozoa incubated in the rat and rabbit uterus, were introduced respectively into the left and right Fallopian tubes of female rabbits at a time (4–$4\frac{1}{2}$ h after ovulation) shortly before termination of the fertile life of the ovum. After collection, the rat-incubated spermatozoa were first washed from the rat uterine fluid and resuspended in rabbit uterine flushings; the rabbit-capacitated sperm were concentrated at the same time by light centrifugation. In spite of the fact that a somewhat greater number of the rat-incubated rabbit spermatozoa were introduced into the oviduct some 10 min earlier than the rabbit capacitated sample, it can be seen in Table 5, that the spermatozoa in the latter sample were able to fertilize 45·5 per cent of the tubal ova at this later hour, in contrast to the very low 3·4 per cent fertilization rate achieved in the contralateral oviduct by the rat-incubated spermatozoa. On the whole, rather more spermatozoa were found on the ovum surface in the latter group.

Discussion

The foregoing results, taken together with the failure to achieve

16*   489

Table 5

The fertilizing capacity of rabbit spermatozoa incubated in the uterus of the rabbit and the rat

Expt. No.	Sperm inc. in rabbit uterus 13–15 h			Sperm inc. in rat uterus 20–24 h			Control		
	Time of insemination (h post ov. inj.)	Ova recovered	% Ova fertilized	Time of insemination (h post ov. inj.)	Ova recovered	% Ova fertilized	Time of insemination (h post ov. inj.)	Ova recovered	% Ova fertilized
1	14·0	12/14	8·5	13·75	20/21	0	0·5	5/5	80
2	14·25	6/6	100·0	14·0	6/11	33	5·0	5/5	80
3	14·5	5/8	62·5	14·25	10/13	0	1·5	6/9	100
4	14·25	3/4	66·0	14·0	7/8	0	1·0	6/9	66
5	14·25	9/10	44·0	14·0	7/7	0	5·0	4/6	75
6	14·25	9/13	44·0	14·0	8/8	0	3·0	8/9	75
Total	14·25	80%	45·5%	14·0	85%	3·4%	0·5–5·0	80%	80%
	100,000–180,000 sperm/tube			140,000–300,000 sperm/tube					

490

fertilization *in vitro*, and in the rat uterus, indicate that rat-incubated rabbit spermatozoa do not have the same capacity for fertilization as that of spermatozoa incubated in the rabbit uterus. The poorer fertilizing ability of the rat-incubated spermatozoa cannot, however, have been due to loss of their physiological integrity in the foreign environment of the rat uterus, since aliquots from the same samples were later able to fertilize tubal ova after their instillation into the oviducts of other females, several hours before ovulation. It seems clear, therefore, that a significant part of the capacitation process had been accomplished in the rat uterus, but that, even after a period of up to 25 h in this heterologous environment, capacitation of rabbit spermatozoa was probably not complete. Bearing in mind the findings of Noyes *et al.* (1958), and Kirton and Hafs (1965), it is reasonable to conclude from all these results that some important phase of capacitation, temporarily occupying as much as two-thirds of the time required for capacitation in the rabbit, may be brought about by agents which are neither species nor organ specific. The present results make it seem highly likely, however, that complete or functional capacitation also involves a second phase which is species specific. This latter phase of capacitation may also be specific to the environment of the uterus and oviducts. Before this latter proposition can be accepted, however, further experiments must be performed to test the fertilizing ability of spermatozoa incubated in ectopic sites (Noyes *et al.*, 1958) or in β-amylase (Kirton and Hafs, 1965), both *in vitro* and *in vivo* towards the limit of the fertile life of the ovum.

THE INFLUENCE OF THE UTERINE ENVIRONMENT ON THE VIABILITY OF RABBIT SPERMATOZOA

As Chang (1965) has pointed out, there is considerable species variation in the length of the viable life of spermatozoa after entry into the female tract. Noyes and Thibault (1962) also have drawn attention to the fact that, within a species, there are many factors which may influence the effective functional life of spermatozoa in the female, including the endocrine state of the female, the degree of uterine phagocytosis, and time of residence in the female before arrival of the ovum.

The prospective fertility of fresh samples of semen is usually estimated by examining the concentration and particularly the motility of the spermatozoa. Progressive motile spermatozoa are not necessarily fertile, however, especially when nearing the end of their viable life in either the male (Hammond and Asdell, 1926; Young, 1929) or the female tract. In the latter situation, it has been shown in the rabbit (Hammond and Asdell, 1926; Noyes and Thibault, 1962), mouse (Merton, 1939), rat (Soderwall and Blandau, 1941), guinea-pig (Yochem, 1929; Soderwall and Young, 1940), and sheep

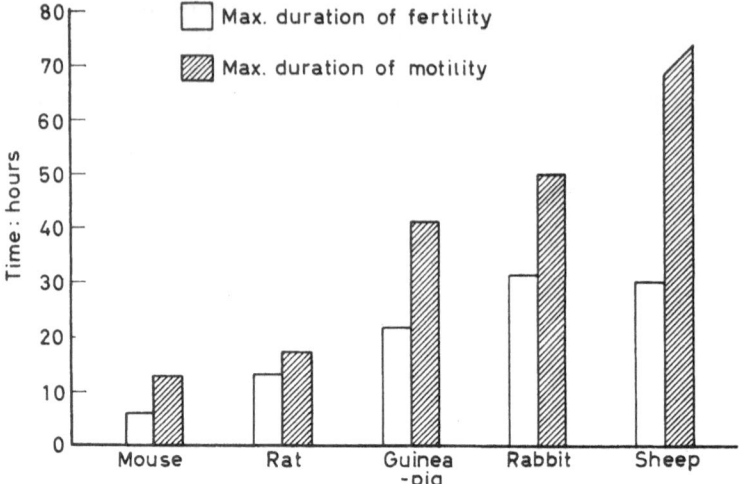

Figure 1. Viability of spermatozoa in the female tract

(Dauzier and Wintenberger, 1952) that the ability to fertilize is retained for a relatively limited period and that this faculty first diminishes, as expressed in litter size in polytocous species, and is then lost completely some hours prior to losing the capacity for progressive motility. The histogram in *Figure 1* shows the temporal relationship between the maximum length of the fertile life and motility in the female tract of these species ('motile life' in the sheep, here refers to spermatozoa found in the cervical canal).

The lesion in 'ageing' though motile spermatozoa which results in an inability to fertilize, has not yet been identified, though Noyes and Thibault (1962), in their study of sperm survival in the female rabbit, have speculated that this situation may come about through what they term 'reversal of capacitation'. In the light of experiments described above, which had shown that rabbit spermatozoa survive well in the uterus of the oestrous rat, it seemed worth while therefore to investigate further the effect of the foreign environment of the rat uterus upon the length of the functional life of rabbit spermatozoa.

Experimental

Epididymal spermatozoa were instilled into rat uteri as described previously, and the spermatozoa were later collected at various times by withdrawal of the uterine fluids. Usually 4–6 oestrous rats were used for incubation of aliquots from a single sperm sample, each horn

Table 6

Survival of rabbit sperm in rat fluid-filled uteri

Time in rat uterus (h)	No motile spermatozoa samples	
–18	32/40	(80%)
18–24	22/30	(73%)
24–30	12/18	(66%)
30–36	9/15	(60%)
36–42	7/18	(39%)
42–46	8/26	(30%)
46–48	· 4/23	(16%)

receiving 4–8 million sperm in 0·02–0·03 ml of Hanks solution, containing streptomycin and/or penicillin. The samples were assessed for motility immediately after collection, and those samples which showed significant numbers of vigorously motile spermatozoa were used in fertility trials. In the motility estimations, samples were considered positive if they contained more than about 15 per cent of vigorously motile spermatozoa, as assessed subjectively. These positive samples were then sometimes used in fertility trials, after light centrifugation at 1,000 rev/min for a few minutes to concentrate the spermatozoa. The fertility trials were carried out in the usual way by insemination of about 0·03 ml containing 150–300,000 spermatozoa into the ampulla of the Fallopian tube, an ovulation injection of HCG (Pregnyl: Organon) being given either before, or more usually some little while after, the operative insemination.

Results: Motility of Spermatozoa

Because previous experience had shown the fluid-filled oestrous uterus to be the most favourable environment, the duration of motility of rabbit spermatozoa in the rat uterus shown in Table 6 is based only on results obtained in those animals having a fluid-filled uterus. Some variation was found between different females which had received aliquots from the same sperm sample. Thus, for instance, at 48 h there was only one animal in each group of 4–6 female rats, which possessed highly motile spermatozoa in the uterine flushings. In the remainder the spermatozoa at this time were either lying free but immotile, or had in some uteri been mostly ingested by uterine leucocytes. Taken overall, there was a definite difference between apparently similar uterine environments in their ability to support the motility of rabbit spermatozoa. There was also wide variation in the degree of local uterine leucocytosis in different individuals. This latter point does not merit further discussion or interpretation, however, since the application of ligatures above the cervix and injection of rabbit spermatozoa must constitute a highly abnormal situation.

The importance of the endocrine state of the animal during this time was clearly, though inadvertently, shown by some experiments performed using ovariectomized or intact females, which had been implanted subcutaneously with 0·5–1·0 mg of stilboestrol. It had become apparent from previous experiments that rabbit spermatozoa survive rather better in the fluid-filled uterus than in that which contains relatively little fluid. As the uteri of rats implanted with oestrogen remain filled with fluid, it was hoped that implanted rats might be used as a ready source of 'fluid-filled uteri'. It was surprising to find, however, that rabbit epididymal spermatozoa would never survive for longer than about 20 h, and often for an even shorter period in the fluid-filled uteri of stilboestrol implanted animals. Aliquots of spermatozoa from the same samples in the uteri of normal oestrous animals, on the other hand, survived for much longer periods, and sometimes for as long as 48 h. Obviously, significant differences exist between the uterine fluid in the naturally oestrous and stilboestrol implanted female, respectively.

The above observations are not emphasized here for their intrinsic interest—the whole is obviously a highly abnormal and contrived situation. The system does seem, however, as though it might lend itself easily to investigation of some of the factors which influence the metabolism of spermatozoa, and which are important for the survival of spermatozoa *in utero*, particularly so as relatively large amounts of these fluids may be obtained readily for analysis.

FERTILITY OF RABBIT SPERMATOZOA AFTER INCUBATION IN THE RAT UTERUS

The extent of the potential fertilizing ability of rabbit spermatozoa, after their previous incubation in the rat uterus, is shown in Table 7.

It seems clear, though perhaps at first sight rather surprising, that the potential fertilizing ability of these spermatozoa incubated in the foreign rat uterus is maintained for a much longer period than the maximum 30–32 h for which rabbit spermatozoa remain fertile in

Table 7

Fertile life of rabbit sperm incubated in the rat uterus

Incubation before fertilization (h)			% Ova fertilized
In rat uterus	Rabbit oviduct	Total	
20	12	32	80
22	11	33	100
24	12	36	80
33	11	44	100
33	11	44	87·5
35	12	47	100
39	11	50	70
39	11	50	87·5
41	11	52	44
42	10	52	10
42	11	53	23
43	11	54	20
44	11	55	87·5
48	11	59	0
48	11	59	0
48	11	59	0
48	3	51	80

the homologous environment of the female rabbit (*Figure 2*). After a total of about 45–48 h the fertility of the inseminated spermatozoa began to fall, as judged by a reduction in the numbers of spermatozoa in association with the fertilized ova. It was not until after 50 h was reached, however, that the fertilization percentage itself was reduced. In one animal in which 7 out of 8 eggs were fertilized, the total time of incubation in the female tract before fertilization was at least 55 h. Beyond this time, at 59 h, however, no ova were fertilized in three separate experiments, even though the spermatozoa showed progressive motility at the time of insemination. Spermatozoa, if motile, may however retain their functional integrity

495

after 48 h in the rat uterus, since one such sample transferred to the rabbit oviduct about 2½ h before ovulation, fertilized most of the tubal ova. It seems reasonable to assume that the maximum limit of the fertile life of rabbit spermatozoa under these conditions is a little short of 60 h.

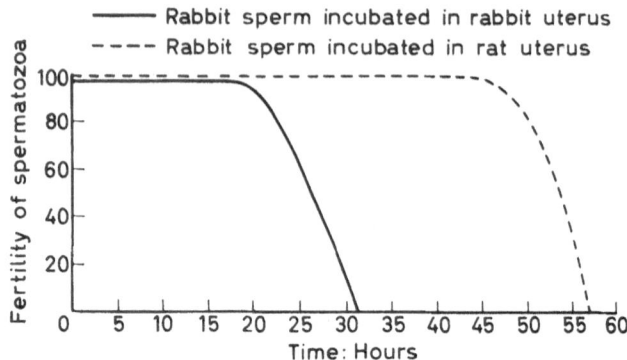

Figure 2. Fertility of rabbit spermatozoa incubated in the uterus of the rat and rabbit (From J. M. Bedford, 1967, by courtesy of the Editor of *Nature*)

Discussion

The experiments outlined above show that the fluid-filled uterus of the oestrous rat is a favourable site for the survival of rabbit spermatozoa; this foreign environment will not, however, support the motility of rabbit spermatozoa for a significantly longer period than will the homologous uterus. As rat spermatozoa survive for a maximum of 17 h only in the oestrous uterus of the rat, it would appear that in this species at least, the viable life of the spermatozoon may be determined primarily by its own characteristics rather than by those of the environment.

The point which stands out in these latter experiments is the finding that rabbit spermatozoa which had been previously confined in the rat uterus retained their fertilizing ability virtually to the end of their motile life, thus almost doubling the length of the fertile life which obtains in the homologous uterine environment alone. This state of affairs contrasts with the homologous situation in the mammals studied so far, in which fertilizing ability is lost before motility. The reason for the prolongation of the fertility of rabbit spermatozoa in the rat uterus cannot be explained with any real confidence, since we have little understanding, as yet, of the important elements about the sperm head which determine the ability to fertilize. It is possible, however, that the delicate systems which are activated on completion of capacitation are the elements in the spermatozoon which often degenerate first, thus producing a failure or degeneration of the capacitated state, as suggested by Noyes and Thibault (1962), to explain the loss of the fertilizing ability of motile spermatozoa in their sperm survival experiments in the rabbit. If capacitation is not completed in the foreign environment of the rat uterus, as was indicated by the earlier experiments reported in this discussion, then the systems activated at complete capacitation would not have been exposed until perhaps 2–4 h after transfer of the spermatozoa to the rabbit oviduct, and thus only a few hours before confrontation with the ovum. Taken in reverse, this phenomenon might conceivably be taken as further circumstantial evidence that capacitation cannot be completed in a definitively heterologous environment.

SUMMARY

Capacitation of rabbit spermatozoa not only permits their passage through the zona pellucida, but apparently is also important for the establishment of contact between sperm and the surface of the zona pellucida, *in vivo*.

Rabbit sperm can be partially capacitated in the rat uterus but appear to require the homologous uterine or tubal environment for completion of the capacitation process. In several mammals, sperm in the female tract retain their motility for longer than their fertilizing ability. In the rat uterus the motile life of rabbit sperm is not significantly extended, but these sperm retain the ability to fertilize throughout their motile life, and thus for almost twice the length of their fertile life in the rabbit. This provides further evidence for the idea that, in the foreign uterus, rabbit sperm cannot reach the same physiological state as that achieved in the homologous female tract.

ACKNOWLEDGEMENT

This study was supported in part by grants from the United States Public Health Service, numbers GM 064089–05, HD 01476–05 and HD 01476–06.

REFERENCES

Adams, C. E. and Chang, M. C. (1962a). *J. exp. Zool.* **151**, 155–158
— — (1962b). *J. exp. Zool.* **151**, 159–166

REPRODUCTION IN THE FEMALE MAMMAL

Austin, C. R. (1951). *Aust. J. scient. Res.* **4,** 581–596
— (1960). *J. Reprod. Fert.* **1,** 310–311
— (1963). *J. Reprod. Fert.* **6,** 313–314
— (1964). *Proc. Vth Int. Congr. Anim. Reprod. Trento,* Vol. III, pp. 7–22
Barros, C., Bedford, J. M., Franklin, L., and Austin, C. R. (1967). *J. Cell.*
Biol. (in press)
Bedford, J. M. (1964a). *J. Reprod. Fert.* **7,** 221–228
— (1964b). *Proc. Vth Int. Congr. Anim. Reprod. Trento,* Vol. VII, pp. 286–288
— (1965). *J. Reprod. Fert.* **9,** 249–256
— (1967) Brook Lodge Symposium on Capacitation and Spermatogenesis,
J. Reprod. Fert., Suppl. 2
— and Chang, M. C. (1962a). *Nature, Lond.* **193,** 898–899
— and Chang, M. C. (1962b). *Am. J. Physiol.* **202,** 179–181
Burkhardt, J. (1949). *J. agric. Sci.* **39,** 201–203
Chang, M. C. (1951). *Nature, Lond.* **168,** 697
— (1952). *J. exp. Zool.* **121,** 351–381
— (1955). In: 'La Fonction Tubaire et ses Troubles': *Colloques sur la Fonction Tubaire,* pp. 40–52. Masson et Cie
— (1957). *Nature, Lond.* **179,** 258–259
— (1958). *Endocrinology* **63,** 619–628
— (1959a). In: *Recent Progress in the Endocrinology of Reproduction,* pp. 131–165. Ed. by C. W. Lloyd. The Academic Press
— (1959b). *Nature, Lond.* **184,** 466–467
— (1965). *J. exp. Zool.* **158,** 87–100
— and Yanagimachi, R. (1963). *J. exp. Zool.* **154,** 175–188
Dauzier, L. and Wintenburger, S. (1952). *C.r. Séanc. Soc. Biol.* **146,** 660–663
Day, F. T. (1942). *J. agric. Sci.* **32,** 108–111
Hadek, R. (1963). *J. ultrastruct. Res.* **8,** 161–169
Hammond, J. and Asdell, S. A. (1926). *J. exp. Biol.* **4,** 155–185
Kirton, K. T. and Hafs, H. D. (1965). *Science, N.Y.* **150,** 618–619
Mattner, P. E. (1963). *Nature, Lond.* **199,** 772–773
Merton, H. (1939). *Proc. R. Soc. Edinb.* **59,** 207–218
Noyes, R. W., Walton, A. and Adams, C. E. (1958). *J. Endocr.* **7,** 374–380
— and Thibault, C. (1962). *Fert. Steril.* **13,** 346–365
Soderwall, A. L. and Young, W. C. (1940). *Anat. Rec.* **78,** 19–29
— and Blandau, R. J. (1941). *J. exp. Zool.* **88,** 55–64
Terner, C. (1965). *Nature, Lond.* **208,** 1115–1116
Wimsatt, W. A. (1942). *Anat. Rec.* **83,** 299–305
— (1944). *Anat. Rec.* **88,** 193–203
Yochem, D. E. (1929). *Biol. Bull.* **56,** 274–297
Young, W. C. (1929). *J. Morph.* **48,** 475–491

DISCUSSION

Dr. A. McLaren (*Edinburgh*)

Dr. Bedford suggested that the loss of fertilizing capacity in ageing sperm might

498

possibly be due to loss of capacitation. I wonder whether any attempts have been made to refresh these sperm by recapacitation.

BEDFORD

I do not know how you would do this. I feel that once sperm have become capacitated this may involve a rather sensitive and delicate step whereby the activated elements of the sperm may then be able to survive for relatively short time. If these are not activated, the sperm will remain potentially fertile until their motility decreases.

Dr. C. POLGE (Cambridge)

Under these circumstances how do you explain the big species variation in the survival of fertilizing capacity of spermatozoa? For instance, in most mammalian species in which the time of capacitation has been studied, it is about 6–10 h, yet you say that rabbit sperm is doomed once it is capacitated and its fertility declines quickly, whereas in other species such as the pig and horse the fertilization capacity may be maintained for 4–5 days yet they have been capacitated after only 6–10 h.

BEDFORD

I cannot give a satisfactory answer to this but it may simply be a question of species differences in the capability of sperm to retain their fertilizing ability after capacitation. Ferret sperm which has to wait a long time before fertilization can retain its integrity for a long period.

THE INFLUENCE OF THE UTERINE ENVIRONMENT ON THE PHAGOCYTOSIS OF SPERMATOZOA

N. B. HAYNES

University of Nottingham, School of Agriculture,
Sutton Bonington, Loughborough, Leicestershire

INTRODUCTION

VARIOUS workers have observed that spermatozoa are removed relatively rapidly from the uterus of many species following copulation and a number of factors appear to play a part in the removal process.

Firstly, in certain species, cervical dilation may be responsible for the clearance of spermatozoa (Blandau and Odor, 1949). In the uterus of the mouse leucocytes appear 4–5 h after insemination and active phagocytosis occurs until the vaginal plug is discharged 15–20 h later and then the contents of the uterus are emptied via the cervix (Dokukin, 1956; Reid, 1965a). A similar situation regarding the removal of spermatozoa by cervical drainage has been observed in the hamster (Yanagimachi and Chang, 1963).

Secondly, it is claimed that spermatozoa may sometimes be removed by penetration of uterine tissue. Spermatozoa were considered by Genin (1953) to actively enter the uterine wall, the tissue cells reacting to break the sperm head. Detection of spermatozoa in the uterine mucous membrane of mice supported the view of an interaction with the uterine mucosa and subsequent fragmentation (Berzedowka, 1955). The uptake of radioactive spermatozoa by cells of the mouse uterus has been reported (Reid, 1965b, c). Spermatozoa have been found in the uterine tube mucosa of the bat (Austin and Bishop, 1959). However, Austin (1960) failed to demonstrate that spermatozoa adhere to the uterine epithelium or that the glands of the rabbit uterus contained spermatozoa. Further, reports of penetration of tissue cells as an artefact of fixation procedures casts doubt on the validity of some cited examples (Vojtiskova, 1956; Posalaky and Töró, 1957).

As a third method of removal, phagocytosis of spermatozoa in the uterus by polymorphonuclear leucocytes has been reported on many occasions (Sobotta, 1920; Yochem, 1929; Merton, 1939; Austin, 1957 and 1964; Pitkjanen, 1960; Menge, Tyler and Casida, 1962; Noyes and Thibault, 1962; Yanagimachi and Chang, 1963). It is not clear whether the actual presence of spermatozoa and seminal plasma is the primary cause of the leucocyte migration into the uterus. Austin (1957) considered uterine distension to be a major factor involved. Howe and Black (1963), on the other hand, observed that spermatozoa and seminal plasma affected leucocyte migration in the cow. In the hamster, semen evoked infiltration of leucocytes into the uterine lumen. Ligation of the uterus and the consequent distension did not, however, give rise to any notable increase in the numbers of intra-uterine leucocytes (Yanagimachi and Chang, 1963). Although the cause of leucocyte invasion brought about by the presence of spermatozoa is not established, there seems no doubt that phagocytosis is a most important method for clearance of spermatozoa from the uterus.

INTRA-UTERINE PHAGOCYTOSIS OF MATERIALS OTHER THAN SPERMATOZOA

Relatively few publications have appeared which describe removal of spermatozoa from the uterus by phagocytosis in a quantitative manner, but insight into the process can be obtained by a consideration of the large amount of work concerned with intra-uterine phagocytosis of other materials.

The mobilization and influx of leucocytes into the uterus under the influence of materials other than spermatozoa, e.g. bacteria, is not constant and is influenced by the ovarian status of the animal. The rabbit is resistant to experimental infection when under the influence of oestrogens but is susceptible to infection when under the influence of progesterone. Polymorphonuclear leucocytes are the agents mainly responsible for this bactericidal activity within the uterus. The differences in activity, under the different hormone conditions, has been shown to correspond in the early stages of infection to different rates of leucocyte infiltration into the uterine lumen (Broome, Lamming and Smith, 1959; Broome, Winter, McNutt and Casida, 1960; Hawk, Turner and Sykes, 1960; Winter, Broome, McNutt and Casida, 1960; Noyes and Thibault, 1962). The differential response is probably a local one restricted to the uterus since the ovarian status did not influence the leucocyte response to induced infection in the peritoneal or the pleural cavities (Hawk, Simon, Cohen, McNutt and Casida, 1955). Killingbeck (1963) has demonstrated that the differential rates of leucocyte migration and phagocytosis in the rabbit is not restricted to bacteria but also applies to inert materials such as starch particles when introduced into the uterus. The slower rate of infiltration of leucocytes into the pseudopregnant uterus does not appear to be the sole reason for a lowered resistance to infection in the rabbit. A factor in the secretion from the uterus under the influence of progesterone seems to inhibit the phagocytic mechanism of those leucocytes present.

Certain materials in the uterine flushings from pseudopregnant rabbits but not oestrous animals have been suggested as being responsible (Heap and Lamming, 1960; Heap, Robinson and Lamming, 1962; Killingbeck, Haynes and Lamming, 1963; Killingbeck and Lamming, 1963; Lamming and Haynes, 1964). Similar materials have been identified in other species in which differential phagocytic activity has not been investigated (Heap, 1962; Haynes and Lamming, 1964). This material appears to exert its effect on the leucocyte rather than on the material being phagocytosed (Killingbeck et al., 1963). Ovariectomized, oestrous and luteal-phase ewes showed differences in the rate of leucocytic response to uterine innoculation with E. coli. The response was slowest in luteal-phase ewes, intermediate in oestrous ewes and fastest in ovariectomized animals. It was suggested that the ovarian hormones of both oestrous and luteal-phase ewes inhibited the leucocytic response, the latter functioning to a greater degree than the former. This differs from results obtained with oestrous rabbits and cattle where endogenous ovarian hormones hasten and intensify the response (Brinsfield, Hawk and Leffel, 1963). The response in ovariectomized ewes was stimulated when they were injected with oestradiol and was inhibited by injection with progesterone or oestradiol and progesterone simultaneously. These results led to the suggestion that the inhibition of induced leucocyte migration in the oestrous sheep may be due to endogenous progestogens (Brinsfield, Hawk and Richter, 1964).

Still other factors have been implicated in the process of uterine defence. The oestrogen-treated rabbit uterus was found to be a most effective substance in stimulating leucocyte migration (Spector and Storey, 1958). Some evidence has been presented for the presence of a bactericidal substance, phagocytin, in the cell free fraction of uterine flushings of rabbits previously inoculated with a bacterial suspension (Hawk, 1958). The phagocytic process and the reticulo-endothelial system in general are modified by hormones. Both natural and synthetic oestrogens stimulate the reticulo-endothelial system (Nicol and Helmy, 1951; Halpern, Stiffel, Biozzi and Mouton, 1960).

INTRA-UTERINE PHAGOCYTOSIS OF SPERMATOZOA

In cases where spermatozoon removal from the uterus at different stages of the cycle has been studied quantitatively, the results bear similarities to the uterine reaction to bacteria. The cow is more resistant to uterine infection at oestrus and susceptible to infection during the luteal phase (Rowson, Lamming and Fry, 1953). Spermatozoa are removed more rapidly from the oestrous than from the luteal-phase uterus in this species and the maximum leucocyte response occurs later in luteal-phase animals (Mahajan and Menge, 1965). Similarly, it is reported that spermatozoa are removed more rapidly from the oestrous than from the pseudopregnant rabbit uterus (Henderson and Mayer, 1965). The leucocytic response to injection of spermatozoa into the uterus of the hamster was significantly greater in ovariectomized oestrogen-treated animals than in ovariectomized progesterone-treated animals. The degree of the leucocytic response and phagocytosis did not appear to be species specific with regard to spermatozoa or dependent on spermatozoa mobility (Marcus, 1966). Bedford (1965) has demonstrated that in the rabbit a difference exists in the phagocytic activity of leucocytes towards spermatozoa in the oestrous or the pseudopregnant uterus. The ingestion of both intact and damaged spermatozoa by leucocytes in the oestrous uterus could take place. However, in the pseudopregnant uterus, only spermatozoa with damaged acrosomes or damaged head membranes were ingested to any extent. It is suggested that the change in spermatozoa, when in the oestrous uterus, which renders them acceptable to leucocytes, may also constitute one phase of the capacitation phenomenon which is known to occur in the oestrous but not the pseudopregnant uterus (Chang, 1958).

A quantitative difference in the rates of phagocytosis of different spermatozoa from ligated segments of oestrous and pseudopregnant rabbit uteri has been demonstrated (Menge *et al.*, 1962). Bull spermatozoa were removed more rapidly from the oestrous rabbit uterus by phagocytosis than from the pseudopregnant uterus. This is in agreement with results obtained from bacterial and inert foreign body inoculations. In the case of rabbit spermatozoa the situation was reversed. These spermatozoa were removed more rapidly from the pseudopregnant than from the oestrous uterus. This difference between bull and rabbit spermatozoa was not a consequence of different numbers of leucocytes infiltrating into the uterus, as these were similar in each case. Alterations in the efficiency of the leucocytes in carrying out phagocytosis was suggested as the factor responsible.

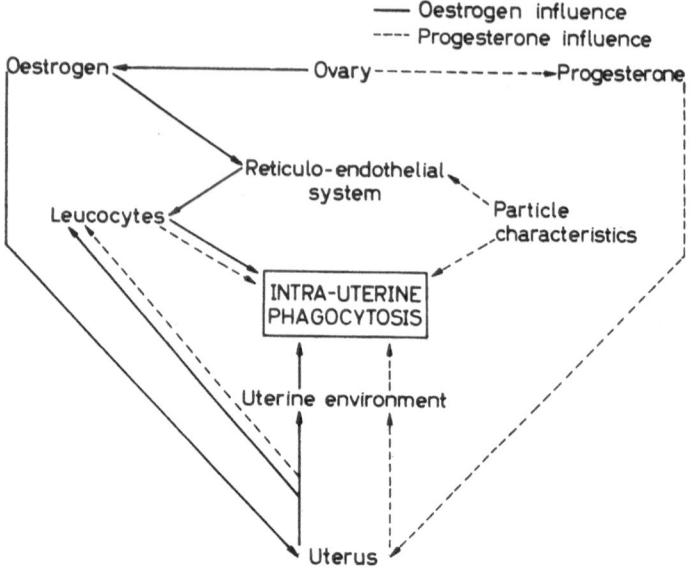

Figure 1. Interrelationships which may affect intra-uterine phagocytosis

All the above concepts point to a complicated series of mechanisms which operate to influence the uterine environment, its resistance to infection, its effect on the removal of spermatozoa and its ability to cause capacitation. It is possible that all these activities may be interrelated. A representation of the factors which may influence intra-uterine phagocytosis is shown in *Figure 1*.

The work of Menge *et al.* (1962) is of particular interest since it implies that the animal is capable of making a distinction between types of spermatozoa with regard to their removal by phagocytosis. It seemed unlikely that the leucocyte is capable of making this distinction in view of the general nature of phagocytosis. A more probable explanation could be that rabbit spermatozoa may be capable of enzymatically destroying material occurring in pseudopregnant rabbit flushings which is preventing phagocytosis. A foreign spermatozoa may, on the other hand, be incapable of doing this. Experiments were designed, therefore, to investigate such a possibility.

MEASUREMENT OF *IN VITRO* PHAGOCYTOSIS OF SPERMATOZOA

Bull or rabbit spermatozoa were incubated with leucocytes and uterine flushings from pseudopregnant or oestrous rabbits *in vitro* and the degree of phagocytosis measured. This has the advantage of removing the factor of different rates of infiltration of leucocytes and the general effects of the reticulo-endothelial system and should allow an assessment of the part played by the uterine secretion. The technique used was similar to that described by Killingbeck and Lamming (1963) for the measurement of phagocytosis of starch particles by leucocytes in the presence of uterine flushings. Freshly obtained spermatozoa, collected by artificial vagina, were used in all the experiments described. The counting technique was that reported by Menge *et al.* (1962). The results as far as comparison can be made between *in vitro* and *in vivo* experiments are at variance with those of Menge *et al.* (1962) in that the rabbit spermatozoa are phagocytosed more rapidly in the presence of oestrous than pseudo-pregnant rabbit flushings or in the presence of material extracted from pseudopregnant rabbit flushings and demonstrated to inhibit the phagocytosis of starch particles. Spermatozoa of the same species, therefore, seem equally to be regarded as a foreign body and to behave similarly to foreign particles. Bull spermatozoa were slightly different in that phagocytosis took place at the same rate in oestrous and pseudopregnant rabbit flushings. However, they be-haved in the same manner as rabbit spermatozoa when incubated in concentrated pseudopregnant rabbit uterine extracts. The reason for this difference has not been investigated further.

The experiments were carried out using both sperm washed free from seminal plasma and untreated spermatozoa and no significant differences were found. The presence or absence of seminal plasma, therefore, played no part in governing the rates of phagocytosis. Washing the spermatozoa resulted in large numbers of immotile or dead spermatozoa and the lack of a significant difference between the results implied that dead and immotile spermatozoa were phagocytosed at the same rate. *Figure 2* presents a composite analysis of results from washed and untreated spermatozoa com-bined.

There are a number of possible reasons why the results are not in agreement with those of Menge *et al.* (1962). Obviously such factors

505

as the mobility of the spermatozoa in the uterus are not reproducible *in vitro*. Further, in the experiments described by these workers, the time of incubation in the uterus was between 14–38 h, presumably to allow time for the infiltration of the large numbers of leucocytes

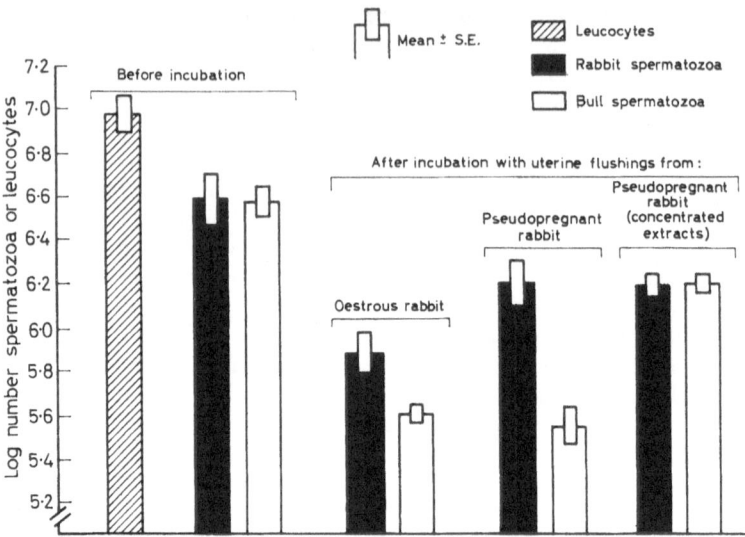

Figure 2. *Recovery of rabbit and bull spermatozoa after* in vitro *incubation with uterine flushings and concentrates from oestrous and pseudopregnant rabbits*

into the uterine lumen for appreciable and measurable phagocytosis. In the experiments described here, the leucocytes were added 'en masse' and incubation carried out for 90 min since considerable phagocytosis occurred within this time. Further, the leucocytes were taken from the peritoneal cavity and not from the uterus. There is, however, no evidence for a difference between uterine and peritoneal leucocytes and both are systemically derived. All the leucocytes used to obtain the results in *Figure 2* were taken from oestrous rabbits and further experiments were carried out to obtain a comparison between rates of phagocytosis of rabbit spermatozoa by leucocytes collected from peritoneal cavities of oestrous or pseudo-pregnant rabbits. Incubations were carried out in modified Krebs Ringer solution. No differences were observed between the rates of phagocytosis (*Figure 3*). Attempts to add a further group of spermatozoa and uterine leucocytes were not successful due to the problem of obtaining leucocytes from the uterus in sufficient quantity. These experiments together with preliminary work incubating rabbit and bull spermatozoa with oestrous rabbit leucocytes in pseudopregnant uterine rabbit flushings and *vice versa*, indicate that the difference

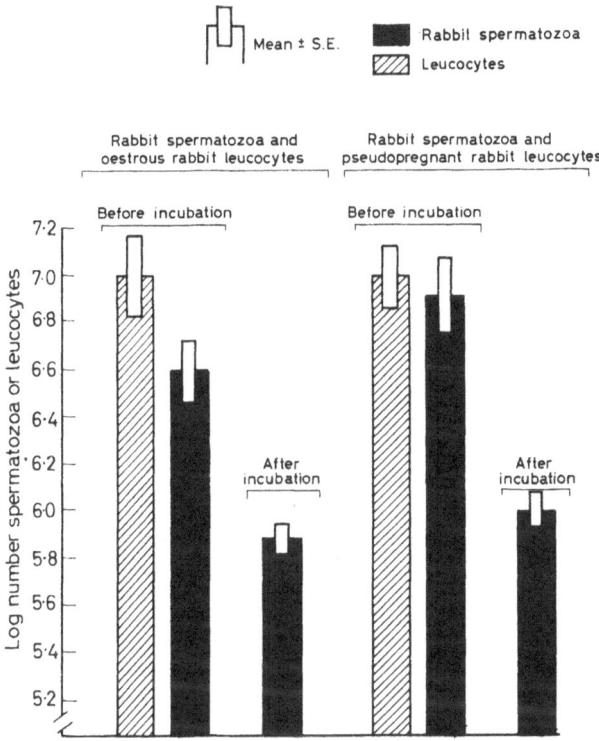

Figure 3. Recovery of rabbit spermatozoa after in vitro *incubation with leucocytes from pseudopregnant or oestrous rabbits*

cannot be explained by a functional difference in leucocytes obtained from animals of different endocrine status.

The difference in the rate of phagocytosis between the incubation of spermatozoa and leucocytes in the presence of oestrous uterine flushings (*Figure 2*) and incubations in Krebs Ringer Phosphate (*Figure 3*) are small, indicating that oestrous flushings do not stimulate phagocytosis and in consequence progestational flushings must have an inhibitory effect.

Research into phagocytosis of bacteria *in vitro* and its inhibition by

detergent compounds revealed that the detergent acted by coating
the leucocyte surface, thus preventing engulfment of the bacteria
(Berry, Starr and Haller, 1949). It has been suggested (Killingbeck
et al., 1963) that material occurring in uterine flushings could act in

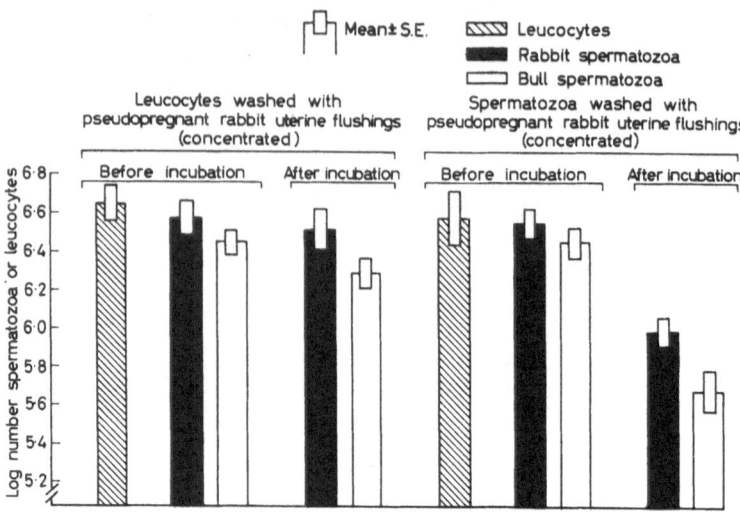

*Figure 4. The effect on incubation of leucocytes or spermatozoa with rabbit uterine flushings
or the* in vitro *phagocytosis of bull and rabbit spermatozoa*

a similar manner and prevent phagocytosis of bacteria by coating
the leucocyte surface. A further series of *in vitro* experiments were
carried out, therefore, in order to determine whether the inhibiting
influence of pseudopregnant uterine flushings on phagocytosis
occurred by an action on the spermatozoa or on the leucocyte.
Leucocytes were washed with pseudopregnant rabbit flushings prior
to incubation with untreated spermatozoa and compared with a
group in which spermatozoa were washed with pseudopregnant
rabbit flushings before incubation with untreated leucocytes. The
phagocytosis of both bull and rabbit spermatozoa was inhibited
when the leucocytes were treated. Treatment of the spermatozoa
on the other hand, gave a much smaller recovery of spermatozoa
(*Figure 4*).

It appears from the above that, under the conditions described,
there is little to indicate that rabbit leucocytes have vastly different
phagocytic abilities towards bull and rabbit spermatozoa. The

trends in phagocytosis were similar to those observed with bacteria and starch particles and the difference occurring in uterine flushings from oestrous and pseudopregnant rabbits was manifested by an action on the leucocytes.

INVESTIGATIONS INTO THE PHAGOCYTOSIS OF SPERMATOZOA *IN VIVO*

Although Menge *et al.* (1962) obtained a constant influx of leucocytes in all treatments, the possibility existed that the differences in the speed of removal of rabbit and bull spermatozoa could lie in their ability to cause leucocyte infiltration at different rates which were not detected.   The long incubation period could have resulted in maximal infiltration being reached in all cases.   In order to investigate this possibility, *in vivo* experiments were performed using the technique of Menge *et al.* (1962) with the intention of incubating for a shorter time.   A surprising result was obtained in that very few spermatozoa could be recovered from ligated uteri after an incubation of more than 10 h.   The leucocyte infiltration was maximal at this time also.   Histological examination of some uteri failed to reveal the presence of spermatozoa adhering to the uterine wall and it was assumed that removal had taken place by phagocytosis.

In further experiments a time interval for incubation of $5\frac{1}{2}$ h indicated that already considerable phagocytosis had taken place.   After this time both rabbit and bull spermatozoa were phagocytosed more rapidly from the oestrous than from the pseudopregnant uterus, rabbit and bull spermatozoa being removed at a similar rate in each case.   Differences were observed in the rate of infiltration of leucocytes.   These correlated inversely with the number of spermatozoa recovered, the infiltration being more rapid in the oestrous than in the pseudopregnant rabbits.   This is in accordance with the findings of Chang (1958) but does not substantiate the similarity in numbers of leucocytes between groups observed by Menge *et al.* (1962).   The slower infiltration of leucocytes into the progestational lumen does not, however, appear to be the only factor responsible for the lowered rate of spermatozoa removal since a visible difference was also apparent.   After incubation of spermatozoa in pseudopregnant uteri, many leucocytes were surrounded by up to 20 spermatozoa which appeared to be attached to the leucocytes by their heads whilst still exhibiting tail movements.   Active phagocytosis did not appear to be taking place.   This was more apparent with rabbit spermatozoa than bull spermatozoa and only occurred on incubation

509

in pseudopregnant uteri. This phenomena was not observed when leucocytes were examined during *in vitro* experiments.

Two rabbits out of nine did not show the expected behaviour of the remainder when rabbit spermatozoa were incubated in oestrous

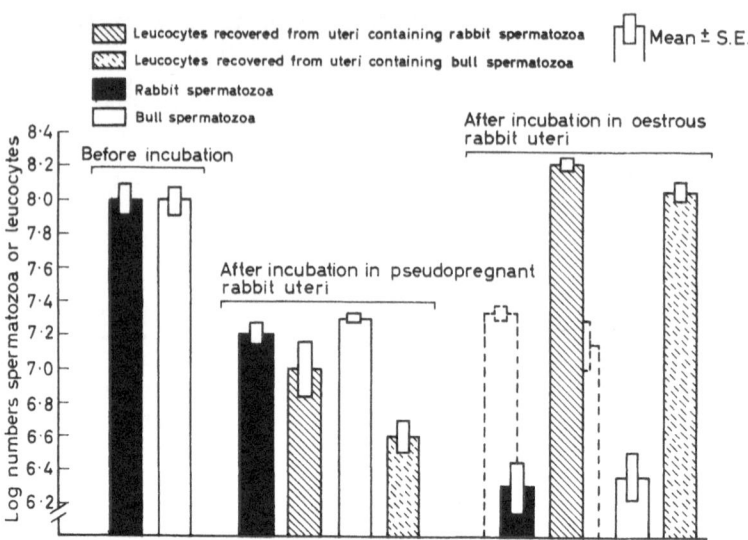

Figure 5. Recovery of rabbit and bull spermatozoa and leucocytes after incubation in oestrous or pseudopregnant rabbit uteri

rabbit uteri. The infiltration of leucocytes in these was much less and correspondingly more spermatozoa were found after incubation. The results from these 2 animals in terms of numbers of leucocytes and spermatozoa disappearance were similar to those obtained by incubation in the pseudopregnant uteri. These are not included in the overall analysis shown in *Figure 5* but are indicated by the additional data adjacent to the oestrous rabbit–rabbit spermatozoa group. This could have been due to an abnormal ovarian status of these rabbits although gross observation showed no evidence for this.

In summary, the results are suggestive that the removal of spermatozoa from the uterus is closely allied to the removal of bacteria and starch particles by phagocytosis. It appears to take place rapidly and leucocyte infiltration into the uterus is appreciable within 5 h. There are no large differences between the two species of spermatozoa used. The slower rate of phagocytosis from the pseudopregnant rabbit uterus could be the result of two factors; a slower rate of infiltration of leucocytes into the lumen and a material present in the uterine secretion of the progesterone-dominated animal which exerts its effect by causing alterations in the ability of the leucocytes to function. Spermatozoa from the same or a different species are removed from the uterus in fact, in a manner similar to other foreign particles.

ACKNOWLEDGEMENT

The author wishes to thank the Milk Marketing Board Cattle Breeding Centre, Sutton Bonington, for providing the samples of bull spermatozoa.

REFERENCES

Austin, C. R. (1957). *J. Endocr.* **14,** 335–342
— (1960). *J. Reprod. Fert.* **1,** 151–156
— (1964). *Proc. Vth. Int. Congr. Anim. Reprod. and A.I.*, Trento **3,** 7–22
— and Bishop, M. W. H. (1959). *J. Endocr.* **18,** viii–ix
Bedford, J. M. (1965). *J. Reprod. Fert.* **9,** 249–256
Berry, J. L., Starr, R. W. and Haller, E. C. (1949). *J. Bact.* **57,** 603–611
Berzedowka, B. (1955). *Pam. Inst. zootech. Polsce.* 39–45
Blandau, R. J. and Odor, D. L. (1949). *Anat. Rec.* **103,** 93–109
Brinsfield, T. H., Hawk, H. W. and Leffel, E. C. (1963). *J. Reprod. Fert.* **6,** 79–86
— — and Richter, H. F. (1964). *J. Reprod. Fert.* **8,** 293–296
Broome, A. W. J., Lamming, G. E. and Smith, W. J. (1959). *J. Endocr.* **19,** 274
— Winter, A. J., McNutt, S. H. and Casida, L. E. (1960). *Am. J. vet. Res.* **21,** 675–682
Chang, M. C. (1958). *Endocrinology* **63,** 619–628
Dokukin, A. V. (1956). *Byull. eksp. Biol. Med.* **42,** 61–63
Genin, D. I. (1953). *Zh. obshch. Biol.* **14,** 441–451
Halpern, B. N., Stiffel, C., Biozzi, I. and Mouton, D. (1960). *C.r. Séanc. Soc. Biol.* **154,** 1144
Hawk, H. W. (1958). *J. Anim. Sci.* **17,** 416–425
— Simon, J., Cohen, M., McNutt, S. H. and Casida, L. E. (1955). *J. Am. vet. med. Ass.* **126,** 268–270
— Turner, G. D. and Sykes, J. F. (1960). *Am. J. vet. Res.* **21,** 644
Haynes, N. B. and Lamming, G. E. (1964). *Proc. Vth. Int. Congr. Anim. Reprod. and A.I.*, Trento **2,** 335–341
Henderson, V. and Mayer, D. L. (1965). *Fedn Proc. Fedn Am. Socs. exp. Biol.* **44,** 450
Heap, R. B. (1962). *J. Endocr.* **24,** 367–378
— and Lamming, G. E. (1960). *J. Endocr.* **20,** 23

Heap, R. B., Robinson, D. W. and Lamming, G. E. (1962). *J. Endocr.* **23,** 351–356

Howe, G. R. and Black, D. L. (1963). *J. Reprod. Fert.* **6,** 305–311

Killingbeck, J. (1963). *Ph.D. Thesis,* Nottingham

— Haynes, N. B. and Lamming, G. E. (1963). *Nature, Lond.* **199,** 255–256

— and Lamming, G. E. (1963). *Nature, Lond.* **198,** 111–113

Lamming, G. E. and Haynes, N. B. (1964). *Proc. Vth Int. Congr. Anim. Reprod. and A.I.,* Trento **2,** 355–360

Mahajan, S. C. and Menge, A. C. (1965). *J. Anim. Sci.* **24,** 924

Marcus, S. L. (1966). *Fert. Steril.* **17,** 212–220

Menge, A. C., Tyler, W. J. and Casida, L. E. (1962). *J. Reprod. Fert.* **3,** 396–365

Merton, H. (1939). *Proc. R. Soc. Edinb.* **59,** 207–217

Nicol, T. and Helmy, I. D. (1951). *Nature, Lond.* **167,** 199–200

Noyes, R. W. and Thibault, C. (1962). *Fert. Steril.* **13,** 346–365

Pitkjanen, I. G. (1960). *Zh. Obshch. Biol.* **21,** 28–33

Posalaky, Z. and Töró, I. (1957). *Nature, Lond.* **179,** 150–151

Reid, B. L. (1965a). *Aust. J. Zool.* **13,** 189–199

— (1965b). *J. Anat.* **99,** 947

— (1965c). *Expl Cell. Res.* **40,** 679–683

Rowson, L. E. A., Lamming, G. E. and Fry, R. M. (1953). *Nature, Lond.* **171,** 149

Spector, W. G. and Storey, E. (1958). *J. Path. Bact.* **75,** 381–387

Sobotta, F. (1920). *Arch. Anat.* **XCIV,** 165

Vojtiskova, M. (1956). *Folia biol.* **2,** 239

Winter, A. J., Broome, A. W. J., McNutt, S. H. and Casida, L. E. (1960). *Am. J. vet. Res.* **21,** 668

Yanagimachi, R. and Chang, M. C. (1963). *J. Reprod. Fert.* **5,** 389–396

Yochem, D. E. (1929). *Biol. Bull.* **56,** 274–295

THE REACTION OF THE UTERUS
DURING IMPLANTATION
IN THE MOUSE

C. A. FINN

Department of Physiology, Royal Veterinary College, London N.W.1

INTRODUCTION

IMPLANTATION is the first stage in the formation of the placenta when blastocysts are attached to the wall of the uterus in a sufficiently intimate manner to permit the exchange of nutrients and waste materials between the mother and the offspring. The extent to which changes occur in the uterus to allow this exchange varies considerably between species. In the pig, for example, the histological changes are minimal, whereas in rodents the implantation reaction in the uterus is considerable. The mouse provides a useful experimental animal for the study of these changes and this paper is largely concerned with the changes induced in the mouse uterus during implantation.

If fertilized eggs are removed from the Fallopian tubes of a mouse and placed in other situations in the body, for example under the capsule of the kidney or testis (Kirby, 1960), the trophoblast grows rapidly, invading and destroying the tissue of the host organ. The embryonal area of the blastocyst, however, does not usually grow. If, on the other hand, the eggs are transplanted after they have reached the uterus then the embryo continues its development (Kirby, 1962). These experiments show that in the absence of uterine stimulation the trophoblast is able to grow and invade adult tissue, but for growth of the embryo the blastocyst has to be exposed to the influences of the uterus.

In lactating pregnant mice, fully developed blastocysts remain in the uterus in a quiescent condition for several days and development does not proceed until implantation occurs. A similar condition of delayed implantation occurs if pregnant non-lactating mice are ovariectomized on the third day of pregnancy and treated with progesterone (Canivenc and Laffargue, 1957; Cochrane and Meyer, 1957). Implantation and continued development of the blastocyst can be induced in both these cases by the injection of small quantities

17+   513

of oestrogen. It is apparent therefore, that whereas the blastocyst will grow rapidly if removed from the uterus and transplanted in another organ, it is in some way restrained when in a uterus in which implantation cannot take place.

It is apparent, therefore, that the uterus influences the development of the blastocyst both by inhibiting the invasiveness of the trophoblast and by stimulating the embryonal area of the egg to further development. Implantation is thus a complex process involving endocrine sensitization of the uterus by hormones secreted by the ovary, stimulation of changes in the sensitized uterus by the blastocyst and control of the development of the blastocyst by the uterus. I shall consider first the changes induced in the uterus during implantation and then the endocrine control of uterine sensitivity. Very little is at present known about the mechanism whereby the uterus exerts a controlling influence over the development of the blastocyst.

THE CHANGES INDUCED IN THE MOUSE UTERUS DURING IMPLANTATION

Macroscopic examination of the uterus of a mouse on the sixth day of pregnancy reveals small beadlike swellings along the length of the uterus. These swellings are due to the development of decidual cells in the stroma of the endometrium which is one of the most obvious early changes of implantation in the mouse. A typical decidual cell has been described by Amoroso (1952) as an endometrial connective tissue cell which has enlarged and become rounded or polyhedral due to the accumulation of glycogen or lipoids in its cytoplasm. Recently, Jollie and Benscosme (1965) have described the ultrastructure of decidual cells in the rat as seen with the electron microscope. Glycogen and fat are abundant and the mitochondria become more numerous and smaller. The stroma appears epithelioid with desmosomes between the cells. They also make the very interesting observation that the stromal cells of unstimulated pseudopregnant rats resemble embryonic fibroblasts.

According to Amoroso (1952) decidual cells originate from undifferentiated perivascular mesenchymal cells. The function of the cells is unknown although it seems likely that they may be a mechanism whereby the uterus protects itself against the invasion of the trophoblast once the blastocyst has been stimulated to further development. Possibly the cells are an evolutionary adaptation of the granulation tissue reaction to foreign bodies. It may be relevant that some substances which induce the granulation tissue reaction, for example carraghenin or oil, can also induce decidual cell formation in the rat (Finn and Keen, 1962). Furthermore, mechanical injury to the endometrium of a sensitized uterus, such as scratching or crushing, will cause a massive decidual cell reaction. This was shown originally in guinea-pigs by Loeb (1908) and since then in mice and rats by Parkes (1929) and Selye and McKeown (1935) respectively.

Kirby (1965) has reported some interesting experiments which provide direct evidence for the protective function of decidual cells. He transplanted pieces of ectoplacental cone from 7½-day mouse embryos into the uteri of pseudopregnant or non-pseudopregnant mice. In the former case, a decidual cell reaction was induced and the implanted tissue did not invade or destroy the wall of the uterus. In the latter case, in which the uteri were not in a hormonal condition to allow the formation of decidual cells, the piece of ectoplacental cone grew rapidly causing considerable destruction of the uterus. These results demonstrate very convincingly the barrier presented by decidual tissue to an invading trophoblast. The morphological changes during decidualization, as seen with the electron microscope, are also consistent with such an hypothesis. As mentioned above, Jolie and Benscosme described decidual tissue in the rat as appearing epithelioid with desmosomes between cells. I have found a similar ultra-structural arrangement in the decidual area around blastocysts on the sixth day of pregnancy in the mouse. The decidual cells appear to be joined together at numerous points by desmosomes so that there is very little intracellular space left. The blastocyst is thus surrounded by a more or less solid mass of cells which presumably provides a much more effective barrier to invasion than the loose connective tissue normally found in the uterine stroma.

The fact that the uterus will respond to an artificial stimulus makes it possible to study the capacity of the uterus for morphological change in the absence of a blastocyst and thus differentiate the uterine and embryonic components of the reaction. Trauma, however, is a poor stimulus because the damage inflicted on the uterus makes histological interpretation of the results difficult. Krehbiel (1937) overcame this difficulty by inducing deciduoma in rats with the passage of an electric current across the uterus. In his classical paper he describes the histological changes which follow electrical stimulation of the pseudopregnant uterus and compares these with those which occur during implantation of a blastocyst. The changes in both cases were very similar, suggesting that the uterus is 'an organ of specific potentialities and that the inherent nature of the

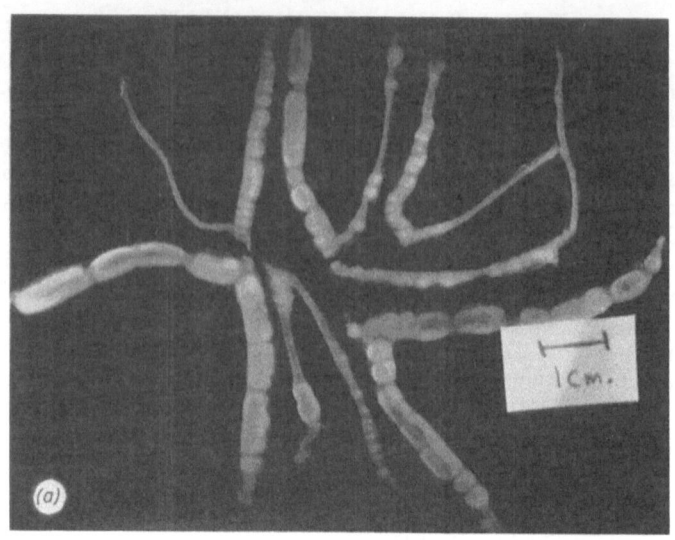

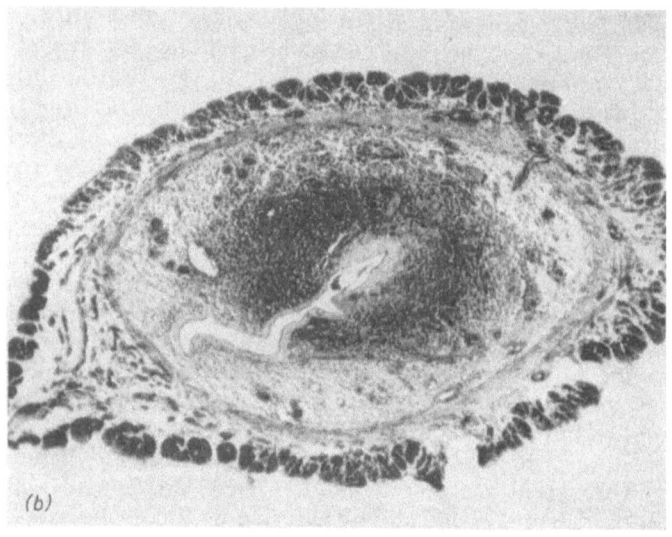

PLATE I

a. *The uteri from a group of pseudopregnant mice injected with oil on* day 4 *and killed on* day 7. *The uteri have been fixed and cleared in benzyl benzoate to demonstrate the decidual swellings*

b. *T.S. of uterus on fifth day of pregnancy, treated to demonstrate alkaline phosphatase* (From C. A. Finn and J. R. Hinchliffe, 1964, by courtesy of the Editor, *Journal of Reproduction and Fertility*)

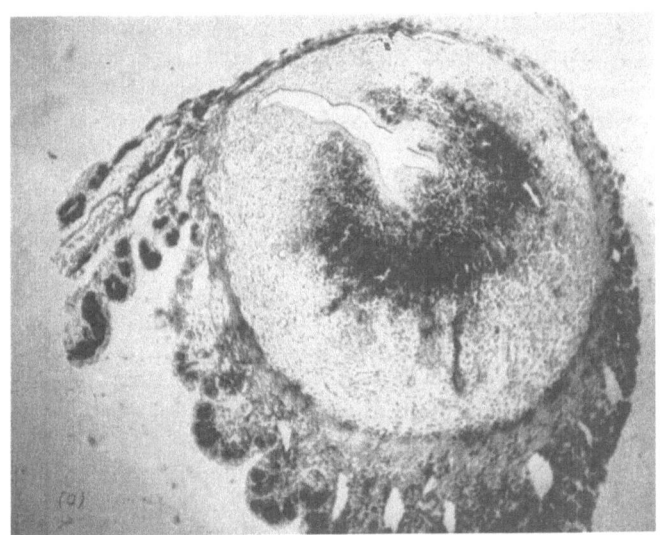

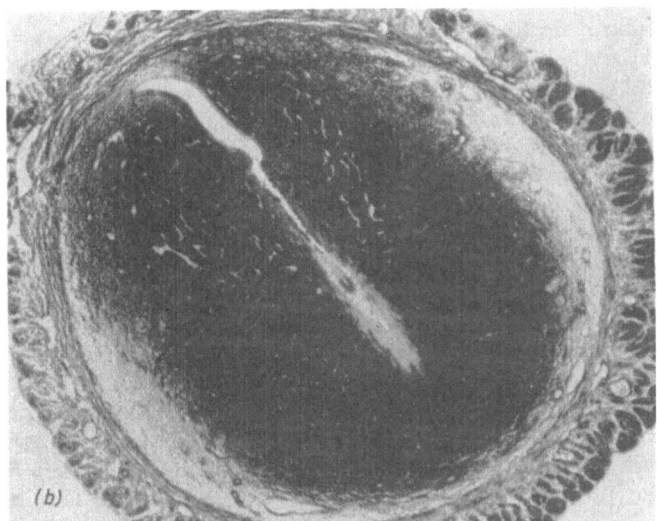

PLATE II

a. T.S. of uterus on fifth day of pseudopregnancy after installation of oil on the afternoon of the fourth day, treated to demonstrate alkaline phosphatase (From C. A. Finn and J. R. Hinchliffe, 1964, by courtesy of the Editor, *Journal of Reproduction and Fertility*)

b. T.S. of uterus on sixth day of pregnancy treated to demonstrate alkaline phosphatase (From C. A. Finn and J. R. Hinchliffe, 1964, by courtesy of the Editor, *Journal of Reproduction and Fertility*)

517

sensitized uterus merely requires some slight stimulus, or irritation, to initiate activity which continues for at least the first half of pregnancy when the proper hormonal environment is provided'.

Recently, Finn and Keen (1962) showed that the intraluminal injection of oil-induced deciduoma formation in rats and Finn and Hinchliffe (1964, 1965) studied the reaction of the pseudopregnant mouse uterus to intra-uterine oil. The histological and histochemical changes occurring in the oil stimulated uterus were compared with those occurring during pregnancy and pseudopregnancy over an equivalent period. In these experiments 0·03 ml arachis oil was injected into the uteri of the pseudopregnant mice on the afternoon of day 4 (the day on which a copulation plug is found is day 1). When this is done it is usual to get several discrete swellings along the uterus which are just visible on day 6 and easily seen on day 7. *Plate Ia* shows the uteri of a group of pseudopregnant mice injected with oil on day 4 and killed on day 7. The uteri have been fixed and cleared in benzyl benzoate according to the method described by Orsini (1962).

Histochemical examination of the uteri of pregnant animals on day 5 showed an accumulation of alkaline phosphatase in the stroma around the blastocyst (*Plate Ib*). A similar area of alkaline phosphatase activity also appeared in the stroma of the oil-injected uteri at this time (*Plate IIa*), but not in the pseudopregnant non-injected uteri. It was interesting that the enzyme always appeared on the antimesometrial side of the uterus and was more or less continuous along the length of the injected horn. Blastocysts, of course, always implant on this side in the mouse as do pieces of tumour or beads (Wilson, 1963a; Blandau, 1949). However, in these cases it was assumed that there was some locating mechanism directing the objects antimesometrially, which can hardly be so in the case of a liquid such as oil, and it suggests, therefore, that the antimesometrial side of the uterus is preferentially sensitive to the implantation stimulus. It is particularly noticeable that there is no breakdown of epithelium, which is free of alkaline phosphatase, at this time, either in the pregnant or oil-injected uteri, indicating that the initial stimulus to the uterus is not injury by the blastocyst.

On the following day (day 6) the alkaline phosphatase in the stroma of the pregnant animals almost circumscribes the lumen and the blastocysts are embedded in a chamber which is an indentation of the lumen at the centre of the implantation swelling (*Plate IIb*). The uterine epithelium at the base of the chamber is in the process of breaking down and reacts positively for alkaline phosphatase. In

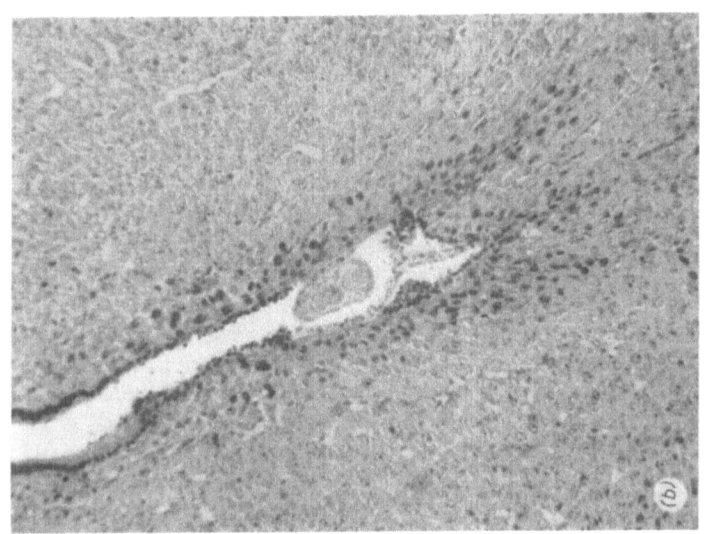

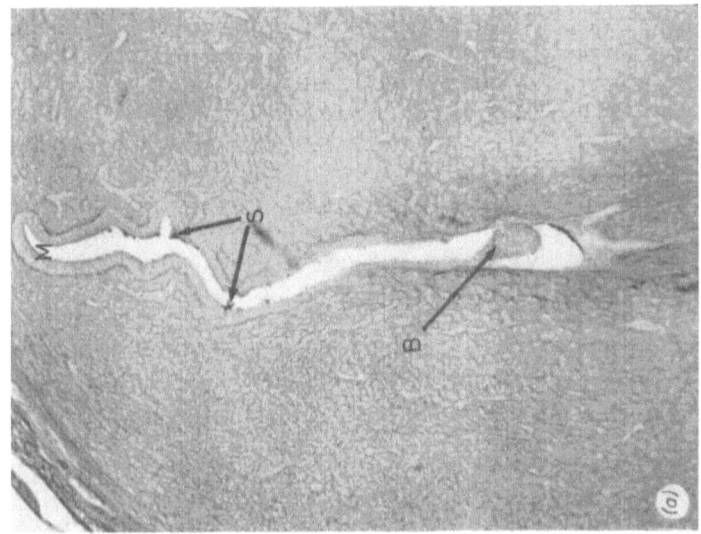

a. T.S. of uterus containing a 6-day *implanting blastocyst (B). PAS positive material can
be seen in the stroma around the blastocyst and in the lumen attached to the epithelial cells (S)*
(From C. A. Finn and J. R. Hinchliffe, 1965, by courtesy of the Editor, *Journal
of Reproduction and Fertility*)

b. T.S. of implantation chamber on the sixth day of pregnancy stained with iron haematoxylin a small area around the base of the chamber the stroma is free of alkaline phosphatase. This area also shows other characteristic features. *Plate IIIa* shows a section through a chamber stained by the PAS reagent. The area around the base of the chamber stains more strongly than the remainder of the section. High power examination reveals that there are granules of PAS positive material within the cells and non-granular material around the outside of the cells. The former material is digested by diastase and is presumably glycogen whilst the latter is not and is probably a mucopolysaccharide. Also present in the section is a small amount of a PAS positive material in the uterine lumen. This is not found in pseudopregnant unstimulated uteri on day 6 and is not the same as uterine milk found earlier in pregnancy or pseudopregnancy.

When similar sections are stained with iron haematoxylin or Toluidine Blue the cells around the base of the chamber are seen to have prominent large nucleoli (*Plate IIIb*). The Toluidine Blue staining is ineffective if the slides are previously incubated in ribonuclease, indicating that the staining is due to a high content of RNA.

By day 6 the oil-stimulated uteri of pseudopregnant animals have developed several discrete swellings along the length of the uterus. Serial sections through these show that they have an internal structure closely resembling that found in a normal implantation swelling, with a central chamber, although not surprisingly this is not as well defined as an implantation chamber. The characteristic features described earlier are all present (*Plate IVa* and *b*), the only major difference being that the PAS positive intraluminal fluid is present in much larger quantities. This may indicate that it is a nutrient substance which is normally taken up by the blastocysts but accumulates when the latter are absent. The epithelium on the antimesometrial side is thin and in the process of breaking down, suggesting that the breakdown of the epithelium during implantation is due to intrinsic autodestructive changes in the uterus rather than to any phagocytic activity of the blastocyst.

From this study it is apparent that most of the early implantation changes occurring in the mouse uterus will take place without the presence of a blastocyst. It is, of course, necessary for the uterus to be stimulated and this is normally done by the blastocyst, but it seems that once stimulated the uterus undergoes a series of reactions with the formation of an implantation chamber. Apparently the mouse uterus when sensitized by the hormones of pregnancy has the competence to react to an appropriate stimulus with the formation

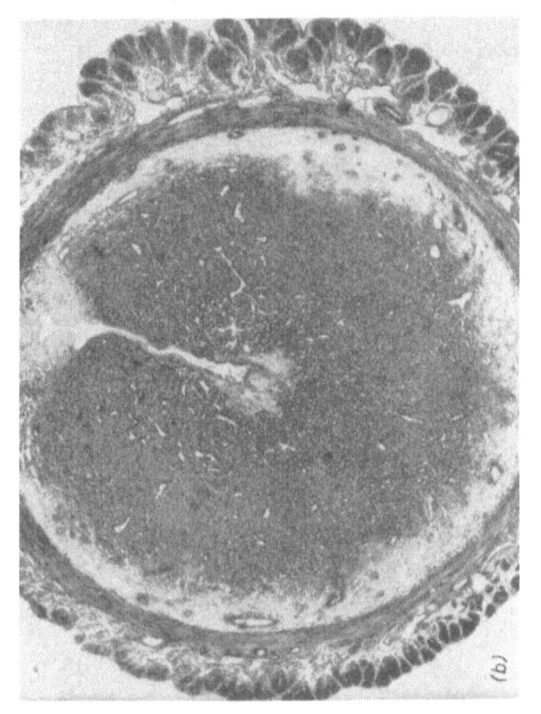

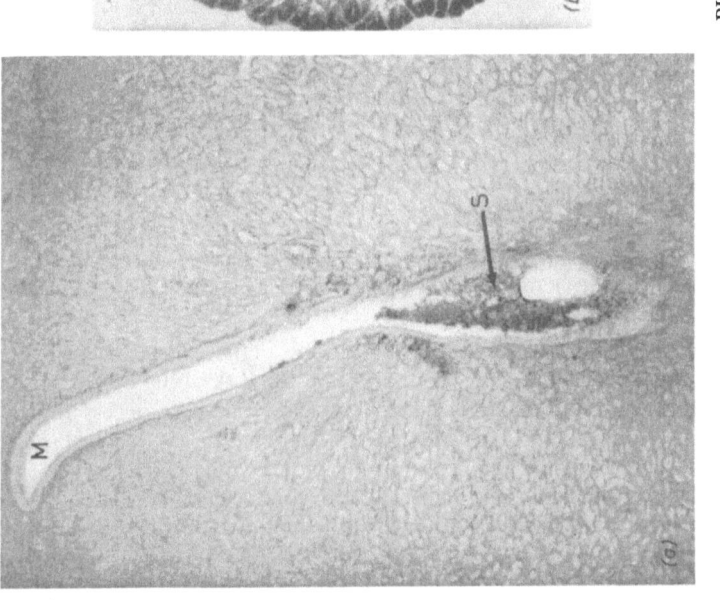

PLATE IV

a. T.S. through the centre of a 6-day oil-induced decidual swelling showing secretion (S) and PAS positive material round the antimesometrial end of the lumen (M-mesometrial end) (From C. A. Finn and J. R. Hinchliffe, 1965, by courtesy of the Editor, *Journal of Reproduction and Fertility*)
b. T.S. through centre of a 6-day oil-induced decidual swelling, treated to demonstrate alkaline phosphatase (From C. A. Finn and J. R. Hinchliffe, 1964, by courtesy of the Editor, *Journal of Reproduction and Fertility*)

of a well defined morphological structure in which the blastocyst can implant.

Psychoyos (1960) has shown that one of the earliest changes which can be detected in the uterus of a rat during implantation is increased permeability of the blood vessels supplying the uterus at the site of an

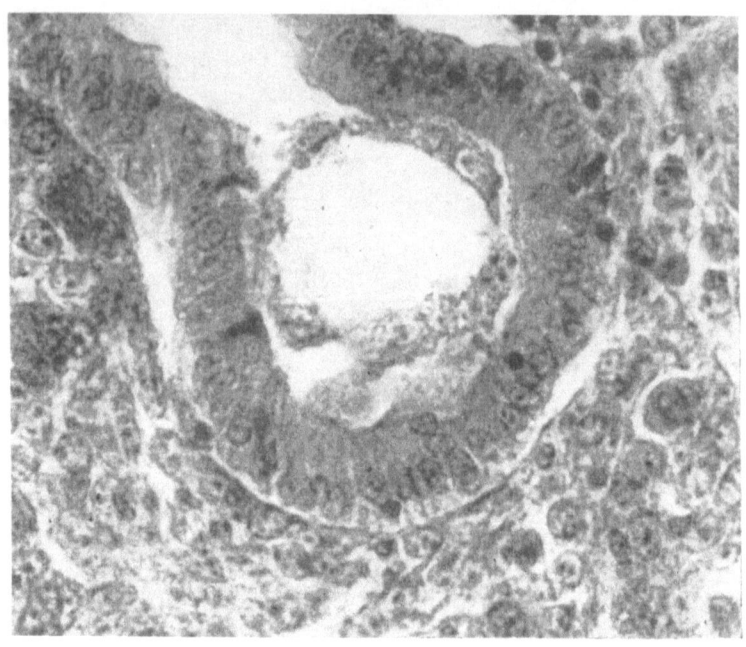

PLATE V

T.S. through uterus at approximately 100 h *after copulation showing blastocyst with three basic cells passing into the uterine epithelium*

implanting blastocyst. This can be clearly demonstrated by the intravenous injection of a dye such as pontamine sky blue or Evans blue. Following such an injection blue bands appear across the uterus where blastocysts are present. This reaction also occurs in mice in which it can be produced on the fifth day of pregnancy and thus provide a simple method for determining the position of implanting blastocysts before any other change is visible macroscopically.

Wilson (1963b) has shown another interesting early reaction of implantation in the mouse; the passage of basically staining cells from the blastocyst through the maternal epithelium (*see Plate V*). Wilson states that not more than six are found in one implantation site and that they pass across about 100 h after copulation.

There are thus three clearly discernible changes in the uterus during the early stages of implantation. The appearance of alkaline phosphatase in the stroma, increased permeability of blood vessels and the passage of basophilic cells from the blastocyst to the uterus. Dr. McLaren and I have recently studied the appearance of these changes in a group of mice in order to determine in which order they occur. We found that increased permeability of the blood vessels invariably occurred in advance of the migration of the basophil cells and this was then followed by the appearance of alkaline phosphatase in the stroma. All three events occurred within a period of a few hours during the night between day 4 and day 5. The function of the basic cells is completely unknown at present. It is not likely that they are involved in stimulating the uterine reaction since they do not pass across into the uterus until after the first uterine reaction to implantation has occurred. The increased permeability of the blood vessels is very interesting as this is also one of the first reactions during an inflammatory response, again suggesting the possibility that the implantation reaction may be related to the granulation tissue reaction.

THE CONTROL OF THE SENSITIVITY
OF THE UTERUS BY HORMONES

Since the discovery by Loeb (1908) that the maternal reaction to implantation could be elicited by traumatization of the uterus of a pseudopregnant guinea-pig, the traumatic decidual cell reaction (DCR) has been widely used in the study of the endocrine factors controlling uterine sensitivity during implantation. The reaction can be elicited in the uterus of pregnant or pseudopregnant animals of several species and the formation of a DCR has been widely accepted as evidence of uterine sensitivity to implantation. Using the traumatic DCR of pseudopregnant mice and rats it has been shown that there is a rather restricted period of pseudopregnancy during which a DCR can be formed, with maximum sensitivity on days 4 and 5 in the mouse and rat respectively.

It is, of course, well known that the corpus luteum plays an important role in the maintenance of pregnancy, the original discovery being by Fraenkel and Cohn (1901) following evidently a suggestion by Gustav Born. Later, several groups of workers succeeded in isolating a steroid from the corpus luteum which was called progesterone in view of its importance in pregnancy. Removal of the ovary or corpus luteum results, in most species, in failure of implantation, thus demonstrating the necessity for progesterone during early pregnancy (for further details *see* Amoroso and Finn, 1962). Furthermore, a DCR can be initiated in ovariectomized unmated rats or mice given injections of progesterone (Rothchild and Meyer, 1942; Greenwald, 1958), indicating that the maternal side of the reaction is dependent on a supply of the hormone. However, removal of the ovaries from pregnant rats on the fourth day of pregnancy (or on the third day in mice) prevents implantation occurring even though exogenous progesterone is given, although the blastocysts remain alive for several days (Canivenc and Laffargue, 1957; Cochrane and Meyer, 1957). Thus progesterone alone is not sufficient to sensitize the uterus to implantation or to the initiation of the DCR *by the blastocyst*. If ovariectomy is delayed until the following day, implantation is not prevented provided exogenous progesterone is given. This has led to the suggestion that another hormone is released from the ovary at some time between the limits defined by the ovariectomy experiments. As oestrogens are able to precipitate implantation in rats and mice showing delayed implantation due to lactation (Krehbiel, 1941) this seemed the likely hormone. It has not so far been possible to demonstrate the secretion of oestrogen at the appropriate time, and it should be remembered that there is a high level of circulating progesterone which will tend to mask most of the secondary oestrogenic responses. It has however, been conclusively shown that the injection of a small quantity of an oestrogen into ovariectomized rats and mice kept in a state of delayed implantation with progesterone, will precipitate implantation, thus providing good indirect evidence for a surge of oestrogen being responsible for sensitizing the uterus to implantation (Mayer, 1959, 1963; Psychoyos, 1961). The situation is complicated, however, by the fact that it is possible to stimulate the DCR by trauma in ovariectomized unmated rats or mice given progesterone without an oestrogen surge, or in pregnant rats ovariectomized on day 4 and given progesterone only, or even in pregnant rats kept in a state of delayed implantation by lactation (Rothchild and Meyer, 1942; Greenwald, 1958; Meyer and Cochrane, 1962; Lyon and Allen, 1938). Obviously, therefore, there is a difference in the hormone requirements for implantation or the induction of the DCR by the blastocyst and for the induction of the traumatic DCR. Two main possibilities can be put forward to account for this difference. Either there is a difference in the nature of the stimulus given to the uterus in the two cases or the oestrogen surge is necessary to activate the blastocyst in some way so that it is able to stimulate the uterus. The latter possibility does not seem very likely in view of the fact that inert objects placed in a properly sensitized uterus are capable of stimulating a DCR.

In the first part of this paper it was shown that the injection of oil into a pseudopregnant uterus stimulated the formation of structures in the uterus which closely resemble natural implantation chambers. As the oil must be acting, at least initially, on the surface of the luminal epithelium, it seems likely that the stimulus given by it to the uterus would resemble more closely the stimulus given by the blastocyst than would trauma, and would thus provide a physiological stimulus for testing whether the first possibility is true. Experiments were therefore performed to determine whether the oestrogen surge was necessary for the oil-induced decidual cell reaction in pseudopregnant mice ovariectomized on day 3 or day 4 and given exogenous progesterone. When the oil was injected into mice ovariectomized on day 3 no deciduomata were induced unless a small quantity of oestradiol was given about 7 h before the oil injection on day 4. No injection of oestradiol was necessary when the oil was injected into animals ovariectomized on day 4. Trauma induced a good DCR in both cases without the necessity for any oestradiol (Finn, 1965).

The decidual stimulus given to the uterus by oil thus appears to require the same endocrine conditions to be effective as are required for implantation of blastocysts. Both blastocysts and oil act initially by making contact with the luminal surface of the uterine epithelial cells and it is possible that the oestrogen surge is necessary for the sensitization of the epithelial cells to this initial surface reaction. Trauma perhaps by-passes this initial reaction and acts directly on the stromal cells. If this is correct, it would suggest that the stromal cells are brought into a condition in which they can be transformed into decidual cells by progesterone alone. Before transformation can take place, however, they must be stimulated. Under normal circumstances this stimulus is received on the luminal surface of the uterine epithelial cells and a message transmitted to the stroma. The oestrogen surge is possibly concerned in sensitizing the epithelial cells to react to this stimulus.

I have recently used the intraluminal injection of oil as a decidual stimulus to get some more detailed information about the endocrine factors controlling uterine sensitivity. In this study (Finn, 1966), unmated ovariectomized mice were used and all hormones were given by injection. Both the quantity of oestradiol given prior to the decidual stimulus (nidatory surge oestrogen) and the timing of the injection of the oil relative to the surge and to the length of progesterone treatment were found to be important.

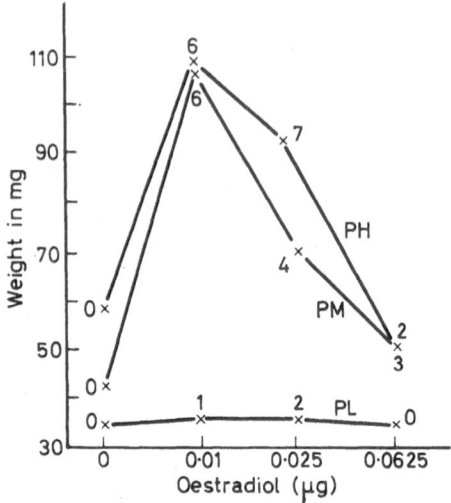

Figure 1. The variation of decidual response with varying dosages of progesterone and nidatory oestrogen surge. Progesterone levels—PH=4 mg, PM=1 mg, PL=0·25 mg. The figures at each point represent the number of animals showing a positive response (group size—8) (From C. A. Finn, 1966, by courtesy of the Editor, Journal of Endocrinology)

Initially, a factoral experiment was performed in which groups of ovariectomized mice were treated with three levels of progesterone (0·25, 1 and 4 mg/day), and four of nidatory surge oestrogen (0, 0·01, 0·025, 0·0625 μg oestradiol) after pretreatment with oestradiol (*see* later). The progesterone was given by daily subcutaneous injection and the oestradiol injected on the morning of the fourth day of progesterone treatment, 7 h before the intraluminal injection of the oil as a decidual stimulus. The animals were killed on the eighth day of progesterone treatment and the decidual response assessed both quantitatively on the number of uteri showing a positive response and on the weight of the oil-injected horns. The results are shown in *Figure 1.*

Little response was obtained when only 0·25 mg progesterone was used whilst with higher doses of 1 and 4 mg the response was more or less equal. Varying the dose of oestradiol had interesting effects. As expected no decidual response was obtained when no nidatory surge oestrogen was given. With 0·01 or 0·025 μg a good response occurred, but with the highest dose, 0·0625 μg, the response was inhibited. This indicates a double threshold for sensitivity to nidatory surge oestrogen, low doses sensitizing and higher doses inhibiting the response. A similar double threshold theory has been postulated by Folley and Malpress (1947) to account for the action of oestrogens on milk secretion. Analysis of variance of the results showed that whereas both progesterone and oestradiol had very significant effects there was no interaction between the two hormones, indicating that in this situation they are acting independently.

In these experiments the nidatory surge oestrogen was given on the fourth day of progesterone treatment 7 h before the intraluminal injection of the oil decidual stimulus. This corresponds as far as could be ascertained with the times on which optimum sensitivity can be obtained in ovariectomized pseudopregnant animals, and in pilot experiments good results were obtained. In the next experiments information was obtained on the temporal limits of these two parameters, firstly by varying the interval between giving the oestradiol and the oil on the fourth day and then by keeping this constant at 7 h but inducing sensitivity and injecting the oil on various days of progesterone treatment. In the first case, maximum response was obtained when the interval between giving the oestrogen surge and the decidual stimulus was 4 or 8 h. When the interval was 2 or 16 h, the decidual response was significantly reduced. In the other case, with a constant interval of 7 h but inducing sensitivity and applying the decidual stimulus on various days of progesterone treatment, maximum response was obtained on the fourth and fifth days of progesterone treatment with diminished response on the third, sixth and seventh days and no response on the second day. It thus appears that there is an optimum time interval for the oestradiol to induce sensitivity and that the progesterone preparation of the uterus reaches a maximum on the fourth and fifth days and then declines.

Following the period of sensitivity the uterus appears to go into a refractory condition when it will no longer respond to nidatory surge oestrogen. This was suggested by Finn and Emmens (1961) to explain the implantation inhibiting effect of small doses of oestradiol and good experimental evidence for such a state has been produced by Psychoyos (1963) in rats. As sensitivity and the oil DCR can be induced equally well on days 4 or 5 of progesterone treatment in ovariectomized mice, they are useful experimental animals for testing the hypothesis. Two groups of mice were both given an oestrogen surge on the morning of the fifth day followed by the intra-uterine injection of oil 7 h later. The mice in one group were also given an oestrogen surge on the morning of the fourth day. It was found that the premature surge on the fourth day almost completely inhibited the response to the oil. This result supports the hypothesis of a refractory period following the period of sensitivity.

Another factor in the sensitization of the uterus is the oestrogen secreted by the ovary during the oestrus preceding implantation. This, of course, plays an important part in the transport of ova along the Fallopian tubes and in the behavioural changes associated with oestrus, but does it also play an important part in the sensitization of the uterus to implantation? The results of experiments reported in the literature are indecisive. Greenwald (1958) and Rothchild and Meyer (1942) have reported experiments on the induction of traumatic deciduomata in rats and mice respectively, in which some of the animals were subjected to a period of oestrogen priming prior to being given progesterone and others were not. Both groups of workers were able to induce deciduomata in the latter animals although the magnitude of response was considerably lower than in those animals which had been pretreated with oestrogen. To differentiate this period of oestrogen secretion from that associated with nidation later in pregnancy, I refer to it as the period of oestrous oestrogen. In the experiments referred to earlier, a period of oestrous oestrogen consisting of $0 \cdot 1$ μg oestradiol daily for 3 days had been included. In the following experiments I kept the other parameters optimal and varied the quantity of oestrous oestrogen. When no oestrous oestrogen was given, no DCR to oil was obtained and even with $0 \cdot 05$ μg oestradiol daily for 3 days the level of response was very poor. Apparently oestrous oestrogen does play a part in the sensitization of the uterus to implantation, although like nidatory surge oestrogen it is not essential for the induction of the traumatic DCR. I did, in fact, include a group in which progesterone only had been given and was able to induce a DCR to crushing of the uterus, although in agreement with the results of previous workers the response was rather poor.

To summarize, it appears that there are three hormone stages in the sensitization of the uterus to the oil DCR; *oestrous oestrogen, pro-*

gesterone and *nidatory surge oestrogen.* It is likely that the endocrine requirements for implantation are similar.

ACKNOWLEDGEMENT

I would like to thank Professor E. C. Amoroso for help in the preparation of the manuscript and Mr. Kenneth Barber for technical assistance. The histological work described was assisted by a grant from the M.R.C. and the electron microscope work by U.S. public health grant HD 01476/06.

REFERENCES

Amoroso, E. C. (1952). 'Placentation.' In: *Marshall's Physiology of Reproduction.* Ed. by A. S. Parkes. London; Longmans, Green
— and Finn, C. A. (1962). 'Ovarian Activity during Gestation, Ovum Transport, and Implantation.' In: *The Ovary.* Ed. by S. Zuckerman. New York; The Academic Press
Blandau, R. J. (1949). *Anat. Rec.* **104,** 331–359
Canivenc, R. and Laffargue, M. (1957). *C.r. hebd. Séanc. Acad. Sci., Paris* **245,** 1752–1754
Cochrane, R. L. and Meyer, R. K. (1957). *Proc. Soc. exp. Biol.* **96,** 155–159
Finn, C. A. (1965). *J. Endocr.* **32,** 223–229
— (1966). *J. Endocr.* **36,** 239–248
— and Emmens, C. W. (1961). *J. Reprod. Fert.* **2,** 528–529
— and Hinchliffe, J. R. (1964). *J. Reprod. Fert.* **8,** 331–338
— — (1965). *J. Reprod. Fert.* **9,** 307–309
— and Keen, P. M. (1962). *J. Reprod. Fert.* **4,** 215–216
Folley, S. J. and Malpress, F. H. (1947). Abstract of communication, XVII *Int. Physiol. Cong.,* p. 340
Fraenkel, L. and Cohn, F. (1901). *Anat. Anz.* **20,** 294–300
Greenwald, G. S. (1958). *J. Endocr.* **17,** 24–28
Jollie, W. P. and Benscosme, S. A. (1965). *Am. J. Anat.* **116,** 217–236
Kirby, D. R. S. (1960). *Nature, Lond.* **187,** 707
— (1962). *J. Embryol. exp. Morph.* **10,** 496–506
— (1965). 'The "Invasiveness" of the Trophoblast.' In: *The Early Conceptus, Normal and Abnormal.* Ed. by W. W. Park. University of St. Andrews
Krehbiel, R. H. (1937). *Physiol. Zoöl.* **10,** 212– 234
— (1941). *Anat. Rec.* **81,** 381–392
Loeb, L. (1908). *J. Am. med. Ass.* **50,** 1897–1901
Lyon, R. A. and Allen, W. M. (1938). *Am. J. Physiol.* **122,** 624–626
Mayer, G. (1959). *Mem. Soc. Endocr.* **6,** 76–83
— (1963). 'Delayed Nidation in Rats: a Method of Exploring the Mechanisms of Ovo-implantation.' In: *Delayed Implantation,* pp. 213–231. Ed. by A. C. Enders. University of Chicago Press

REPRODUCTION IN THE FEMALE MAMMAL

Meyer, R. K. and Cochrane, R. L. (1962). *J. Reprod. Fert.* **4,** 67–79
Orsini, M. W. (1962). *J. Reprod. Fert.* **3,** 283–287
Parkes, A. S. (1929). *Proc. R. Soc.* B. **104,** 183–188
Psychoyos, A. (1960). *C.r. Séanc. Soc. Biol.* **154,** 1384
— (1961). *C.r. hebd. Séanc. Acad. Sci., Paris* **253,** 1616–1617
— (1963). *C.r. hebd. Séanc. Acad. Sci., Paris* **257,** 1153–1156
Rothchild, I. and Meyer, R. K. (1942). *Physiol. Zoöl.* **15,** 216–223
Selye, H. and McKeown, T. (1935). *Proc. R. Soc.* B. **119,** 1–31
Wilson, I. B. (1963a). *Proc. zool. Soc. Lond.* **141,** 137–151
— (1963b). *J. Reprod. Fert.* **5,** 281–282

DISCUSSION

Dr. L. Martin (*London*)

Mr. Humphreys, working in Sydney, has been able to confirm some of the results you have reported here. He found that injecting 10 μl of peanut oil, gives a response along the whole length of the uterus, but a smaller volume gives a discrete response. Do the droplets of oil need to stay in the uterus for a long time and if you inject a larger amount do you get the response along the whole length of the uterus?

Finn

I have tried to determine how long the oil stays in the uterus without success. I have injected lard, which is liquid at body temperature but solid at room temperature, but this experiment was not successful in showing where the oil was located and I am now using radio-opaque oil, followed by x-rays. If you inject 0·03 ml, the initial stimulus on day 5 appears along the length of the uterus, but on day 6 you get discrete areas, with some as large chambers. I think most of the oil disappears from the uterus for usually there is oil in the vagina half an hour after injection. I suggest that most of the oil is removed from the uterus. The uterus quickly organizes with stimulation and possibly inhibition of decidual cell formation occurring at the same time, and gradually some sites get ahead and form swellings whereas others are inhibited.

Dr. M. C. Chang (*Worcester Foundation, Boston*)

Has experimental work shown that there is an oestrogen surge?

Finn

All the available evidence is indirect but I think it is good evidence. In the human there is a secondary peak of oestradiol in the urine, about the time of implantation and this may indicate the surge. I am hoping to use a method to determine exactly whether there is an oestrogen surge.

Dr. B. J. Weir (*London*)

Are you sure the cells you showed are moving from the blastocyst to the uterus. Have you found them in the blood? In man and in the chinchilla there is spontaneous deportation of trophoblast cells which pass via the blood to form choriocarcinoma in the lungs.

THE UTERUS DURING IMPLANTATION

FINN

Dr. I. Wilson is convinced that they are passing in this direction, but we do not know of their occurrence in blood.

DR. A. McLAREN (*Edinburgh*)

The invading trophoblast cells shown in humans and hamsters are from the embryonic end of the blastocyst at the ecto-placental core stage. These particular Wilson cells come from the abembryonic end at a much earlier stage.

WEIR

In the hamster it is an active process via the arteries but in man and the chinchilla it is a passive process whereby the cells are moved via the venous system.

DR. M. J. K. HARPER (*Alderley Edge, Macclesfield*)

Concerning your reference to implantation of blastocysts being analogous to an inflammatory reaction, we have tried to block implantation in the rat using paramesazone which is a potent anti-inflammatory steroid. It is quite effective in preventing the occurrence of oedema in the rat following the injection of killed TB organisms but it had no effect at all on implantation, although we cannot be sure it reached the uterus in the correct concentration. On the question of the oestrogen surge, we have used a compound related to triphenylethylene which is anti-oestrogenic. It will prevent the effect of oestrogen in the vagina and the uterus of immature rats. The effectiveness of this compound given orally or intravenously in preventing implantation is closely related to the time at which the oestrogen surge probably occurs in our rats. We consider this good evidence for the oestrogen surge, for we can get implantation if we inject a massive dose of oestrogen (250 mg/rat) at the same time as this compound. If we delay application of the oestrogen we can still get implantation of a few blastocysts even up to day 11, but smaller doses of oestrogen are required.

OVARIAN CONTROL OF EARLY EMBRYONIC DEVELOPMENT WITHIN THE UTERUS

C. E. ADAMS

A.R.C. Unit of Reproductive Physiology and Biochemistry,
University of Cambridge

THE PREIMPLANTATION period consists of two phases, tubal and uterine, and although this communication is primarily concerned with the uterine phase, some reference must be made to the former since the two are so intimately connected. Moreover, so far as the early embryo and uterine environment are concerned, the most significant change in ovarian function, that from oestrogen to progesterone dominance, takes place whilst the egg is still in the Fallopian tube.

In most mammals the eggs spend 3 or 4 days in the tubes. However, they remain in the tubes up to 7 days in the dog and cat but only a day in the opossum. In general, the tubal phase appears to be independent both of the length of gestation and of the dimensions of the tube (Boyd and Hamilton, 1952). The first half of the eggs' journey through the ampulla is rapid, e.g. in the rat (Andersen, 1927), guinea-pig (Squier, 1932), sheep (Winterberger, 1953), and rabbit (Harper, 1961), in some cases taking only a few minutes. Subsequently, it appears as if the eggs are held captive in the isthmus, probably under the influence of progesterone, before being released into the uterus. There are several reports that low doses of oestrogens accelerate egg transport in both spayed and intact animals (Noyes, Adams and Walton, 1959; Greenwald, 1961; Deanesley, 1963a). Spaying does not prevent the eggs entering the uterus in the rat (Alden, 1942), rabbit (Adams, 1958), or guinea-pig (Deanesley, 1960). The timing of the eggs' entry into the uterus is quite critical because prematurity leads to expulsion from the uterus, whilst an extended stay in the tube sooner or later leads to degeneration, usually at the early blastocyst stage (*see* page 539).

As the eggs of different species cleave at different rates, it follows that they will enter the uterus at different stages of development. Thus, that of the rabbit, which is fast cleaving, enters the uterus as an early blastocyst, those of the sheep and cow as 8–16 cell morulae and that of the pig at the 4-cell stage. Boyd and Hamilton (1952) have summarized in tabular form, according to species, the time and stage of the eggs' entry into the uterus. The period during which the mammalian blastocyst remains free in the uterine lumen varies from species to species, and even according to conditions within a species. In the rabbit this period extends over 4 days; it is shorter in the rat, mouse and guinea-pig, several days longer in the cat and longer still in the ungulates; e.g. the sheep embryo first attaches about day 18. Most remarkable are those species, including roe deer, badger, certain martens, weasels and bears, in which delayed implantation occurs naturally, for in these the blastocyst may persist unattached for a period of many months. In the rat and mouse, delayed implantation occurs when pregnancy is initiated *post partum* concurrent with lactation, and it can also be induced experimentally.

Cleavage stages normally associated with the uterus are not absolutely dependent upon the uterine environment, as shown by the following observations:

1. Eggs prevented from entering the uterus by tubal ligation may continue to develop until the early blastocyst stage is reached, e.g. in the mouse, Kirby (1965); in the rabbit (Pincus and Kirsch, 1936; Adams, 1958) and in the sheep (Wintenberger-Torres, 1956). In the human where tubal ectopic pregnancy is not uncommon special conditions seem to exist, as tubal ectopics have been scarcely reported in any other species.

2. Eggs may be cultivated *in vitro* to early blastocysts in the mouse (Whitten, 1957; Brinster, 1963; Mulnard, 1964; Cole and Paul, 1965) and rabbit (Pincus and Werthessen, 1938a; Adams, 1965a). Although the number of species whose eggs have been successfully cultivated *in vitro* is limited, this may simply reflect a lack of systematic study.

3. Eggs of one species may develop in another species. Sheep eggs can develop in the rabbit's genital tract for 4 or 5 days, from early morulae into blastocysts (Averill, Adams and Rowson, 1955). Moreover, equally good results were obtained in the ligated tubes of follicular phase does (Adams, Rowson, Hunter and Bishop, 1961), indicating a lack of specificity of species, organ, and ovarian hormone status. Under similar conditions cow and pig morulae (8–16 cell) may also develop to the early blastocyst stage, but so far practically no success has been obtained with 2–4 cell stages.

It seems significant that in all of the above experimental situations degeneration occurs at the early blastocyst stage—a stage that is always normally associated with the uterus.

MYOMETRIAL ACTIVITY AND RETENTION
OF EGGS IN THE UTERUS

When considering the uterine environment, attention is usually focused on the endometrium almost to the exclusion of the myometrium. That the unattached embryo must somehow be retained in the uterus prior to implantation is all too frequently taken for granted. It has long been recognized that before the eggs enter the uterus, the endometrium undergoes preparation for their reception. Simultaneously, myometrial activity, which is high at oestrus, subsides as luteal activity increases (Schofield, 1957). Though less

Table 1

Recovery of eggs 3 days after transfer to the uteri of does in which the number of corpora lutea was reduced to 2, 4 or 8 by semi-spaying, or 0 by spaying 15 h *post coitum*

No. of corpora lutea	No. of eggs		Recovery (%)
	transferred	recovered	
0	89	4	4·5
2	123	49	39·8
4	125	62	49·6
8	121	100	82·6

appreciated this is significant because under conditions of oestrogen dominance eggs are quickly expelled into the vagina (Chang, 1950; Noyes *et al.*, 1959, for the rabbit; Doyle, Gates and Noyes, 1963, for the rat and mouse).

Recent work on the rabbit (Adams, 1965b) has revealed the existence of a quantitative relationship between the number of corpora lutea and the efficiency of retention of eggs within the uterus. If the number of corpora lutea is progressively reduced, without interfering with the number of mature follicles, an increasing proportion of eggs is expelled from the uterus (Table 1), due it is assumed to an increasing oestrogen/progesterone ratio. It is particularly noteworthy that whereas one corpus luteum, or possibly even a fraction of a corpus luteum, suffices to stimulate endometrial proliferation, nearly the full complement of corpora lutea is required to ensure that a high proportion of eggs is retained in the uterus.

After spaying during the follicular phase the myometrium continues to behave as if it were oestrogen dominated for some time

(14–21 days) before entering a 'non-reactive' state, as recorded by isometric techniques (Schofield, 1954). Table 2 summarizes the results of an experiment in which eggs were transferred to the uteri of does spayed for 3–6 weeks. It will be seen that a high proportion of eggs may be retained within such uteri for 48 h, which is in keeping with Schofield's findings, but by 96 h most have been expelled. In this respect, the atrophic uterus behaves similarly irrespective of whether spaying takes place during the follicular or luteal phase.

Table 2

Egg recovery after transfer of 60 h morulae to uteri of does spayed for 21–40 days

Interval (h) from egg transfer to autopsy	No. of does	Control uterine horn			Ligated uterine horn		
		No. of eggs		%	No. of eggs		%
		transferred	recovered		transferred	recovered	
48	6*	25	19	76	30	23	77
72	5†	45	27	60	10	10	100
72	6*	78	35	45	0	0	
96	5†	35	5	14	15	7	47
168	2†	24	2	8	0	0	

* Spayed during follicular phase (oestrogen dominance).
† Spayed during pseudopregnancy (d.6) (progesterone dominance).

PROGESTATIONAL PROLIFERATION AND EMBRYO DEVELOPMENT

It is now well recognized that following ovulation the endometrium primed with oestrogen, responds to progesterone secreted by the developing corpora lutea, and undergoes characteristic changes, called progestational proliferation after Corner and Allen (1929). The extent of these changes varies from one species to another but they are especially well defined in the rabbit, the species most studied by the early investigators in this field. The link between endometrial proliferation and the presence of corpora lutea was first noted by Bouin and Ancel (1910) and the quantitative aspects of the relationship were studied by Joublot (1927); Corner (1928) and Brouha (1934), who established that only one corpus luteum, or perhaps even less, was required to induce a full response, as judged histologically. Wu and Allen (1959) concluded that the histological appearance of the endometrium does not provide a reliable index of the adequacy of the endometrium to nourish an embryo, but their results were based upon progesterone treated, spayed rabbits. Though it has been reported that implantation may occur when only one or two corpora lutea are present quantitative data on the numbers of eggs surviving and implanting are not available.

For this reason Adams (1965b) examined the development of 60 h morulae in the uteri of recipient rabbits in which the number of corpora lutea had been reduced to 2, 4 or 8 by semi-spaying shortly after ovulation. The reason for using egg transfer was to avoid any possible complicating effect of spaying on egg movement through the tubes and on cleavage. At recovery 3 days after transfer, 94·5 per cent of the eggs had developed into blastocysts, whose mean diameter was 1·33 mm (range 0·5–1·95 mm). Neither the number of eggs transferred, 5 or 20 per uterine horn, nor the number of the corpora lutea affected blastocyst size significantly. This finding is in keeping with the results of Hafez (1964), who found in rabbits spayed 2–4 days *post coitum* that increasing the dose of progesterone above the threshold (1 mg) had no significant effect on blastocyst size or viability. On the other hand, Wintenberger-Torres (1964) reported that in superovulated and/or progesterone-treated sheep embryonic development was accelerated from day 8 to 11, though before day 8 the eggs developed at the same rate as controls.

INFLUENCE OF SPAYING ON EGG DEVELOPMENT

In Eutheria, with the notable exceptions of the guinea-pig and armadillo (*Dasypus novemcinctus*), the ovaries appear to be essential for the maintenance of the early stages of pregnancy. Spaying during the preimplantation period quickly terminates pregnancy in the cat, hamster, mouse, rabbit, rat and others (*see* Amoroso, 1955 and Deanesley, 1960). Fraenkel and Cohn (1901) and Fraenkel (1903), 1910) working with the rabbit, appears to have been the first to show that spaying or destruction of the corpora lutea interfered with early pregnancy. However, it was Corner (1928), who firmly established the dependence of early embryonic development upon the ovaries, particularly the corpora lutea. He found that egg development was arrested at the early blastocyst stage, which was later confirmed by Adams (1958).

In the guinea-pig, implantation takes place at the normal time 6½ days *post coitum* after spaying on days 3–5, though loss of the ovaries on day 2 interrupts pregnancy (Deanesley, 1960, 1963b). In the Armadillo, a species in which the blastocyst is normally quiescent for several months, bilateral spaying towards the end of the delay period does not interfere with implantation either (Buchanan, Enders and Talmage, 1956). However, removal of the corpus luteum (= one ovary) for the most part resulted in loss of the blastocyst and implantation failure. This could have been due to expulsion of the blastocyst (*see* page 534).

Hartman (1925), working with the Opossum (*Didelphis virginiana*), found that spaying led to uterine involution and embryonic death—an observation which for long remained the only one on the effect of spaying in marsupials. According to Tyndale Biscoe (1963), it led to the supposition that marsupials are equally dependent upon ovarian secretions for pregnancy maintenance, as are Eutheria. However, Tyndale Biscoe's own results on the Quokka (*Setonyx brachyurus*), a representative of the most advanced group of marsupials, show that this is not necessarily the case for he found that blastocyst development resumed in each pregnant quokka spayed on day 4 after removal of the pouch young, but development did not occur after spaying on days 0–2.

It is possible that early spaying may also interfere with egg development by disturbing the tubal environment. To exclude this possibility, Adams (1965b) transferred 60-h morulae to the ligated uteri of does spayed 45 h previously. After 3 days it was once again found that development ceased at the early blastocyst stage (200–270 μ). However, in recipients that had been spayed for 3 weeks morulae were able to develop beyond the '84 h' early blastocyst stage, which had proved vulnerable in freshly-spayed does. Thus, in 6 does autopsied 72 h after transfer, the mean diameter of 28 blastocysts was 528 μ (195–973 μ), equivalent in size to blastocysts recovered 96–108 h *post coitum*. In a further experiment similar sized blastocysts were recovered 4–7 days after transfer, showing that the marked growth characteristic of days 5–7 had not occurred (*see Plate I, a–d*). It was wondered, therefore, whether a 'delayed implantation' condition had been induced, as the blastocysts' development was about 2 days retarded. To test their viability, representative blastocysts were transferred to the uteri of three synchronized recipients, ovulated by LH injection 93, 117 and 137 h earlier. None became pregnant and it therefore seems likely that pregnancy failed at or before implantation. Consequently, it will now be necessary to recover the eggs earlier in order to determine just how long they remain viable. It is worth recording that delayed implantation does not occur naturally in the rabbit and so far has never been induced experimentally. One suspects that this condition may demand a certain specialization of the blastocyst which that of the rabbit does not possess. Furthermore, the possible lack of correlation between appearance and developmental potential of early blastocysts has already been demonstrated with rabbit eggs following cultivation *in vitro* (Adams, 1965a).

MAINTENANCE OF PREGNANCY IN SPAYED ANIMALS

Since Allen and Corner (1929, 1930) succeeded in demonstrating that pregnancy could be maintained in spayed rabbits by treatment with extracts of corpora lutea, and later with progesterone (Corner, 1937, Pincus and Werthessen, 1938a and b), this subject has been intensively studied, particularly as a result of the synthesis of many new compounds suspected of having progestational activity. It is noteworthy that progestational activity of different compounds may vary markedly according to the parameter under test. For example, the relative activity of 20α and 20β-hydroxy pregnen-3-one, as judged by the induction of implantation, was estimated at one tenth that of progesterone, whereas they surpass progesterone in the Hooker Forbes or Clauberg bio-assays (Rennie and Davies, 1965). In some species, e.g. hamster (Orsini and Psychoyos, 1965), rabbit, cow (Hawk, Brinsfield, Turner, Whitmore and Norcross, 1963) and quokka (Tyndale Biscoe, 1963) progesterone alone is sufficient to support embryo development and permit implantation. However, in others notably the rat and mouse, whereas progesterone will maintain a quiescent blastocyst for several weeks as in naturally occurring delayed implantation, a single small dose of oestrogen is necessary to induce implantation (Canivenc and Meyer, 1956; Cochrane and Meyer, 1957). If implantation occurs, it follows that one or more embryos must have survived throughout the preimplantation period. Up to now little attention has been given to the evaluation of embryonic survival and results are not easily related to what might have happened in the intact animal. In fact, of many species it is still true to say that analyses of preimplantation mortality are not available. In future, a more critical quantitative approach is required.

LOSS OF THE ZONA PELLUCIDA

In many species, e.g. mouse, rat, hamster, sheep, cow and pig, the zona pellucida is shed from the blastocyst as a preliminary to its attachment to the uterus. It appears to be almost exceptional for the zona to be retained as in the rabbit. So far, little is known concerning the mechanisms responsible for the escape of the blastocyst from the zona, but ovarian hormones may be involved. In the hamster, Orsini (1963, 1965) reported that the loss of the zona may be progesterone dependent and she suggested that the hormone acts via the uterus. According to Mintz (1965) there are two hatching mechanisms in the mouse, which can be independent of each other. One is subject to an external influence known to be uterine in origin whereas the other is of a mechanical nature. The latter mechanism is held to be responsible for the hatching of mouse blastocysts *in vitro* and in blastocysts held in the tube by ligation. Cole and Paul (1965) have filmed cultured mouse blastocysts escaping from the zona.

THE 'QUALITY' OF PREIMPLANTATION DEVELOPMENT

Recently, it has been recognized that the 'condition' of the egg before implantation may influence later stages of development. Noyes, Doyle and Bentley (1961) confirmed a preliminary observation made by Dickmann and Noyes (1960) that the preimplantation development of rat blastocysts at implantation affected foetal weight near term, on day 18. When 4- and 5-day eggs were transferred to a 4-day recipient the 5-day eggs gave rise to significantly heavier foetuses, in spite of the fact that each lot was implanted simultaneously. A similar finding was recorded in mice (Noyes, Doyle, Gates and Bentley, 1961) after 3- and 4-day eggs were transferred to a 3-day uterus. Later, Doyle et al. (1963), re-examining this question, confirmed that foetuses derived from 'older' eggs were significantly heavier than those from eggs transferred in synchrony with the uterus, although both groups were implanted at the same time. Noyes and his co-workers have concluded that the older ova 'mature' in some way, as yet unspecified, during the additional day that they remain free in the uterus.

This is an appropriate juncture to recall the so-called 'uterine factor' described by Kirby (1962; 1965), which influences the development of mouse eggs. He found that tubal eggs or tube-locked blastocysts, when transplanted beneath the kidney capsule, give rise to trophoblast and extra embryonic membranes only, whereas about 30 per cent of equal aged uterine blastocysts placed in a similar position develop into embryos. Kirby concluded that a mouse blastocyst, to realize totipotency, must experience the uterine environment, $3\frac{1}{4}$ to $4\frac{1}{4}$ days *post coitum*. Whether this applies to other species is unknown. The occurrence of tubal pregnancy in the human suggests that here the tubal egg is totipotent; however, migration back from the uterus into the tube cannot be excluded

(Iffy, 1963). It has also been suggested that a tube that permits implantation is probably always an abnormal one (Lennox, 1965).

SYNCHRONIZATION OF ENDOMETRIUM
AND EGG DEVELOPMENT

Chang (1950) was the first to investigate systematically the development and fate of eggs or blastocysts in relation to the ovulation time of the recipient. Working with the rabbit, he firmly established the significance of co-ordinating the stage of egg development with the luteal phase of the recipient, and this synchronization of donor and recipient is an essential preliminary to egg transfer. McLaren and Michie (1956) found in the mouse that much better results were obtained if the egg's developmental stage was 24 h in advance of that of the endometrium than in the reverse combination, with synchronous transfers giving intermediate results. A likely explanation of this finding is that mouse eggs may suffer a delay in development of about 12 h as a result of transfer (Tarkowski, 1959). This would mean that the combination 'egg in advance of endometrium' becomes more synchronous, whereas with the 'endometrium in advance of egg' a somewhat unfavourable situation is exaggerated.

In the rat, too, Noyes and Dickmann (1960) found that the best results were obtained when the egg was in advance of the endometrium. However, in this case, the explanation may be different from that offered for the mouse. In the latter species, younger eggs appear to fail shortly after transfer (Doyle et al., 1963), whereas in the rat they may survive until implantation, when they are destroyed by changes that take place in the environment (Dickmann and Noyes, 1960). On the other hand, 'older' eggs are able to survive in a less advanced uterus until it is ready for implantation. Noyes and Dickmann were led to conclude that 'when ova first contact the uterine environment their chance for survival depends not only on the absolute stage of maturation of the ova and uterus, but also upon a particular relationship between the stages of development of the ova and uterus'.

Until recently, studies on synchronization have been exclusively based upon three laboratory species, namely rat, rabbit and mouse, in which implantation occurs early on days 5–7. This has led to the belief that the necessity for close synchronization between donor and recipient is governed by implantation requirements. For example, Tyndale Biscoe (1963) wrote: 'The close synchronization is undoubtedly related to the intimate association that develops between ovum and uterus during implantation in eutherian mammals.'

540

Whereas this is apparently the case in the rat (Dickman and Noyes, 1960), it is not necessarily so in the mouse and rabbit, in both of

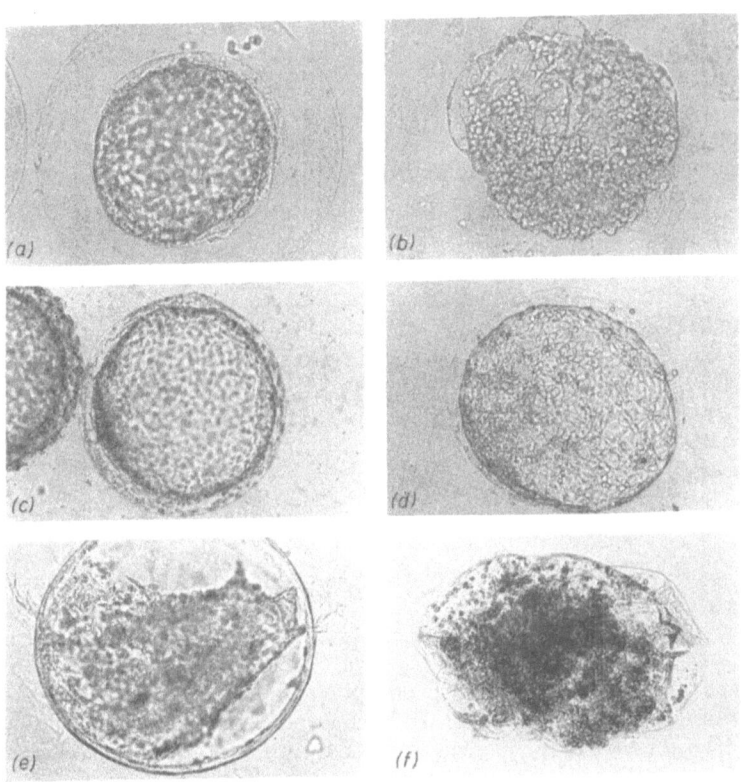

PLATE I
(Reduced ⅔ on reproduction)

Sixty-hour morulae were transferred to the uteri of recipients, which were either spayed 28 days previously (a–d) or pseudopregnant following LH treatment 108 h before transfer (e–f), and recovered 72 h later

a. *Atypical early blastocyst: note ruptured zona pellucida.* × 192
b. *Abnormal early blastocyst.* × 154
c and d. *Blastocysts.* × 96
e and f. *Degenerate blastocysts.* × 120 and × 96

which eggs, if transferred asynchronously, degenerate before the implantation stage is reached. Thus, Doyle *et al.* (1963), reported that 'younger mouse eggs often cease cleaving or begin to fragment before the time of implantation and unlike the rat ova cannot be

541

recovered, damaged or otherwise, after the period of normal implantation'. Similar results have been obtained by Adams in the rabbit. He transferred 60-h morulae to the uteri of recipients that were given LH at the same time as, or 24 and 48 h in advance of the donors. The results, based upon findings at autopsy 72 h after transfer, appear in Table 3.

It will be noted that with increasing asynchrony (1) the proportion of transferred eggs recovered fell steeply, (2) the proportion of normal blastocysts fell even more sharply and (3) there was a corre-

Table 3

Fate of 60 h eggs transferred to the uteri of does given LH 60, 84 or 108 h before transfer

Recipient's luteal stage (h)	No. recipients	No. eggs		Recovery (%)	Blastocysts			Late morulae and early blastocysts
		transferred	recovered		normal	collapsed	degenerate	
60	4	43	37	86·0	33 (89·2%)	3 (8·1%)	1 (2·7%)	0
84	4	37	25	67·6	14 (56%)	4 (16%)	4 (16%)	3 (12%)
108	5	56	20	35·7	1 (5%)	4 (20%)	13 (65%)	2 (10%)

sponding increase in the proportion of degenerate morulae and blastocysts (*Plate I, e* and *f*). Furthermore, the mean size (surface area) of the blastocysts fell from 18·2 mm² (60 → 60) to 9·3 mm² (60 → 84) and 4·3 mm² (60 → 108).

In a still more asynchronous combination, e.g. 60-h eggs into an 8-day uterus, no egg development occurred beyond the early blastocyst stage, corresponding with 4–4½ days *post coitum* under normal

PLATE II
(Reduced ⅔ on reproduction)

Sixty-hour morulae were transferred to the uteri of pseudopregnant recipients, treated with LH to induce ovulation 8 days before transfer. The eggs were recovered after 3 days (a–f), 4 days (g), or 13 days (h)

a. degenerate native egg.   × 154   b. very early blastocyst.   × 192
c. degenerate early blastocyst.   × 154   d. degenerate morulae.   × 192
e and f. degenerate blastocysts.   × 75 and 120
g and h. degenerate eggs. It is not possible to identify them as native or alien.   × 75 and × 192

conditions, and the majority were very degenerate in appearance. In some cases degeneration was so advanced as to render identification of 'natives' and 'aliens' speculative, so that no reliable estimate can be given for the loss of transferred eggs *in utero* (*Plate II*).

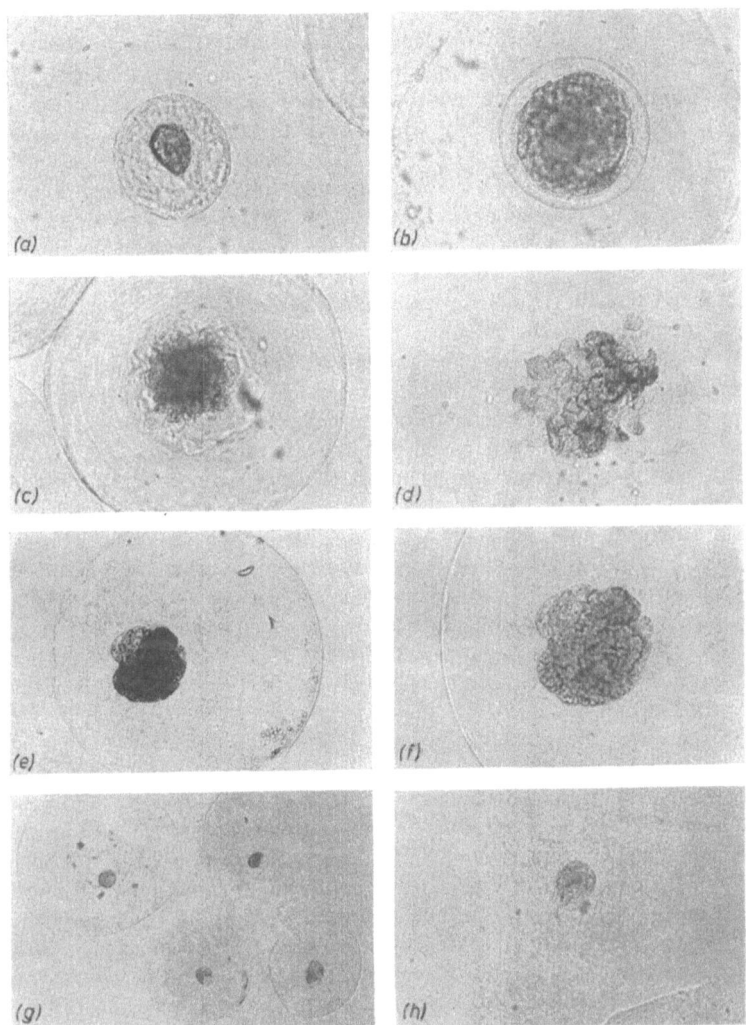

PLATE II

These experiments clearly show that for successful embryo development the limits of asynchrony between egg stage and endometrium are very narrow. For 50 per cent survival it is estimated that the luteal stage should be no more than 24 h different (recipient in advance of donor) whilst 48 h appears to be almost the extreme limit of tolerance. It is remarkable that an endometrium that appears perfect structurally can prove such a hostile environment for an embryo. It fully supports Moor's suggestion that 'changes in the function of the corpus luteum are not in themselves the most critical determinants of embryonic survival following asynchronous transfer'.

In the sheep, although there is no early attachment of the embryo as in rodents or lagomorphs, there is, nevertheless, still a need for synchronization of donor and recipient (Moore and Shelton, 1964; Moor, 1965). The limits of synchronization compatible with embryo survival are a little wider than in the rat, rabbit, or mouse, but this may merely reflect a more gradual change in the endometrium and its secretions consequent upon an extended pre-attachment period. Whilst the limits of asynchrony apparently remain relatively constant from the second to the twelfth day after oestrus, Moor observed a significant shift in the compatibility of different 'egg/endometrium' combinations relative to the age of the egg. Thus, whereas the '5-day' egg fared better in synchronized or less advanced recipients than in a more advanced recipient, 9-day blastocysts survived better in the synchronous or more advanced combinations than in a less advanced recipient. Information is not available concerning the stage at which the transferred sheep eggs died, as Moor was primarily interested in the fate of the corpus luteum. This ruled out an early examination before the end of the normal cycle.

From the foregoing analysis of the embryo's fate in a variety of uterine environments the picture that emerges is one of stringent conditions being imposed upon the embryo, which may only survive if these are satisfied. This is in marked contrast to the position in certain ectopic sites (Kirby, 1965)—a subject discussed by McLaren (1965). One cannot escape the conclusion that we need to know far more about the uterine environment, especially the secretions. Biochemical studies of the secretions should aid the experimental embryologist in his endeavours to explore either the control or augmentation of fertility. For control in particular, the hostility of the non-synchronous uterus coupled with the vulnerability of early embryonic stages should offer considerable scope.

REFERENCES

Adams, C. E. (1958). *J. Endocr.* **16,** 283-293
— (1965a). In: *Preimplantation Stages of Pregnancy*, pp. 345-373. Ed. by G. E. W. Wolstenholme and M. O'Connor. London; Churchill
— (1965b). *J. Endocr.* **31,** 29-30
— Rowson, L. E. A., Hunter, G. L. and Bishop, G. P. (1961). *Proc. IVth Int. Congr. Anim. Reprod.*, pp. 381-382. The Hague
Alden, R. H. (1942). *J. exp. Zool.* **90,** 159-170
— (1942). *J. exp. Zool.* **90,** 170-182
Allen, W. M. and Corner, G. W. (1929). *Am. J. Physiol.* **88,** 340-346
— — (1930). *Proc. Soc. exp. Biol. Med.* **27,** 403-405
Amoroso, E. C. (1955). *Br. med. Bull.* **11,** 117-125
Andersen, D. H. (1927). *Am. J. Physiol.* **82,** 557-569
Averill, R. L. W., Adams, C. E. and Rowson, L.E.A. (1955) *Nature, Lond.* **176,** 167
Boyd, J. D. and Hamilton, W. J. (1952). *Marshall's Physiology of Reproduction 2*, 3rd edn. Ed. by A. S. Parkes. London; Longmans, Green
Bouin, P. and Ancel, P. (1910). *J. Physiol. Path. gén.* **12,** 1-16
Brouha, A. (1934). *Arch Biol. Liège* **45,** 571-609
Brinster, R. L. (1963). *Expl. Cell. Res.* **32,** 205-208
Buchanan, G. D., Enders, A. C. and Talmage, R. V. (1956). *J. Endocr.* **14,** 121-128
Canivenc, R. and Mayer, G. (1956). *C.r. Séanc. Soc. Biol.* **150,** 2208-2212
Chang, M. C. (1950). *J. exp. Zool.* **114,** 197-216
Cochrane, R. L. and Meyer, R. K. (1957). *Proc. Soc. exp. Biol. N.Y.* **96,** 155-159
Cole, R. J. and Paul, J. (1965). *Preimplantation Stages of Pregnancy*, pp. 82-112. Ed. by G. E. W. Wolstenholme and M. O'Connor. London; Churchill
Corner, G. W. (1928). *Am. J. Physiol.* **86,** 74-81
— (1937). *Cold Spring Harb. Symp. quant. Biol.* **5,** 62-64
— and Allen, W. M. (1929). *Am. J. Physiol.* **88,** 326-339
Deanesley, R. (1960). *J. Reprod. Fert.* **1,** 242-248
— (1963a). *J. Reprod. Fert.* **5,** 49-57
— (1963b). *J. Reprod. Fert.* **6,** 143-152
Dickmann, Z. and Noyes, R. W. (1960). *J. Reprod. Fert.* **1,** 197-212
Doyle, L. L., Gates, A. M. and Noyes, R. W. (1963). *Fert. and Steril.* **14,** 215-225
Fraenkel, L. (1903). *Arch. Gynaek.* **68,** 438-545
— (1910). *Arch. Gynaek.* **91,** 705-761
— and Cohn, F. (1901). *Anat. Anz.* **12,** 294-300
Greenwald, G. S. (1961). *Fert. and Steril.* **12,** 80-95
Hafez, E. S. E. (1964). *J. Reprod. Fert.* **7,** 241-249
Harper, M. J. K. (1961). *Proc. IVth Int. Congr. Anim. Reprod.*, pp. 375-380. The Hague
Hartman, C. (1925). *Am. J. Physiol.* **71,** 436-454

Hawk, H. W., Brinsfield, T. H., Turner, G. D., Whitmore, G. E. and
Norcross, M. A. (1963). *J. Dairy Sci.* **46,** 1397–1401
Iffy, L. (1963). *J. Obstet. Gynaec. Br. Commonw.* **70,** 996–1000
Joublot, J. (1927). *Archs Anat. Histol. Embryol.* **7,** 437
Kirby, D. R. S. (1962). *J. Embryol. exp. Morph.* **10,** 496–506
— (1965). In: *Preimplantation Stages of Pregnancy,* pp. 325–339. Ed. by
G. E. W. Wolstenholme and M. O'Connor. London; Churchill
Lennox, B. (1965). *The Early Conceptus, Normal and Abnormal,* p. 38.
Edinburgh; Livingstone
Magnus, V. (1901). *Norsk Mag. Lægevidensk.* **62,** 1138–1145
McLaren, A. (1965). *The Early Conceptus, Normal and Abnormal,* pp. 27–33.
Edinburgh; Livingstone
— and Michie, D. (1956). *J. exp. Biol.* **33,** 394–416
Mintz, B. (1965). In: *Preimplantation Stages of Pregnancy,* p. 167. Ed. by
G. E. W. Wolstenholme and M. O'Connor. London; Churchill
Moor, R. (1965). *Ph.D. Thesis.* Cambridge University
Moore, N. W. and Shelton, J. N. (1964). *J. Reprod. Fert.* **7,** 145–152
Mulnard, J. (1964). *C.r. hebd. Séanc. Acad. Sci., Paris* **258,** 6228–6229
Noyes, R. W., Adams, C. E. and Walton, A. (1959). *J. Endocr.* **18,** 108–117
— and Dickman, Z. (1960). *J. Reprod. Fert.* **1,** 186–196
— Doyle, L. L. and Bentley, D. L. (1961). *J. Reprod. Fert.* **2,** 238–245
— — Gates, A. H. and Bentley, D. L. (1961). *Fert. and Steril.* **12,** 405–416
Orsini, M. W. (1963). In: *Delayed Implantation.* Ed. by A. C. Enders.
Chicago; Univ. Chicago Press
— (1965). *Preimplantation Stages of Pregnancy,* pp. 162–167. Ed. by
G. E. W. Wolstenholme and M. O'Connor. London; Churchill
— and Meyer, R. K. (1959). *Anat. Rec.* **134,** 619 Abstr.
— and Psychoyos, A. (1965). *J. Reprod. Fert.* **10,** 300–301
Pincus, G. and Kirsch, R. E. (1936). *Am. J. Physiol.* **115,** 219–228
— and Werthessen, N. T. (1938a). *J. exp. Zool.* **78,** 1–18
— — (1938b). *Am. J. Physiol.* **124,** 484–490
Rennie, P. and Davies, J. (1965). *Endocrinology* **76,** 535–536
Schofield, B. (1954). *Endocrinology* **55,** 142–147
— (1957). *J. Physiol.* **138,** 1–10
Squier, R. R. (1932). *Contr. Embryol. Carneg. Instn.* **23,** No. 137, 225–250
Tarkowski, A. (1959). *Acta theriol.* **2,** 251–267
Tyndale Biscoe, C. H. (1963). *J. Reprod. Fert.* **6,** 25–40
Whitten, W. K. (1957). *Nature, Lond.* **179,** 1081
Wintenberger, S. (1953). *Annls Zootech.* **3,** 269–273
— -Torres, S. (1956). *3rd Int. Congr. Anim. Reprod.,* pp. 62–64. Cambridge
— - — (1964). *Vth Int. Congr. Anim. Reprod., Trento* **2,** 414–418
Wu, D. H. and Allen, W. M. (1959). *Fert. and Steril.* **10,** 439–460

DISCUSSION

Dr. A. McLaren (*Edinburgh*)

I have found recently that in delayed implantation in mice whether experimentally induced by bilateral spaying or physiologically during lactation there is some delay in blastocyst development as judged both by the timing of loss of zona pellucida or by a delay in trophoblast giant cell transformation. I think Dr. Adams implied that the greater necessity for synchronization in the rabbit by not having the egg older than the uterus, might be connected with the fact that mice and rats normally have some delay while rabbits do not. Are rabbit blastocysts more difficult to store *in vitro* than those of the mouse or the rat?

ADAMS

I have not worked with mouse or rat eggs *in vitro*. We can take rabbit eggs to an early blastocyst stage *in vitro* but they die if the period of culture is extended. Pincus and Werthessen in 1938 obtained some large blastocysts in culture but they did not test their viability. However, on the basis of my own observations of these stages I should be surprised if they were still normal.

DR. MARY HAY (*Cambridge*)

We have completed a few experiments culturing blastocysts and in the few cases where we have transferred them back into a rabbit, we have had a limited success with 5-day blastocysts cultivated for 24 h but not with 6-day blastocysts cultivated for 24 h. The 6-day blastocysts continue to develop in culture but not when transferred into the uterus which may indicate some difficulty with implantation in the recipient.

DR. J. M. BEDFORD (*London*)

It seems that the zygote is rather more sensitive to its environment than the male gamete. In the 1961 experiments where sheep morulae were transferred to rabbits, was the rate of development normal or was development of the morula retarded in any way?

ADAMS

The observations we made suggested that the rate of development was normal. Ultimately, if we left the eggs in the rabbit for longer than 4 to 5 days we did get retardation and presumably a loss of viability again at the early blastocyst stage.

DR. C. A. FINN (*London*)

I am particularly interested in the growth of blastocysts *in vitro*. As in your experiments, Biggers of Philadelphia can culture eggs as far as the blastocyst stage but no further. It appears that they require a stimulus to develop further and it would be extremely interesting to find what this is. There appears to be a species variation for those species which have delayed implantation, e.g. the mouse, eggs can be kept at this stage, but in those species that do not, e.g. the rabbit, this does not occur.

ADAMS

The rabbit blastocyst does not go into quiescence, it either develops or it dies, and this is one of the problems in trying to induce delayed implantation experimentally in the rabbit. It seems that we are searching for a uterine factor which is missing in culture.

547

REPRODUCTION IN THE FEMALE MAMMAL

Dr. M. J. K. Harper (*Macclesfield*)

There seems to be an interesting species difference, for in the rat and the mouse, implantation is delayed, whereas in the hamster or the rabbit there is no apparent delay. In cultural conditions *in vitro* rabbit and mouse eggs are easily cultured, yet no one seems to have had success with rat and hamster eggs.

Adams

I should have referred to that in my paper. Historically, the rabbit egg seems to be the one that is easy to cultivate almost irrespective of treatment, at least through the cleavage stages. Many workers have attempted culture *in vitro* of the eggs of other species but with little success, except in the mouse. We need more systematic work on the culture of eggs.

Author's note added in proof:

Professor Weiert Velle has drawn my attention to a little known paper in Norwegian by Vilhelm Magnus whose observations on the effect of spaying or destruction of the corpora lutea on pregnancy appear to have gone unrecognized in the English literature. Using rabbits, Magnus demonstrated the essential role played by the ovaries or corpora lutea in maintaining the early stages of pregnancy. As Magnus' paper appeared in 1901, it therefore ranks historically with those of Fraenkel: it is interesting that Magnus, like Fraenkel, credits Gustav Born for having given him the idea.

VIII. THE SYMPOSIUM IN PERSPECTIVE

THE SYMPOSIUM IN PERSPECTIVE

PROFESSOR E. C. AMOROSO

When in the summer of 1965 Professor Lamming approached me on the matter of acting as co-ordinator, I accepted, perhaps more from a sense of duty than anticipation of a new and exciting venture. I had been aware only that the Easter School had become a recurrent activity of the Faculty of Agricultural Science. Also, I was assured that the group would consist of friends of long standing, interested in developing more intense interests in the biology of reproduction. Now, in retrospect, I shall always regard this Symposium as a very rewarding personal experience for which I am most grateful to those many who made it possible. To the organizers as well as to the participants we give our sincere thanks.

You will recall that the conventional purpose of a Chairman's summing up is to bring into relief the main points of agreement and disagreement that have emerged during previous discussions, and to underline those points of difference whose resolution would definitely advance the subject. It would be impossible, however, to give an adequate summary of a Symposium so rich and varied in content as this one. Hence little but folly and conceit can be attributed to anyone who chose to embark on such an enterprise. Besides, some of the most valuable features of this Symposium would be lost in a summary, however adequate this might be.

Indeed I am reminded at once of such episodes as the relative merits of quantitative versus semi-quantitative assays which became the basis of a preliminary skirmish, early in our proceedings between Dr. Williams and Dr. Van Rees, and which was echoed again in Dr. Loraine's and Dr. Hartree's insistence on the value of biological and biochemical standards. And here too, I must myself chide Dr. Van Rees for his remark that 'the use of quantitative assays is more or less a question of fashion'. Surely the thing to remember is that statistical procedures have advanced beyond the point when it was only necessary to calculate the standard error of a mean or the variance and to compare one mean with another.

In reviewing the Symposium in perspective I propose to indicate to you to what extent my education on the biology of reproduction in the female mammal has been improved and to what extent it has,

551

perhaps, not proceeded as far as it might well have done. Accordingly, if any part of this review fails in its allusion to major points whilst emphasizing minor ones unduly, this should be ascribed to my own shortcomings and not to any departure from the strict neutrality that is customarily associated with a Chairman's function.

The first point I should like to make is that it can be no accident that so large a part of our discussions has been focused on the area where neurology and endocrinology meet. Indeed, the manner whereby the nervous and endocrine systems of the body become integrated, continues to constitute a bewildering series of problems about which no common explanation is yet at hand. But alas, after years of inquiry all we know as fact is that, of the endocrines in the higher vertebrates, *if we except the adrenal medulla*, only the pituitary has an important regulatory link with the nervous system in addition to a vascular connection. It is not surprising, therefore, that in the opening paper Dr. Donovan set the stage with a balanced account of gonadal-hypophysial interactions during infancy. In this he cited recent evidence that in the infant, just as in the adult, gonadal activity and pituitary function lie in a delicate state of equilibrium and that the mode of action of the adult hypothalamo-hypophysial system, be this cyclic as in the female or acyclic as in the male, is determined during the immediate post-natal period. But we should not, as Professor Folley reminded us, forget that experimental proof for such a notion was provided by Hohlweg and Junkman more than 30 years ago when they suggested that a steroid-sensitive sex centre in the hypothalamus mediated the effects of sex steroids on the hypophysis.

These and other facts that Donovan presented, pose fascinating problems and indicate that despite the apparent inactivity of the gonad-hypophysial axis in infancy, a variety of data make it clear that the control of gonadal function before puberty is not as simple as it sometimes seems or is made out to be.

Hypothalamic regulation of hypophysial hormones was a central theme in the contributions of a number of other speakers. Thus, from Dr. Rothchild's analysis it emerges that the neural mechanisms through which ovulation is prevented during the luteal phase by progesterone, and during lactation by suckling, constitute a complex of inhibitions involving the influence of the ventro-medial nucleus of the hypothalamus over the far-lateral area.

Linked with these observations are those of Dr. Van Rees, who postulates a dual control of the secretion of LH and possibly also of FSH. In the first place there are the mechanisms regulating basal or tonic secretion of LH and FSH, and secondly, those which in the female rat cause rhythmic peaks in the secretion of the two hormones, thus leading to ovulation and the formation of corpora lutea.

At this point I should digress to mention the pleasure Dr. Roth-child expressed when he believed that Dr. Van Rees had stated that progesterone inhibition of ovulation was a direct effect on the pituitary, but which was denied by Dr. Van Rees, who would only admit that he claimed that the pituitary became insensitive.

Professor McCann next introduced us to a fascinating new world of endocrinological mechanisms, in which he showed that the stimulating effects of the hypothalamus on the secretion of FSH and of LH, on the one hand, and its inhibitory effect on the secretion of prolactin, on the other, is mediated by specific polypeptides secreted into the hypophysial veins. The FSH releasing factor and LH releasing factor are small polypeptides that have been separated chemically and are dissimilar from vasopressin and oxytocin. As yet, however, the existence of a separate prolactin releasing factor remains equivocal. Whatever the way these separate components become articulated, however, from the neural point of view, into specific patterns of response, we have learned quite definitely that these patterns are activated or powered by hormones.

Proceeding from the notion that a functional anterior pituitary is essential for the process of normal milk secretion, Dr. Tindal argued that the central connecting link in the whole story of the fine control of the neuro-humours and hence the modulation of pituitary function is significantly under limbic influences. This, of course, is not surprising, since the limbic system, the most distinctive anatomical feature of which is the massive richness of its interconnecting pathways, is an outgrowth of the olfactory system—the rhinencephalon and its extensions. And is it not true that among the lower vertebrates and mammals also that it is still the olfactory cue which, for prey and predator alike, most often signals the requirements for survival? Similarly, the olfactory sense guides the mating behaviour of most mammals, and a well supported perfume industry bears familiar testimony to its lingering relevance in the sexual activities even of microsomatic man. Seemingly, therefore, the phylogenetic heritage of limbic functions survive unchanged in the face of the diminished importance of the olfactory system per se.

All this has relevance to the points raised by Tindal when he referred to electrical stimulation of the amygdala and to Dr. Roth-child's obvious pleasure when he learned that mechanisms which

19+   553

increase appetite in goats and sheep, are also connected with pro-lactin secretion and this, in addition to the suckling stimulus. On the other hand, it may be well to remember that bilateral ablation of the amygdala makes the most savage animal docile, whereas de-struction of the ventro-medial nucleus of the hypothalamus converts a previously tame animal into a dangerously vicious one. In the male cat too, loss of the periform cortex produces a sexual athlete, that attempts to copulate with anything that moves or stands.

Having said all this I must express surprise that no mention of the pineal was made, since there is now considerable evidence that the mammalian pineal exerts an inhibitory influence on the incidence of vaginal oestrus in rodents, the oestrus inhibiting effect being blocked by pinealectomy.

The material of our discussion has been derived both from clinical investigations and from laboratory experiment, and in the second session of our proceedings we found ourselves faced by questions involving separation, purification and characterization of the gonadotrophins.

Contrary I imagine to Dr. Loraine and Dr. Hartree, who, because of their authority insist on still more purified hormones, I believe a reasonably good case for relative chemical purity of human FSH and for LH has been advanced by Dr. Butt and Professor Reichert. Those who have been following them in this particular line of work will wish them increasing success because on this depends our clearer appreciation of the full value of these hormones in clinical practice. At the same time it is greatly to be regretted that there was no mention at all of pregnant mares serum gonadotrophin. Here I had hoped that information might be forthcoming of attempts to purify and further characterize this biologically active material, thus enabling the site of hormone production to be identified. Because of the lack of facts, our discussions did not, however, dwell on these possibilities.

This brings me naturally to the immunological methods of gonadotrophin assay, discussed by Dr. Butt and Mr. Saji. Here I must admit my shock of surprise to learn that owing to antigenic impurities, assay procedures involving haemagglutination-inhibition and complement fixation reactions may not be all that could be wished for. I hope, however, it will not be long before this question and others like it are resolved, for until they are it will be difficult to realize the full value of these immunological procedures in clinical practice.

The clinical implications of issues which have been raised during this Easter School are legion, and have been dominated—in spite of Professor Gemzell's illuminating account—by the shadow of impurities in his preparations, shadows which it is to be hoped may grow smaller as the months pass and we begin to know more about the limitations to the use of these protein hormones of pituitary origin. Much work remains to be done on this problem.

The early phases of our discussions were dominated by references to pituitary-gonadal interrelations in the rat. But birds, too, as Professor Folley reminded us when opening our discussions, have a pattern of sexual activity strongly influenced by the hypothalamus, and can be included quite naturally in the same general category. In this connection Professor Nalbandov has proved that the small band of workers who study reproductive mechanisms in birds or, as he put it, rats with feathers, yield nothing in interest to those who work on mammals, and the same may be said of Dr. Lake's contribution. There were several points in Dr. Nalbandov's papers, such as his demonstration of Substance X, and the place of oestrogen in the synthesis of progesterone, which struck me as important and which I had hoped would receive more detailed treatment in discussion.

Variations in reproductive processes are revealed not only in their normal external processes but also in the influence which environmental factors have upon the sensitivity of tissues to different forms of hormonal stimulation. This, in my view, was the clear inference in Dr. Jöchle's account of the circadian rhythm in which the chicken loomed large, and which I would have liked to have seen developed in discussion. To be true, there were sceptics, but this is the kind of generalization which is so useful in science; for it lends itself immediately to experimental verification.

Cyclic variations of gonadotrophic secretion and release were the subject of papers by Dr. Robertson and Mr. Crighton. For the first time a differential release of FSH and LH has been demonstrated to occur around the time of the onset of oestrus, the essential feature of which is that the release of FSH which commences 12 h before oestrus may perhaps be responsible for the initiation of the state of sexual receptivity. The release of LH which commences at the onset of oestrus and is completed 6 h later may be the true ovulatory discharge. The first ovulatory discharge of gonadotrophins which occurs at the commencement of a new breeding season may take place at a fixed time of day, i.e. it may be dependent upon some circadian rhythm.

Pharmacological control of reproductive processes was the central theme in the papers of Professor Robinson, Professor Hansel and

555

Dr. Schilling, and it is hardly necessary for me to allude to the economic and potential benefits to be derived from successful control of oestrus and ovulation. While no new principles have been revealed by these investigations, they are none-the-less a beginning and should be explored still further.

Here I must mention Professor Hansel's attempts to schematize the mechanism which controls the life span of the corpus luteum in cattle. This was raised to the fore in the light skirmish which followed between himself and Dr. Short. You will recall that Dr. Short's objection was that he would prefer the subjunctive rather than the imperative mode, whereas it was said that LH was the triggering mechanism. Dr. Short would rather have us say that it might be.

Our attention has also been focused on the question of the role of the uterus and its effect on the life span of the corpus luteum. The hysterectomy experiments of Dr. Donovan make it clear, however, that the solution of this problem has not advanced beyond the point where Loeb left it nearly 50 years ago. On the other hand, there are glimmerings which suggest that the uterus may have an important endocrine function in regulating the oestrous cycle, but whether it does this through an oestrogen sparing action, or through some unrecognized metabolite remains unknown. You may remember that Donovan suggested the possible implication of relaxin. The ubiquitous nature of relaxin makes it highly tenable. Relaxin occurs more plentifully in rabbit than in any other species, being found not only in the blood and endometrium but also in the ovary.

Linked to the studies of Donovan were those of Professor Melampy, who suggested that the uterus is the source of luteolytic activity during the progestational phase of the cycle in the unmated sow. This luteolytic action is blocked by the conceptus and is absent following hysterectomy. Hence the corpora lutea persist and the cycle is inhibited. All of which may be paraphrased by stating that 'Nature abhors a non-pregnant uterus'.

Professor Nalbandov dealt with a different aspect of ovarian function. He adduced evidence of the possible implication of oestrogen in the synthesis of progesterone, either by direct action on the corpus luteum or by eliciting the release of a luteotrophic substance, which in turn controls progesterone synthesis and release. This material of Dr. Nalbandov is so new indeed that it should cause little surprise if many of the interpretations of the phenomena described are modified as new knowledge becomes available.

Professor Karg's paper was the only one which bore reference to the products of conception in a straightforward way. He demonstrated that whereas the gonads and the pituitary are well equipped before birth to execute their endocrinological functions, the testes are not yet competent to discharge their gametogenic functions.

In their turn Dr. Chang, Dr. Bedford and Mr. Finn have shown that the theme which gives this rich array of facts derived from studies of spermatozoa and the implanting ovum its enduring interest, and, which binds them into an understandable and continuous pattern is primarily biological, and this may also be said of the last paper of the Symposium given so elegantly by Dr. Adams. None-the-less, I would caution Mr. Finn that he may be dealing with activities in which many more physiological factors may be concerned since there is the likelihood that some of the effects we have been debating are secondary and not primary events.

We have covered a great deal of ground during the last 4 days, but we have not covered all the ground we could have covered under our general title. Clearly it is not enough to determine that a mechanism is a necessary precursor for a second one; it must also be ascertained whether it is a sufficient one. Nor is it enough to identify the components of a physiological mechanism and describe the course of events from the stimulus to the result. One can be sure that an explanation is incomplete if one has not also explained how the mechanism is turned off.

But having suggested that we have not covered all the ground that we might have done, let me hasten to point out that we all know and appreciate that science never advances on an even front. Science we might even dare to define as 'errors brought up-to-date'.

This conference has greatly widened our knowledge and experience, but at the same time the wealth of information with which we have been regaled, at all points, has also revealed how extensive are the gaps which remain. In the next Symposium that is organized on this subject I trust that we shall be able to sacrifice expanse for the sake of depth, and that we shall be ready to approach the problem of the way the hormones modulate the activities of the pregnant female and thus provide a cogent answer to Professor Nalbandov.

LIST OF MEMBERS

ADAMS, Dr. C. E.	A.R.C. Unit of Reproductive Physiology and Biochemistry, Animal Research Station, Huntingdon Road, Cambridge
AMOROSO, Prof. E. C.	Royal Veterinary College, Royal College Street, London, N.W.1
BAIER, Prof. W.	Gynaekologische Tierklinik, Universitaet München, Koeniginstrasse 12, München 22, Germany
BEAUMONT, Dr. H. M.	Department of Anatomy, University of Birmingham Medical School, Birmingham 15
BEDFORD, Dr. J. M.	Royal Veterinary College, Royal College Street, London, N.W.1
BENNETT, Dr. J. P.	Biological Research Department, British Drug Houses Ltd., Borough Road, Godalming, Surrey
BENSON, Dr. G. K.	Department of Veterinary Anatomy, The University, Liverpool 3
BERCHTOLD, Dr. M.	Gynaekologische Tierklinik, Universitaet München, Koeniginstrasse 12, München 22, Germany
BETTERIDGE, Mr. K. J.	Department of Anatomy, University of Birmingham Medical School, Birmingham 15
BLAND, Mr. K. P.	Department of Neuroendocrinology, Institute of Psychiatry, Maudsley Hospital, Denmark Hill, London, S.E.5
BRADFIELD, Mr. P. G. E.	Grassland Research Institute, Hurley, nr. Maidenhead, Berks.
BRADSHAW, Dr. S. D.	Department of Zoology, University of Sheffield, Sheffield 10
BREED, Mr. W. G.	Department of Agriculture, University of Oxford, Parks Road, Oxford
BUTT, Dr. W. R.	The United Birmingham Hospitals, Department of Clinical Endocrinology, The Birmingham and Midland Hospital for Women, Showell Green Lane, Sparkhill, Birmingham 11
BUTTLE, Mr. H.	Animal Breeding Research Organisation, Kings Buildings, West Mains Road, Edinburgh 9
CAMPBELL, Mr. C.	U.S. Feed Grains Council, Locomotive House, 30/34 Buckingham Gate, London, S.W.1
CARLES, Mr. A. B.	Faculty of Veterinary Science, University College of Nairobi, P.B. Kabote, Kenya
CALLEAR, Mr. J. F. F.	Abbott Laboratories Ltd., Queenborough, Kent

559

CLARK, Dr. J. B. K.	Twyford Laboratories Ltd., Twyford Abbey Road, London, N.W.10
CLARKE, Dr. J. R.	Department of Agriculture, University of Oxford, Parks Road, Oxford
CHANG, Dr. M. C.	Worcester Foundation for Experimental Biology, 222 Maple Avenue, Shrewsbury, Massachusetts, U.S.A.
CHANNING, Dr. C.	School of Veterinary Medicine, University of Cambridge, Madingley Road, Cambridge
CHOW, Dr. W. F.	Department of Pharmacology, University of Bristol Medical School, Bristol 8
COOPER, Mr. K. J.	Department of Physiology and Environmental Studies, University of Nottingham School of Agriculture
COUDERT, Dr. S. P.	School of Veterinary Medicine, University of Cambridge, Madingley Road, Cambridge
COUTTIE, Mr. M.	Milk Marketing Board, Thames Ditton, Surrey
COLEMAN, Dr. J. R.	Department of Pharmacology, University of Bristol Medical School, Bristol 8
COX, Mr. J. E.	Department of Surgery, Royal Veterinary Field Station, Hawkshead Lane, North Mimms, Hatfield, Herts.
CRIGHTON, Dr. D. B.	Department of Physiology and Environmental Studies, University of Nottingham School of Agriculture
CROSS, Dr. B. A.	Sub-Department of Veterinary Anatomy, School of Veterinary Medicine, University of Cambridge, Tennis Court Road, Cambridge
CROWLEY, Mr. J. P.	The Agricultural Institute, Thorndale, Beaumont Road, Dublin, Eire
CULLEN, Dr. R.	Wellcome Veterinary Research Station, Ely Grange, Frant, nr. Tunbridge Wells, Kent
CUNNINGHAM, Mr.	Reading
CUMMINS, Mr. J. M.	Department of Zoology, University College of North Wales, Bangor, Caernarvonshire
DAS, Mr. R. M.	Department of Veterinary Anatomy, The University, Liverpool 3
DAY, Prof. B. N.	Department of Animal Husbandry, University of Missouri, Columbia, Missouri, U.S.A.
DEANSLEY, Dr. R.	University of Cambridge, School of Agriculture, Downing Street, Cambridge
DENNISON, Miss M.	Department of Zoology, University College of North Wales, Bangor, Caernarvonshire
DONOVAN, Dr. B. T.	Department of Neuroendocrinology, Institute of Psychiatry, Maudsley Hospital, Denmark Hill, London, S.E.5

DORRINGTON, Dr. J.	Department of Zoology, University of Sheffield, Sheffield 10
DRAPER, Mr. S. A.	Department of Physiology and Environmental Studies, University of Nottingham School of Agriculture
ECKSTEIN, Dr. P.	Department of Anatomy, University of Birmingham Medical School, Birmingham 15
EDWARDS, Dr. J.	Milk Marketing Board, Thames Ditton, Surrey
FALCONER, Dr. I. R.	Department of Applied Biochemistry and Nutrition, University of Nottingham School of Agriculture
FINDLAY, Mr. A. L. R.	Sub-Department of Veterinary Anatomy, School of Veterinary Medicine, University of Cambridge, Tennis Court Road, Cambridge
FINK, Dr. G.	Department of Human Anatomy, University of Oxford, South Parks Road, Oxford
FINN, Mr. and Mrs. C. A.	Department of Physiology, Royal Veterinary College, Royal College Street, London, N.W.1
FITZPATRICK, Prof. and Mrs. R. J.	Faculty of Veterinary Science, The University, Liverpool 3
FOLLEY, Prof. and Mrs. S. J.	National Institute for Research in Dairying, Shinfield, Reading, Berks.
FORBES, Mr. J. M.	Department of Agriculture, University of Leeds, Leeds 2
FORD, Miss C.	Ministry of Agriculture, Fisheries and Food Veterinary Investigation Centre (Nottingham University School of Agriculture)
FORTEATH, Mr.	Milk Marketing Board Cattle Breeding Centre, Chase Farm, Little Harwood, Bletchley, Bucks.
FURR, Mr. B. J. A.	Department of Physiological Chemistry, University of Reading, Reading, Berks.
GEMZELL, Prof. and Mrs. C.	Akademiska sjukhuset, Uppsala, Sweden
GERARD, Mme. M.	Station de Recherches de Physiologie Animale, C.N.R.Z. 78, Jouy-en-Josas, Seine et Oise, France
GLOVER, Dr. T.	The University, Liverpool 3
GODING, Dr. J. R.	A.R.C. Institute of Animal Physiology, Babraham, Cambridge
GOOD, Dr. B. F.	Department of Human Anatomy, University of Oxford, South Parks Road, Oxford
GORDON, Dr. I.	Faculty of Agriculture, University College, Dublin, Eire
GREENHALGH, Dr. P. M.	Department of Physiology and Environmental Studies, University of Nottingham School of Agriculture

GREIG, Miss M. A. F.	A.R.C. Institute of Animal Physiology, Babraham, Cambridge
GROVES, Mr. T. W.	I.C.I. Pharmaceutical Division, Alderley Park, Alderley Edge, Macclesfield, Cheshire
GUPTA, Mr. S. K.	Department of Physiology, National Institute for Research in Dairying, Shinfield, Reading, Berks.
HANCOCK, Dr. J. L.	Animal Breeding Research Organisation, Kings Buildings, West Mains Road, Edinburgh 9
HANLY, Mr. S.	University College Dublin, Faculty of Veterinary Medicine, Veterinary College of Ireland, Ballsbridge, Dublin, Eire
HANSEL, Prof. W. M.	Department of Animal Husbandry, Cornell University, Ithaca, New York, U.S.A.
HARPER, Dr. M. J. K.	I.C.I. Pharmaceuticals Division, Alderley Park, Alderley Edge, Macclesfield, Cheshire
HARTREE, Dr. A. S.	Department of Biochemistry, University of Cambridge, Tennis Court Road, Cambridge
HAVARD, Mr. F. W.	Veterinary Sales Department, G. D. Searle & Co. Ltd., Lane End Road, High Wycombe, Bucks.
HAY, Dr. M. F.	A.R.C. Unit of Reproductive Physiology and Biochemistry, 307 Huntingdon Road, Cambridge
HAYNES, Dr. N. B.	Department of Physiology and Environmental Studies, University of Nottingham School of Agriculture
HEALD, Dr. P. J.	Twyford Laboratories Ltd., Twyford Abbey Road, London, N.W.10
HEAP, Dr. R. B.	A.R.C. Institute of Animal Physiology, Babraham, Cambridge
HIGNETT, Mr. P. G.	Department of Veterinary Reproduction, University of Glasgow Veterinary Hospital, Bearsden Road, Bearsden, Glasgow
HOLMES, Dr. M.	Institute of Physiology, University of Glasgow, Glasgow, W.2
HOVELL, Dr. R.	Wellcome Veterinary Research Station, Ely Grange, Frant, Tunbridge Wells, Kent
JARRETT, Dr. A. S.	Syntex Pharmaceuticals Ltd., St. Ives House, Maidenhead, Berks.
JOCHLE, Dr. W.	Director of Veterinary Research, Syntex International A.T.S.A., Apartado M-8797, Mexico 1, D.F.
KARG, Prof. H.	Institut für Tierphysiologie der Universitat München, Veterinarstrasse 13, München 22, Germany
KELLY, Mr. A. W.	Department of Anatomy, University of Birmingham Medical School, Birmingham 15

KENDLE, Mr. K. E.	Biological Research Department, British Drug Houses Ltd., Borough Road, Godalming, Surrey
KILPATRICK, Dr. and Mrs. R.	Department of Pharmacology, University of Sheffield, Sheffield 10
LAKE, Dr. P. E.	A.R.C. Research Centre, Kings Buildings, West Mains Road, Edinburgh 9
LAMMING, Prof. G. E.	Department of Physiology and Environmental Studies, University of Nottingham School of Agriculture
LAND, Dr. R. B.	Institute of Animal Genetics, West Mains Road, Edinburgh 9
LANE, Mr. P. J.	Baywood Chemicals Ltd., P.O. Box 7, Eastern Way, Bury St. Edmunds, Suffolk
LARKIN, Mr. P. J.	Beecham Research Laboratories, Brook Farm, Leigh, nr. Reigate, Surrey
LEDERIS, Dr. K.	Department of Pharmacology, University of Bristol Medical School, Bristol 8
LEIDL, Prof. W.	Gynaekologische und Ambulatorische Tierklinik der Universitat München, Abt. für Andrologie und Kunstliche Besanning, 8, Koeniginstrasse 12, München 22, Germany
LINCOLN, Mr. D. W.	Sub-Department of Veterinary Anatomy, School of Veterinary Medicine, University of Cambridge, Tennis Court Road, Cambridge
LODGE, Dr. G. A.	Department of Agriculture and Horticulture, University of Nottingham School of Agriculture
LORRAINE, Dr. J. A.	M.R.C. Clinical Endocrinology Research Unit, 2 Forrest Road, Edinburgh 1
MARSTON, Mr. J. H.	Department of Anatomy, University of Birmingham Medical School, Birmingham 15
MARTIN, Dr. and Mrs. L.	Imperial Cancer Research Fund, Lincolns Inn Fields, London, W.C.2
MAULEON, Dr. P.	Laboratoire Physiologie Reproduction, C.N.R.Z. 78, Jouy-en-Josas, Seine et Oise, France
McCANN, Prof. S. M.	Department of Physiology, University of Texas, Southwestern Medical School, Dallas, Texas, U.S.A.
McGOVERN, Mr. P. T.	Department of Zoology, University College of North Wales, Bangor, Caernarvonshire
McLAREN, Dr. A.	Institute of Animal Genetics, West Mains Road, Edinburgh 9
MELAMPY, Prof. R. M.	Department of Animal Science, Iowa State University, Ames, Iowa, U.S.A.
MELROSE, Dr. D. R.	Pig Industry Development Authority, PIDA House, Ridgmount Street, London, W.C.1

563

DU MENSIL DU BUISSON, Dr. and Mrs. F.	C.R.V.Z. Tours-Lorfraziere, Nouzilly 37, France
MILLER, Mr. D. J. S.	Animal Health Division, E. R. Squibb & Sons Ltd., Royal House, Twickenham, Middlesex
MILLS, Dr. J. B.	Department of Biochemistry, University of Cambridge, Cambridge
MITCHELL, Mrs. M. E.	A.R.C. Poultry Research Centre, Kings Buildings, West Mains Road, Edinburgh 9
MOON, Prof. R. C.	Department of Physiology, University of Tennessee, Memphis, Tennessee, U.S.A.
MOOR, Dr. R. M.	A.R.C. Unit of Reproductive Physiology and Biochemistry, 307 Huntingdon Road, Cambridge
MORTON, Mr. D. B.	Department of Veterinary Anatomy, The University, Liverpool 3
MOSELEY, Miss S. R.	Department of Physiology and Environmental Studies, University of Nottingham School of Agriculture
MOSS, Mr. J. A.	Avoncroft Cattle Breeders, Bromsgrove, Worcs.
MUELLER, Dr. G. L.	Squibb International, 460 Park Avenue, New York 22, New York, U.S.A.
MUNRO, Mr. I. B.	Ministry of Agriculture, Fisheries and Food, Reading Cattle Breeding Centre, Shinfield, Reading, Berks.
NALBANDOV, Prof. A. V.	Animal Genetics Laboratory, University of Illinois, Urbana, Illinois, U.S.A.
PELLETIER, Dr. J.	Laboratoire de Physiologie de la Reproduction, Jouy-en-Josas, Seine et Oise, France
PERRY, Dr. J. S.	A.R.C. Institute of Animal Physiology, Babraham, Cambridge
PHILLIPPO, Dr. M.	Department of Animal Hygiene, Royal Veterinary College, Boltons Park, Potters Bar, Herts.
PICKARD, Mr. D. W.	Department of Physiology and Environmental Studies, University of Nottingham School of Agriculture
PIKE, Mr. I. H.	Department of Agriculture, University of Leeds, Leeds 2
PINOT, Dr. R	Station de Recherches sur l'Elevage des Ruminants, C.N.R.Z., Jouy-en-Josas, Seine et Oise, France
POLGE, Dr. C.	A.R.C. Unit of Reproductive Physiology and Biochemistry, 307 Huntingdon Road, Cambridge
REED, Mr. H. C. B.	Pig Industry Development Authority, Pig Breeding Centre, Leeds Road, Thorp Willoughby, nr. Selby, Yorks

Reichert, Prof. L. E.	Department of Biochemistry, Division of Basic Health Sciences, Emory University, Atlanta, Georgia, U.S.A.
Rigby, Mr. J. P.	The University, Liverpool 3
Robertson, Dr. H. A.	Division of Agricultural Biochemistry, Department of Biological Chemistry, University of Aberdeen, Aberdeen
Robinson, Prof. J. T.	University of Sydney, Sydney, New South Wales, Australia
Rothchild, Prof. I.	University Hospitals of Cleveland, University Circle, Cleveland, Ohio, U.S.A.
Rowson, Mr. L. E. A.	Cambridge and District Cattle Breeders Ltd., Huntingdon Road, Cambridge
Ryle, Dr. M.	Department of Clinical Endocrinology, Birmingham, and Midland Hospital for Women, Showell Green Lane, Sparkhill, Birmingham 11
Saji, Dr. M. A.	Livestock Experiment Station, Nabisar Road, Lahore, W. Pakistan
Schilling, Dr. E.	Max-Planck Institut für Tierzneht und Tierernahrung, 3051 Mariensee/Winstorf, Germany
Schomberg, Dr. D. W.	School of Veterinary Medicine, University of Cambridge, Madingley Road, Cambridge
Scofield, Mr. A. M.	Department of Zoology, University College of North Wales, Bangor, Caernarvonshire
Shaw, Mr. R.	Milk Marketing Board, Thames Ditton, Surrey
Shearer, Mr. G.	Wellcome Veterinary Research Station, Ely Grange, Frant, nr. Tunbridge Wells, Kent
Short, Dr. R. V.	Department of Veterinary Clinical Studies, University of Cambridge, Madingley Road, Cambridge
Simmons, Mr.	Milk Marketing Board, Thames Ditton, Surrey
Skinner, Mr. J. P.	A.R.C. Unit of Reproductive Physiology and Biochemistry, 307 Huntingdon Road, Cambridge
Smidt, Dr. J. P.	Institute für Tierzucht und Haustiergenetik, 34 Gottingen Albrecht Thaerweg 1, Germany
Smith, Mr. G. F.	Milk Marketing Board, Thames Ditton, Surrey
Spincemaille, Dr. J.	Veterinary School, Casinoplein 21, Ghent, Belgium
Squance, Dr. E. H.	Agricultural Chemistry Department, Queen's University of Belfast, Elmwood Avenue, Belfast 9, N. Ireland
Stack-Dunne, Dr. M.	National Institute for Medical Research, The Ridgway, Mill Hill, London, N.W.7
Stoliaroff, Dr. M.	Laboratories B.Y.L.A., 9 rue Pierre Byla, 94 Gentilly, Seine, France

Swan, Mr. H.	Department of Agriculture and Horticulture, University of Nottingham School of Agriculture
Tindall, Dr. J. S.	Department of Physiology, National Institute for Research in Dairying, Shinfield, Reading, Berks.
Torres, Miss S.	Station de Physiologie Animale, C.N.R.Z. 78, Jouy-en-Josas, Seine et Oise, France
Tripp, Mr. H. R.	Department of Zoology, University College of North Wales, Bangor, Caernarvonshire
Van Rees, Dr. G. P.	Department of Pharmacology, University of Leiden, Wassenaarseweg 62, Leiden, Netherlands
Velle, Prof. W.	Department of Physiology, The Veterinary College of Norway, Oslo, Norway
Vincent, Miss D. S.	Department of Biological Sciences, Wye College, Ashford, Kent
Vinson, Dr. G. P.	Department of Zoology, University of Sheffield, Sheffield 10
Walpole, Dr. A. L.	I.C.I. Pharmaceutical Division, Alderley Park, Alderley Edge, Macclesfield, Cheshire
Waynforth, Mr. A. B.	Courtauld Institute of Biochemistry, Middlesex Hospital Medical School, London, W.1
Weir, Miss B. J.	Wellcome Institute, Zoological Society of London, Regents Park, London, N.W.1
Welch, Mr. R. A.	A.R.C. Unit of Reproductive Physiology and Biochemistry, 307 Huntingdon Road, Cambridge
Wiggan, Mr. L. S.	Department of Animal Husbandry, Royal Veterinary College, Potters Bar, Herts.
Williams, Mr. P. C.	Imperial Cancer Research Fund, Lincolns Inn Fields, London, W.C.2
Willis, Mr. F. J.	N.A.A.S., Shardlow Hall, Shardlow, nr. Derby
Wishart, Mr. D. F.	G. D. Searle & Co. Ltd., Lane End Road, High Wycombe, Bucks.
Wright, Mr. J.	Milk Marketing Board, Thames Ditton, Surrey
Yoshinaga, Dr. K.	A.R.C. Unit of Reproductive Physiology and Biochemistry, 307 Huntingdon Road, Cambridge

AUTHOR INDEX

567

SUBJECT INDEX

ERRATUM

The appointment of Professor G. E. Lamming
is incorrect. He holds the Chair of Animal
Physiology, School of Agriculture, University
of Nottingham.